Basic Instrumentation

Third Edition

by
Bruce R. Whalen

Edited by
Mildred Gerding

Published by
PETROLEUM EXTENSION SERVICE
Division of Continuing Education
The University of Texas at Austin
Austin, Texas
1983

First Edition published 1948. Second Edition 1964
Third Edition 1983. Second Impression 1987
Printed in the United States of America

Catalog No. 1.20030
ISBN 0-88698-003-8

Contents

Preface ... vii
Acknowledgments ... ix
Chapter 1. Introduction to Instrumentation ... 1
The Need for Measurement and Control ... 2
Measuring Methods ... 3
Types of Controls ... 4
Characteristics of Automatic Control Systems ... 8
Modes of Control ... 16
Conclusion ... 21
Chapter 2. The Units and Dimensions of Measurement ... 23
Systems of Units ... 24
The Units of Length Measurement ... 25
The Units of Time Measurement ... 26
The Units of Temperature Measurement ... 26
The Units of Mass, Weight, and Force ... 27
The Units of Work and Energy ... 31
The Dimensions of Various Quantities ... 33
Conclusion ... 36
Chapter 3. Final Control Elements ... 37
Valves ... 38
Actuators ... 50
Controlled Volume Pumps ... 61
Other Final Control Elements ... 63
Chapter 4. Automatic Controls ... 65
Pressure Regulators ... 66
Developing a Pneumatic Controller ... 67
The Controller in Action ... 73
Commercial Pneumatic Controllers ... 78
Volume Booster Relays ... 84
Valve Positioners ... 85
Electronic Controls ... 90
Conclusion ... 103
Chapter 5. Pressure Measurement and Control ... 105
Units of Pressure Measurement ... 106
Mechanical Pressure Elements ... 107
Protective Devices for Pressure Instruments ... 116
High Vacuum Measurement Gauges ... 117
Purposes of Pressure Control ... 123
Pressure Control Devices ... 123

Chapter 6. Temperature Measurement and Control 137
Temperature Scales 138
Liquid-in-Glass Thermometers 140
Filled-System Thermometers 143
Bimetal Thermometers 150
Electrical Measuring Devices 152
Pyrometers 161
Temperature Control 167
Summary 172
Chapter 7. Liquid-Level Measurement and Control 173
Direct Measuring Devices 174
Instruments Using Buoyancy 176
Instruments Using Displacement Elements 179
Instruments Using Hydrostatic Pressure 182
Electric Devices 189
Sonic Devices 193
Vibrator Devices 194
Radiation Devices 194
Chapter 8. Flow Measurement 197
Units and Dimensions 198
Differential Pressure Flowmeters 198
Variable-Area Flowmeters 221
Magnetic Flowmeter 224
Mass Flowmeters 226
Turbine Meters 229
Displacement Meters 230
Chapter 9. Flow Control 235
Flow Control Devices 236
General Considerations of Flow Control 239
Flow Control in Fractionating Columns 244
Installation Arrangements for Flow Measurement and Control 250
Summary 258
Chapter 10. Measurement and Control of Other Variables 259
Humidity and Dew Point 260
Specific Gravity 265
Viscosity 271
pH Factor 277
Conclusion 279
Chapter 11. Transducers, Transmitters, and Converters 281
Transducers 282
Transmitters 286
Converters 292
Conclusion 297

Chapter 12. Recorders and Integrators 299
Types of Recorders 300
Servomechanisms in Recorders 304
Chart Drive Mechanisms 311
Recorder Marking Systems 314
Solid State Devices 317
Integrators 318
Summary 322
Chapter 13. Adjusting Automatic Controllers to a Process 323
Process Reaction Rate 324
Stability of Control 325
Phase Shift and Natural Frequency 326
Making Controller Adjustments 328
Conclusion 334
Appendix A. The International System of Units (SI) 335
The Dimensions, Units, and Symbols of SI 337
Prefixes and Multipliers 337
Rules and Comments 338
Comparing SI with Conventional Units 339
Appendix B. Symbols for Instrumentation and Electronics 343
Instrumentation Symbols 344
Electrical Symbols 347
Greek Alphabet 348
Bibliography 349

Preface

Basic Instrumentation represents an effort to explain the operation of the more common components and the important concepts of measurement and control. It is not a primer or elementary introduction to automation, nor is it a text on advanced process control.

Ideally, the user of this text should have the knowledge that can be gained from high school courses in algebra, chemistry, and physics, plus a knowledge of basic electricity and electronics. Not having such knowledge will merely mean that, in a few instances, the results of derivations and the validity of some statements will need to be accepted by the reader.

Although a liberal mixture of conventional and metric units is used in the text, metric use is limited largely to measurements of force, length, and temperature. The more complex units of force (newton) and pressure (pascal) are generally avoided. Many units in the conventional system of measurements have been metric in nature for many years—all electrical units, for example, and those for viscosity. An appendix is provided to assist in converting between metric and conventional units.

The devices covered in this text include not only modern equipment items but some that have been in service for a number of years. Many components of instrumentation serve industry years after their manufacture has been discontinued in favor of more modern equipment. It is useful to know how this still-serving old equipment functions; thus, the more common of such items are discussed.

Considerable effort has been taken to eliminate typographical and other errors, but in spite of such effort mistakes have a way of cropping up. Petroleum Extension Service will be most grateful to hear from those who detect incorrect statements, typographical errors, or other erroneous content. Errors are usually corrected at each printing of the text.

Bruce R. Whalen

Acknowledgments

The suggestions of many people have been incorporated into this new edition of *Basic Instrumentation.* Serving as content consultants were Willis Finley, Jerry Campbell, and Louis Johnson of the PETEX staff at the Kilgore Training Center; and J. J. (Joe) Sergesketter of Houston, consultant and instructor in PETEX instrumentation schools conducted in Kilgore.

PETEX staff members Jodie Leecraft, Richard House, and David J. Morris assisted in the editing and finalizing of the text. Illustrations and composition were handled by Marty Burns, Deborah Caples, Gale Hathcock, and Denise Sanderson. Their time and talents in their respective areas are greatly appreciated.

In addition, the following companies contributed data and illustrations for this manual, and their assistance is gratefully acknowledged:

American Meter Division
Singer Company
13500 Philmont Avenue
Philadelphia, PA 19116

Analytical Measurements, Inc.
29 Willow Street
Chatham, NJ 07928

Bristol Babcock Instruments/Systems
Bristol Babcock Inc.
40 Bristol Street
Waterbury, CT 06708

C-E NATCO
Division of Combustion Engineering
5330 East 31st Street
Tulsa, OK 74101

The Eppley Laboratory, Inc.
14 Sheffield Avenue
Newport, RI 02840

Esterline Angus Instrument Company
P. O. Box 24000-T
Indianapolis, IN 46224

Fischer & Porter Company
51 Warminster Road
Warminster, PA 18974

Fisher Controls
Marshalltown, IA 50158

The Foxboro Company
38 Neponset Avenue
Foxboro, MA 02035

General Electric Company
One River Road
Schnectady, NY 12305

General Time Controls, Inc.
135 South Main Street
Thomaston, CT 06787

Graphic Controls Corporation
Division of Industrial Products Corporation
One Carnegie Plaza
Cherry Hill, NJ 08003

Halliburton Services
P. O. Drawer 1431
Duncan, OK 73533

Honeywell
Process Controls Division
1100 Virginia Avenue
Fort Washington, PA 19034

Kimray, Inc.
52 Northwest 42nd Street
Oklahoma City, OK 73118

Leeds & Northrup Company
A Unit of General Signal
Sumneytown Pike
North Wales, PA 19454

The Mercoid Corporation
4217 Belmont Avenue
Chicago, IL 60641

Micro Essential Laboratory, Inc.
4206 Avenue H
Brooklyn, NY 11210

Motorola Instrument and Control, Inc.
A subsidiary of Motorola Inc.
P. O. Box 5409
Phoenix, AZ 85010

Rochester Manufacturing Company, Inc.
Rochester, NY 14610

Rockwell International Corporation
400 North Lexington Avenue
Pittsburgh, PA 15208

Taylor Instrument Company
Division of Sybron Corporation
95 Ames Street
Rochester, NY 14692

TXT Texsteam products
Division of Vapor Corporation
320 Hughes Street
Houston, TX 77011

Weston Instruments
Division of Sangamo Weston, Inc.
614 T Frelinghuysen Avenue
Newark, NJ 07114

CHAPTER 1

Introduction to Instrumentation

A century ago musicians were specialists in instrumentation, but the impact of scientific progress on man's way of living and his readiness to adopt new meanings for old words have caused a decided shift in the connotation of the word *instrumentation*. Today, the word is readily associated with measurement and control of processes, although musicians are still specialists in their own field of instrumentation.

Terminology—the use of particular words or phrases (sometimes very ordinary words) to express technical terms associated with an art, science, or business—is a problem in the study of instrumentation, largely because positive standards have not been universally adopted. Terms used in this manual are those most generally accepted by persons and organizations best qualified to determine the needs of the field. Items of terminology will be defined and emphasized in order to acquaint the reader with this important phase of instrumentation.

Since the word *instrument* is so broadly defined as to include "any tool or other device that is useful in accomplishing an objective," instruments are frequently further defined according to the fields they serve—musical, surgical, surveying, and navigating instruments, to name a few. For this field, an instrument is any sensing, measuring, transmitting, or controlling device associated with a process or system. *Instrumentation* shall mean any arrangement of instruments to measure, indicate, record, or control the magnitudes of the variable quantities that might exist in the process. Thermometers, pressure gauges, and control valves are a few of the instruments that make up a system of instrumentation.

The Need for Measurement and Control

Prehistoric man, whose simple club represented an instrument of war and an instrument of survival, existed for many centuries before he devised tools or instruments that were not closely allied with mere survival. Instruments were eventually developed for measuring distances, angles, and time. As man's curiosity about his environment mounted, he devised means of observing stars and other natural phenomena more accurately—not in the hope of controlling these phenomena, but perhaps in the belief that he could adjust his own activity to his advantage. What little control early man achieved was tied closely to his basic needs. Little thought was given to conserving manpower, but recognition of how instruments helped man do a better job motivated the development of tools and standards of measurement.

For many centuries, measuring and controlling have been recognized as necessary functions in the production of acceptable commodities. The brewing and wine-making activities of the Old World needed measurement and control to ensure success in these efforts. Measurements, to be sure, might have been so primitive as to involve only careful observation of the fermenting process, and control was limited merely to the selection of a suitable cellar. Modern industries are bristling with needs for accurate measurement and control. The quality of a chemical product, for example, may depend upon the proper proportioning of ingredients by *weight* or *volume,* maintaining a constant pressure in the reaction vessel for a prescribed *time,* and adjusting the acidity (or pH factor) of the final product by adding a corrective agent.

The economic gains to be achieved through proper measurement and control of processes are of primary importance in the growth of instrumentation. In the field of manufacturing, the savings in material and time, which result from a system of accurate control of product quality, cut overall costs and mean greater profits and lower selling costs. Even in the average modern home, a system of automatic control for heating prevents fuel losses caused by overheating.

The automatic temperature control of a modern home is an example of a desirable and convenient system. Many homes still have no automatic control of temperature, a satisfactory degree of comfort being achieved by manual control. However, it must not be concluded that all forms of automatic control are merely desirable and convenient; some of them are indispensable.

A few years ago officials of a Middle Eastern country contracted with an engineering firm for the design of a modern oil refinery. When the plans were completed and tendered for approval, the officials were alarmed to find that the refinery would possess such a degree of instrumentation that only a small group of personnel would be needed for its operation. The officials considered such extensive automation undesirable because of the enormous supply of cheap manpower available to operate the refinery. The engineering firm was requested to redesign the plant, ridding it of all automatic controls and making provision for manual operation of valves and other equipment. After brief consideration, the firm withdrew its bid to equip the refinery, pointing out that a modern refinery simply cannot be manually controlled, since the coordination needed among the hundreds of individuals to maintain the close tolerances could never be achieved. Thus, since the day James Watt equipped his steam engine with a speed governor—a control that "could do the job better than a man"—industry has advanced to the point of having systems of instrumentation that make processes possible that could not exist without the reliability, stamina, and speed inherent in properly conceived systems.

In modern times, particularly since the beginning of the industrial revolution of the eighteenth century, the need for measurement and control has kept pace with the increase in the number of variable conditions and factors that have evolved with industrialization. Although the number of variables that might be measured and controlled is almost countless, it is heartening to know that for any given process only a few need be considered.

Instrumentation—that is, measurement and control—has found wide use in the petroleum industry; in fact, the chemical and petroleum industries are characterized by the extensive degree to which they have adapted to automation. Of all the variables that might be considered, only a very few are of primary importance in this study. Although pH factor, humidity, and other conditions or properties are important and must be considered, four very basic variables are of principal interest: *temperature, pressure, rate of flow,* and *liquid level.* A variable that must be maintained within specified limits in a system or process is called a *controlled variable.*

Measuring Methods

Human faculties are too insensitive and vary too much among individuals to be useful in most needed measurements. The human body, for example, cannot accurately judge temperature; nor can it estimate distances with acceptable precision by normal visual means. Instruments that augment man's native faculties enable him to conduct measurements with any degree of accuracy desired, although the cost of providing accuracy beyond certain limits becomes prohibitive for practical application. Thus, instruments are used to make measurements or to help man make them.

Some of the variables to be dealt with can be measured directly in units that represent the basic nature of the variable. Thus, liquid level is changed by varying its height above some reference level, and this difference is easily measured directly in feet and inches. Pressure is defined as force per unit-area and could be measured directly by its ability to lift a weight against the force of gravity. Rate of flow is expressed in terms of volume units per minute, hour, day, and so forth and can be measured by noting the rate at which a tank is filled or emptied. Temperature is a different problem and is said to be a measure or, more properly, an indication of molecular activity—a difficult quantity to measure directly. Some of the laws of physics that establish relations between temperature and numerous other physical quantities allow the measurement of temperature by noting the changes it causes in selected mediums that are exposed to its effects. When temperature or any other variable is determined by indirect means, its value is said to be *inferred.* All of the practical means for measuring temperature are *inferential* devices, and most systems for measuring other variables are inferential in nature. The following examples should aid in the understanding of direct and inferential measurements.

A gauge glass having a scale calibrated in inches (fig. 1.1*A*) shows the level of liquid in the tank above the reference level,

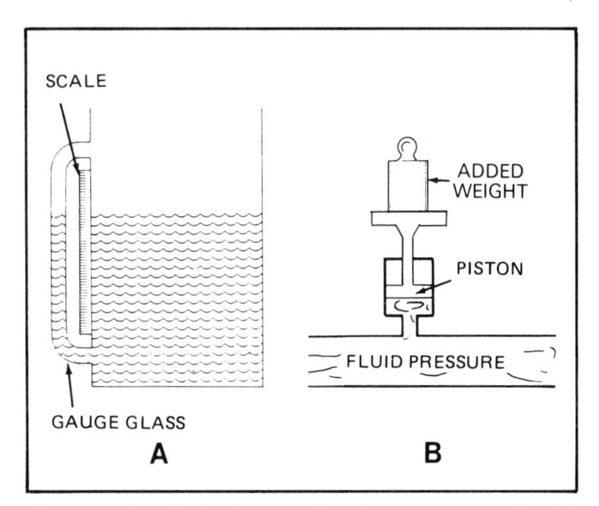

Figure 1.1. Methods for directly measuring liquid and fluid pressure

which is taken to be the bottom of the tank. This is a straightforward measuring device.

The direct measurement of pressure (fig. 1.1*B*) involves the balancing of two forces: (1) the force of gravity acting downward on the piston and the added weight elements and (2) the force acting upward on the piston due to fluid pressure in the line. Assume the piston weighs 1 pound and has a cross-sectional area of 1 square inch. For every 1 pound per square inch of pressure in the line, a force of 1 pound will be exerted against the piston face. Known values of weight units can be added to that of the piston until force caused by line pressure is exactly balanced by the weight of the piston and added weight units.

Although liquid level is easy to measure directly, it also lends itself readily to inferential measurement. Liquid level in the tank (fig. 1.2) can be measured in feet and inches by the gauge glass, but the amount of liquid above the reference level produces a side effect—hydrostatic pressure—that can also be used to infer this height.

A column of liquid having specific gravity G with respect to water exerts a pressure at its base of 0.433 psi $\times$ G for each foot of vertical height. Thus, if a pressure gauge of appropriate range is installed in the tank at the reference level (fig. 1.2), a fairly accurate accounting of the liquid level can be achieved with the equation:

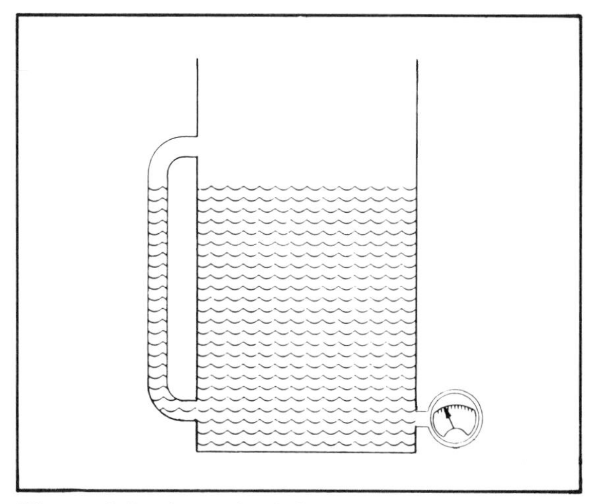

Figure 1.2. A method for inferring liquid level from hydrostatic pressure

$$h = \frac{p}{0.433 \text{ psi} \times G}$$

where

h = the liquid-level height in feet above the gauge;
p = the pressure gauge reading in psi;
G = the specific gravity of the liquid with respect to water.

Clearly, the liquid-level height has been *inferred* from natural physical effects related to density, force of gravity, and so forth.

Types of Control

Methods for conducting measurements can be divided into two broad categories: direct measurements and inferential measurements. Methods for the control of a variable can also be divided into two broad categories: manual control and automatic control.

Manual Control of Variables

The manual control of a single variable ordinarily involves an *operator,* a *measuring means,* and a *final control element.* The measuring means here is a device that indicates to the operator the value of the controlled variable. The final control element is an element of the system, usually a valve, that enables the operator to bring about changes in the controlled variable. The accomplishment of manual control is ordinarily so straightforward and simple that it deserves little study, but its importance cannot be ignored. The manual control of several related variables, as in the control of an aircraft in flight, can become an exacting and complicated task, because of the operator skill required to properly coordinate the adjustments to the related variables.

Two manually controlled systems are (1) a water tank whose liquid level is to be controlled and (2) a hot-water heater requiring temperature control of its contents (fig. 1.3). In each instance there will be a *desired* value for the *controlled variable*—so many feet for the liquid level and so many degrees Fahrenheit for the temperature. In the accepted terminology of instrumentation, the desired value of a controlled variable is called the *set-point* value.

In order to maintain the set-point values of liquid level and temperature in the systems shown in figure 1.3, the adjustments of valves X and Y in each instance must be manually balanced. In the case of liquid-level control, the average rate of flow of water into the tank must be equal to the average rate of flow out of the tank. This last statement implies a fundamental requirement of any successful control system: the set point value of a controlled variable is maintained by achieving a balance between the input and the output of a process. Control of the hot-water system temperature is obtained by balancing the input and the output of heat energy. Steam flows through heating coils and loses a certain amount of heat, which is absorbed by the cold water. It is important to realize, of course, that not all of the energy given up by the steam will eventually be discharged through valve Y. Some of the energy will be lost due to inefficient insulation of the heater and other loss factors.

The measuring means that indicates the liquid level in figure 1.3*A* is a simple system. A float, called the *primary element,* follows the water level and operates the indicator through a set of linkages. The measuring means for the hot-water temperature is an inferential system based upon laws that regulate pressure, temperature, and volume of a gas. In essence, the system operates on the principle that, if a given quantity of gas is confined in a virtually

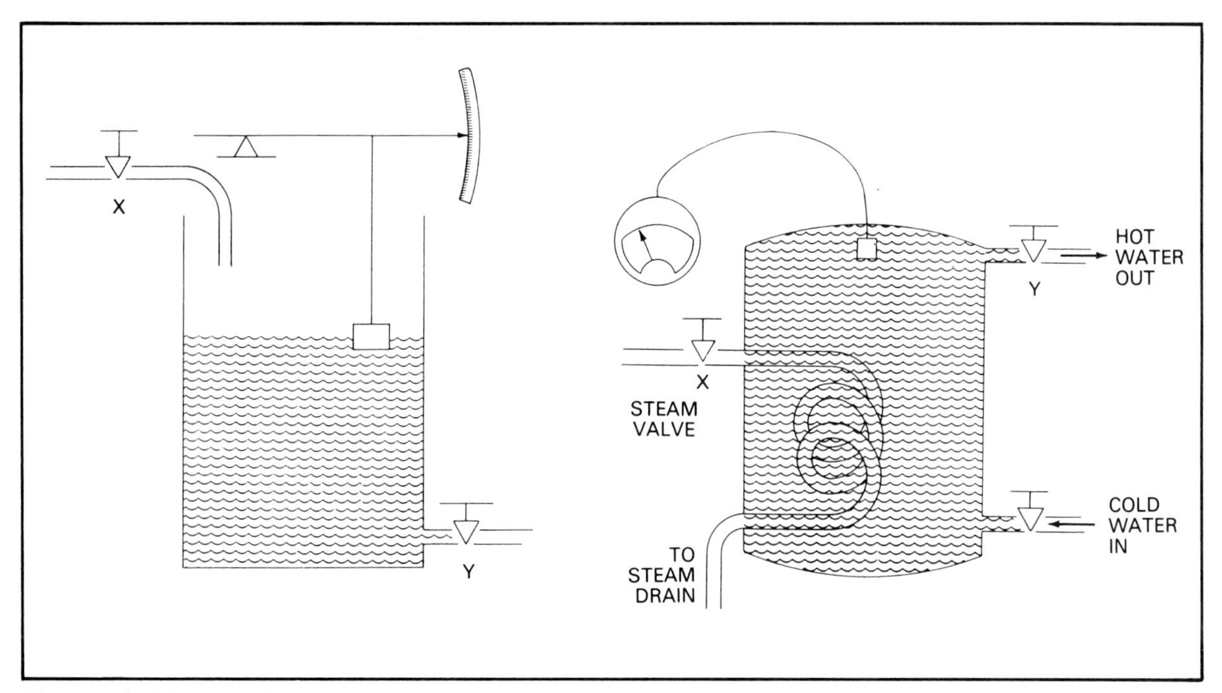

Figure 1.3. Methods for manual control of liquid level (*A*) and temperature (*B*). Valves *X* and *Y* are the means of control.

constant volume, then the absolute pressure of the gas will be directly proportional to its absolute temperature. The *primary element* is a bulb containing a quantity of gas. A metal tube that has a very small inside diameter connects the bulb and a Bourdon tube, a pressure-sensitive device that, like many other pressure-measuring elements, responds to pressure changes by manifesting a change in shape.

Simple Automatic Control

With a few simple additions and changes, the systems in figure 1.3 can be made to provide a reasonable degree of automatic control, as shown in figure 1.4. In addition to operating the liquid-level indicator, the float is made to operate valve *X*. If the rate of discharge through valve *Y* increases and causes the water level in the tank to fall, then the linkages between float and valve *X* will force the latter to open wider to admit more water into the tank. Systems of this sort are in common use for many applications. Note that a turnbuckle has been installed so that the effective length of the link attached to the valve stem can be varied. This action will allow a quick and easy change to be made in the set point: if the link is made shorter, the valve will open wider for a given water level.

Control of the steam flow is now accomplished by a bellows-actuated control valve. The bellows is another pressure-sensitive device, and it responds to pressure changes by varying its length. The pressure changes are proportionally related to temperature changes. The set point can be varied either by the turnbuckle idea or by installing a spring that can be adjusted to vary the tension exerted against the bellows motion. If force of the spring is directed against *expansion* of the bellows in figure 1.4*B*, then greater tension provides for a higher set point.

The feedback principle. All successful automatic control systems are based on feedback. Information relating to the controlled variable is sent back to a controlling means to be checked so that corrective action can be

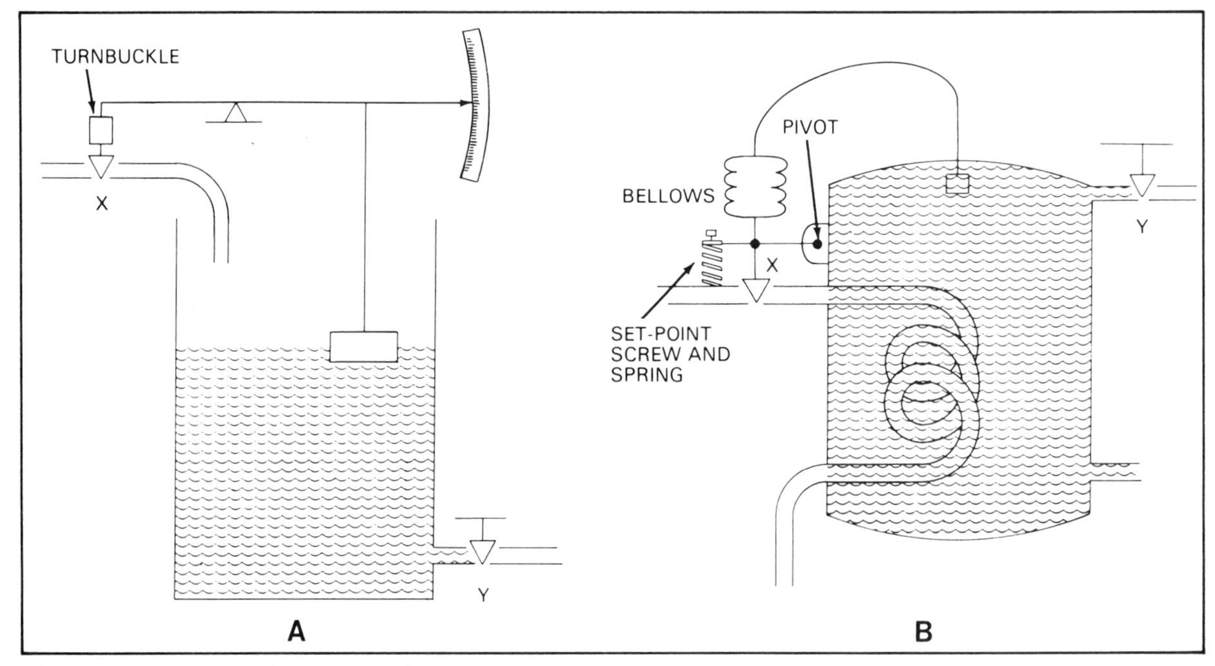

Figure 1.4. Automatic controls for systems in figure 1.3. In *A*, valve *X* is linked to the float; in *B*, valve *X* responds to pressure changes in a bellows.

taken if needed. Action of the controlling means causes the controlled variable to respond, and this response is detected by the primary element that relays the information to the controlling means. Such an arrangement is commonly called a *closed-loop* control system. To emphasize the importance of feedback and the closed-loop system, its performance will be contrasted with another form of control called the *open-loop,* or *open-sequence,* system.

In the open-loop control system shown in figure 1.5, a closed room is heated by a steam radiator, with the primary element of the measuring means located *outside* the room. Steam flow to the radiator is regulated by a

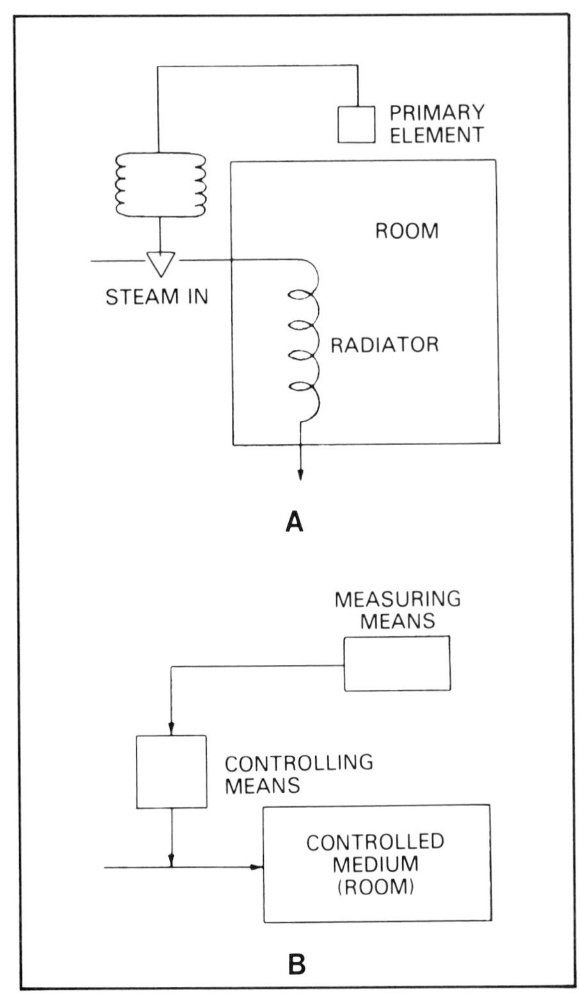

Figure 1.5. An open-loop control system. The primary element is located outside the room.

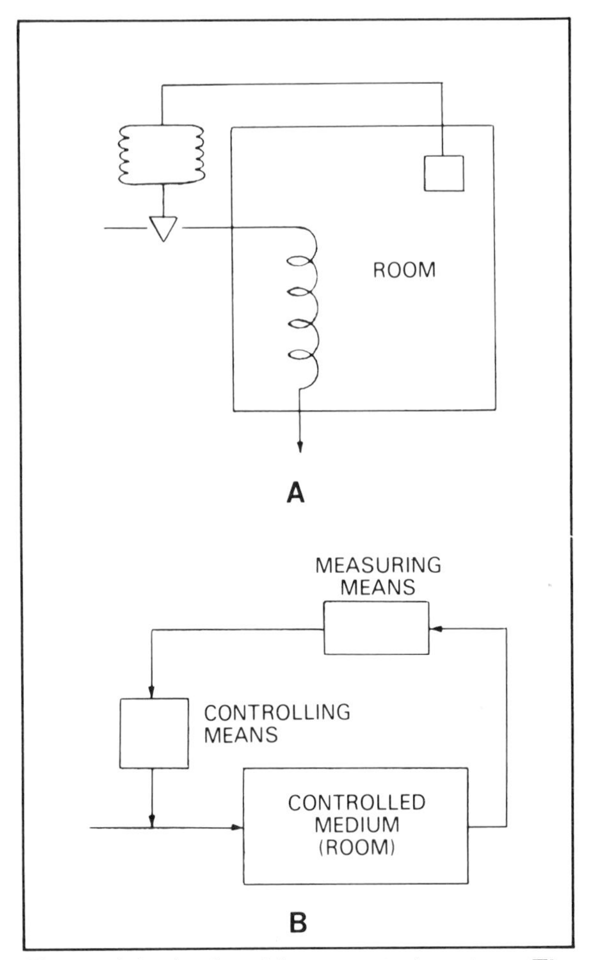

Figure 1.6. A closed-loop control system. The primary element, located inside the room, provides feedback to the controlling means.

bellows-actuated valve similar to that described earlier. Since the primary element is located outside the room, it is exposed to the uncontrolled air surrounding the room. It is probable that the room temperature will vary roughly in proportion to the outside air temperature, but this method of control is unacceptable because the room temperature becomes a function of an uncontrolled environment—the outside air temperature.

The closed-loop system shown in figure 1.6 is similar, but its primary element is located *inside* the room and will therefore respond to action of the radiator. Figure 1.6*B* is a block diagram of the closed-loop system and clearly shows the feedback connection that forms, or closes, the loop. Closed-loop systems of control are useful and popular, while open-loop systems find only limited application.

The simple systems of automatic control are far more interesting than one might suspect. Embodied in these systems is a multitude of considerations that must be taken into account with great seriousness for many critical applications. It must be obvious, for example, that none of the systems will smoothly and automatically maintain a set-point value in the face of varying loads, that is, for various discharge rates from valve *Y* or for changes of temperature in the heated room.

Important items of terminology. In each of the systems studied thus far, the controlled variables—*liquid level, water temperature,* and *room temperature*—were regulated by varying the flow of water or steam. Water and steam are not considered to be variables, but their rates of flow are variables. In these cases, rate of flow is called the *manipulated variable.* The water and steam are called *control agents* because through them the desired values of liquid level and temperature are achieved. Liquid level and temperature are the controlled variables in the systems studied, and the valves controlling the rate of flow of the control agents are final control elements. The float device and thermometer bulb were defined as primary elements. The water tank, hot-water heater, and room are controlled mediums. To fully understand the terms *primary element* and *final control element,* it is necessary to know their full definitions and their relations to other parts of the system. Thus, the following definitions are important:

> The *controlling means* consists of those elements of a controller that are involved in producing a corrective action.
>
> The *final control element* is that portion of the controlling means that directly changes the value of the manipulated variable.
>
> The *measuring means* consists of those elements of a controller that are involved in ascertaining and communicating to the controlling means either the value of the controlled variable or its deviation.
>
> The *primary element* is that portion of the measuring means that first utilizes or transforms energy from the controlled medium to produce an effect in response to a change in the value of the controlled variable. The effect produced by the primary element may be a change of pressure, force, position, electrical resistance, etc.

After careful study, it will be obvious that these definitions are difficult to apply in many instances. It is ironic that the more simple systems of automatic control are the most troublesome in the matter of applying these definitions. Consider, for example, the common thermostat found in automobile engines. It regulates cooling water circulation to achieve proper engine operating temperature and is characterized by its compactness and efficiency. Encompassed in its small size are measuring means, controlling means, final control element, and primary element, but when the definitions are applied, considerable doubt exists about where the measuring means leaves off and the controlling means begins. Fortunately, this will not be a problem in dealing with most of the systems in this study, although careful interpretation is usually required to avoid error.

Characteristics of Automatic Control Systems

Each of the systems of automatic control possesses certain characteristics that affect the quality of the control function. It will be advantageous to learn some of these characteristics and what, if anything, can be done to offset those that are undesirable.

Response Lag

A complete and immediate response by the measuring means to a change in the controlled variable is difficult to achieve in any physical system. The response might start immediately but requires a finite amount of time to complete its effect. This time element is called *lag*. It is the falling behind of one physical condition with respect to a related condition. Thus, a temperature change in the hot-water system (shown in fig. 1.4*B*) is not detected instantaneously. Heat must be transferred through the bulb wall to the gas medium. The gas then expands and operates the control valve. Lag in response to a change in the controlled variable exists whether the value is rising or falling.

The graph in figure 1.7 shows the sort of performance expected from the system shown in figure 1.3*B* if the steam valve is opened and the water temperature allowed to rise at a moderate rate. The steam valve is shut off abruptly after *x* minutes, and the rate of temperature rise begins to decline immediately. Measured values lag true values until the two eventually coincide after the rate of increase in true values declines sufficiently. The difference between the true and measured values of temperature in this case is called a *dynamic error*. It is an error caused by changing conditions, not a result of any constant inaccuracy associated with the measuring means.

Lag is not a property that belongs entirely to primary elements or measuring means. Lag occurs throughout the closed loop of a control system. In some systems a large lag is acceptable, while in others special efforts must be made to reduce it to the barest minimum.

Cycling

Any change in value of the controlled variable (temperature) in figure 1.4*B* brings about corrective action by the controlling means—the final control element and its actuating bellows. A gradual deviation above and below the set-point temperature, coupled with the

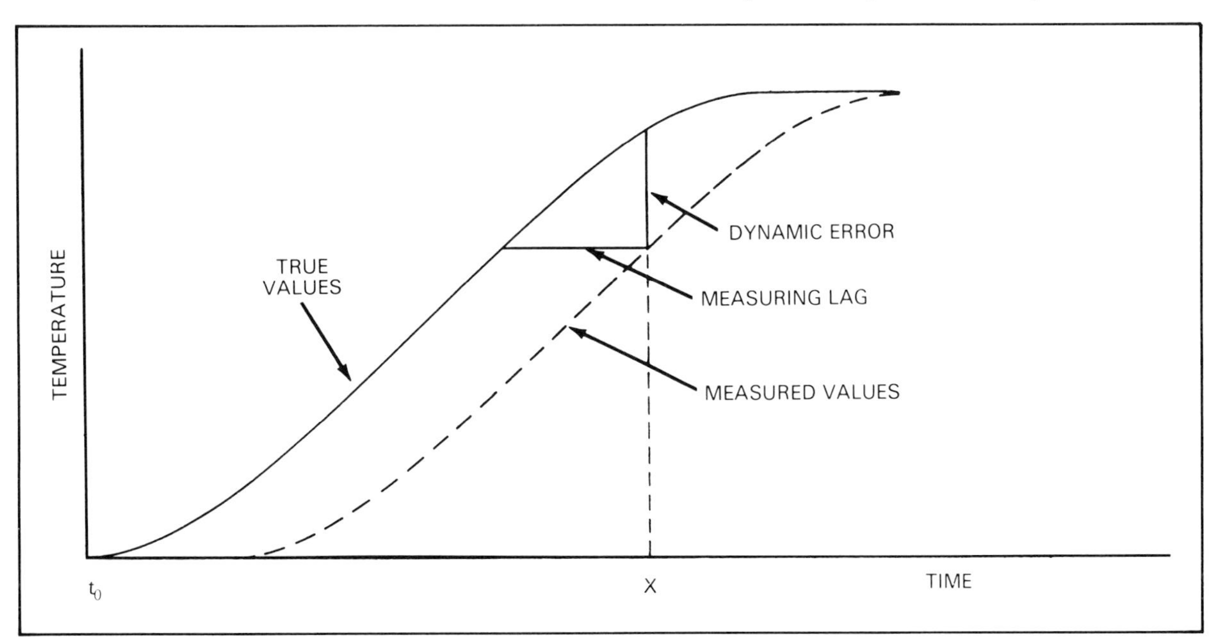

Figure 1.7. Typical lag between measured values and true values of temperature in a manual control system.

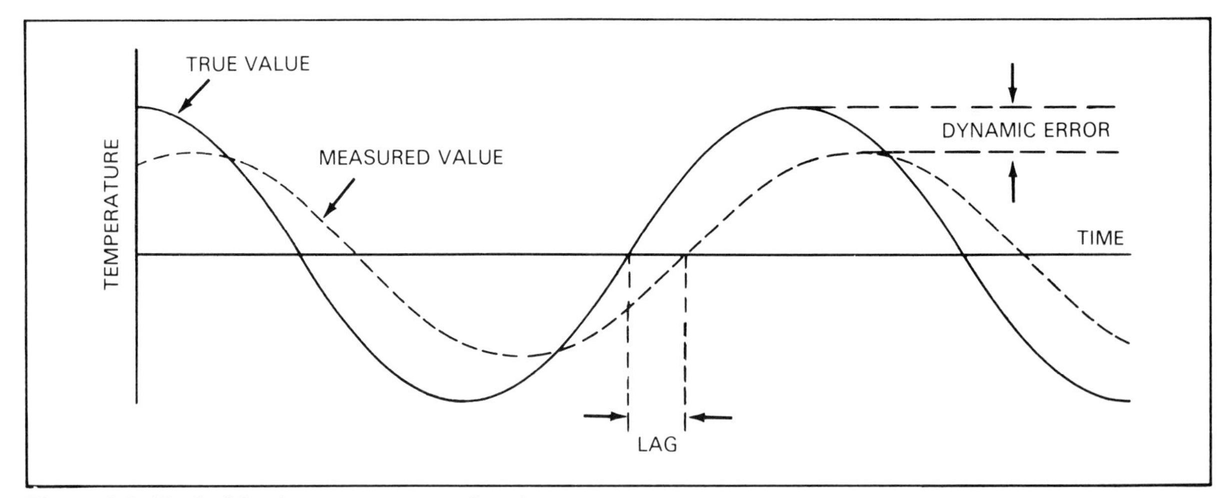

Figure 1.8. Typical lag between measured and true values of temperature in an automatic control system that oscillates about the set point.

existence of lag in the measuring means and controlling means, results in an oscillation of true and measured values of temperature about the set-point value. Figure 1.8 shows a possible form of deviation to be encountered in such a system.

The presence of considerable lag and dynamic error results in a true temperature deviation somewhat *greater* than is accounted for by the measuring means. The periodic deviations above and below the set-point value represent a condition called *hunting,* or *cycling.* Any system or process that is automatically controlled must allow some tolerance for deviations of its controlled variables. In some cases the tolerance is extremely fine, while in others it might be quite broad. While it is desirable to maintain a controlled variable strictly at the set point, cycling to small values above and below the set point is frequently accepted.

Dead Zone

In figure 1.4*A* a direct mechanical linkage connects the primary element (float device) with the control valve. Any movement of the float will be instantly transmitted to the control valve, with the rate of flow adjusted accordingly. The mechanical linkage in this case represents the transmission system between the primary and final control elements. Such mechanical systems, although satisfactory for many applications, suffer deficiencies that are as troublesome as cycling. Unless the working joints in the linkage are designed with close tolerances, a significant amount of motion in the movement will be lost; that is, the float can move vertically a certain distance without moving the valve stem. Other factors to be considered are inertia and friction.

If the linkage is made with tight-fitting joints and the valve stem packed adequately to avoid leaks, the system will suffer a serious delay in responding to level changes. This delay is not a function of time, but it is related to the amount of departure from the set point and the degree of friction in the moving parts. The action is simple and logical. The static friction of the moving parts resists the force exerted by the float until the force reaches a critical value, at which time the static friction is overcome and is replaced by sliding friction—a resisting force of somewhat smaller value. The valve quickly opens or closes, and the inertia of the system in all probability will cause the valve to overshoot its mark; that is, the valve will be opened or closed an amount in excess of that needed for proper correction.

This action will contribute to the extent of the deviations noted earlier as cycling. The lost motions in the joints of the linkage and the friction in the overall system will result in a condition that will permit the float device to move a certain distance above and below the set point without operating the control valve. The range of values for which this condition exists is called the *dead zone.* All automatic controllers have dead zones, although in many cases the zones are remarkably small. Dead zone is closely related to the sensitivity of a controller: a controller having a very narrow dead zone is said to be sensitive.

Capacity and Resistance

It should be clear that each of the systems studied thus far, whether manually or automatically controlled, possesses at least one element that is capable of storing energy. The tank whose liquid level has been controlled stores energy in the form of the weight and hydrostatic pressure of the water it contains, while the water heater stores heat energy. Those parts or agents of a system or process capable of storing energy are called *capacities.*

Figure 1.9 shows two water tanks, each manually controlled as to its inflow and discharge. Both tanks have the same height (*h*) and the same liquid level (*l*), but the diameter of the tanks differs considerably: one is 2 feet in diameter, and the other is 4 feet. The hydrostatic pressure developed at the discharge of each tank will be identical, since the factors that account for this pressure—liquid level, density of the water, and so forth—are identical. However, the capacities of the tanks and thus the amount of energy each can store are very different.

What applies to energy storage in the water tanks also applies to hot-water systems. The greater the volumetric capacity of the hot-water tank, the greater will be its heat energy storage capacity for a given water temperature. The hot-water system, however, differs importantly from the liquid-level system in one respect: it is a *multiple-capacity* system, while the liquid-level system is virtually a *single-capacity* system. In the liquid-level system the only significant storage of energy is represented by the weight of water and its height. In the hot-water system the bulk of the water within the heater tank stores energy in the form of heat, but so do the heating coils and the metal walls of the tank. Clearly, there is more than one capacity to deal with; and, if the tank and the heating coils are made of very thick metal, these additional capacities

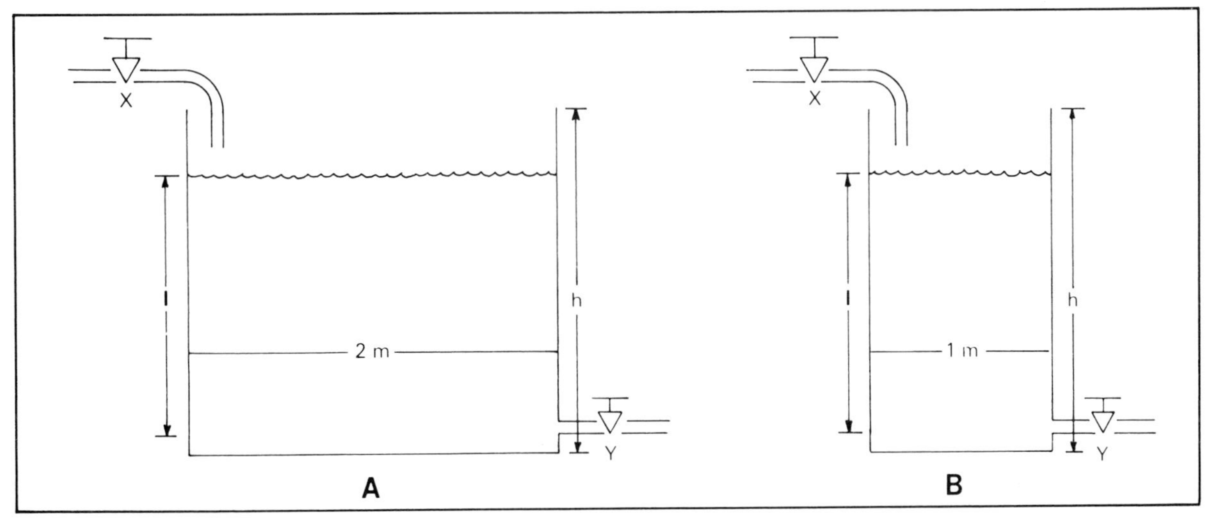

Figure 1.9. A comparison of energy-storing capacity

will become tremendously important factors in the problem of control.

The *resistance* encountered in a control problem comes into play when it is necessary to have a transfer of energy from one capacity to another in the system or process. For example, in the case of the hot-water system, heat energy is transferred from the steam coils to the water. The transfer of energy between two capacities is never instantaneous. The transfer is always resisted to some degree by one or more components of the system—for example, the walls of the steam coils and layers of steam and water on either side of the coils. Any component, or part, of the system that opposes the free transfer of energy between two capacities is called a *resistance.*

An example of a liquid-level system containing two capacities and a resistance connecting them is shown in figure 1.10. The resistance in this case is a restricted passage connecting two tanks near their bases. The effects of this resistance will be quite similar to the resistance existing in the hot-water system between the steam coils and the water in the heater tank.

Process Reaction Rate and Other Characteristics

Assume that the liquid-level control system of figure 1.11*A* is stabilized and therefore operating with a steady liquid-level height. At an arbitrary time referred to as t_0, the final control element, valve *X,* is suddenly operated to a slightly greater opening. The change in the valve setting is assumed to take place instantaneously, and this sort of action is called a *step change.* The rectilinear curve of figure 1.11*B* showing the rate of flow through the final control element resembles a step at time t_0.

The liquid level, as indicated by the curve of figure 1.11*C,* begins an immediate response to the change in input to the tank. The response is rapid at first, then tapers off as it

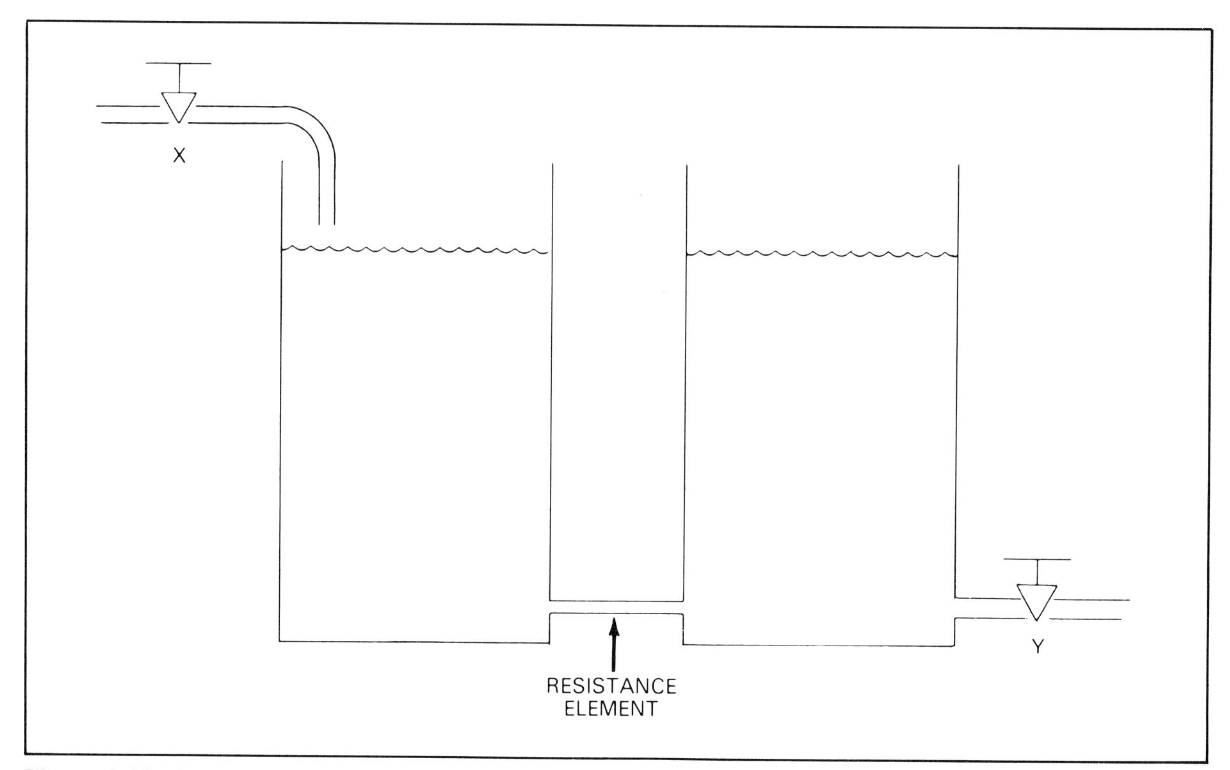

Figure 1.10. An example of resistance in transfer of energy from one capacity to another

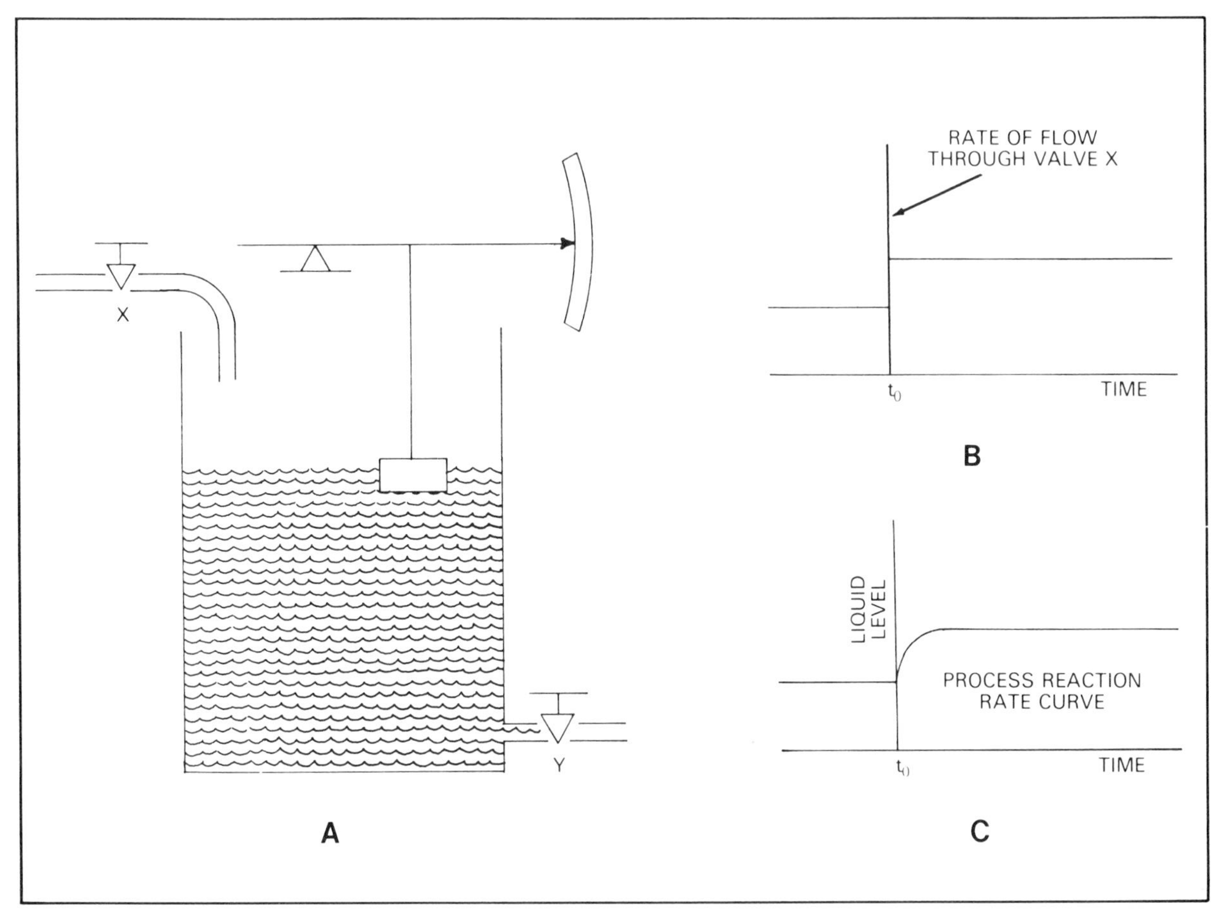

Figure 1.11. Reaction to a sudden operation (step change) of a control valve to a greater opening in a liquid-level control system

nears the new stable value of liquid-level height. The fact that the level begins to rise simultaneously with the step change in valve *X* indicates there is virtually no delay and no resistance to a transfer of energy to the system. The steep slope of the curve immediately after the step change in valve *X* indicates that energy is being stored in the tank at a rate somewhat greater than it is being dissipated through valve *Y*. At the new stable value the energy stored in the tank is at a new high but has ceased to increase. As noted earlier, the ability of a component in a process to store energy is called capacity.

The rate of change in the controlled variable that took place following the step change in valve *X* is called the *process reaction rate* of the system. One can learn much about the characteristics of a process and its control system by observing the process reaction curves. In order to obtain the curves for an automatically controlled system, certain parts of the automatic control must be temporarily disabled; that is, the system must not be allowed to correct the "upset" condition that is brought about by the change in setting of the final control element. A study of the process reaction curve will aid in determining the most simple sort of control system that will satisfactorily meet the needs of the process.

The system diagrammed in figure 1.11 is simple to control, largely because only a single capacity is involved and little or no resistance

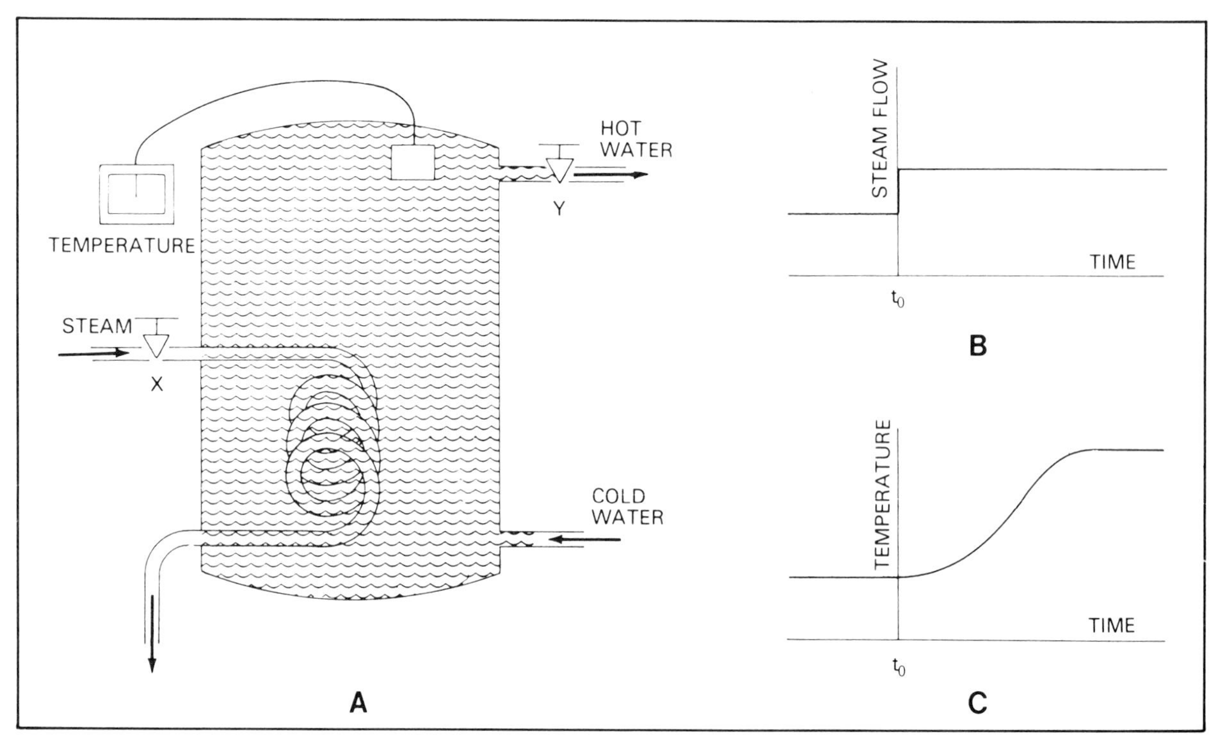

Figure 1.12. Reaction to step change in the control valve in a water-temperature control system

complicates the problem. Figure 1.12 depicts a more complicated system, and the rate of change in controlled variable following a step change in valve X follows a curve different from that shown for the system in figure 1.11. Note that the temperature change shown by the indicator does not begin simultaneously with the sudden additional opening of valve X. Once the controlled variable begins to change, it accelerates for a while; then, as it approaches the new stable value, its rate of increase declines.

The delay in response of the indicator to a change in the final control element setting is caused by several factors. For one thing, it is necessary to have a transfer of energy from one capacity to another—from heating coils to water. This form of delay is called *transfer lag*. A delay is also caused by the fact that a certain amount of time is consumed in carrying the temperature change from the steam coils to the location of the primary element. This delay is called *transportation lag,* or *dead time.* Another delay will be involved in the primary element responding to the change in temperature of the water.

The rate of change in the controlled variable that occurs as a result of a step change in the final control element—that is, the *process reaction rate*—has a curve shape that is determined largely by the size of the capacities involved in the process. Figure 1.13 shows two tanks of very different capacities, and the reaction curves that might apply to them with respect to liquid-level heights. The small-capacity tank has a very rapid reaction rate, and a new stable value of liquid level is attained in relatively short time. In the larger tank, the reaction rate is much slower for a given change in the setting of valve X. Figure 1.14 contains two hot-water systems and their reaction rate curves. The system in figure 1.14A contains a small-volume tank and very light heating coils, while the system in figure 1.14B contains a moderate-volume tank and very heavy heating coils. The point to notice

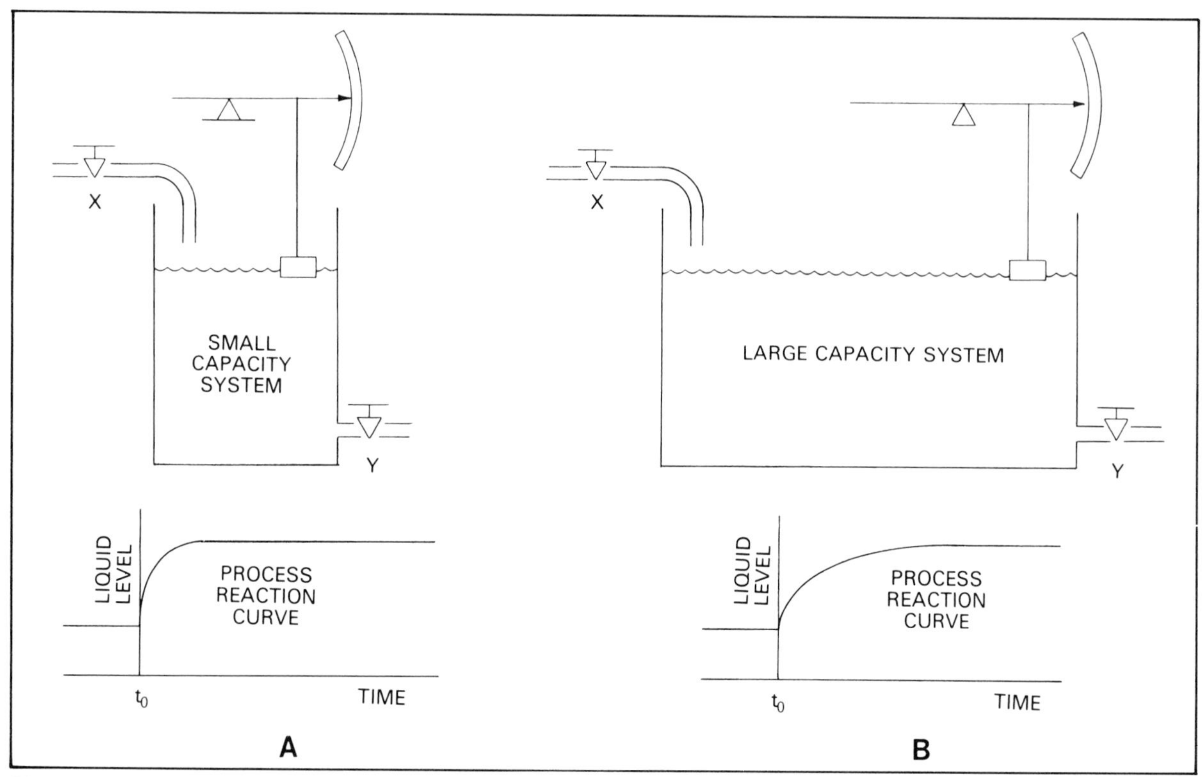

Figure 1.13. Relation of process reaction rate to capacity in liquid-level systems

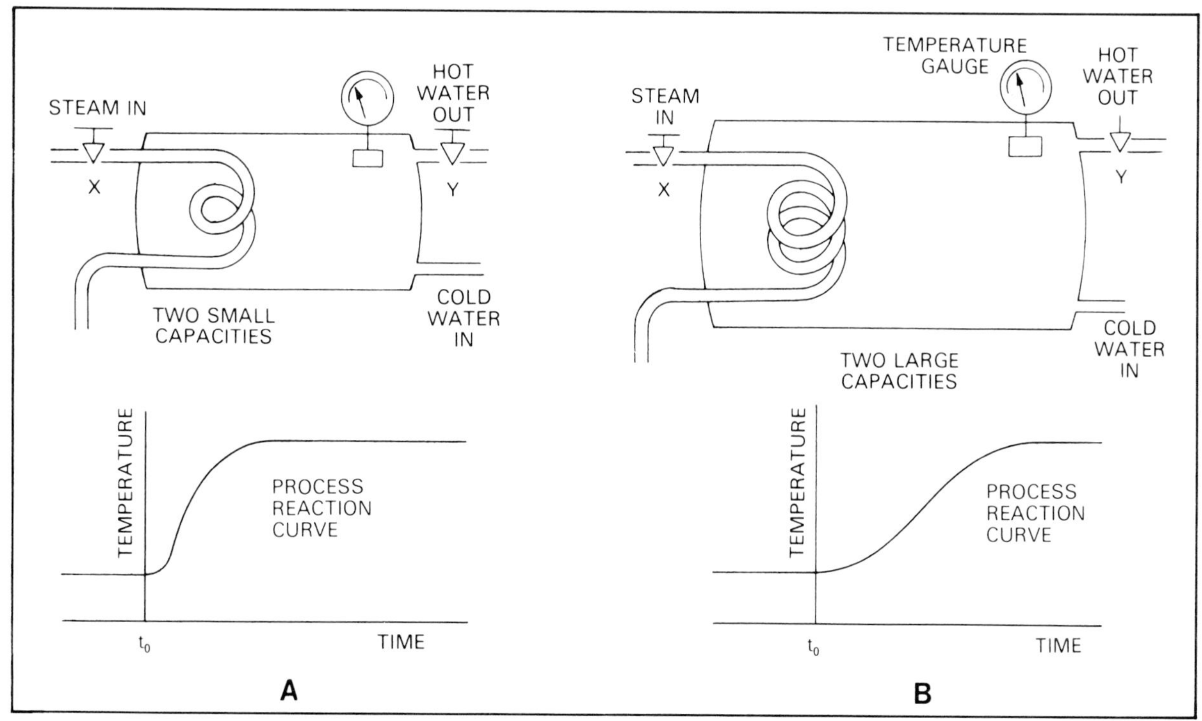

Figure 1.14. Relation of process reaction rate to capacity in water temperature control systems

here is that the larger volume of water, which represents a larger capacity, causes a much gentler sloping of the reaction curve—a slower reaction rate. The dead time is greater due to a greater distance between the area at which the heat is introduced and the point at which the measurement is made. The gentle convex sloping of the curve that begins after dead time is a result of the resistance to the energy transfer between steam coils and the water.

The study of reaction curves is of importance to engineers charged with the responsibility of designing or improving control systems. The practical application of the curves is based on many empirical equations, that is, equations derived from actual situations and experience. It is enough to learn some of the terminology and the fact that most processes and their control systems possess such properties as slow reaction rates, dead time, and transfer lag.

Modes of Control

The systematic action that the controller follows in response to changes in the controlled variable is a *mode*. Many modes are possible; a few of them are proportional, floating, proportional plus reset, and on-off, or two-position.

Proportional Control

The tanks shown in figure 1.15 illustrate proportional control. In *A*, the control valve allows water to flow from the tank at a rate of 50% of maximum flow. If the discharge valve is open enough to require half again as much flow (75%), the level in the tank will begin to fall as the new demand is supplied (*B*). The control valve will open wider, but the level will decline until such time as the inflow equals the outflow. If the system is a means of liquid-level control, the point at which the system stabilizes will be below the original set point for 50% of maximum flow. A similar situation will exist if the discharge rate falls somewhat below 50% of maximum. In this case the liquid level will stabilize above the set point.

Assume that the tank in figure 1.15 is 100 inches deep, that the float can move up and down over this entire distance, and that the control valve is completely closed only when the tank is full and is completely open only when the water level in the tank falls to zero on the liquid-level scale. Also, assuming that the rate of flow through the control valve *X* is a straight-line function of its opening (meaning that for each incremental movement of the valve stem, the flow changes in exact proportion to that movement), then for each change in flow from the discharge valve *Y*, a proportional change in opening of valve *Y* will occur. For this situation, a maximum change in the controlled variable (liquid level) produces a maximum change in the control valve *X*. Any controller for which there exists a continuous linear relation between the value of the controlled variable and the full range of positions of the final control element is called a *proportional controller*.

The one-to-one ratio that exists between changes in a controlled variable and the final control element brings up another bit of terminology—*proportional band*, sometimes called the *throttling range*. The proportional band is the range of values of a controlled variable that corresponds to the full operating range of the final control element. In the example above, the proportional band is 100% because a full range in a controlled variable produces a full range in setting of the final control element.

By changing the lever action of the linkages between primary and final control elements in figure 1.15, the control valve can be made to operate from fully closed to fully open with only a comparatively small change in controlled variable. Assume further that the leverage system has been changed so that if the water rises in the tank to 55 inches (this

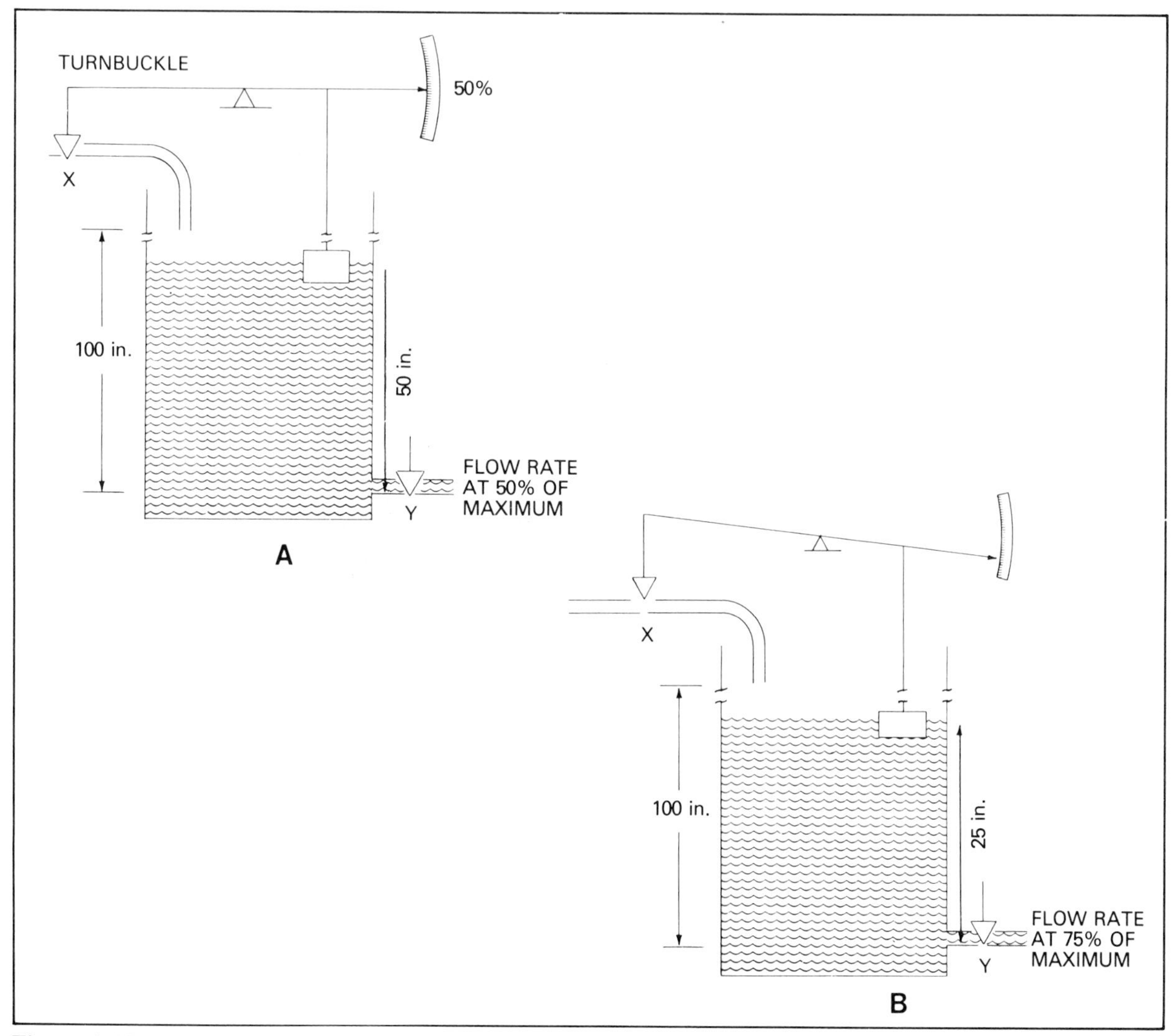

Figure 1.15. A system using proportional control with a proportional band of 0 to 100 percent

will be a rise representing 5% of the overall range of the controlled variable), the control valve will be fully opened. In this case the proportional band would be said to be 10%, because a change of 10% in the overall range of the controlled variable (from 55 to 45 in.) would produce a full range of openings in the control valve. Obviously, by proper choice of valve for the final control element and/or changes in the lever action of the linkage system, practically any proportional band desired can be obtained—from perhaps a few percent to more than 100%.

Generally speaking, proportional controllers are suitable for use in systems having a slow reaction rate, load changes that are neither large nor rapid, and small transfer lag and dead time. The offset deviation can be reduced by narrowing the proportional band, but a limit is soon reached if there are appreciable dead time and lag present. Exceeding this limit by narrowing the band still more causes an unceasing oscillation about the set point, as shown in figure 1.8. The curves of this figure might exist for some particular system but are not general in nature.

The Floating Mode

An arrangement in which an electric motor opens and closes a control valve through a worm-gear drive is shown in figure 1.16. The motor revolves in one direction when energized across contacts *1* and *2* and in the opposite direction when energized across contacts *1* and *3*. Contacts *2* and *3* are stationary, while contact *1* is fitted to the end of a length of spring steel that moves up and down by action of the primary element (float) and its connecting linkages.

When the liquid level in the tank shown in figure 1.16 changes enough to cause a meeting between contact *1* and either *2* or *3*, the motor will revolve at a constant speed and move the valve stem in a way that will tend to restore the liquid level to the set point. Once the float is restored to the desired position, the contact connection is broken and the motor stops.

The system shown in 1.16*A* is called a single-speed floating control. In this form of control the final control element slowly moves toward either the open or the closed position when the controlled variable deviates a predetermined amount from the set point, and the rate at which it moves is a single speed for all deviations. So long as the controlled variable stays within its prescribed differential gap—the range of values existing for a no-contact relation between contacts *1, 2,* and *3*—the valve "floats" in a partially open position.

For the single-speed system, the rate at which the valve is opened or closed must be slow enough to prevent serious overshooting. High-speed response of the valve, coupled with a narrow differential gap, can cause the controlled variable to cycle continuously from one side to the other of the set point. A possible form of floating control with two speeds is shown in figure 1.16*B*. Note that the motor can now be energized from either of two contacts for a given direction of rotation. For slight deviations, the motor is energized

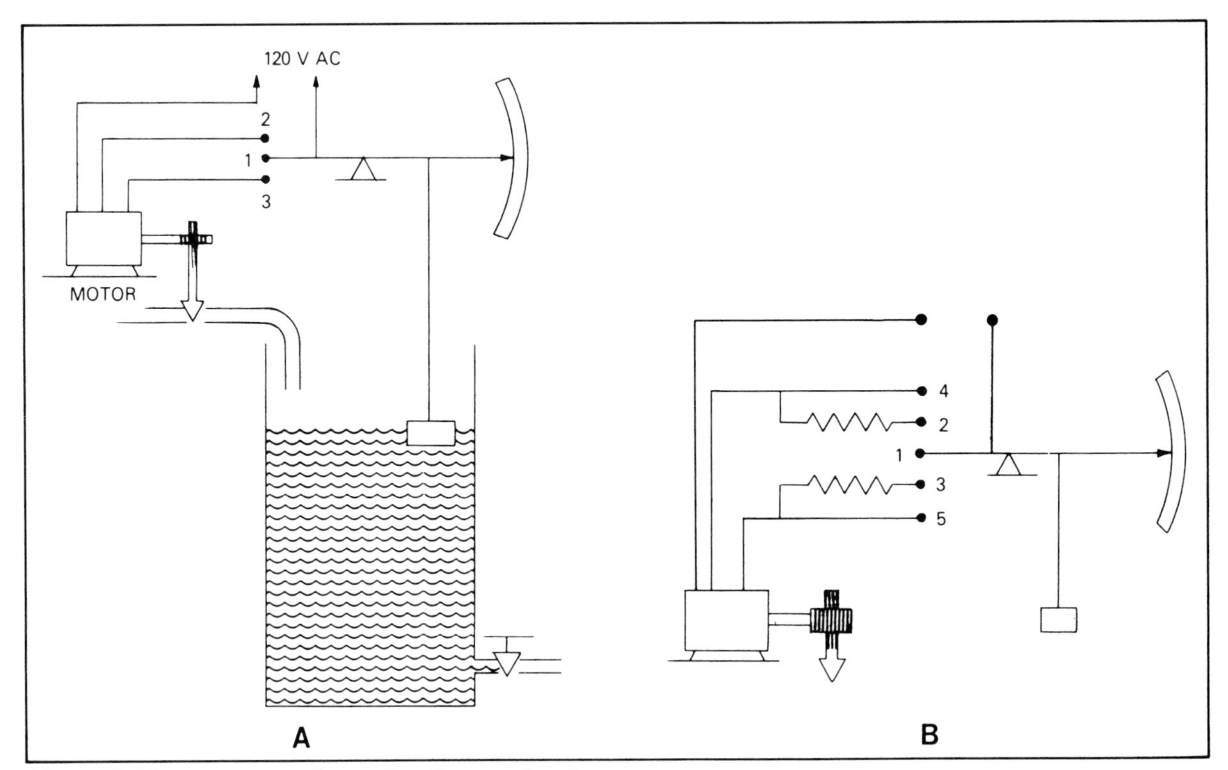

Figure 1.16. A system using a floating mode of control: *A*, single-speed and *B*, two-speed.

through a resistor that causes it to turn at a relatively low speed. For large variations, the resistor is bypassed, and the motor turns at high speed. In effect, the controlled variable has two differential gaps.

The ultimate form of floating control is one called *proportional-speed floating control* (fig. 1.17). In this form, the speed at which the motor operates to position the final control element is in proportion to the amount of deviation of the controlled variable beyond the differential gap. The motor is energized for each direction of rotation through a variable resistor whose value becomes less in proportion to the amount of deviation. As the resistance value decreases, the motor speed increases, and vice versa.

The floating mode of control is quite useful and provides best results where the amount of transfer lag and dead time are small. Such control can adequately deal with processes having fast reaction rates, but the single-speed mode cannot cope with rapid load changes.

Proportional Plus Reset Mode

A turnbuckle is added to the linkage system to aid in changing the set point of the controller. This turnbuckle idea might be adapted to an automátic system for reestablishing the set point of a proportional controller in the face of varying discharge rates.

Remember that the system shown in figure 1.15 was assumed to be stable when the set point was 50 inches of water and input-output rates were 50% of maximum flow. Also, recall that when discharge reached 75% of maximum flow, the water level dropped until the final control element opened to 75% of maximum. Now, the level can be readjusted to the 50-inch mark by *manually* adjusting the turnbuckle to open the valve wider. This action will cause the water level to rise in the

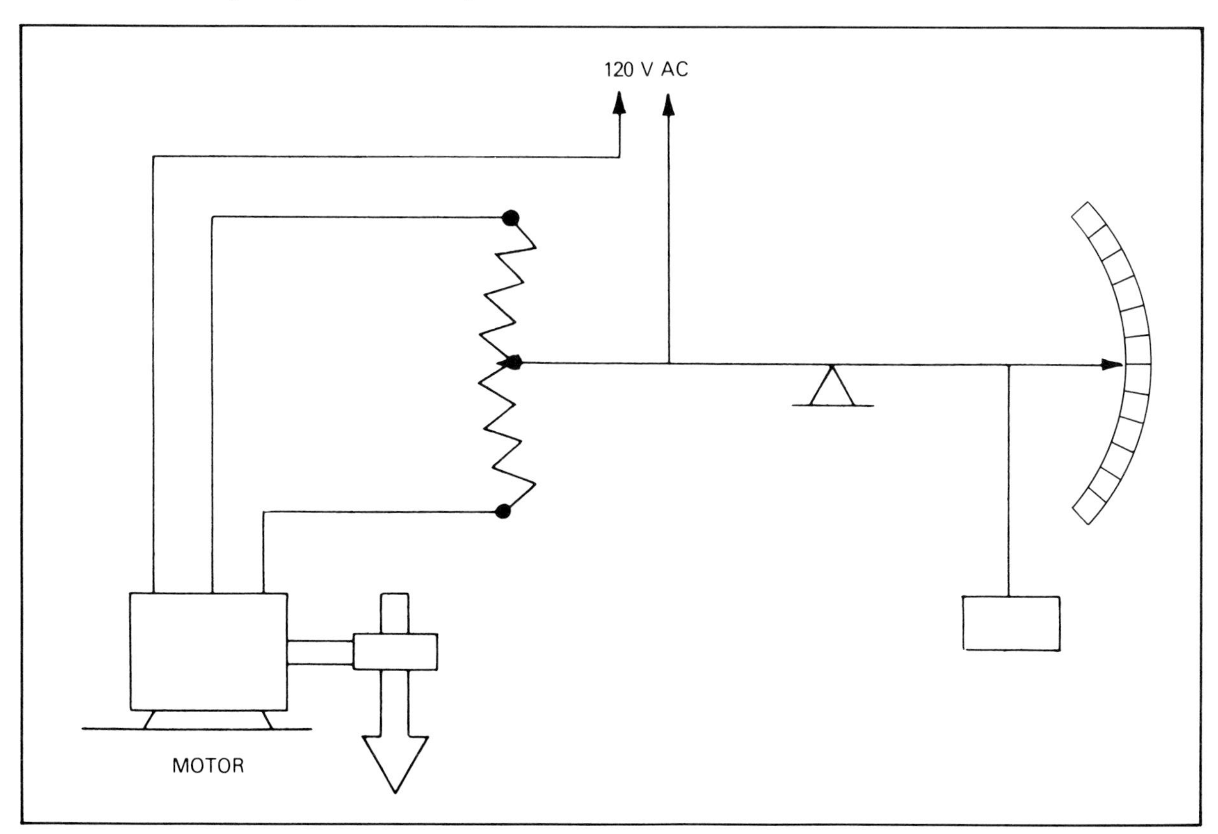

Figure 1.17. A proportional-speed floating control system

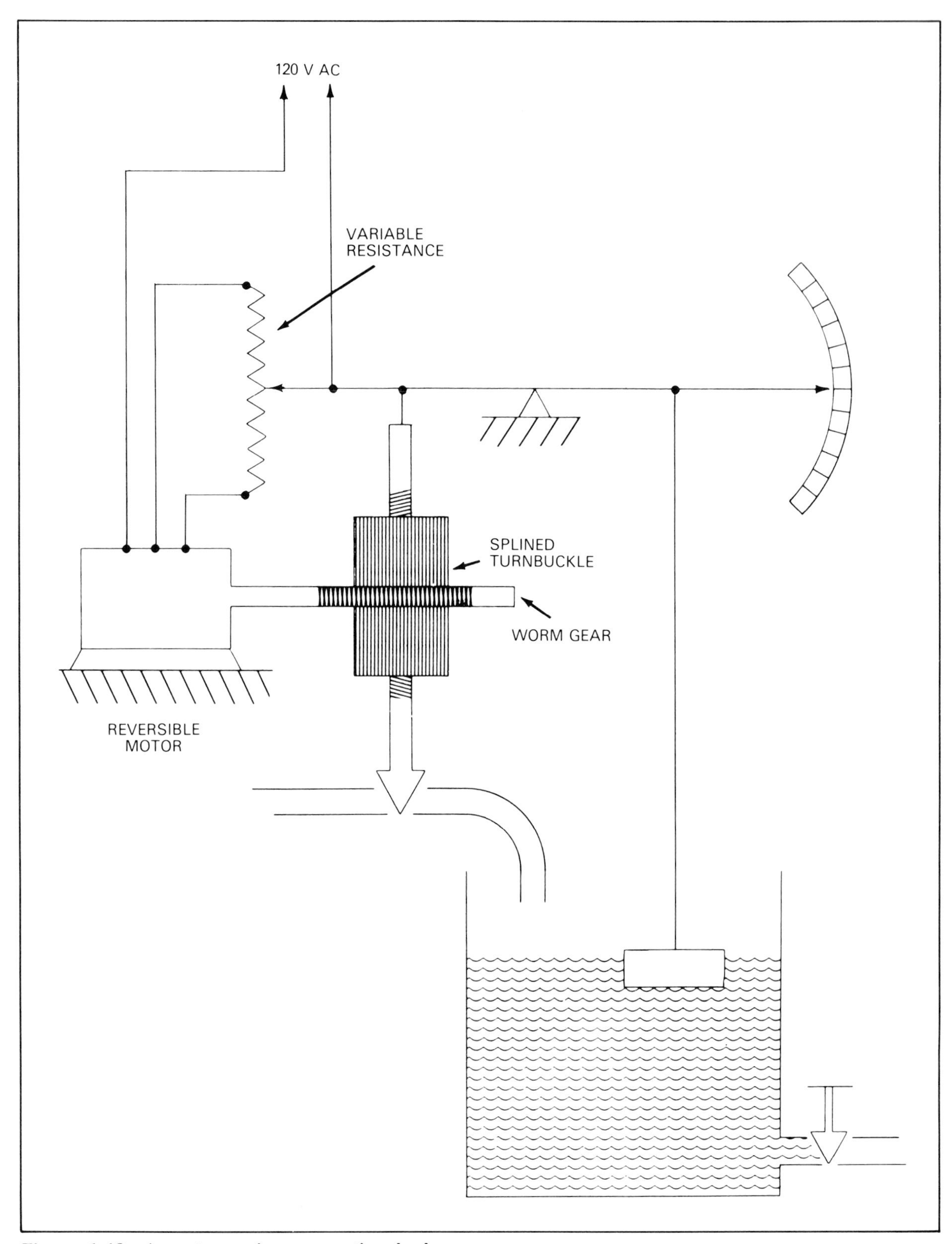

Figure 1.18. A system using proportional plus reset mode of control

tank, because the rate of inflow will exceed the discharge rate. Eventually the original set point will be reached, and the inflow and discharge rates can be equalized.

It is conceivable that a mechanical arrangement could be incorporated into the system for automatically correcting the offset that occurs in a straightforward proportional controller. An alternative way of operating the turnbuckle is with a reversible electric motor very similar to the one used in the single-speed floating control. The arrangement (fig. 1.18) simply combines a proportional-speed floating mode with a proportional controller. This system, called proportional plus reset, combines the great stabilizing features of a proportional controller with the added advantage of the reset capability offered by the floating mode.

Proportional plus reset controllers can be used to control processes that have a large amount of dead time or transfer lag; but, if these factors become very large, it will be necessary to use a broad proportional band and a slow reset rate. An additional feature is sometimes incorporated with the proportional plus reset controller that will largely offset the effect of dead time and transfer lag. It is called *rate response*—a term derived from the fact that the action produced by the feature is based on the rate at which the controlled variable is changing value.

On-Off, or Two-Position, Mode

Many systems of automatic control—for example, domestic hot-water heaters or central heating systems—employ final control elements that are characterized by the fact that the flow through them is either maximum or zero. This mode of control is called *two-position,* or *on-off,* mode. The actuating mechanism for a two-position final control element can be an electrically operated, a pneumatically operated, or a self-operated device. A self-operated control element is one that uses energy from the controlled medium for its operation.

All forms of on-off controls use appreciable differential gaps to prevent excessive cycling, but this gap can be reduced to a small value if the capacity of the controlled medium is large and the reaction rate is slow.

An on-off type of control is used in the hot-water system shown in figure 1.19. Steam is the control agent, and its flow is controlled by a solenoid valve, which is open when energized and closed at all other times.

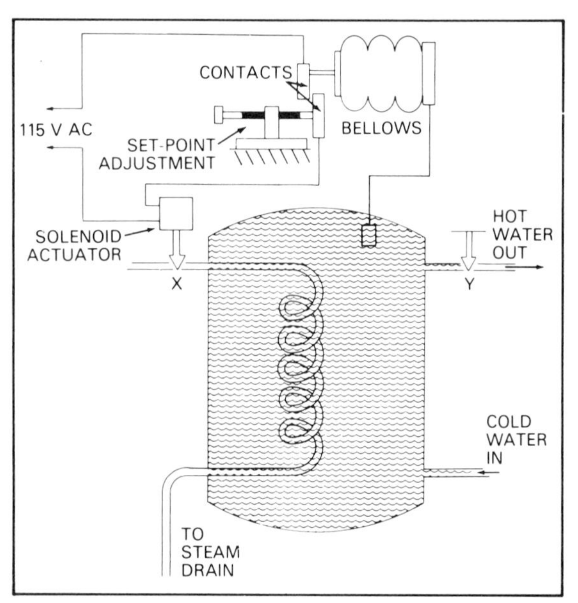

Figure 1.19. A system using on-off mode of control

Conclusion

Measuring and controlling are necessary functions in the efficient use of natural resources and in the production of man-made commodities. Understanding the terminology of instrumentation is vital to the study of measuring and controlling methods. Thus, armed with a better understanding of the phraseology that will beset you throughout your study, you should be able to proceed to learn the simple details that make up the complex overall subject of measurement and control.

CHAPTER 2

The Units and Dimensions of Measurement

The study of instrumentation principally concerns the measurement of only a few fundamental quantities—length, mass, time, and temperature. To a lesser extent, electric current, which is also a fundamental quantity, is involved in several aspects of instrumentation. A *quantity* is something that can be measured. A *fundamental quantity* is one that cannot be subdivided into other quantities. Speed is a physical quantity because it can be measured, but it is not fundamental. Speed is comprised of two fundamental quantities—length and time. Speed is defined as distance divided by time and is expressed in feet per second, miles per hour, and so forth.

Temperature, electric current, luminous intensity, and amount of substance are also regarded as fundamental quantities. Temperature can be a measure of the heat energy that an amount of substance contains, and heat energy has dimensions of length, mass, and time. However, treating the above four quantities as fundamental is quite useful.

All quantities, have *dimensions*. The most commonly known dimension, of course, is length, but it is also proper to regard mass and time as dimensions. For example, speed has dimensions of length and time, and force has dimensions of length, time, and mass.

A *unit* is a standard measure of a quantity. Units of measurement are established by law or adopted through common usage. Units are used for measuring quantities of any size, with the measurement always expressed in terms of the chosen unit.

Systems of Units

Two centuries ago those nations of the world having industrial and other developed technology enjoyed little agreement among themselves concerning systems of measurement. Each nation or empire had its own system. Standards of measurement, if they existed at all, were poorly conceived, and conversion of units among the various systems was onerous and unpredictable. Of course, commerce and communication among the nations of that day were neither as abundant nor as important as they are today. Today, the world is well on the way to adopting a single set of measurement standards common to all nations.

Two systems of measurement units are used in this book: the *conventional system* and the *International System of Units (SI)*. The first is also commonly known as the British, Imperial, or English system. The second is an outgrowth of the metric system. Both systems are used to some extent, although the conventional system is predominant in this text.

The Conventional System of Measurement Units

The system of measurement generally used for trade and commercial purposes in the United States is referred to as the *conventional system*. It has been in use for a long time, but it is characterized by ambiguity and a poor mathematical relation between one unit and another used to measure the same quantity. For example, the unit of mass in the conventional system is the pound, which in the United States is defined in terms of the kilogram, an SI unit. The pound is divided into ounces, drams, grains, and other units, each bearing some clumsy relation to the pound. The yard is the standard of length in the system, and it is subdivided into feet and inches. The yard is defined in terms of the metric (SI) system.

In addition to the poor mathematical relations existing among the units of measurement in the conventional system, another serious shortcoming is the fact that a given name for a unit quantity can have several different meanings or values. For example, at least three kinds of miles are used in the system: nautical, geographical, and statute. The same is true for the volume measurement unit, the gallon. Each of the several different gallons differs at least superficially from the others.

The International System of Units (SI)

The International System of Units, known worldwide as Le Système International d'Unités (SI), is an outgrowth of the metric system that was developed by a body of French scientists in the 1790s. The metric system represented a remarkable advance in the science of weights and measures, and its adoption by many nations proceeded rapidly. SI retains the standards set up by the metric system but encompasses many improvements. Almost totally unambiguous, calculations involving its units are facilitated by the logical relations existing among quantities and the powers-of-ten multiples used to measure a given quantity.

The initial metric system recognized only three fundamental quantities—mass, length, and time. SI has added four others—temperature (kelvin), electric current (ampere), amount of substance (mole), and luminous intensity (candela).

The use of SI requires that strict attention be given to symbols and typographical style. Abbreviations are not used in SI—only symbols. The unit of length in SI is the metre, and its symbol is m, with no period. The unit of force is the newton, and its symbol is N, with no period. The SI temperature scale is kelvin, and its symbol is K with no degree mark (°). The Celsius scale (°C) is not strictly an SI entity but is used for most nonscientific needs. The appendix contains a more extensive discussion of SI.

Table 2.1 is a tabulation of conventional units and symbols and their relation to SI units and other quantities.

The Units of Length Measurement

Length is one of the fundamental quantities, and a score or more units in the various systems of measurement are used for its expression. Feet and inches and millimetres and metres are the units that are of greatest importance in instrumentation. Length is used to measure more than just *distance*. It is used to measure *volume* and *area* as well.

Area Measurement

If two dimensions form a product, that is, if the two are multiplied together, the result is in effect an area. Area is expressed in square units, such as square feet (ft^2) or square metres (m^2). If a plot of land is 90 feet long

TABLE 2.1
CONVENTIONAL AND SI UNITS OF MEASUREMENT

Quantity Name and Symbol	*Dimension (Relation to Other Quantities)*	*Conventional Unit and Symbol*	*SI Unit and Symbol*
Length (L)	Fundamental	foot (ft)	metre (m)
Mass (M)	Fundamental	pound (lb)	kilogram (kg)
Time (t)	Fundamental	second (sec)	second (s)
Temperature (T)	Fundamental	degree Fahrenheit (°F) degree Rankine (°R)	degree Celsius (°C)* kelvin (K)
Electrical current (I)	Fundamental	ampere (A)	ampere (A)
Amount of substance (Mol)	Fundamental	mole (mol)	mole (mol)
Luminous intensity (I)	Fundamental	candela (cd)	candela (cd)
Acceleration (a)	L/t^2	feet per second per second (ft/sec^2)	metre per second per second (m/s^2)
Area (A)	L^2	square feet (ft^2)	square metre (m^2)
Density (ρ)	M/L^3	pounds per cubic foot (lb/ft^3)	kilogram per cubic metre (kg/m^3)
Energy (E), Work (W), Heat (Q)	$M \cdot L^2/t^2$	foot-pound force (ft-lbf)	joule (J)
Force (F)	$M \cdot L/t^2$	pound force (lbf)	newton (N)
Frequency (f)	$1/t$	hertz (Hz)	hertz (Hz)
Power (P)	$M \cdot L/t^3$	watt (W) horsepower (hp)	watt (W)
Pressure (p)	$M/(L \cdot t^2)$	pound force per square inch (psi)	pascal (Pa)
Velocity (v)	L/t	foot per second (ft/sec)	metre per second (m/s)
Volume (V)	L^3	cubic foot (ft^3)	cubic metre (m^3)

*Although not a part of SI, the Celsius temperature scale is used with SI.

and 40 feet wide, then the area of the land is calculated to be

$$90 \text{ ft} \times 40 \text{ ft} = 3{,}600 \text{ ft}^2.$$

Notice the implication that feet times feet equals square feet, just as $x \times x = x^2$. Notice also that the names of the quantity units can be treated as mathematical terms.

Frequently an area measurement turns up unexpectedly, as in the expression for kinetic energy: one-half mass times velocity squared ($\frac{1}{2}Mv^2$). Since velocity is defined as length divided by time (L/t), velocity squared is length squared divided by time squared ($v^2 = L^2/t^2$), or area divided by time squared (A/t^2). Such expressions have solid foundations in physics and mathematics and should be no cause for wonderment.

Volume Measurement

Three dimensions of length taken as a triple product form a measurement of volume. Thus,

$$a \text{ ft} \times b \text{ ft} \times c \text{ ft} = abc \text{ ft}^3.$$

Here again the unit name *foot* is treated as a mathematical entity. Scores of units for expressing volume exist, but in instrumentation such terms as *cubic feet* (ft^3), *cubic inches* (in.^3), *cubic centimetres* (cm^3), *barrels, gallons,* and *litres* are by far the most common.

Terms that express volume come about quite naturally sometimes. Consider the result of multiplying area times velocity, for example:

$$A \times v = L^2 \times L/t = L^3/t.$$

Simply, the result is volume per unit of time.

The Units of Time Measurement

Time is measured in seconds, minutes, hours, days, and so on, and these units have been almost universally adopted. In SI measurements the second is the standard unit.

The second was formerly defined in terms of rotation of the earth about its axis. The year 1900 was chosen as the standard base for the invariable unit of time, the *ephemeris second.* (*Ephemeris* is a word that describes a chart, table, or almanac related to astronomy, specifically the orbital motions of the planets. The use of such an obscure word here seems unfortunate. The words *universal* or *standard* would have been better understood.)

Today, the second is intended to be exactly equal to the ephemeris second but is not defined in terms of planetary motion. It is defined in terms of the frequency of radiation from the cesium 133 atom. Atomic clocks controlled by cesium radiation are extremely accurate and provide a time base adequate to meet the exacting demands of modern technology. Such clocks are expensive, but their time signals can be made available through radio signals to laboratories and other activities.

The Units of Temperature Measurement

Temperature is regarded as a dimensionless quantity, although the presence of temperature in an object is clear indication that the object contains energy. As noted earlier, temperature is considered to be a fundamental quantity.

Three temperature scales are of importance in instrumentation: the Celsius scale (°C), formerly called the centigrade scale; Fahrenheit scale (°F), a part of the conventional system of units; and kelvin scale (K). Foundation of the Celsius scale is the choice of 0° and 100° as the freezing and boiling points of pure water at standard atmospheric pressure. The Fahrenheit scale is arranged to have equivalent points at 32° and 212°. The span of 1°C is exactly equal to the span of 1 K.

The kelvin scale is more adequately described as one of thermodynamic temperature. Kelvin can be a measure of the temperature, or it can be an interval of temperature. For example: 273 K − 250 K = 23 K, but not 23°C. Also, 38°C − 13°C = 25°C or 25 K. In the latter example, the 25 K is an interval of temperature.

The kelvin scale hypothesizes the existence of an absolute zero temperature, meaning that as temperature is lowered to ever smaller values, a point will eventually be reached at which it is impossible to go lower. This will be the point at which absolutely all heat energy in an object will have been removed, and all molecular motion within the object will cease. Absolute zero temperature has not been achieved, but its existence is quite accurately established. Temperatures within a fraction of a kelvin have been attained. On the basis of the Celsius scale, absolute zero temperature is *−273.15°*C.

The relation between the Celsius scale and the kelvin scale is a simple one:

$$K = °C + 273.15°.$$

An absolute temperature scale based on Fahrenheit measurements is called *Rankine* (°R). The relation between the Fahrenheit and the Rankine scale is

$$°R = °F + 459.6°.$$

Most applications of temperature measurement in this book are concerned with the Fahrenheit scale. Use of Celsius values and the absolute scales (kelvin and Rankine) are used where appropriate. The behavior of gases under various conditions of temperature and pressure is more easily understood when dealing in terms of absolute pressure and absolute temperature.

The relation existing between Celsius and Fahrenheit temperatures is more involved than the relation between Celsius and kelvin:

$$°F = \frac{9}{5}(°C) + 32°;$$
$$°C = \frac{5}{9}(°F - 32°).$$

The factor $\frac{9}{5}$ comes from the fact that 1°C represents a spread of $\frac{9}{5}$°F. The factor $\frac{5}{9}$ comes from the reverse relation; that is, 1°F is only $\frac{5}{9}$ of 1°C in spread. There are 100°C between freezing and boiling temperatures of water, and 180°F between the same points.

The Units of Mass, Weight, and Force

The concepts of mass, weight, and force, and the relations that exist among them are poorly understood ideas and require some discussion.

Mass and Weight

Mass is sometimes thought to be another word for weight, but in the strictest sense of meaning it is different. For all practical purposes the *mass* of a given quantity of something remains constant regardless of its location in the universe. The *weight* of an object, in the ordinary sense of meaning, is apt to vary according to the physical conditions that form its environment. For example, the astronauts are said to become "weightless" in orbit. The condition of weightlessness comes about from the fact that the force of attraction (gravity) the earth has for the astronaut is exactly counterbalanced by the centrifugal force resulting from the orbiting motion.

Mass can be defined as the measure of the inertial properties of a quantity of substance. This statement can be partly clarified by noting that inertia is the resistance the substance shows to a change in its state of motion. If it is standing still, it will resist an effort to move it. If it is moving, it will resist an effort to slow it down, speed it up, or change its direction of travel. The effort that is resisted by the inertia of the substance is properly termed a *force,* and the amount of force that is required to effect a given change in the state of motion is a measure of the mass of the substance.

Another way to define *mass* is to say simply that it is a measure of the amount of any given substance.

Force

It can be gathered from the discussion about mass and weight that force is closely related to each of them. In fact, *force* is defined as the product of mass and acceleration and therefore has dimensions of mass times length divided by time squared (ML/t^2).

For many practical purposes, weight and force can be considered synonymous. If, for example, the weight of an astronaut is said to be 180 pounds, in effect his mass is being attracted to the earth's mass by a force of 180 pounds-force.

Gravitational force. Two bodies of mass will interact with one another. They will be drawn toward one another, and the force of attraction between them can be used as a measure of their masses. The force of attraction acting between any two bodies is a gravitational attraction, and for pairs of small masses the force is extremely feeble and difficult to detect without special instruments. For those instances where one (or both) of the masses is very large (for example, if one of the masses is the earth), the force of attraction will be quite significant.

Every physical force has *direction* as well as magnitude. The force of attraction between two bodies of mass will have a definite direction. It will act along an imaginary line that connects the two centers of mass. When the astronaut is launched into orbit and becomes "weightless," forces are still acting on his mass. The two main forces, however, are acting in exactly opposite directions, so the force of gravity that pulls him toward the earth is neutralized by the equal and opposite centrifugal force that tends to drive him farther out into space. The net force acting on the astronaut is virtually zero, so his weight is also zero.

Magnetic field. Two magnets, whether of the permanent or the electromagnetic type, will be attracted strongly to one another if their unlike poles are brought close together. This is not a gravitational attraction, but it is a force that has direction and magnitude. A magnetic field is a force field, meaning that, if a magnetized particle of mass is placed in the field, the particle will be accelerated in some definite direction.

Electric field. Electric fields are also force fields in which an electrically charged particle of mass will be accelerated with a definite magnitude and in a definite direction. Electric and magnetic fields are related and similar. Their directions of force, however, may have different effects on a given particle of mass. An electrically charged particle will be accelerated in a given direction if placed in an electric field; when the same sort of particle is placed in a parallel magnetic field, it will be accelerated in a direction that will be perpendicular to that observed for the electric field.

Springs. A very common object that produces force is the spring, and it can take many forms—coil springs, spirals, leaf springs, and the like. In fact, almost anything that returns to its original shape after it has been stretched, compressed, bent, or twisted can be considered a form of spring.

Other forces. In addition to these forces, there are contained forces that result in pressure. When an airtight container is pumped up to a pressure somewhat greater than the surrounding outside air, forces act on the inner walls of the container, trying to push them out. These forces are resisted by the opposing stresses in the material of the container.

Mass and Force

In the original metric system the basic unit for mass was the gram. It represented the mass of 1 cubic centimetre of water at a temperature of 4°C, this being the temperature at which water is most dense. The gram may be expressed in powers of ten to arrive at milligrams, kilograms, and so on. In SI units, the kilogram is the base unit of mass. In the United States, the U.S. pound-mass is defined in terms of the kilogram (1 lb = 0.453592427 kg).

It has been noted that force is the product of mass and acceleration ($F = Ma$). In conventional units the unit of force is the pound-force, equal to the product of pound-mass and acceleration in feet per second per second. In SI units, force is expressed in newtons. A newton is the amount of force needed to accelerate a mass of 1 kilogram at 1 metre per second per second ($N = kg \cdot m/s^2$).

The measurement of mass and force in the conventional systems of weights and measures is made clumsy by the poor mathematical relation existing among units. For example, pound, instead of being expressed in multiples of the powers-of-ten, is expressed in *ounces, drams, grains, tons,* and other terms. Each of these units is related to the others in ratios that are troublesome to calculate—12 or 16 ounces per pound, 16 drams or 437.5 grains per ounce, for example.

A reasonable amount of care is required to avoid confusion from the fact that *pound* is used to represent mass in some instances and force in others. It is fairly common practice to indicate mass by the expression *pound-mass* and force by *pound-force.*

In order to measure anything, scales and fundamental units must first be established. These scales and units can be balanced against the unknown quantities, and the values of these quantities can then be determined in terms of the chosen units.

Determining mass by balancing gravitational forces. Once a standard of mass has been established, the best way to use it in determining unknown quantities of mass is to set up a balance of forces. On the one hand, a gravitational force between the earth's mass and the standard mass exists; on the other hand, a similar force exists between the earth and the unknown mass. By choosing the proper number of standard mass units, a balance of forces is attained (fig. 2.1).

Since the gravitational forces existing between earth and standard mass on the one hand and earth and unknown mass on the other are equal, then the masses are equal. Force (F) is the product of mass (M) and acceleration (a), or in equation form,

$$F = M \times a.$$

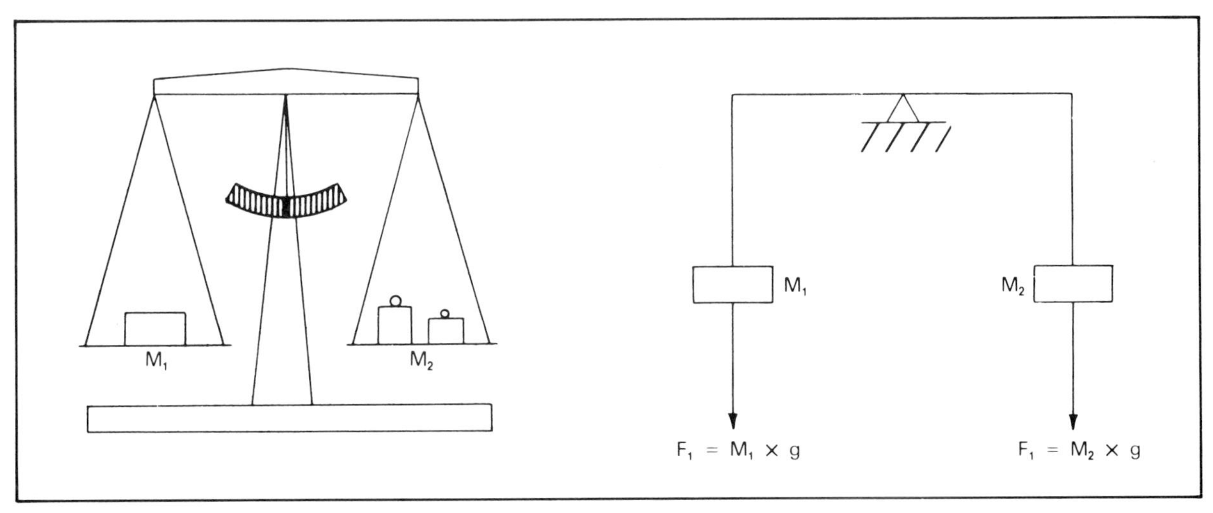

Figure 2.1. Balance scale for determining mass. Unknown mass M_1 is determined by balancing with standard mass M_2. Gravitational forces F_1 and F_2 are then equal.

For a condition of equilibrium, the balance of forces for weighing is

$$M_1 \times g = M_2 \times g,$$

where

M_1 = unknown mass;
M_2 = standard mass;
g = acceleration due to gravity at the earth's surface.

If both sides of an equation are divided by the same number, the result is still a valid equation; thus, the factor g can be removed by dividing it through the equation. The result is simply $M_1 = M_2$, showing that the two masses are equal for a balanced equation.

The factor g, acceleration due to gravity, is equal to about 32.2 ft/sec^2 or about 9.8 m/s^2 at the earth's surface. It varies slightly from point to point; for example, it is 32.155 ft/sec^2 at Washington, D.C., and 32.118 ft/sec^2 at Pike's Peak, Colorado. This factor g was determined quite accurately even in the time of Sir Isaac Newton.

Spring scales for measuring mass. The balance (fig. 2.1) is capable of determining mass without suffering an error due to a deviation in the assumed value of g. A spring scale is also used for weighing (fig. 2.2). The spring scale works on the principle of applying weight or *force* and measuring the distance the weight or force causes the spring to stretch. Within limits the spring will obey certain laws of physics and will stretch an amount that is in linear relation to the applied force; that is, it will stretch a certain amount if one unit of force is applied, twice that

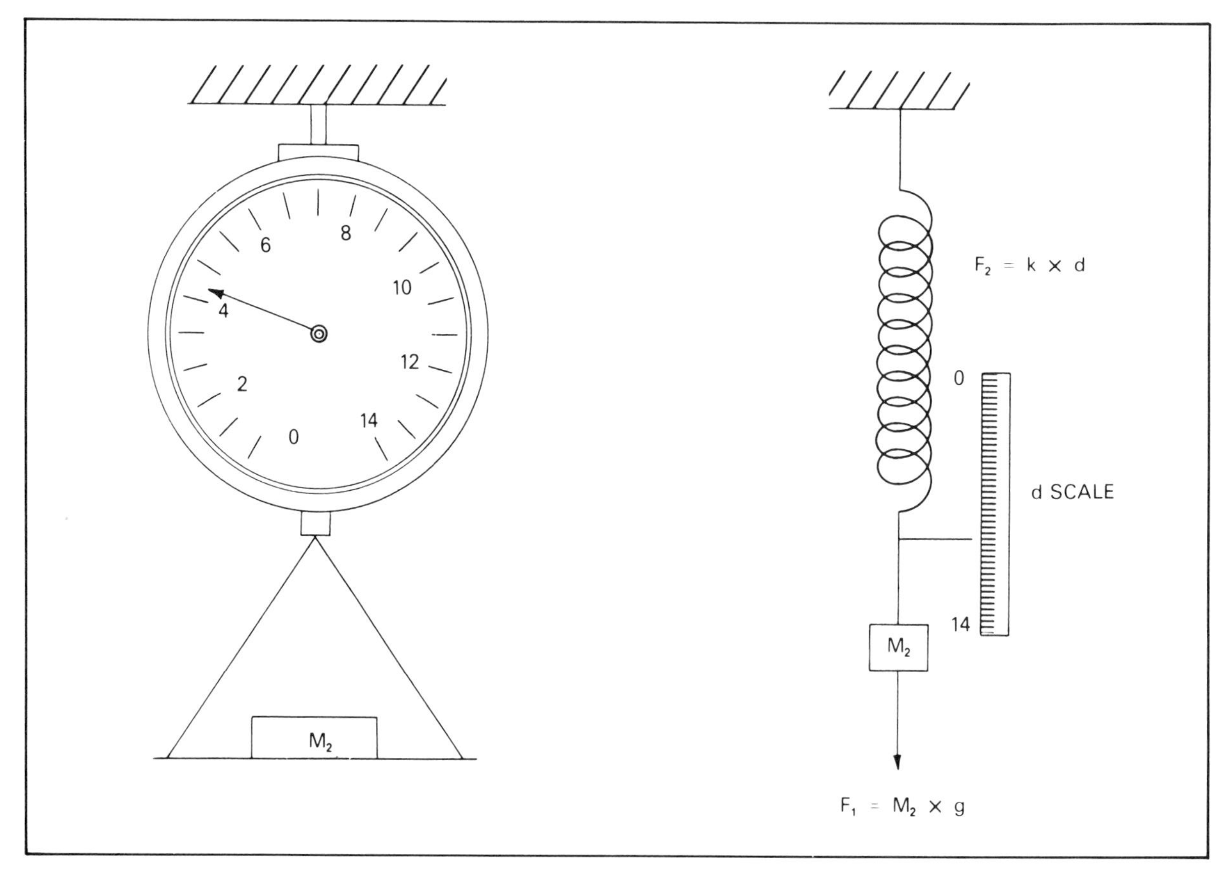

Figure 2.2. A spring scale determines mass by measuring the distance the spring is stretched. F_1 is the force exerted by the mass against the upward force F_2 of the spring.

amount if two units are applied, and so forth. The equation of the spring for weighing mass is

$$F = k \times L,$$

where

F = force;
k = force constant for the spring;
L = distance spring stretches for the applied force.

The spring constant k is determined by the properties of the particular spring. A strong, stiff spring has a large k, while a limber spring has a small k. Using a spring scale to weigh mass is again a balancing of forces. The force $k \times L$ is balanced against the gravitational force that is attracting the mass M being weighed, which is $M_2 \times g$. In equation form the balance of forces is

$$k \times L = M_2 \times g.$$

Notice that any change in the value of g is not compensated and that an error will arise. Spring scales are usually not legal for trade purposes, but as far as the error caused by deviation in value of g is concerned, they usually will indicate 0.99 absolute accuracy.

The Units of Work and Energy

Work and *energy* are sometimes used as synonymous terms. The words have very different meanings when used to discuss physical concepts on the one hand and nonscientific concepts on the other.

Work

In everyday use, the word *work* denotes any kind of mental or muscular effort. In science, however, the term *work* has a restricted and explicit meaning. In physical mechanics the accomplishment of work means that a *force* has acted to move an object. The amount of work accomplished is determined by (1) the distance the object is moved along a line parallel to the direction of the force and (2) the amount of force applied. This gives a simple definition of *work*: work W is the product of force F and length L; that is, $W = F \times L$. It is important to note, however, that movement of the object must be parallel to the direction of the force. For example, if a force acts to push an object toward the east, and the object actually moves in a northeasterly direction, the work done is the product of the total force applied and the total distance the object moves in an *easterly* direction.

As noted earlier, force, in addition to having magnitude, has *direction*. An object that is being moved by a force is obviously moved along a line that also has direction. However, even though work is the product of two quantities—force and distance—that are described as having direction, work has only magnitude. A quantity having both magnitude and direction is called a *vector* quantity, while one having only magnitude is called a *scalar* quantity.

Energy

Energy is another word that has different meanings in scientific and nonscientific usage. In everyday use the word *energy* can denote a capacity for activity of a mental, physical, or even social nature. In physics the term *energy* has an explicit and limited meaning. It denotes the *capacity to do work*. Energy, like work, has only magnitude and is measured in the same quantities—force and length.

Two kinds of energy are *potential* and *kinetic*. If an object or body is capable of doing work because it possesses a certain position, or because it is existing in a stressed condition, it is said to be a source of *potential energy*. A weight suspended above the floor has potential energy, and this is an example of energy by virtue of position. A tightly wound clock spring also has potential energy, and this is an example of energy by virtue of a stressed condition. If a massive object is in motion, it possesses *kinetic energy* because of this motion and is capable of doing work. Spinning flywheels, projectiles in flight, and automobiles in motion are examples of kinetic energy. Kinetic energy, just as in the case of potential energy, is measured in the same quantities as work—force and distance.

The Equations of Work and Energy

Although work and energy are measured in the same quantities, the equations that represent them sometimes have forms that require a small amount of interpretation in order to arrive at the basic product of force and distance. The equation for work is straightforward—the product of force and distance:

$$W = F \times L,$$

where

W = work;
F = force;
L = the lineal measurement between the beginning and ending points that represent the path over which the work is being computed.

Potential energy in the form of a suspended weight, or water trapped behind a dam, has the same equation as that for work. In this case, however, F is the gravitational force, or weight, and L is the vertical distance between a reference point and the point at which the weight is located, or the surface of the water behind the dam.

The potential energy stored in a compressed or stretched spring can be written as

$$E_p = \tfrac{1}{2}k \times L^2,$$

where

E_p = potential energy;
k = the force constant of the spring;
L = the distance the spring is stretched or compressed from its reference position.

It is known that $k \times L$—that is, the force constant times the distance the spring is stretched—is equal to force. Therefore, this equation is easily resolved into the form

$$E_p = \tfrac{1}{2}F \times L,$$

where $F = k \times L$.

Kinetic energy, the capability for doing work by virtue of a moving mass, is expressed in equation form as

$$E_k = \tfrac{1}{2}M \times v^2,$$

where

E_k = kinetic energy;
M = mass;
v = velocity.

Recalling the dimensions of force ($M \times L/t^2$) and velocity (L/t), the kinetic energy equation can be written as follows:

$$E_k = \tfrac{1}{2}M \times L^2/t^2.$$

$$E_k = \tfrac{1}{2}F \times L,$$

which is the same equation as that shown for the compressed or stretched spring. Manipulation of the last two equations, those for potential and kinetic energy, demonstrates that energy and work can be resolved into two dimensions—force and distance.

Several other forms of energy equations relate to rotating masses and spiral springs. These equations can be resolved into the products that involve only force and distance and perhaps a constant factor such as the $1/2$ noted in equations for kinetic energy and forms of potential energy.

The Relation of Power to Work and Energy

Work and energy are expressions for a product of force and distance. The *time rate* at which work is accomplished is defined as *power*. Power (P) is made up of force (F), length (L), and time (t) and is written in equation form as

$$P = F \times L/t.$$

In the conventional system of units, the work might be expressed in foot-pounds, and the time in seconds or minutes. A common unit of power in this system of measurement is the *horsepower,* which is defined to be 550 foot-pounds per second, or 33,000 foot-pounds per minute.

Electric power is usually measured in watts or kilowatts. In terms of watts, a *horsepower* is defined to be 746 watts, or 0.746 kilowatt.

The common practice followed by utility companies is to measure electric energy in kilowatt-hours. This is equivalent to saying *power* times *time,* or the product of power and time. The result is, in fact, energy or work.

The Dimensions of Various Quantities

In terms of science and instrumentation the word *dimension* means more than a measurement of length, or distance. All fundamental quantities are treated as dimensions. An interesting technique called *dimensional analysis* establishes certain physical quantities as fundamental, and from these few quantities all other compound quantities are derived in mathematical fashion. This sort of analysis is used to study cumbersome physical systems that do not lend themselves to easy, straightforward mathematical solution.

Pressure

Pressure (p) is defined as force (F) per unit area (A) and can be written as

$$p = F/A, \text{ expressed in lbf/in.}^2\text{, or psi.}$$

Obviously, any units of force and area can be used to express pressure, but in the conventional system the combination of pounds force and square inches is the most common. In SI, newtons and square metres are used to express pressure in pascals (N/m², or Pa).

Two other expressions for pressure are used quite widely in particular applications. They deserve some discussion because of their apparent serious departure from the basic dimensions of pressure—force and area. A U-tube manometer (fig. 2.3) is a device in

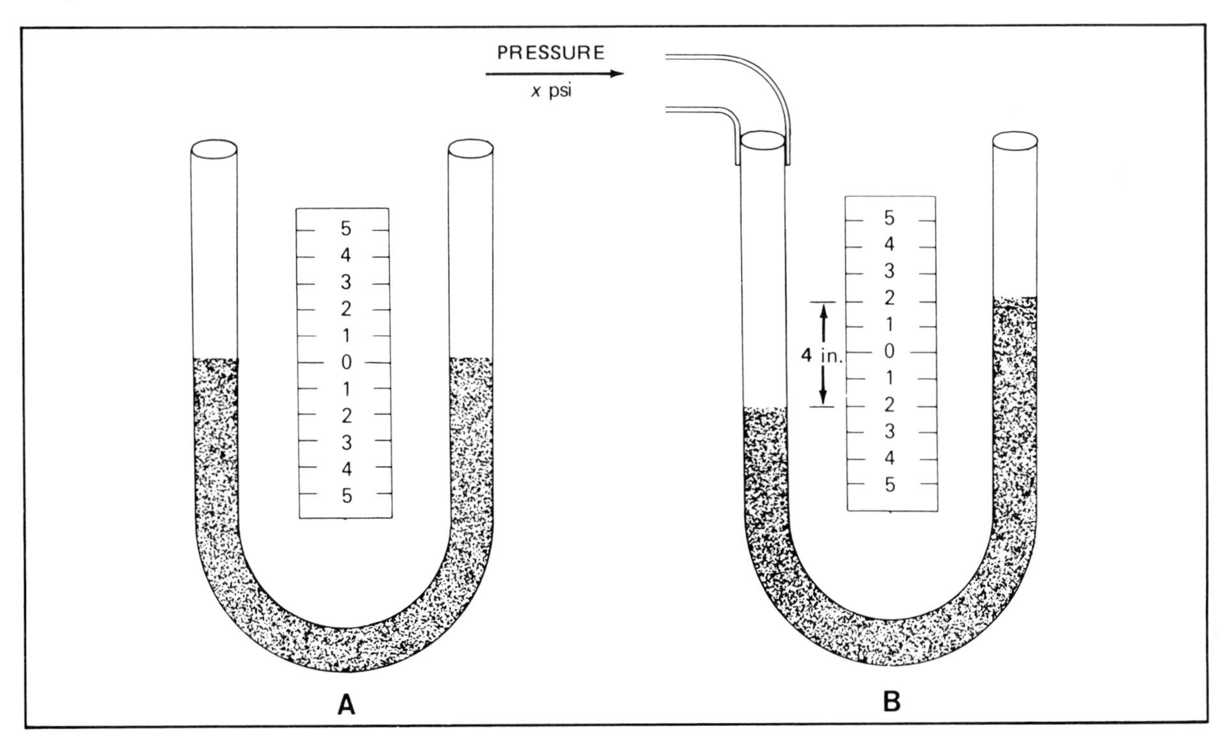

Figure 2.3. A U-tube manometer is a common device for measuring pressure.

common use for performing pressure measurements. The manometer tube contains mercury, which weighs about 0.5 pound per cubic inch. With both tube ends open to the atmosphere, the surface of the left and right columns of mercury are at equal levels, and the gauge pressure reading is 0, as shown in *A*. If now a pressure of x psi is applied to the left column, as shown in *B*, the surface of the mercury in the left and right columns will fall and rise, respectively. With the gauge scale graduated in inches, the left scale reads 2 inches below the zero mark, and the right scale reads 2 inches above the zero mark. The difference in level of the two surfaces is therefore 4 inches; thus, the column height is 4 inches.

Carrying out some basic calculations, the diameter of the mercury columns in the U-tube is found to have no effect at all on how high the mercury will rise or fall in the two branches of the manometer for any applied pressure. In fact, the calculations reveal that the applied pressure, x psi, is simply the product of column height in inches and the weight of the mercury per cubic inch; that is,

$$\begin{aligned} x &= 4 \text{ in.} \times 0.5 \text{ lb/in.}^3 \\ &= 2 \text{ lb/in.}^2\text{, or 2 psi.} \end{aligned}$$

Since the applied pressure is directly proportional to the column height, it is a matter of convenience to express it in terms of *inches of mercury*. This method of pressure measurement in terms of length began in laboratory work but became widely used in industry to express generally low values of pressure. Most are familiar with the fact that meteorologists express atmospheric pressure in inches, although pressure is usually given as a barometer reading and not strictly as one of pressure.

Another commonly used measure of pressure is directly related to this method. It involves substituting water for the mercury in the manometer tube. Imagine the U-tube in figure 2.3 using water instead of mercury. Since the density of water is approximately $1/13.6$ of the density of mercury, the manometer is now much more sensitive; that is, the value of pressure that will cause the mercury column to rise 1 inch will cause the water column to rise 13.6 inches. The term *inches of water* is a popular expression for pressure in the gas industry where fairly low values of gauge pressure are commonplace.

In summary, several popular expressions exist for pressure—*pounds per square inch; newtons per square metre;* and *inches, centimetres, or millimetres of mercury or water.* Some of these are not true pressure dimensions, but from the standpoint of convenience they are extremely useful and have a proper place in instrumentation.

Rate of Flow

A natural assumption is that rate of flow can be measured only by using units of length and time. Rate of flow of fluids is usually measured by noting the time required for a given volume of fluid to pass a certain point. There have been scores of volume units and time units available for measuring the rate of flow; gallons per minute or hour, barrels per day or hour, cubic feet or cubic metres per minute or hour are examples. The following equation shows that volume rate of flow has dimensions of length (L) and time (t):

$$\text{rate of flow} = L^3/t.$$

Rate of flow is also measured in terms of mass per unit of time. Mass flowmeters are enjoying considerable popularity today and probably will continue to do so. The conversion from mass to volume is a simple one if density of the gas or liquid is known, as it certainly will be in all cases of importance. The relation between volume and mass is

$$\text{volume} = \frac{\text{mass}}{\text{density}}.$$

Clearly, two logical sets of dimensions are available for expressing rates of flow: *mass per unit of time* and *volume per unit of time.*

Other Variables

Specific gravity. *Specific gravity* may be defined in several ways. Most commonly it is said to be the ratio of the mass of a given volume of substance to the mass of an equal volume of water. The resulting quotient is said to be the specific gravity of the substance. Assume that *a* grams of substance has a volume of *b* cubic centimetres. The specific gravity of the substance can be determined from the equation:

$$\text{specific gravity} = \frac{a\ \text{g} \times b\ \text{cm}^3}{c\ \text{g} \times b\ \text{cm}^3}$$

where *c* equals the number of grams of water. Remembering that grams and centimetres are treated as mathematical entities, all terms in the equation's right member cancel except a/c, which is a pure number; that is, it has no dimensions of length, mass, or other fundamental quantity.

In the International System of Units, specific gravity is called *relative density*.

pH factor. By virtue of its definition and use, pH factor is treated as a pure number, that is, dimensionless. The quantities from which it is derived, however, have a dimension of mass. pH factor is arrived at by the following equation:

$$\text{pH factor} = -\log_{10}[\text{H}^+].$$

The factor $[\text{H}^+]$ designates the hydrogen ion concentration in moles per litre of a solution. In pure water, which is neither acid nor base, the concentrations of hydrogen and hydroxyl ions are equal and amount to 10^{-7} mole per litre. The pH factor for pure water is found to be

$$\begin{aligned} \text{pH} &= -\log_{10}[\text{H}^+] = -\log_{10}[10^{-7}] \\ &= 7 \log_{10} 10 \\ &= 7 \end{aligned}$$

Decreasing values of pH indicate increasing values of acidity, and vice versa.

Viscosity. Like the term *mile, viscosity* needs an adjective to identify the particular kind to be discussed. *Absolute viscosity* is expressed in a basic unit called a *poise.* The poise has dimensions of length, mass, and time. As a standard of viscosity, poise was established in the era when the centimetre-gram-second (cgs) formed one subsystem of the metric system of units. The poise is therefore best expressed in cgs units. The International System (SI) discourages the use of those units, however.

Consider the following arrangement. A solid smooth surface of indefinite length and width is covered with a layer of viscous liquid 1 centimetre (cm) thick. Another solid surface having a unit area of 1 square centimetre is on the surface of the liquid. Now, if a force of 1 dyne (d) (the unit of force in the cgs system) is required to move this unit-area surface at a velocity of 1 centimetre per second, the absolute viscosity of the liquid is equal to 1 poise. Absolute viscosity in poises can be expressed as follows:

$$\begin{aligned} \text{poise} &= \frac{\text{d} \times \text{sec}}{\text{cm}} = \frac{M \times L \times t}{L^2 \times t^2} \\ &= \frac{\text{g}}{\text{cm} \times \text{sec}} \end{aligned}$$

In its simplified form, 1 poise equals 1 gram per centimetre per second.

Kinematic viscosity is obtained by dividing the absolute viscosity by the density of the fluid in grams per cubic centimetre. The resulting dimensions are arrived at as follows:

$$\begin{aligned} \text{kinematic viscosity} &= \frac{\text{g/cm} \times \text{sec}}{\text{g/cm}^3} \\ &= (\text{g/cm} \times \text{sec}) \times \text{cm}^3/\text{g} \\ &= \text{cm}^2/\text{sec}. \end{aligned}$$

Kinematic viscosity has units called *stokes* with dimensions of centimetres squared per second.

Several other forms of viscosity are in use and will appear in the chapter on measurement and control of other variables.

Conclusion

In instrumentation, only a very few fundamental quantities are dealt with. It is the myriad grouping of these quantities that results in the somewhat larger number of problem-causing variables.

An important concept to remember is that the very names of the units of measurement are treated as mathematical entities. When a given unit appears in both numerator and denominator of an expression, sometimes a cancellation step can be carried out, just as with equal numerical factors. Each variable measured and controlled is really a quantity made up of a particular grouping of one or more of the fundamental quantities.

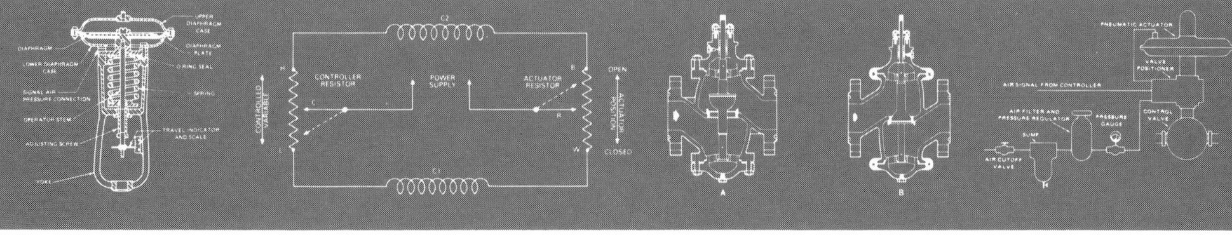

CHAPTER 3

Final Control Elements

The final control element performs a necessary function in the fluid flow process: it regulates the rate of flow. Valves are the predominant form of final control elements (the two terms might be considered synonymous); however, controlled-volume pumps and other final control elements are also used in the petroleum industry.

A final control element usually consists of a valve, an actuator, and the proper design, size, and piping arrangements. The main components of a valve are its body, plug, guides, and seat. The design and arrangement of these components determine the function and capabilities of a valve. An actuator provides the force that actuates the valve, determining the rate of flow of an agent through the valve. Means used to operate the actuator may be mechanical, pneumatic, electrical, hydraulic, or a combination of two of these methods.

A controlled-volume pump delivers a definite and predetermined volume of liquid with each stroke, or cycle; thus, its use in forcing chemicals into lines and vessels is important to the petroleum industry.

Valves

Many terms are used to describe parts of a valve (fig. 3.1); however, most of them are not ambiguous and not apt to be a stumbling block in the study of controls. The more important parts of valves are the bodies, plugs, guides, and seats. The port area in the valve body, the position of the plug in the valve, the trim, and other design details determine the characteristics of a valve.

Valve Bodies

Most valve bodies used in control applications are *globes,* a term derived from the global shape of the body. Of course, many other body types are important for specific applications.

Single-ported globe bodies. Single-ported valves have a single path for passage of fluid. The two forms of single-ported valves shown in figure 3.2 are similar but differ somewhat.

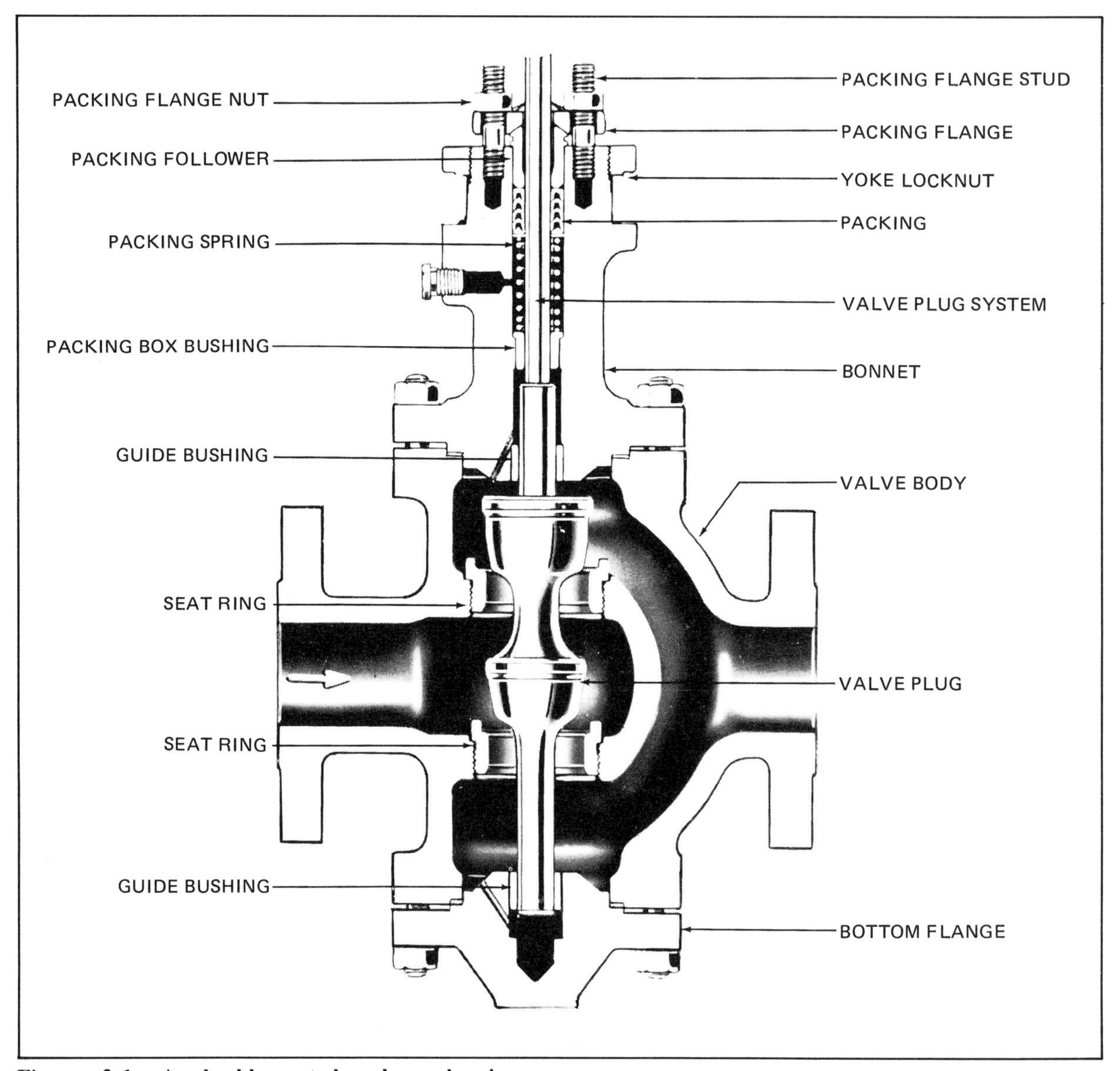

Figure 3.1. A double-ported valve, showing nomenclature commonly used for various parts *(Courtesy Fisher Controls)*

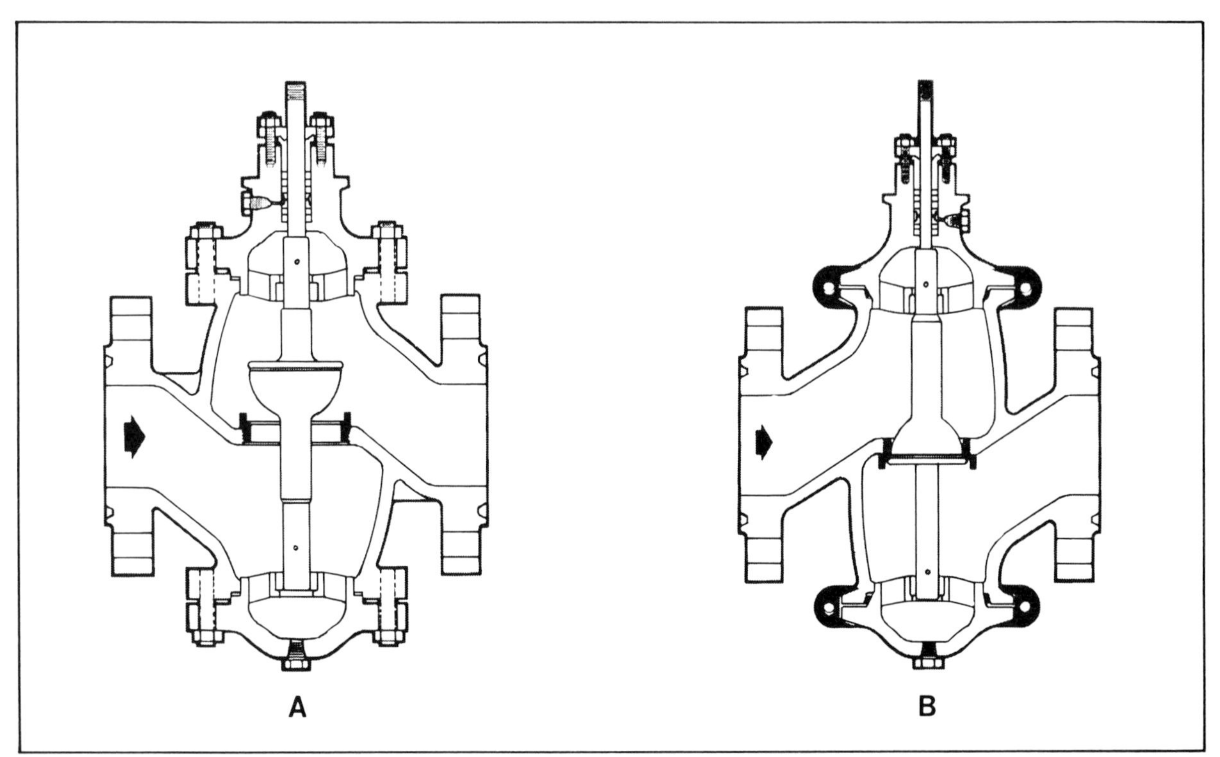

Figure 3.2. Two types of single-ported valves. *A*, direct-acting; *B*, reverse-acting.

One obvious difference is the manner in which the bonnets and bottom flanges are attached to the valve bodies. In one, they are attached by studs and nuts, while the other uses a clamp ring to secure the parts to the main valve body. Another difference is that the plugs are reversed with respect to one another. Pushing down on the valve stem in *A* causes the valve to close, while the same action on the valve stem in *B* causes it to open. The valve in *A* is called a *direct-acting valve,* and that of *B,* a *reverse-acting valve.* Each of the forms finds wide use in control application.

An important point is the fact that the valve plugs shown in figure 3.2 can be reversed, thus changing the valve from direct-acting to reverse-acting, and vice versa. Of course, the valve seats must be reversed, too, but reversal is easily accomplished. The greatest advantage of this feature is the reduced manufacturing cost, as the producer needs to make only a single set of parts to obtain either a direct- or reverse-acting valve.

Single-ported valves cost less, are easier to maintain, and are more resistant to leakage when fully closed than the double-ported valves. A major disadvantage of single-ported valves is dealing with the fluid pressure of the line, which bears heavily against the plug in the closed position. However, such valves are usually installed in the line so that the fluid pressure tends to force the plug away from the seat. This action results in smoother operation and deters the plug's tendency to slam shut when the valve is installed in the reverse direction.

Double-ported globe bodies. Double-ported valves (fig. 3.1) are the most popular types for automatic control applications. The original idea in the development of the double-ported valve was to provide a unit that would require little force to position its plug to any point between fully opened and fully

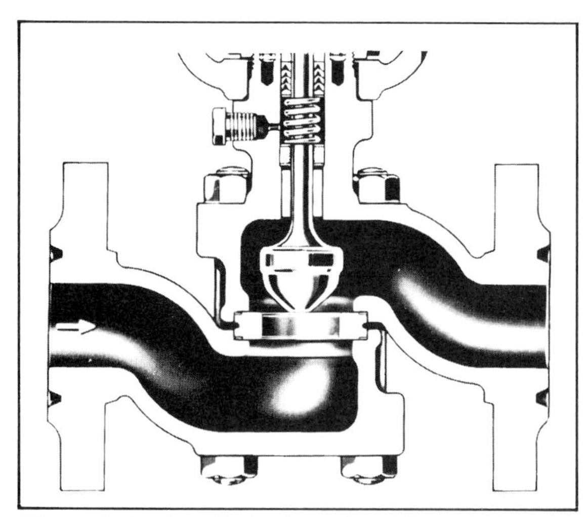

Figure 3.3. A split-body valve *(Courtesy Fisher Controls)*

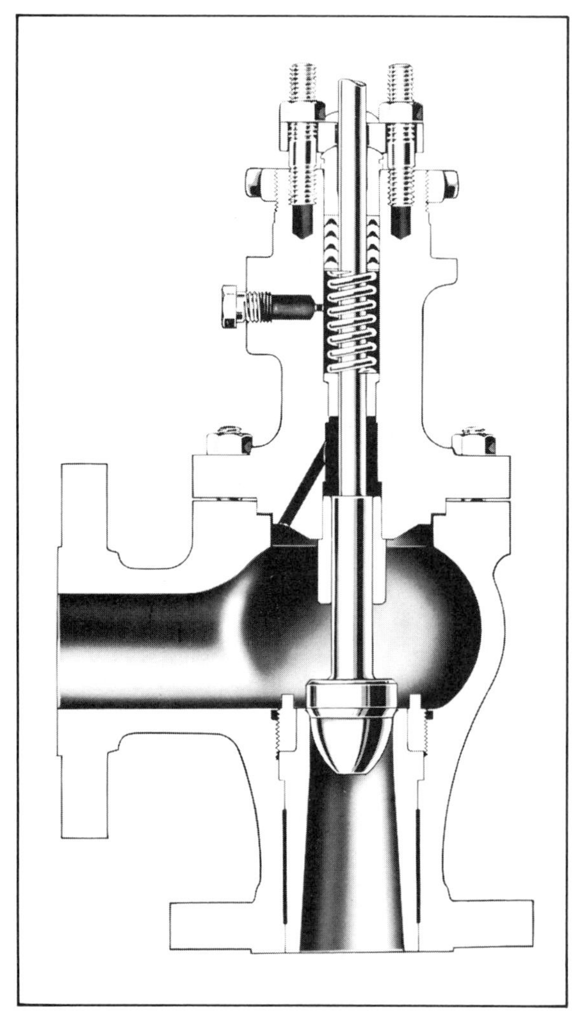

Figure 3.4. A typical angle-body valve

closed. In practice this idea does not work out perfectly for a number of reasons, but the amount of force required to position the plug is indeed quite small as compared to single-ported valves.

Split-body valves. Split-body valves (fig. 3.3) are made up of two major components joined by a system of bolting that permits the forming of either a straight-through or right-angle valve. Other advantages are easy servicing and reasonable freedom from sediment collection.

Angle-body valves. Angle-body valves are usually single-ported styles and are used where lines need regular draining. A typical right-angle valve (fig. 3.4) is comparatively free from pockets and cavities, thus affording good drain qualities. An angle valve usually is installed in such a way that flow is into the side port and out the bottom since this minimizes body erosion, but the valve tends to slam shut as the plug nears the seat.

Bodies for three-way valves. Two forms of three-way valves are shown in figure 3.5: an adaptation of a single-ported valve (*A*) and a double-ported valve (*B*). These three-way valves are constant-flow devices; that is, the position of the plug does not affect the total flow through the valve. The single-ported body style is used for mixing service, the *U* and *L* connections being inlets and the *C* connection being the mixed outlet. The double-ported style is used for proportioning service, having one inlet at *C* and two outlets, *U* and *L*.

Valve bodies for small flow rates. The need for low-capacity control valves has brought about the development of several designs that perform special services, such as injecting inhibitors into lines or processes and setting up laboratory or experimental arrangements. The body design seems considerably less important than the design of the seats, plugs, and actuating mechanisms.

Butterfly valve bodies. The bodies of butterfly valves are best described as cylindrical

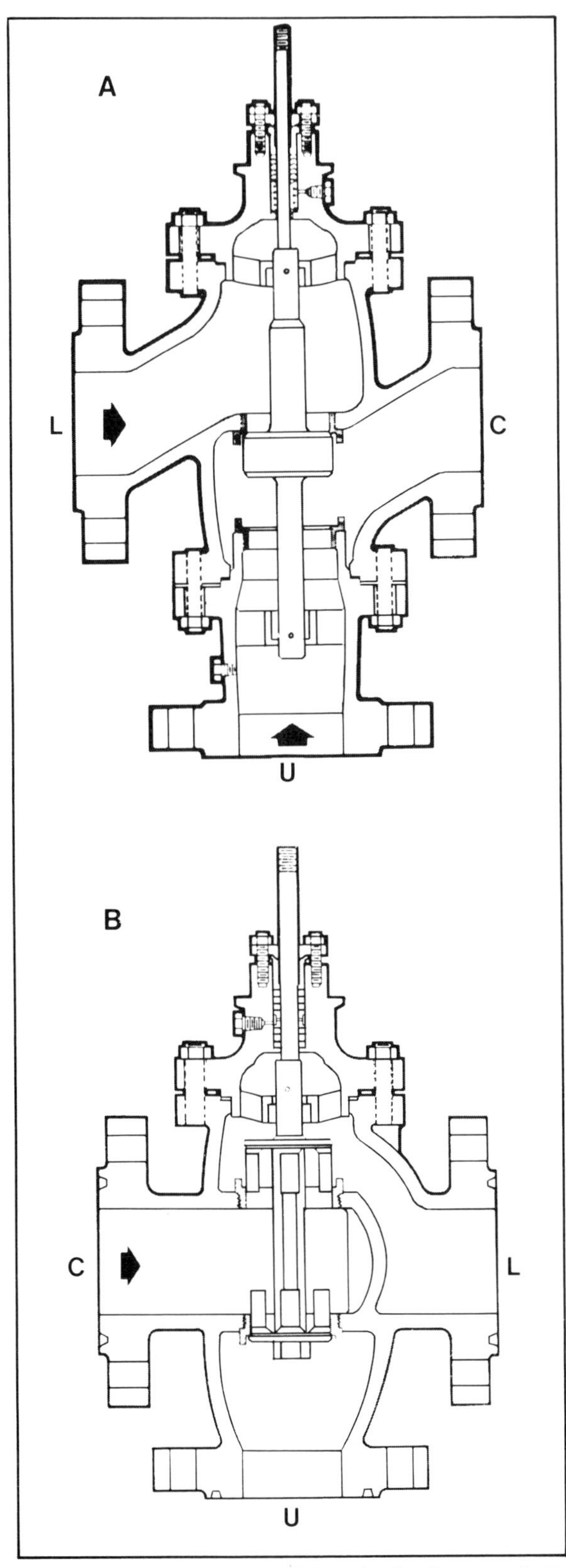

Figure 3.5. Three-way valves. *A*, single-ported; *B*, double-ported.

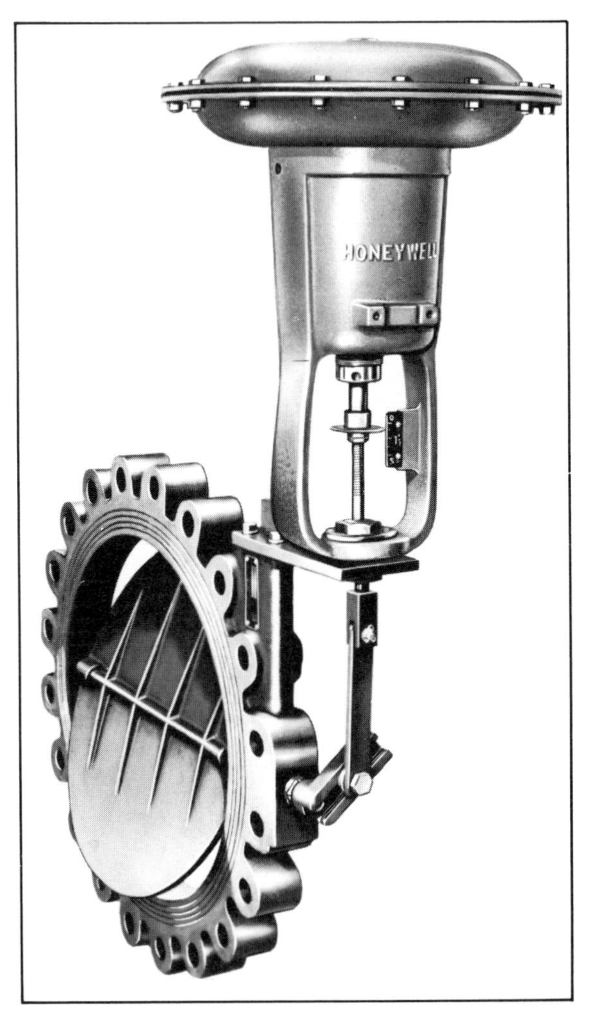

Figure 3.6. A butterfly valve *(Courtesy Honeywell)*

forms (fig. 3.6). Butterfly valves are characterized by their large capacity. In the wide-open position there is a straight-through flow with only the small area of the edgewise surface of the disc or vane presenting an obstruction.

Although butterfly valves have high capacity for a given size and are usually less expensive to build than plug-type globe valves, their use in control functions is limited by such factors as leakage and the considerable force required to operate the vane against any but the lowest pressures. They have a flow that is logarithmic (equal percentage) for low values of flow and lift, and linear flow for high lift and flow. The need for this sort of characteristic is not widespread.

Gate valve bodies. Gate valves find extensive use in pipeline service but are not suited to throttling service typical of process control applications. Gate valves are characterized by their full flow capability, opening to the same inside diameter of the pipe in which they are installed (fig. 3.7). This feature permits easy use of scrapers and other full-diameter devices that must be sent through the pipeline from time to time. The shutoff capability of gate valves is quite good.

Valve Characteristics

Various factors make one type of valve better suited for a given control problem than others. Of greatest importance, of course, are flow characteristics and flow capacity. Other considerations assume importance in specific applications: ease and smoothness of operation, leakage character, ease of maintenance, and self-cleaning features, to name a few.

Flow characteristic. The flow characteristic of a control valve describes the relation between the amount of fluid flow through the valve and the extent to which the valve is open. This characteristic is usually expressed in a graph for which maximum flow is plotted as the abscissa and maximum lift (opening) as the ordinate. The characteristic curves of several control valves using typical forms of plugs are shown in figure 3.8. The curves are

Figure 3.7. A pipeline gate valve

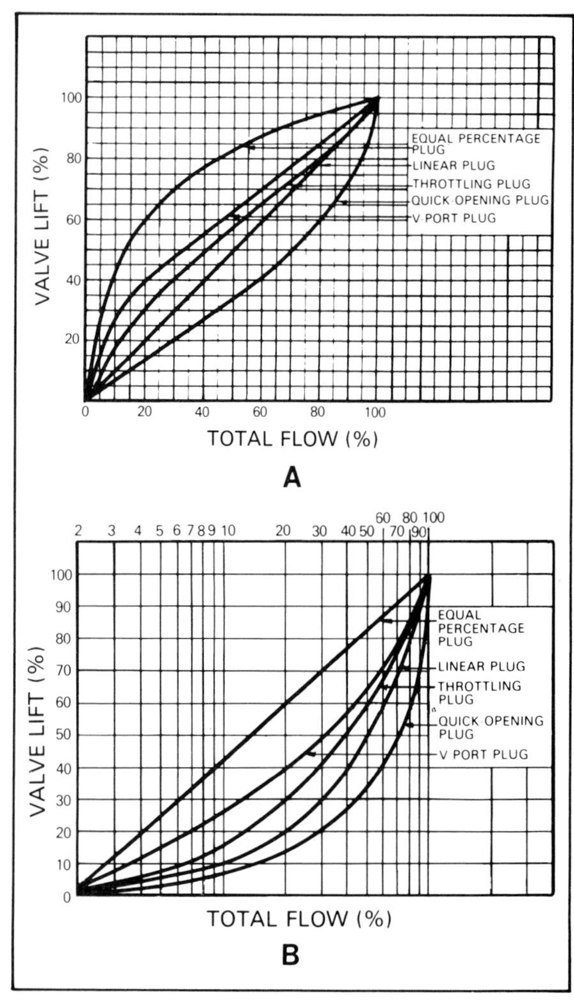

Figure 3.8. Flow characteristic curves plotted on graph paper. *A*, linear; *B*, linear-log. *(Courtesy Fisher Controls)*

drawn on linear graph paper in *A* and on linear-log, or semilogarithmic, paper in *B*.

Flow capacity. The term *flow capacity* describes the maximum volume of fluid that can be put through a wide-open valve in a unit of time. This is the maximum flow rate of the valve, but several factors become involved in its measurement: (1) pressure drop across the valve; (2) energy difference between inlet and outlet; and (3) viscosity, density, and type of fluid.

In the conventional United States measurement system, the *flow coefficient* (C_v) has been a convenient factor to express the flow characteristics of a control valve. C_v is the volume of water in gallons that will pass through the wide-open valve in 1 minute with a pressure drop of 1 pound per square inch across the valve. This factor is still in widespread use about the world. As of now, no equivalent flow coefficient is available in SI.

A few valve manufacturers in western Europe have used a flow coefficient K_v, based on flow per hour in cubic metres and a pressure drop measured in bars. The bar is a pressure unit equal to 100 kPa (14.5 psi).

A useful equation can be set up for approximating the rate of flow through a valve:

$$Q = k\sqrt{\Delta p}$$

where

Q = rate of flow;
k = a constant;
Δp = pressure drop across valve (or other restriction in the line).

The simple equation becomes complicated as more considerations are taken into account, but for water it may be expressed

$$Q = C_v\sqrt{\Delta p}.$$

It is not the purpose of this manual to study the extensive considerations involved in the process of selecting valves for special applications; this is a function of engineers who design control systems. However, understanding the fact that every control valve is chosen on the basis of having met many critical specifications allows appreciation of the exacting requirements that go into the design of satisfactory control systems.

Valve Plugs

It is difficult to discuss valve flow characteristics without closely associating this information with the components that are most important in shaping them—*valve plugs*, sometimes called *inner valves*—and their performance. Valve plugs (fig. 3.9) can be

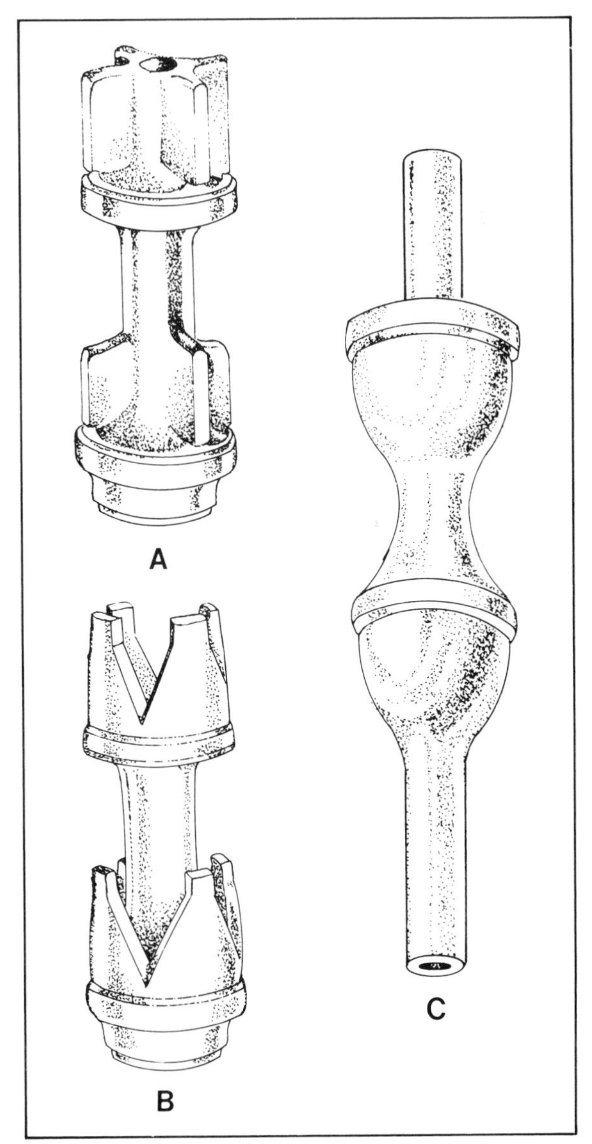

Figure 3.9. Types of valve plugs. *A*, quick-opening; *B*, V-port; *C*, throttling.

designed to produce all forms of flow from ordinary on and off service to any desired characteristic of throttling action.

Quick-opening plugs. The intake and exhaust valves of most internal-combustion engines are examples of quick-opening valves. Sometimes called *poppets,* they are characterized by their ability to provide full flow with comparatively small stem movement. Valves of this type are used in control service where on and off flow is required but are not particularly well suited to throttling service. Valve plug *A* (fig. 3.9) is a double-ported quick-opening type. It is port-guided; that is, the wings that radiate from the axis of the plug ride against the seat rings and thus keep the plug aligned. The flow characteristic of the plug is shown in the curves of figure 3.8.

Modified linear, or V-port, plugs. *Modified linear* and *V-port* are names that enjoy equal popularity. One company calls its version of this plug *characterized V-port* because of the V-shaped slots cut into the skirts of the plug. The term *modified linear* can be misleading, but the plug does have a linear characteristic for all high values of lift and flow. Plug *B* (fig. 3.9) is a double-ported V-port plug; the graph of its flow characteristic is shown in figure 3.8. This plug is also port-guided.

Throttling plugs. The throttling plug (fig. 3.9*C*) is a superior plug for use in handling slurries and other fluids having any solid content. Its flow pattern is similar to that of the V-port plug. Its physical differences are a result of its being stem-guided and lathe-turned. *Stem-guided* plugs are held in alignment by having their stems ride in special bushings installed in the valve body. The term *lathe-turned* means the plug can be machined from round stock in a lathe, while quick-opening and V-port plugs require more extensive techniques.

Equal percentage plugs. Valves with an equal percentage characteristic can best be described as those having a logarithmic flow pattern (fig. 3.10). Assuming that the pressure drop across the valve is held constant, this type of characteristic will produce equal percentage changes in flow for equal changes in stem lift; that is, each time the stem lift is changed an equal percentage of the total lift, the flow through the valve changes 50% from the flow existing before the change in lift. As an example, suppose a valve is such that its flow changes 50% for each change of 10% of total lift. Now, with the valve wide open—lift

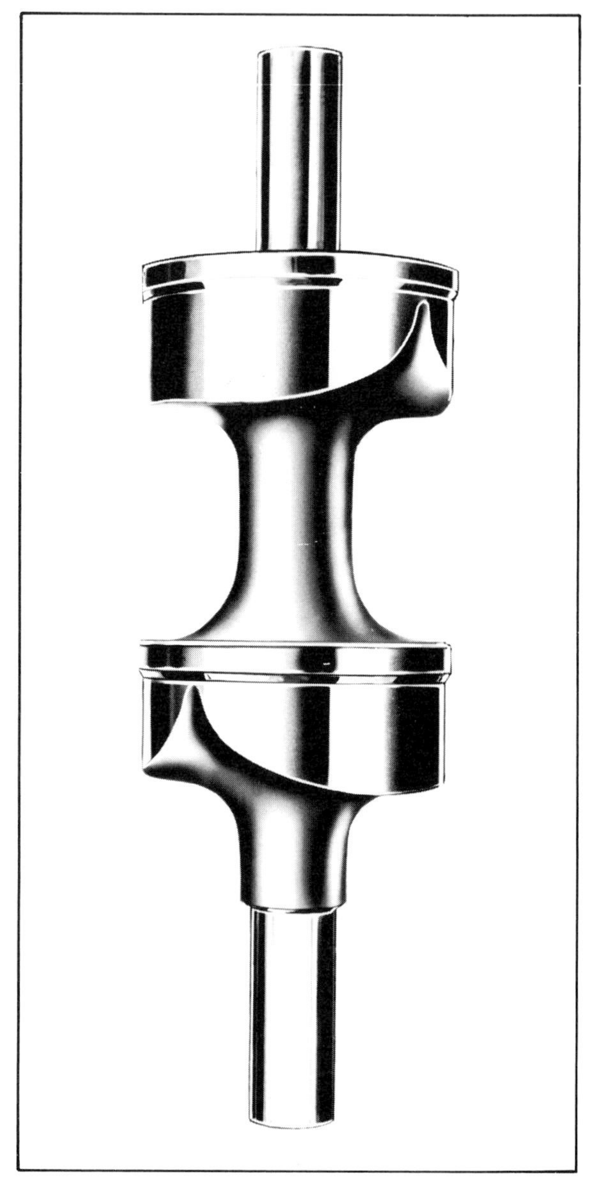

Figure 3.10. Typical equal percentage plug *(Courtesy Fisher Controls)*

and flow at maximum—the lift is reduced to 90% of maximum. This produces a 10% change in lift. Flow will change by 50%. Another reduction of lift—to 80% of maximum—will reduce flow to 25% of maximum, and so on until flow and lift reach 0.

The equal percentage plug in figure 3.8 clearly indicates that, for low values of flow, considerable stem movement is required for appreciable changes in flow, but the opposite is true for large values of flow. Such flow characteristics are desirable on applications where a large portion of the pressure drop is

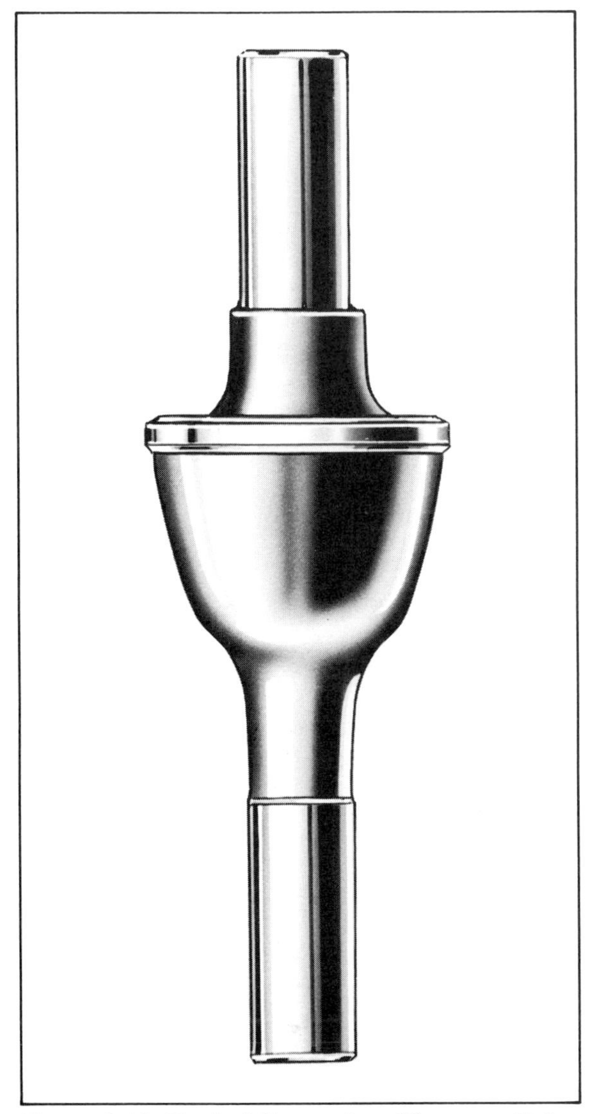

Figure 3.11. Typical linear plug *(Courtesy Fisher Controls)*

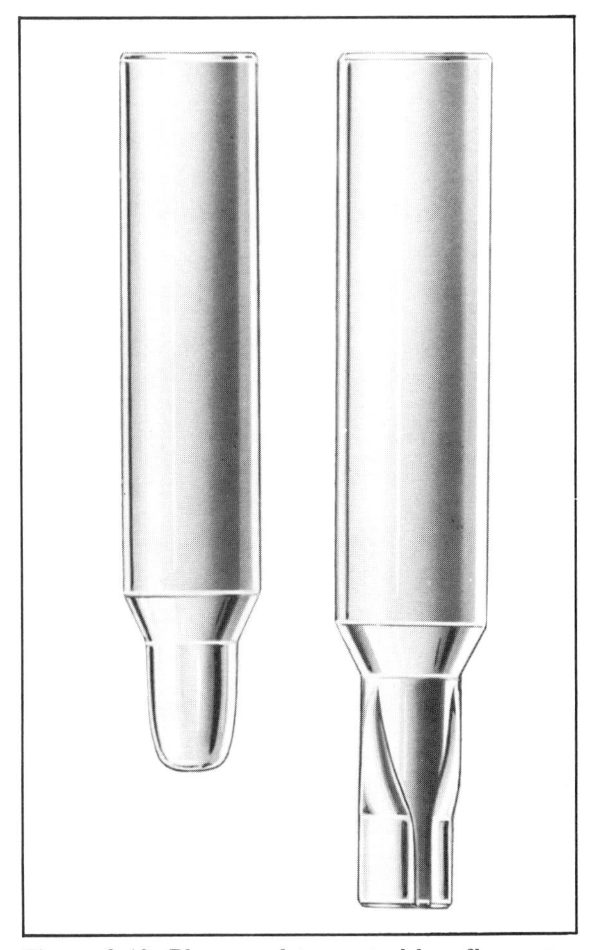

Figure 3.12. Plugs used to control low flow rates *(Courtesy Fisher Controls)*

normally absorbed by the system itself, with only a relatively small amount available at the control valve. These characteristics are also useful where highly varying pressure-drop conditions are expected.

Linear plugs. Linear plugs (fig. 3.11) are made in lathe-turned and skirted types, as well as for single- or double-ported service. The lathe-turned types are stem-guided, while the skirted plugs may be either stem- or port-guided. Port-guided plugs have two or more rectangular slots cut in the skirts. The flow characteristic of linear plugs is a straight line from zero to maximum flow and lift.

Plugs for low flows. Some plugs are designed to control low flow rates (fig. 3.12). The flow characteristics of such plugs can be tailored to fit any pattern desired. Physical characteristics of low-flow plugs include the usual condition that the guide stem is of considerably greater size than the controlling part of the plug. These plugs almost always have single ports and top guides.

Valve Guides and Seats

Proper mating of the valve plug and seat requires a steady and durable means of maintaining correct alignment of the moving parts relative to the valve seat. This function is performed by guides, of which there are several forms.

The V-port (*B*) and quick-opening (*A*) plugs shown in figure 3.9 are skirt-guided and wing-guided, respectively. This form of plug is called a port-guided plug. Valves with port-guided features are less expensive to build and are satisfactory for some services although not for wide ranges of throttling. At positions of low flow any misalignment of the plug will distort the flow in the annular space between the seat ring and the plug, causing the plug to bind.

The stem-guided plug shown in figure 3.9*C* has top and bottom guides and represents the most desirable method of guiding used in control valves. Stem-guided, double-ported units may have either top and bottom guides or top guides only. Such guiding results in valves that are less subject to vibration and binding.

Seats for control valves are usually metallic components that provide for a metal-to-metal contact with the valve plug. If positive shutoff is required, such seats are not satisfactory for extended periods of time, and composition seats that include a resilient compound are used. Of course, where complete shutoff is never necessary, the advantages of suitable all-metal seats are desirable.

Single-seated valves give better shutoff than double-seated units, because they are easier to keep aligned and their wear is more uniform. However, because of the difficulty of overcoming the forces that affect a single-ported valve near the closure point, the double-ported valve is still favored for use in such throttling ranges.

Valve Trim

Trim is a term that refers to the kind of material used to make the internal valve parts. It logically includes the plug, seats, valve stem, valve stem guide bushings, and internal parts of the stuffing box. The type of trim will depend almost entirely on two factors: (1) pressure drop across the valve and (2) the corrosive and erosive qualities of the fluid being controlled.

Stainless steel is the most popular trim material, and one form or another of it can be used for all parts. This trim is not only mechanically tough and durable, but it resists most corrosive fluids. Many special forms of trim have been developed for valves and bear registered trade names, for example, Stellite, Colmony, Stoody, and Monel. Stainless steel with chrome plating has been used with success as trim for large-size valves handling erosive material.

Because of its low cost, bronze trim is used in a few instances of mild service, but the trend is away from using such soft materials for trim in control valves.

Valve Design Details

Although many design details exist, the more important are stuffing boxes (valve stem seals), bonnets, and end connections.

Stuffing boxes. The valve stem must transmit motion from the actuator to the plug through a seal capable of withstanding the operating pressure in the line. Such seals are usually called *stuffing boxes*. Good seals must not only be leak-resistant, but they must possess a minimum of friction so that positioning the valve plug can be done smoothly by the actuator. One form of stuffing box uses a number of rings of composition material and a system for lubricating the stem

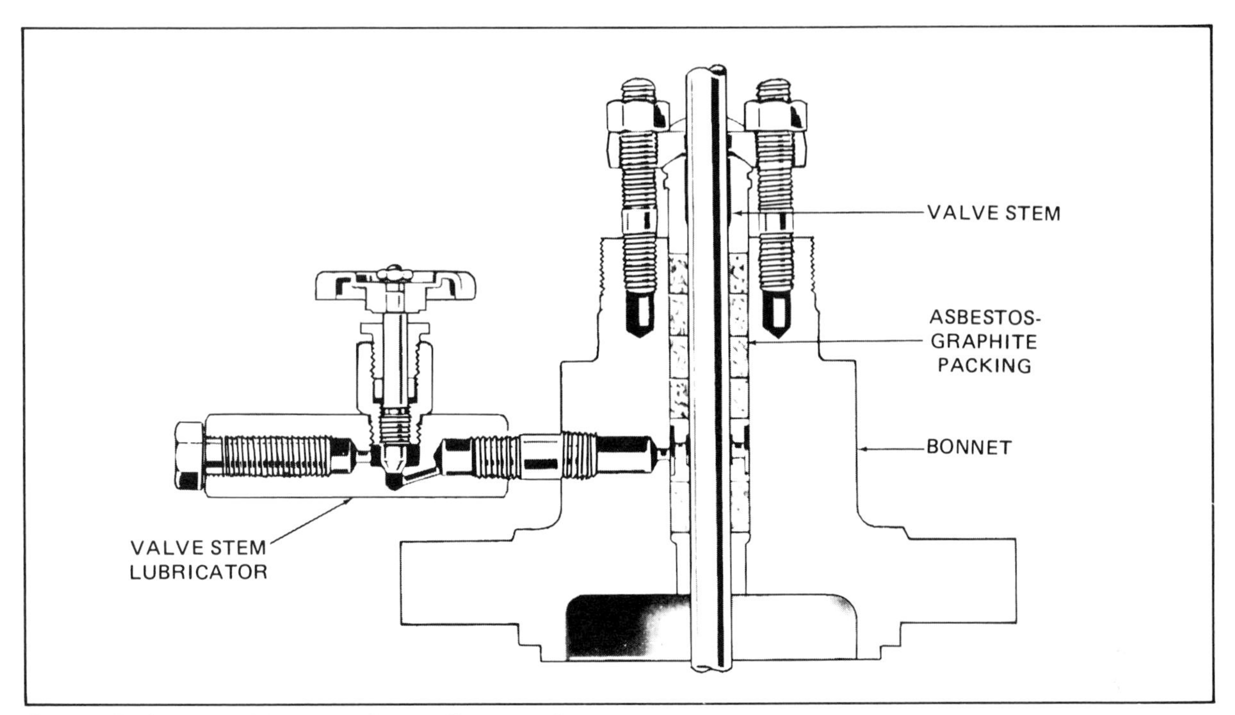

Figure 3.13. A stuffing box for sealing a valve stem. Note the composition packing and lubricating system. ***(Courtesy Fisher Controls)***

(fig. 3.13). The material might consist of asbestos, mica, or graphite mixed with artificial rubber.

Teflon, a particular form of plastic, is popular as a packing material since it is inert to most chemicals, has a wide temperature range, and is self-lubricating. Its most satisfactory form is that of molded V-rings, or chevron-type packing (fig. 3.14). Valves with molded Teflon packing require careful maintenance, and procedures recommended by the valve manufacturer should be learned and followed if difficulties are to be avoided.

Bonnets and bonnet extensions. The bonnet of a valve is fastened to the main valve body by any of several methods: (1) direct mating of internal- or external-threaded joints, (2) the fastening together of flanges with studs and nuts, or (3) the clamping together of flanges by a clamp ring. A bonnet provides a mounting support between the main body and the actuator mechanism, and it contains the stuffing box.

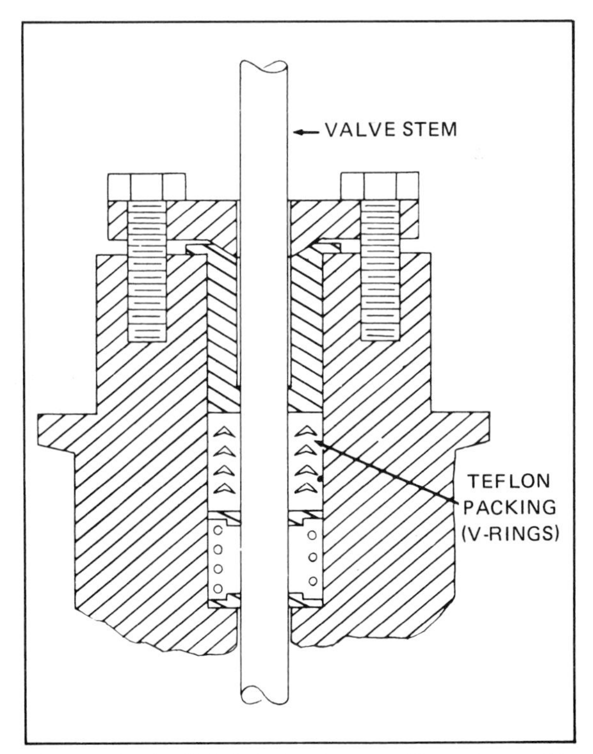

Figure 3.14. Self-lubricating stuffing box with Teflon packing

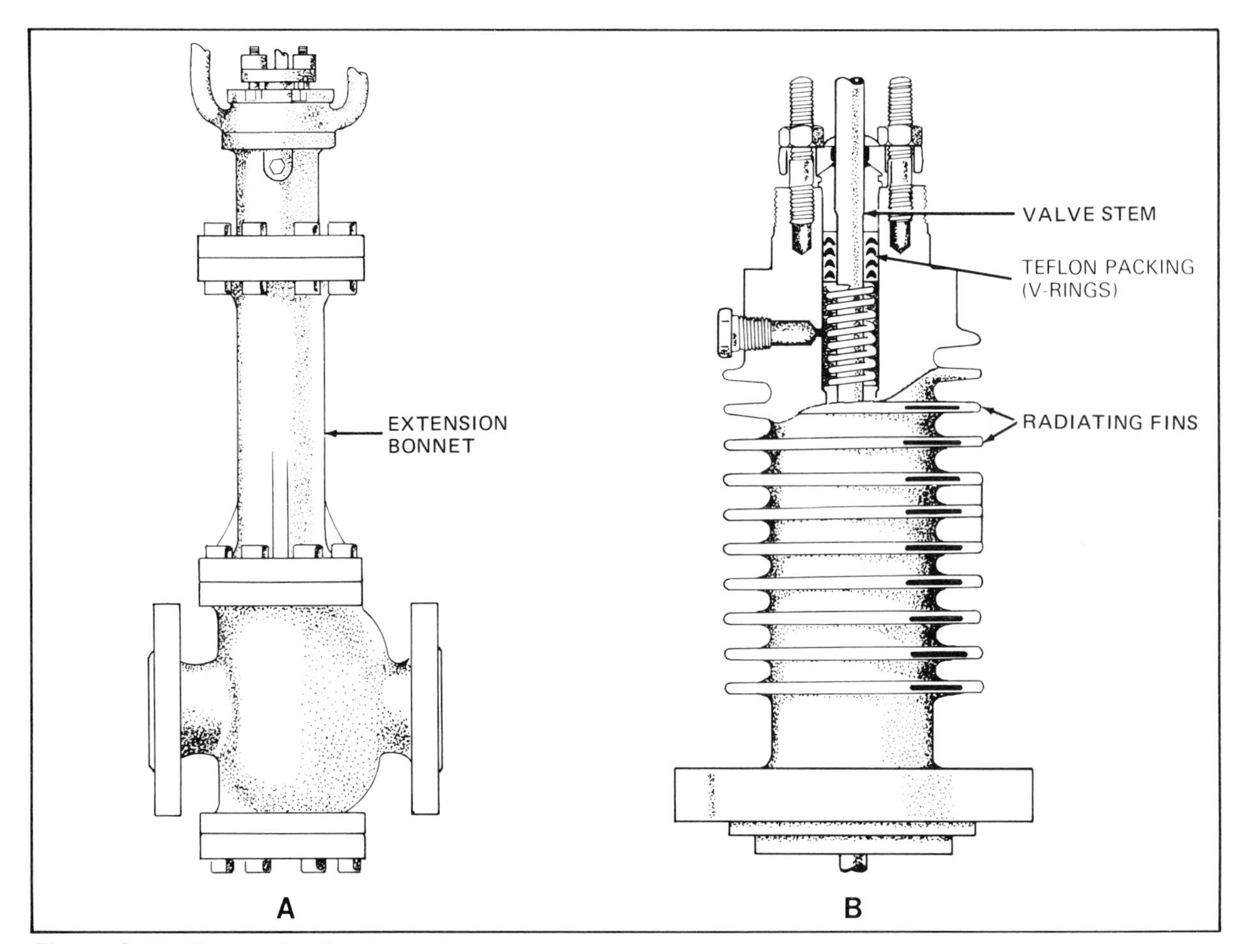

Figure 3.15. Types of valve bonnets and extensions. *A*, extension for low-temperature service; *B*, finned bonnet for high-temperature service. *(Courtesy Fisher Controls)*

Sometimes bonnets are provided with radiating fins in order to carry away heat of the process fluid, since this heat might cause serious deterioration of the packing material. These finned units (fig. 3.15) are usually elongated bonnets or actual extensions fitted between bonnet and body. Extensions without fins are used for low-temperature service to help insulate packing and actuator parts from the low temperature.

End connections. Valves are usually attached to piping and equipment by mating flanges, screwing together threaded fittings, or welding. Each method possesses an advantage for a given application.

Threaded fittings are used in applications requiring valves of less than 2 inch size, although high-pressure or high-temperature service or unusually corrosive conditions might dictate the use of *flanged fittings.* Threaded fittings are used in large-sized valves for low-pressure gas service, but the threading of valves and pipes beyond certain diameters becomes impractical. The valve shown in figure 3.15*A* has flanged ends.

Welded ends find prevalent use in processes that treat dangerous or valuable products where leakage cannot be tolerated. If the permanency of the installation can be assured, welded connections offer real advantages. They are capable of withstanding maximum pressure and are inexpensive.

Sizing and Piping Arrangements

Selecting the right type and size of control valve for some processes and systems can be very difficult. It involves the expert use of all known flow data of the process or system and also the use of assumptions that might not be justified by theory but that have been deemed necessary by previous experience. Valve sizing will not be considered except to point out that *throttling* valves are almost always considerably smaller than the line in which they are installed.

A rough criterion for throttling valves is that not less than one-third of the pressure drop in the system should occur across the valve during maximum flow conditions. There are notable exceptions to this idea, but it remains workable in most instances. If a control valve of large size is used—one that produces only a small part of the overall system pressure drop—then its ability to control flow near the maximum rate of the system will be seriously impaired. Such a valve will probably be operating at less than half-open positions for all conditions.

In spite of the relatively large pressure drop that is desired across a control valve, the valve must still be able to pass a quarter to a half more flow than the maximum rate ever demanded by the process or system.

In the typical piping arrangement for the installation of control valves (fig. 3.16), the block valves and a bypass line are provided so that repairs, cleaning, or complete removal of the control valve can be carried out without

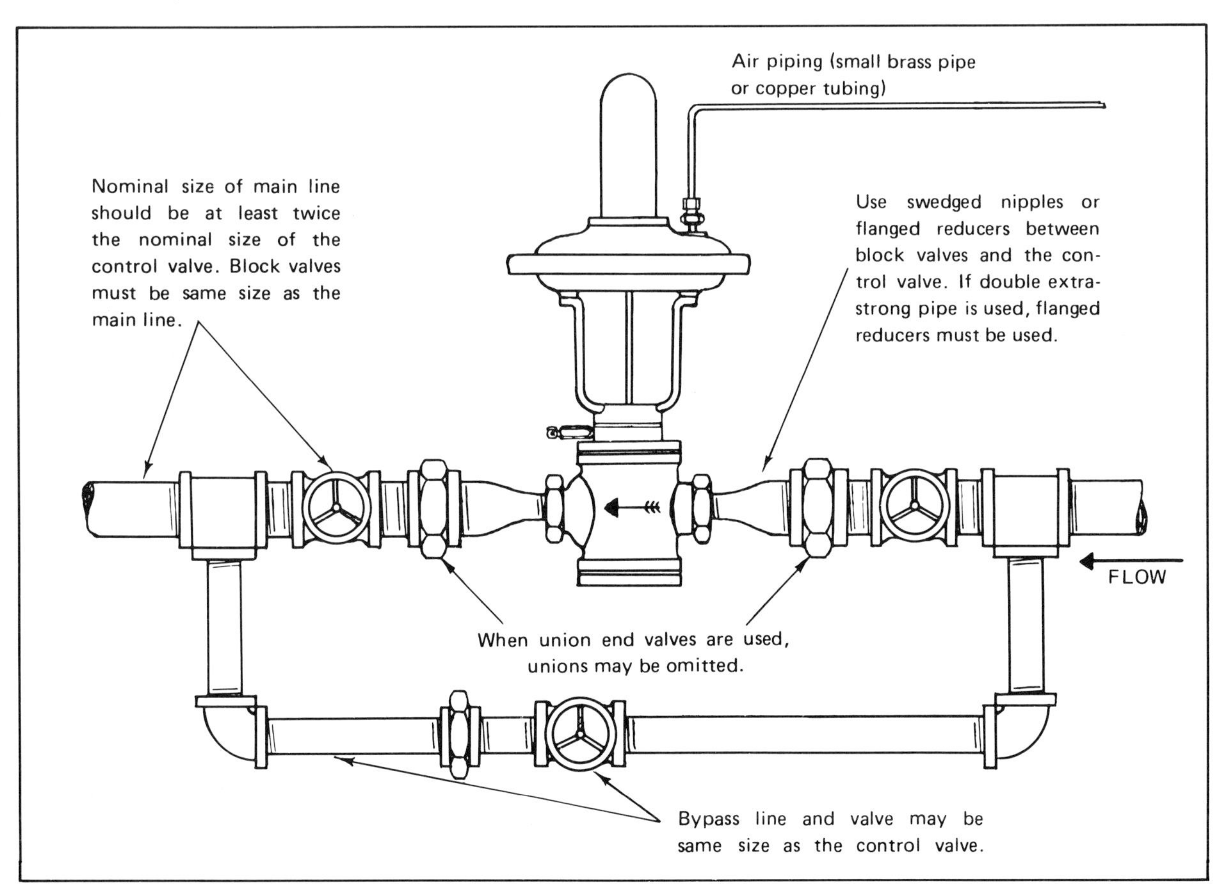

Figure 3.16. Typical piping arrangement for a throttling valve. The valve should be half as large as the main line or smaller. ***(Courtesy Foxboro)***

putting the line out of service. The control valve size is considerably smaller than the line in which it is installed, and the bypass line and valve have a similar relation to the main line.

Actuators

An actuator (also called operator) is a device that provides the force to vary the orifice area through which the control agent flows. It accomplishes this by positioning the valve stem (or other driven element) in the valve (or other final control element).

Actuators are classified according to the form of input signal and output power used. Thus, actuators can be mechanical, pneumatic, electric, hydraulic, or a combination. For example, an electrohydraulic actuator is one that receives an electrical signal and uses a hydraulic system to perform the mechanical motion.

Mechanical Actuators

Mechanical actuators use a mechanical linkage to transmit motion between a sensing device and the final control element. Mechanical actuators are used extensively for liquid-level control systems, although they are not sensitive enough to serve some processes that require extremely accurate control.

Pneumatic Actuators

The use of pneumatic actuators is widespread in the petroleum and chemical industries. The popularity is due to a number of factors, such as safety, simplicity, and reliability. Two major types of pneumatic actuators are *diaphragm* and *piston.* A diaphragm actuator usually contains a spring that opposes the air pressure applied against the diaphragm, although springless types in which controlled air pressure can be applied to either side of the diaphragm are quite common. A piston actuator is usually springless.

Spring-loaded diaphragm actuator. A spring-loaded diaphragm actuator (fig. 3.17) has characteristics and action that deserve attention. In a *direct-acting* actuator, air pressure is applied to the inlet at the very top of the diaphragm case, and downward motion of the diaphragm and actuator stem is opposed by a heavy spring. The application of air pressure tends to drive the actuator stem down and thus *closes* a direct-acting valve. A *reverse-acting* actuator is quite similar, but there are important differences. Close study of the drawing in figure 3.18 will show that air pressure is applied to the lower portion of the diaphragm case and that this application of pressure will cause the actuator stem to be driven up, thus simultaneously compressing the actuator spring and *opening* a direct-acting valve.

Two ideas concerning force are important in the operation of the valve actuator shown in figure 3.17: (1) the force caused by air pressure bearing against the diaphragm is equal to the product of the area of the diaphragm and the pressure, and (2) the force needed to compress a spring is equal to the product of the distance the spring is compressed and some particular constant number that can be determined by experiment. Thus, in one case, *force* equals *area* times *pressure,* and for the spring, *force* equals *distance* times a constant k. In acceptable abbreviations,

$$F = Ap = kx, \text{ or simply } Ap = kx,$$

where

F = force;
A = area in square inches;
p = gauge pressure in psi;
k = a constant;
x = linear distance in inches.

If the spring rests against the diaphragm assembly with virtually zero force, the actuator stem will begin to move once the gauge pressure exceeds zero. Most, if not all, pneumatic actuators have the springs compressed a small amount so that the valve is

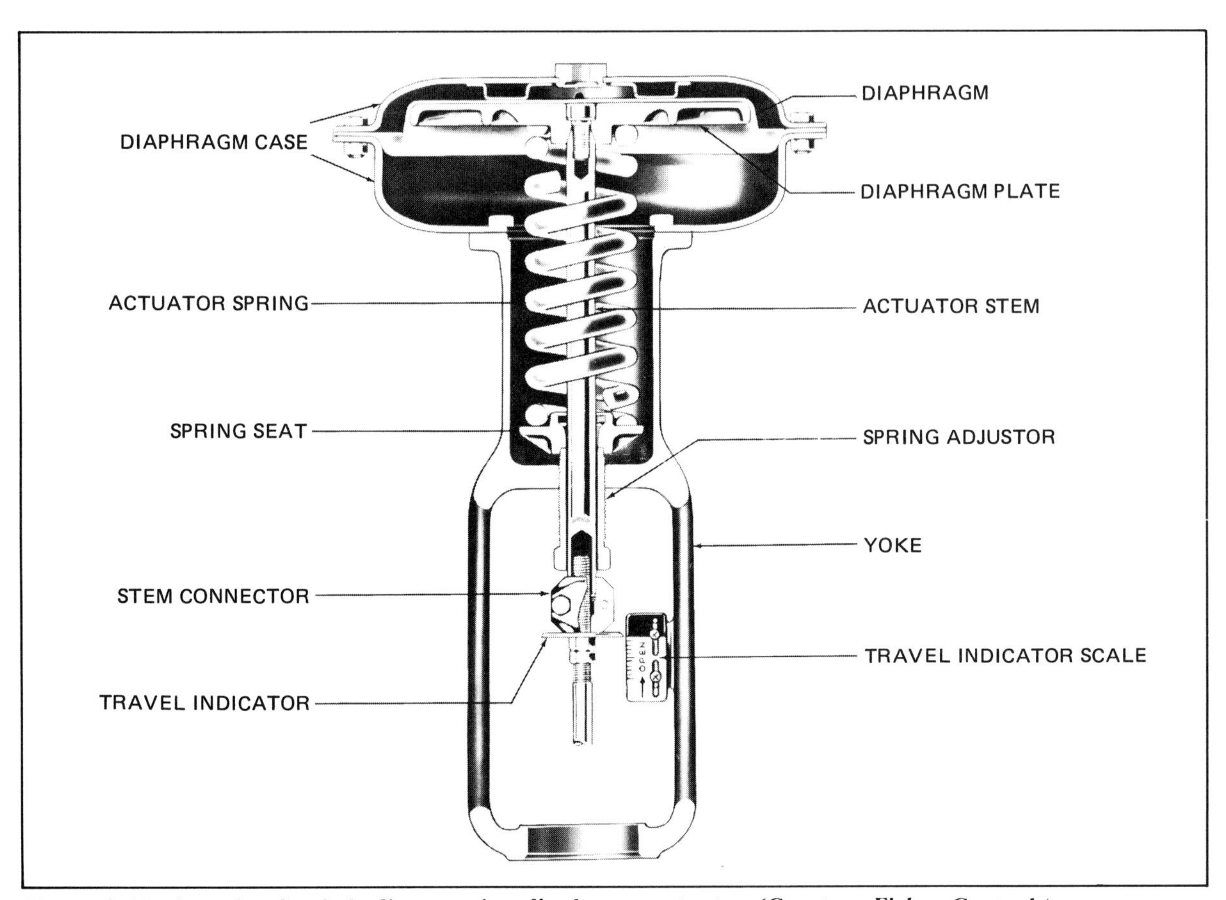

Figure 3.17. A spring-loaded, direct-acting diaphragm actuator *(Courtesy Fisher Controls)*

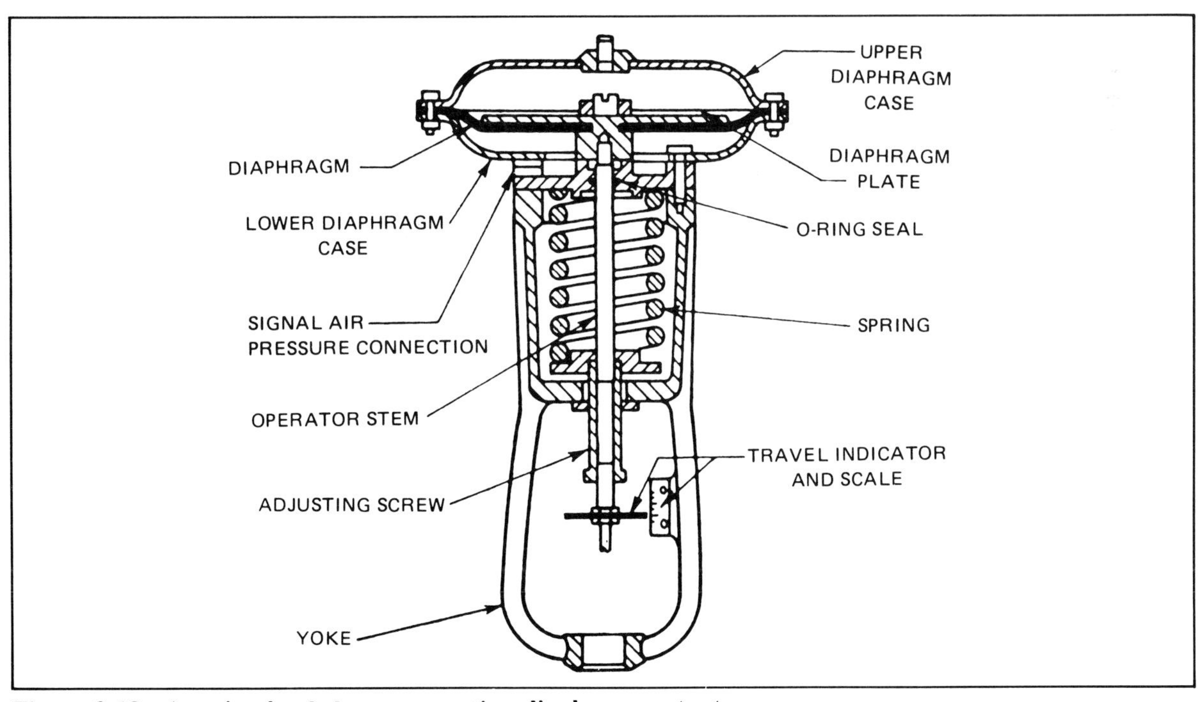

Figure 3.18. A spring-loaded, reverse-acting diaphragm actuator

held firmly open or closed with zero gauge pressure applied to the diaphragm. In such instances, it is necessary to apply some given minimum pressure to the diaphragm before it can overcome the force caused by the slightly compressed spring. Taking this initial compression factor into account, this equation can be rewritten as follows:

$$Ap = kx + F_i,$$

where F_i represents the initial force that must be overcome before stem movement occurs.

To understand the relationship between applied air pressure and actuator stem displacement, it is useful to know that manufacturers and control system engineers have agreed on certain standards, one of which is the use of a specific range of air pressure values for control purposes. This range is from 3 to 15 psi, which means that when 3 psi is applied to the diaphragm, the actuator stem will just begin to move; when 15 psi is applied, the stem will be fully extended in its travel.

For example, suppose a valve actuator using the standard range of pressure values has a total stem travel of 1 inch and a diaphragm area of 75 square inches. At 3 psi pressure the stem is virtually "weightless," and at 15 psi pressure it has moved 1 inch. Expressing length and area in inches and square inches, the following equations can be set up:

For 3 psi,

$$3 \text{ psi} \times 75 \text{ in.}^2 = kx + F_i;$$

however, since there is no stem movement at this pressure ($x = 0$),

$$F_i = 225 \text{ lb.}$$

For 15 psi,

$$15 \text{ psi} \times 75 \text{ in.}^2 = kx + 225 \text{ lb.}$$

Here, 225 pounds has been substituted for F_i. Since stem movement is 1 inch ($x = 1$), the value of k can be determined by substituting 1 inch for x:

$$k = 15 \text{ psi} \times 75 \text{ in.}^2 - 225 \text{ lb.}$$

$$k = 900 \text{ lb.}$$

The following equation can be used to determine the stem movement x in the above problem for any applied pressure between 3 and 15 psi:

$$x = \frac{P - 3 \text{ in.}}{12}$$

where P is applied pressure in psi.

This equation reveals that the actuator stem can be made to assume any position within its limits of travel merely by adjusting the air pressure applied to the diaphragm. In theory, the movement should be a predictable function of the applied pressure; but in practice, a number of factors arise that tend to cause such difficulties as the friction between guides and packing of the valve to which the actuator is attached and the effects of flow through the valve. In instances where these factors might be unacceptably troublesome, an auxiliary device is used with the actuator to ensure that the valve is positively positioned at the setting required by the process under control. Such devices are called *positioners*.

Air-loaded diaphragm actuator. The spring-loaded diaphragm is replaced with an air-loaded version in the air-loaded diaphragm actuator (fig. 3.19). The air-loaded and spring-loaded diaphragm actuators are not directly interchangeable, and, unless it is used with a valve positioner, the air-loaded actuator is suitable only to on and off service. For on and off service a constant air pressure of a few psi, depending on the particular type and use of the valve, is applied to one side or the other of the diaphragm, and air pressure from the controller is applied to the opposite side.

In many applications requiring the use of a valve positioner, the air-loaded diaphragm actuator is superior to the spring-loaded type because the force exerted by a spring is constant at any given degree of compression. In

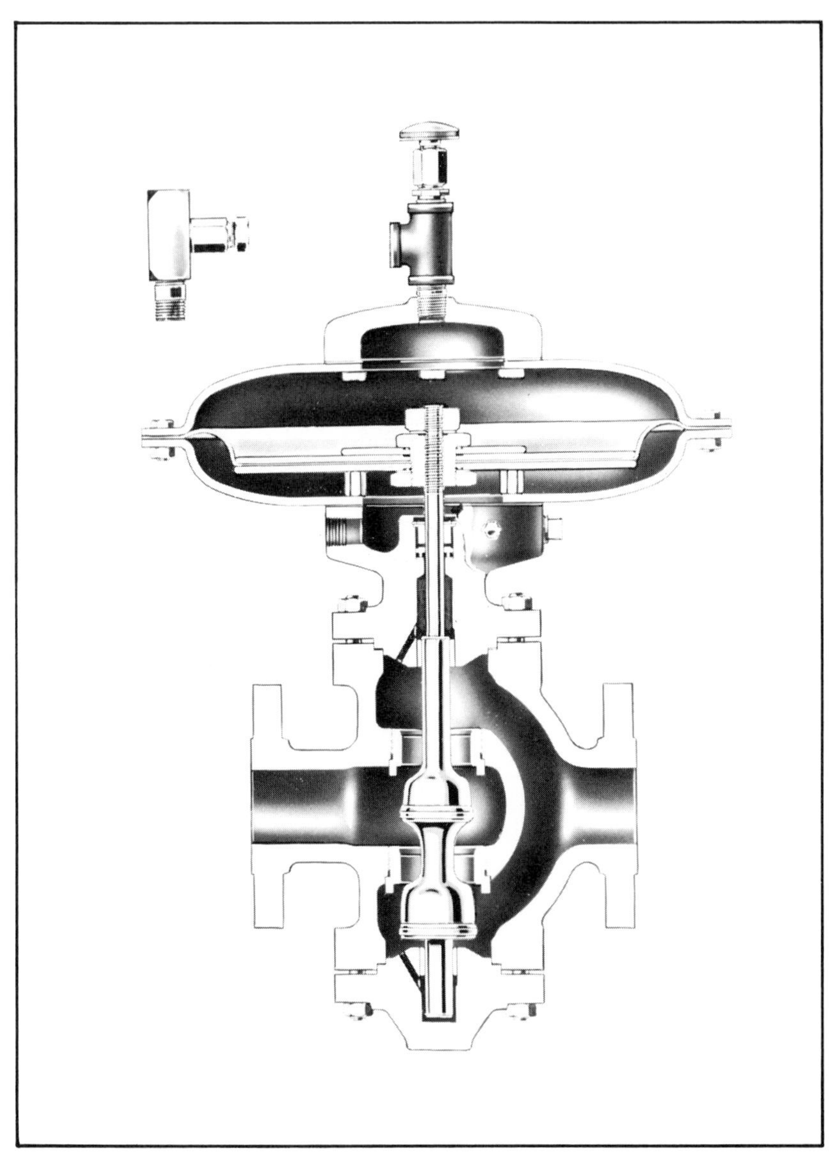

Figure 3.19. An air-loaded diaphragm actuator *(Courtesy Fisher Controls)*

applications in which the force required to position the valve stem varies widely for different positions, the spring action is apt to be erratic. The air-loaded actuator used with a proper positioner is capable of exerting a force in either direction of travel equal to the product of the effective area of the diaphragm and the applied air pressure.

Piston actuator. The use of pneumatic- or hydraulic-driven piston devices to actuate valves is an old idea and one that finds specialized use today. Piston actuators possess two advantages that make them desirable for some applications:

1. They are of rugged design, capable of handling high operating pressures. This enables them to provide rapid response and enormous linear forces.
2. They can be made to deliver very large linear movement.

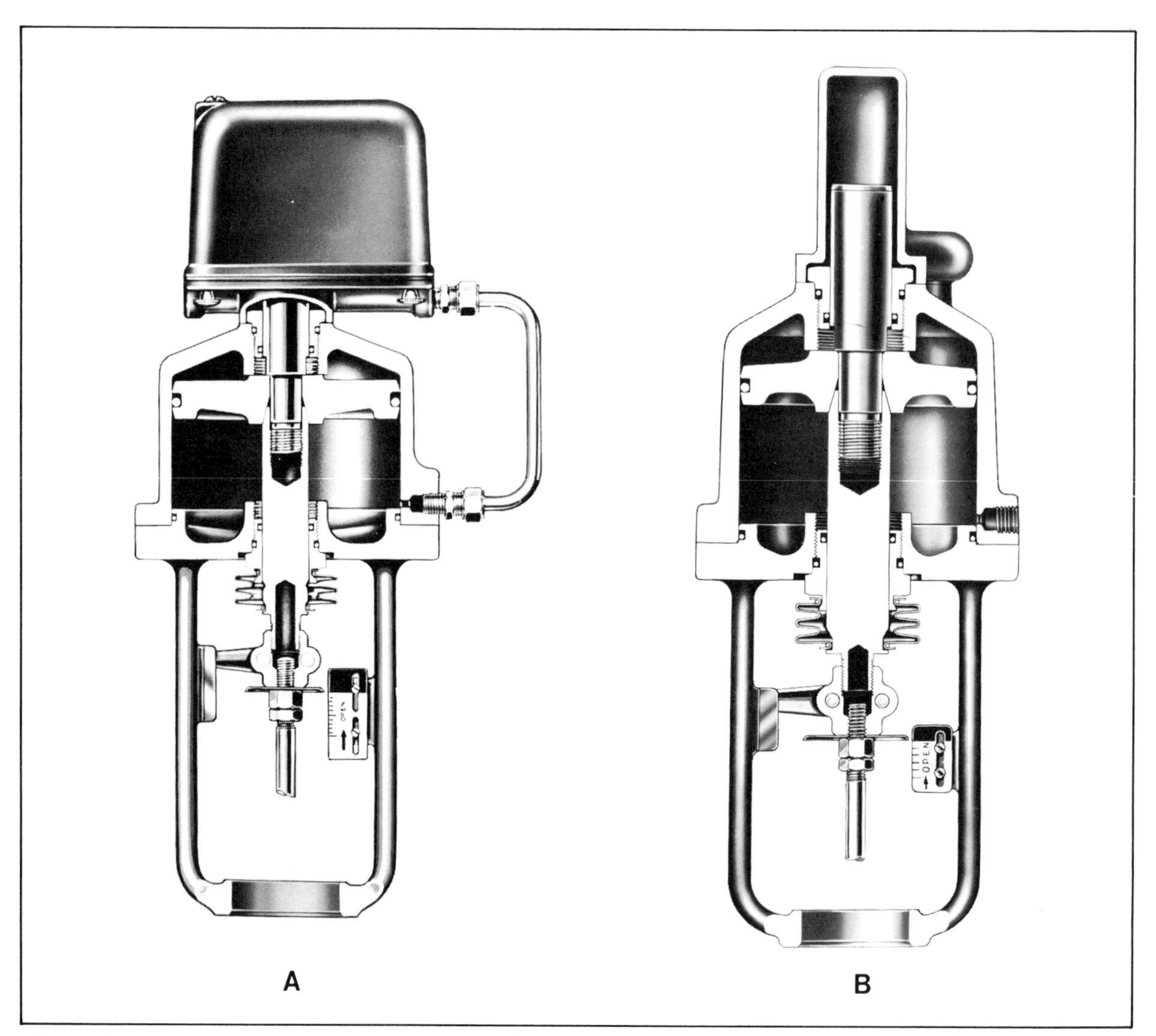

Figure 3.20. Pneumatic piston actuators. A valve positioner is attached to actuator *B*. (*Courtesy Fisher Controls*)

In the piston actuators depicted in figure 3.20, a positioner assembly rides atop actuator *A*, while *B* is for on and off service. The on-off actuator requires that the cylinder be loaded and unloaded through the operation of a solenoid valve, pneumatic switching valve, or similar equipment. A *valve positioner* is an attachment that fits on the valve actuator. Its purpose is to assure that valve stem movement follows the demands of the controller with great accuracy. It accomplishes this by acting as a force amplifier between the controller and the actuator. Positioners are used where friction occurs between valve stem and packing. They are also used when the force of flowing fluid on the valve plug causes unsatisfactory response by unaided actuators.

Air supply to pneumatic actuators. Air for all control purposes should be as free of moisture, oil, and other contaminants as practicable. Sometimes the air necessary to operate an actuator comes directly from the controller. In this case a system of piping or tubing exists between the two components, and necessary filtering is accomplished ahead of the controller.

Many control valve actuators using positioners or volume boosters receive air from a source separate from that supplying the controller. In such instances a filtering system might be installed at the valve (fig. 3.21). The sump collects condensed moisture that can be drained out through the bottom petcock. The filter, which is often a part of a pressure-reducing regulator, prevents the passage of viscous oils and particles of dirt.

Electric Actuators

Electrically powered devices used to position final control elements may be broadly classified into two categories: (1) solenoid-operated and (2) electric motor operated. Manufacturers usually provide the actuator and valve as an integrated unit.

Solenoid actuators. Solenoid-operated valves enjoy an enormous popularity in many automatic control systems that can function with a two-position, or on and off, flow control. They are employed as fuel flow control devices on many automatic central heating systems and in industrial processes where on and off action is acceptable.

Although solenoid actuators are as reliable as their source of power and are among the most inexpensive types to manufacture, their use for operating large valves at even moderate pressures is not feasible because of heavy construction and power consumption needed to obtain the required actuating force. As an example, solenoid valves capable of handling flow in 1-inch lines at 600 psi

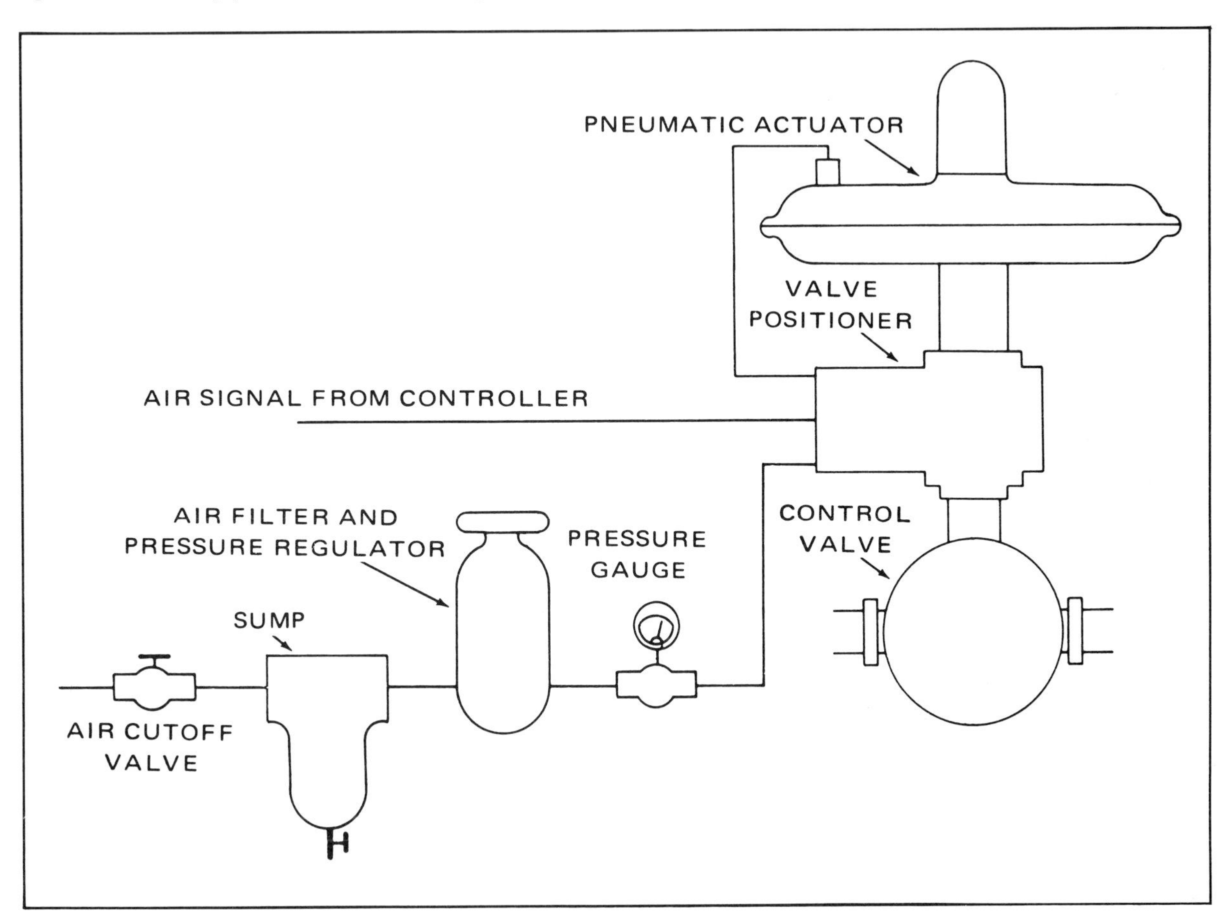

Figure 3.21. Arrangement of a typical air supply for pneumatic actuators

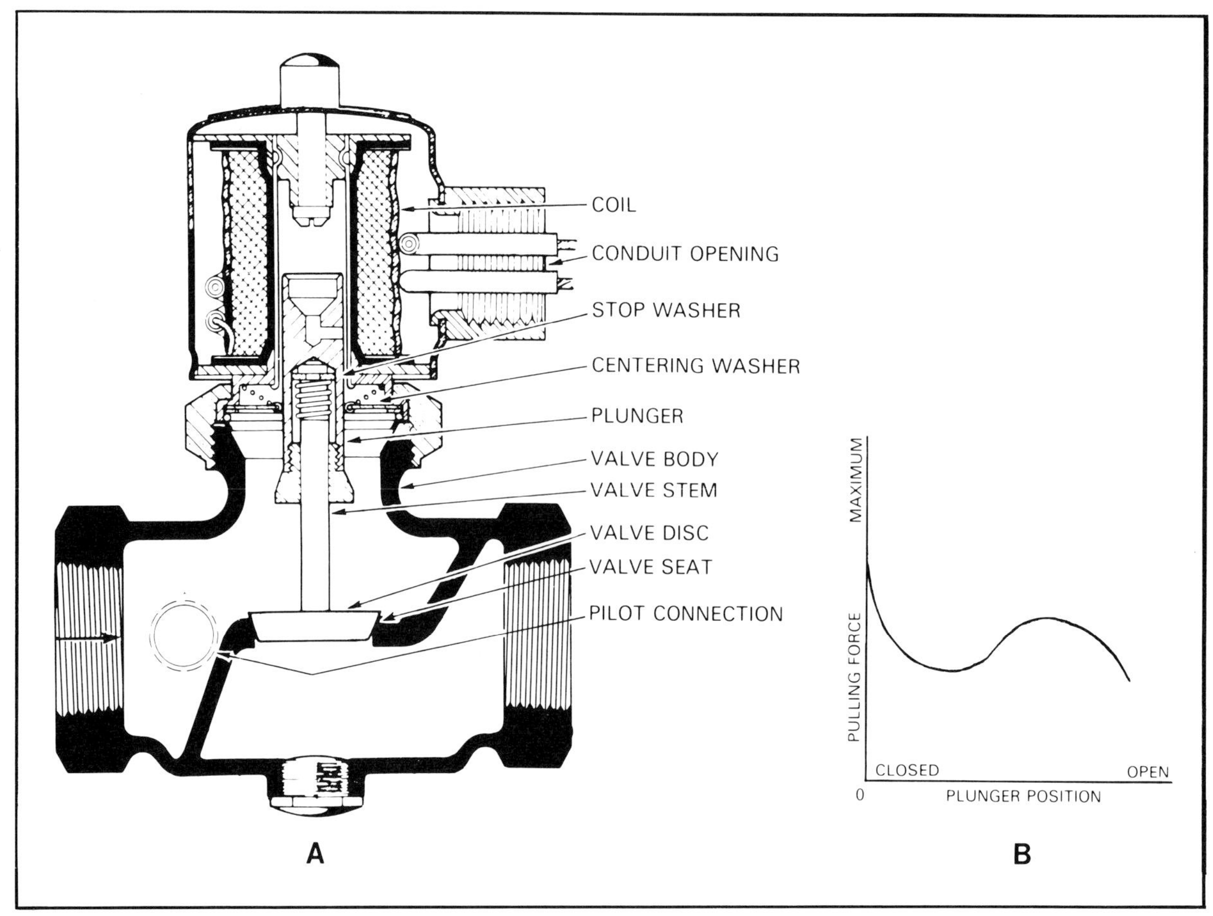

Figure 3.22. A solenoid actuator and valve (*Courtesy Honeywell*)

pressures are not common. On the other hand, solenoid valves for control of a 6-inch line at low pressures up to 10 psi are available.

In the sectional view of a typical solenoid-operated valve (shown in fig. 3.22*A*), the valve stem is attached to the soft iron plunger that rides in the hole centered on the axis of the solenoid coil. The valve is in the deenergized position, and, if flow is from left to right as it must be for proper operation, the valve disc is held firmly in its seat. In the deenergized position the plunger is positioned, as nearly as possible, to experience the maximum pull when the coil is energized. This position enables the actuator to overcome the seating force caused by the line pressure bearing on the disc. The pull force varies from a maximum in the seated position (*B*) to lesser values for the remainder of the plunger travel.

The closing action of the valve in *A* is achieved by the weight of the plunger, valve stem, and disc. Once the disc nears its seat, flow will snap the valve tightly shut. This sort of closing action cannot tolerate the friction effects inherent in having the valve stem pass through packing material; therefore, "packless" valves are in common use. This feature is attained by providing a fluid-tight cylinder for the solenoid coil. If such a liner is made of nonmagnetic material, the efficiency of the unit is virtually unaffected. Sometimes exposure of the soft iron plunger to the fluids in the valve causes corrosion or other unacceptable conditions. In these cases valve stem packing must be used, and springs are provided to assist in overcoming the friction forces that affect the closing motion.

Electric motor operated actuators. Electric motors have been used to open and close valves for many years because they offer an excellent means for remote control of valves. The exacting demands of automatic process control have resulted in the development of electric motor actuators that are capable of any desired mode of control.

The motors used for powering actuators can be reversing or unidirectional types. They are available in a multitude of power and voltage ratings. Motors are almost invariably connected to the valve stems through a system of reduction gears, and such arrangements obviously reduce the speed with which a valve can respond to a control signal. The time required for such actuators to move from one extreme position to the other may be as low as 2 seconds or as high as 240 seconds.

The rotary motion of a reversible electric motor can be converted into linear motion suitable for operating a valve; the result is the problem of how to control the action of the motor itself. One solution is to provide floating control of the motor as shown in the circuit diagram in figure 3.23. The mercury switch acts as a single-pole, double-throw switch capable of assuming a neutral or no-contact center position. Movement of the switch can be effected by very feeble forces such as a Bourdon spring. If the instrument makes contact from *C* to *L*, the Open winding of the motor is energized and the actuator will continue to move toward its open position until the result of its control action or some other influence causes the contact between *C* and *L* to be broken. When the contact is broken, the motor will stop and the valve will remain set until contact is made again from *C* to *L* or from *C* to *H*. If contact is made from *C* to *H*, a similar action will occur, except the actuator will move toward the closed position. Obviously, as long as the mercury switch stays

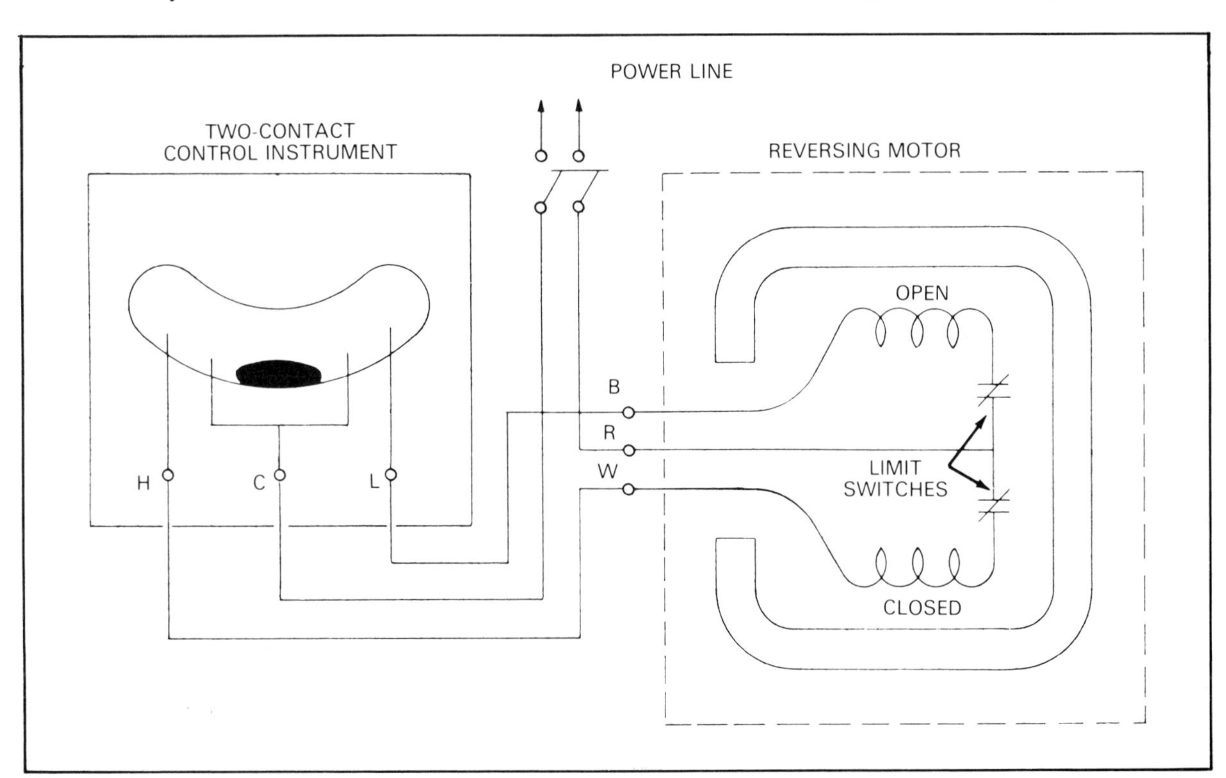

Figure 3.23. Circuit diagram for floating control of reversible motor

in its neutral position, the actuator will not move and the valve will simply float in its set position.

Electric motor operated actuators usually require a means for preventing overtravel of the valve stem or other device being positioned by the actuator. Limit switches that open the motor circuit at each extreme of travel are in common use for this purpose. Such switches may form a part of the actuator mechanism or may be a separate assembly actuated by the valve stem.

Of course, this type of floating control is not suitable for many applications because the amount of process deviation permitted by the floating band might be in excess of critical demands of control. A proportional control system (fig. 3.24) permits positioning the valve specifically in accord with the amount of deviation detected by the measuring means of the system. The reversible motor in this system can be assumed to be identical to the one described in figure 3.23. The 235-ohm resistor that forms a part of the proportional control instrument is positioned by the measuring means, and the other resistor is directly connected to the actuator mechanism for positioning purposes. A brief study of the circuit will disclose that motion of the motor and the direction of that motion are controlled by the position of the balancing relay. Thus, it becomes a matter of determining the action that controls the position of the relay.

Two facts about relay operation and motor rotation should be noted (fig. 3.24):

1. If more current flows in coil *C1* than in *C2,* the center contact of the relay will close with relay contact *3,* thus completing the motor circuit, and driving the actuator toward the open position.

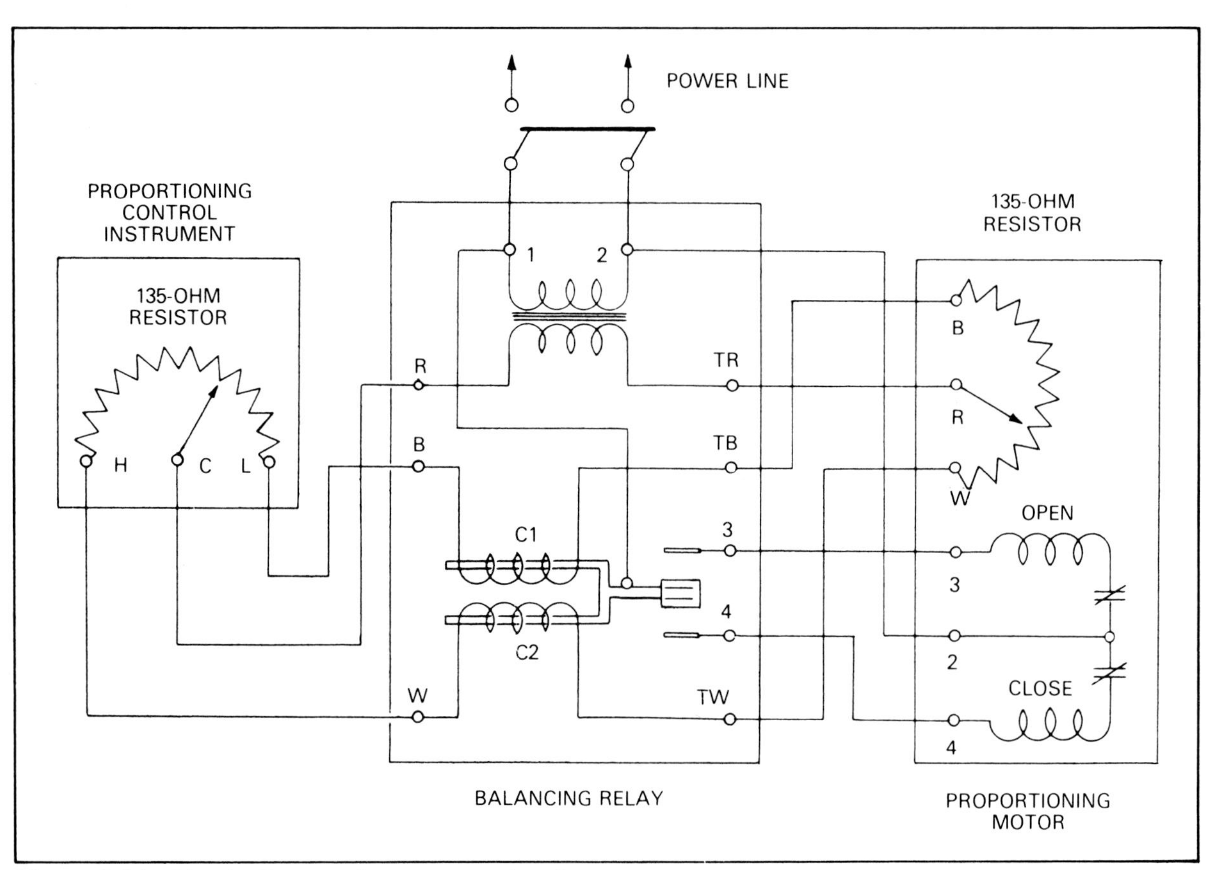

Figure 3.24. Circuit diagram for a proportional control system *(Courtesy Honeywell)*

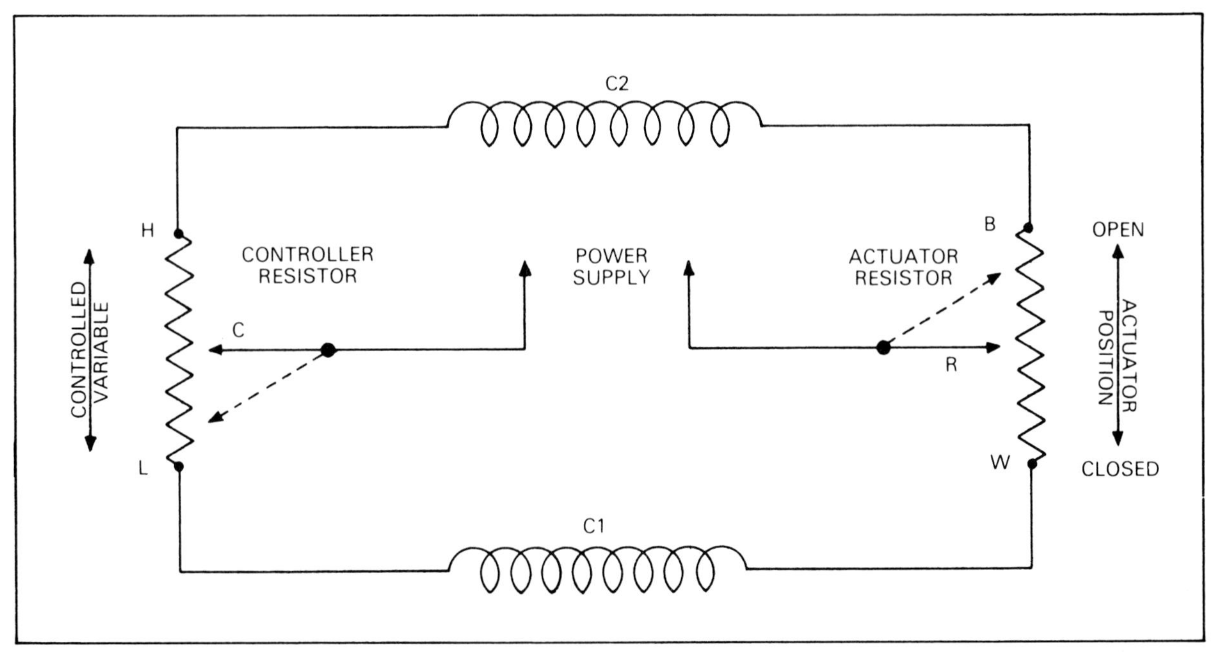

Figure 3.25. Schematic of the balancing relay shown in figure 3.24 ***(Courtesy Honeywell)***

2. If coil *C2* carries heavier current than *C1,* the relay will operate to drive the actuator closed. Now, to simplify circuitry and bring about a better understanding of how the balancing relay operates, the schematic in figure 3.25 should be considered.

With the sliding arms or movable contacts of the two resistors resting in their center positions, equal currents will flow through coils *C1* and *C2,* and the relay center contact will not move. A decrease in the controlled variable will drive the controller resistor movable contact toward *L.* This will add resistance to the circuit containing *C2* and subtract a like amount from that containing *C1,* resulting in a lessening of current flow through *C2* and an increase of current through *C1.* This unbalanced condition will cause the relay to close the circuit to the motor in such a way as to drive the actuator to the open position. The actuator-resistor movable contact will begin travel to a new position, simultaneously reducing the unbalanced condition caused by movement of the controller-resistor movable contact. Eventually the currents in *C1* and *C2* will be balanced and the motor brought to a stop. The action described provides a linear relation between the controlled variable and the position of the actuator.

Hydraulic Actuators

The use of hydraulic pressure in automatic process control does not enjoy the wide popularity associated with electrical or pneumatic systems, but hydraulic systems are not without advantages. The principal advantage lies in the fact that liquids are virtually incompressible. This enables the transmission of pressure changes almost instantaneously, while in a pneumatic system the considerable delay in transmission resulting from compressibility of air is sometimes a significant factor. Right now, however, the interest is in hydraulic pressure as it relates to use in an actuator.

A simple speed control for a steam turbine is a type of hydraulic actuator. The system shown in figure 3.26 is in a balanced position with the steam valve passing the correct quantity of steam to maintain desired speed. The entire governor block is revolving with its weights (shown as two round balls diametrically situated) compressing the spring. The central governor shaft protruding from the center of the spring revolves with the governor, and its vertical motion is transmitted to the stationary shaft *AB* through the ball bearing at *A*. The rigid bar *BCD* has pivots at the points shown. The tension of the speed-adjustment spring keeps the rotating and stationary shafts firmly against the ball bearing.

If the turbine speeds up, the weights compress the governor spring and raise the rigid bar. The pilot valve piston moves upward and admits oil to the top of the actuator piston, at the same time opening the drain to the oil trapped below the actuator piston. The actuator then moves in response to the hydraulic pressure and reduces the steam flow through the valve. As soon as the actuator piston begins its movement, the rigid bar repositions the pilot valve and shuts off the flow of oil to the actuator cylinder. The action that takes place if the turbine loses speed for some reason is similar, except that the actuator opens the steam valve to admit more steam to the turbine.

Combination Actuators

Only a small step separates a hydraulic or pneumatic actuator from one that can be designated as electrohydraulic or electropneumatic. In such actuators the advantages of using hydraulic or pneumatic pressure to

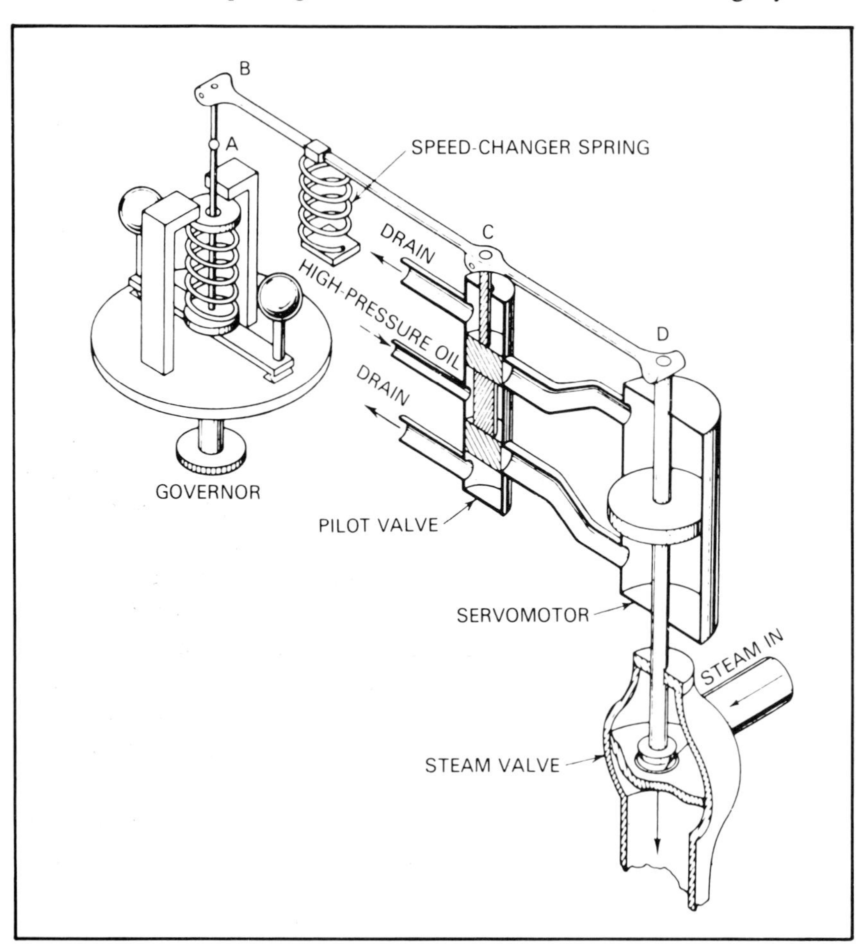

Figure 3.26. A speed control system for a steam turbine

drive the actuator mechanism and the relative ease of producing and transmitting electrical or electronic signals to trigger the response of the actuator are combined. The principal difference between electrohydraulic and electropneumatic actuators and the straightforward types is the means used to operate the actuator.

Controlled-Volume Pumps

A controlled-volume pump is one in which each cycle, or complete stroke, delivers a definite and predetermined volume of liquid. It is a positive-displacement pump, but, when used as a final control element, it usually contains refinements that set it apart from mud pumps, well pumps, and most other common positive-displacement devices. The principal differences, however, are capacity and the accurate metering capability of the pumps used as final control elements.

These pumps are used as injectors for forcing emulsion breakers, inhibitors, and other chemicals into process lines, tanks, and other vessels. They are capable of delivering enormous pressures—perhaps as high as 50,000 psi. On the other hand, they are more likely to be found delivering very small quantities of liquid to the process to which they are attached; for example, they may be designed to produce flows as low as 2 cubic inches per day.

The liquid ends of controlled-volume pumps are made in several forms, but the reciprocating piston type is the most common. Most controlled-volume pumps are provided with an adjustment that varies the length of the piston stroke to achieve a change in liquid volume delivered per stroke. Capacity is known to be a function of displacement and speed, and displacement is the product of piston area and stroke.

A cutaway view of chemical injection equipment that can be used at a pumping wellhead is shown in figure 3.27. The injection pump associated with this arrangement has an adjustable stroke length and rachet mechanism. The rachet mechanism is actuated by the operating lever, which can be linked to the walking beam of the well pump. Each well pump stroke delivers a constant volume of oil to the line, while at the same

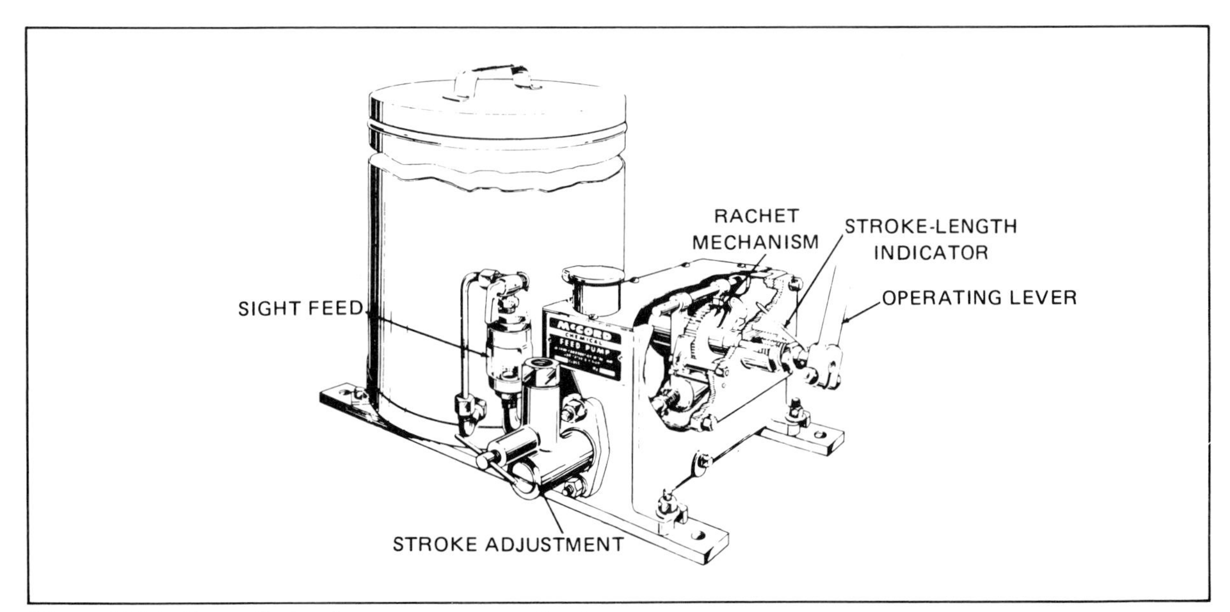

Figure 3.27. Controlled-volume pump as a final control element

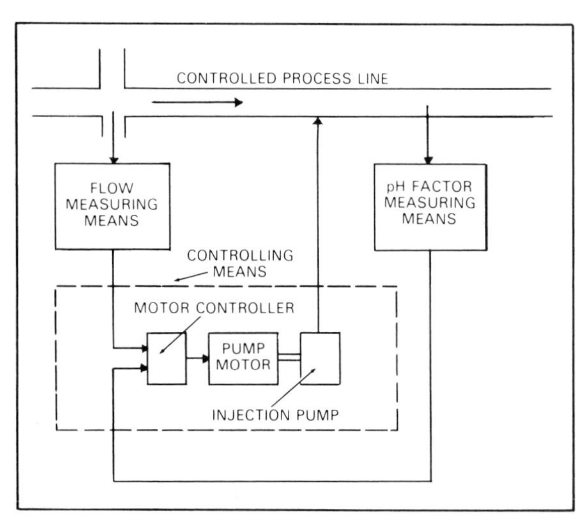

Figure 3.28. A combination open-loop and closed-loop system with a controlled-volume pump

time the walking beam actuates the injection pump. Such a system lends itself to varying the piston stroke to attain the proper quantity of chemical additive. This is an open-loop system of control since there is no feedback of information from the final product. It is an example of one of the few practical applications of the open-loop system.

The combination open- and closed-loop system (fig. 3.28) employs a controlled-volume pump. The pump is driven by a variable-speed electric motor, which is subjected to control by two variables—rate of flow and pH factor. The rate of flow controls the motor speed within certain very wide limits, allowing greater speeds with increasing flow values. The rate of flow has a governing effect on the injection of corrective chemical into the line, but no mutual adjustment exists between the *correct* amount of injection and the rate of flow. In other words, this part of the system is an open loop. Now study the effect on the motor caused by the pH measuring means and its connection to the controlling means. Here a pH detector transmits corrective information to the controller; this in turn adjusts the speed of the pump to bring about a proper value of pH factor. This part of the system is a closed loop.

In the example (fig. 3.28) the capacity of the pump was varied by regulating the motor speed. The accurate control of injection could be accomplished equally well by having the feedback loop actuate a mechanism that varies the piston stroke length, leaving the open-loop portion of the system to control the motor speed.

Stroke length variation is often accomplished automatically by an actuator that is powered by an air signal whose pressure is

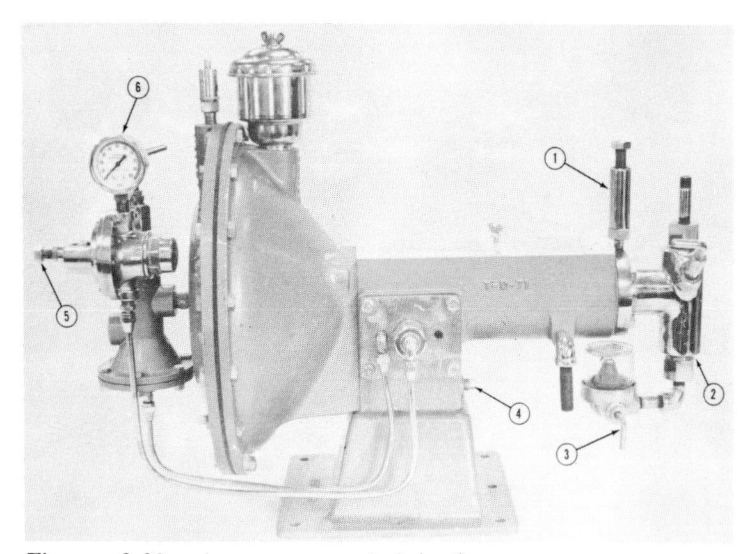

1. GREASE CHAMBER
2. CHECK VALVE
3. CHEMICAL INLET VALVE
4. DRAIN PLUG
5. REGULATOR ADJUSTMENT
6. REGULATOR GAUGE

Figure 3.29. A gas-powered injection pump *(Courtesy Texsteam)*

established by the measuring means. Some reciprocating pumps are air- or gas-operated, and in such pumps the capacity is partially controlled by a pneumatic positioner that limits the piston travel (fig. 3.29).

Capacity regulation by pump speed control assumes one of several forms:

1. The pump has a fixed stroke length and is driven by a variable-speed rotary motor.
2. The pump has a fixed stroke length and is driven by a reciprocating pneumatic or hydraulic actuator.
3. A constant-speed electric motor drives the pump whose speed is regulated by mechanical speed drives installed between the motor and the pump.

Each of these means is useful and practicable for some purpose; however, the variable-speed motor is not satisfactory over very low ranges. For example, its delivery would probably be erratic at low ranges if designed to cover the range of one to a hundred strokes per minute. The pneumatic and hydraulic drives are ideally suited to applications requiring, say, one stroke per minute or less.

Other Final Control Elements

In addition to valves and pumps, *louvers* and *dampers* occasionally function as final control elements of an automatic control system. One additional form of final control element that finds frequent application is the *electric switch.* In figure 3.30 the automatic hot-water system's heating element is electric, and its final control element is simply a heavy-duty, remote-controlled switch that controls current to the heating element.

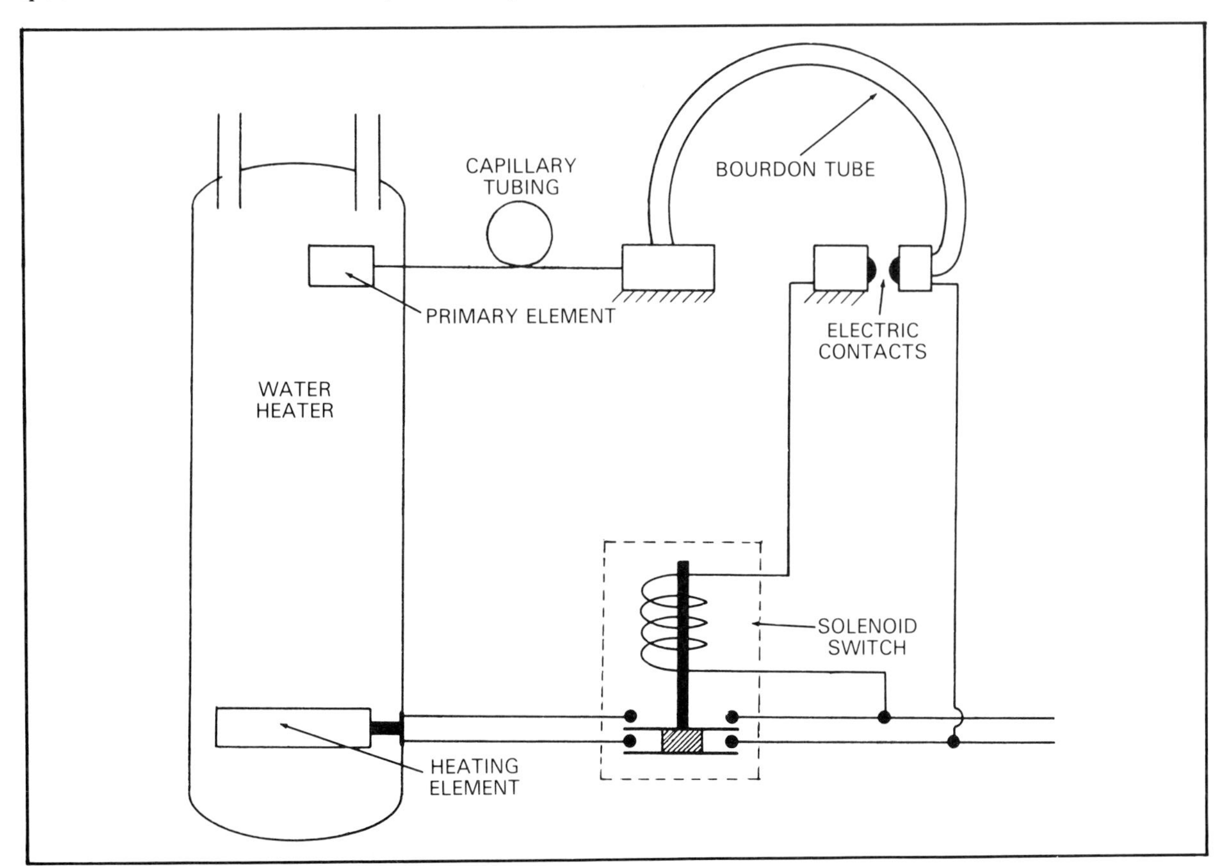

Figure 3.30. An electric switch as the final control element in an electric hot-water system

CHAPTER

4

Automatic Controls

Automatic control. The term implies that with the setting of a few dials and the pushing of a start button a complex system can be made to perform with near perfection, requiring only the casual attention of a human operator. Modern techniques of instrumentation have not achieved perfection, but the great strides made in recent years and those to be made in the near future are sure to provide a level of performance undreamed of before the advent of solid-state electronics.

This chapter will begin with a development of pneumatic control concepts. Pneumatic control enjoys widespread use in instrumentation. That makes it important. In addition to its importance, it is an ideal vehicle for developing a good understanding of automatic control principles, particularly those that apply to the modes of control.

Electronic control forms the latter part of the chapter. The comprehension of the subject is made less difficult if a grasp of pneumatic principles has been achieved. The study is based on the use of operational amplifiers to obtain the various signal-producing functions, because these remarkable devices spare the student the need to be concerned with the enormous complexity of detailed circuitry.

Pressure Regulators

Before studying the methods for accurately varying the pressure applied to an actuator, it is necessary to understand how air pressure is maintained at a given constant value. For example, a source of air pressure of 100 psi might be available for control applications, but this value is too great for direct application to a control system, so a means for reducing this pressure to a safer value of perhaps 20 psi is desirable. Such a means, of course, must be capable of maintaining this reduced pressure regardless of severe fluctuations in demand.

Weight-Loaded Regulators

A self-contained, weight-loaded regulator (fig. 4.1) is a simple means of reducing and regulating pressure. It is a double-ported valve with its poppet-type plug actuated by a diaphragm that supports a specific constant weight. The weight tends to force the valve open. On the other hand, pressure buildup on the outlet side of the valve has an open path to the lower side of the diaphragm, tending to lift it against the weight and thus closing the valve. By choosing the correct value of mass to bear down on the diaphragm, the desired pressure can be obtained at the valve outlet. The regulator is called *self-contained* because the controlling weight and the diaphragm pressure connections are internal features of the unit.

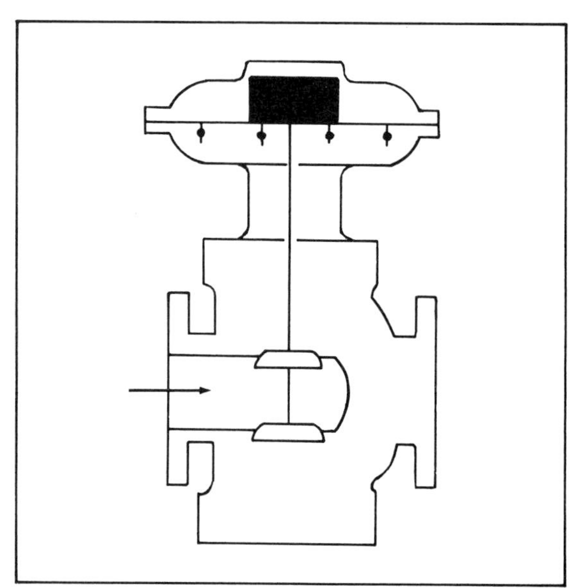

Figure 4.1. A self-contained, force-loaded regulator

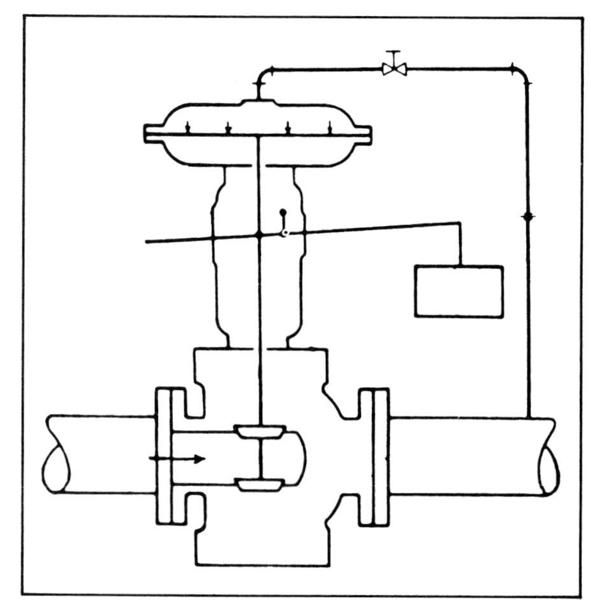

Figure 4.2. A force-loaded regulator with an external-pressure connection

A weight-loaded regulator with an external-pressure connection to the diaphragm (fig. 4.2) has a levered system with weights external to the main valve assembly. Such an arrangement offers more versatility for some applications than the self-contained type but is of little use in the petroleum industry.

Spring-Loaded Regulators

Spring-loaded units are the most popular form of regulators in reducing and controlling pressure in pneumatic controllers. They are characterized by simplicity, light weight, small size, and reliability. The small, self-contained, spring-loaded regulator (fig. 4.3) finds wide use as a reducer for control purposes. The compressed spring tends to drive the valve plug open, while pressure existing at

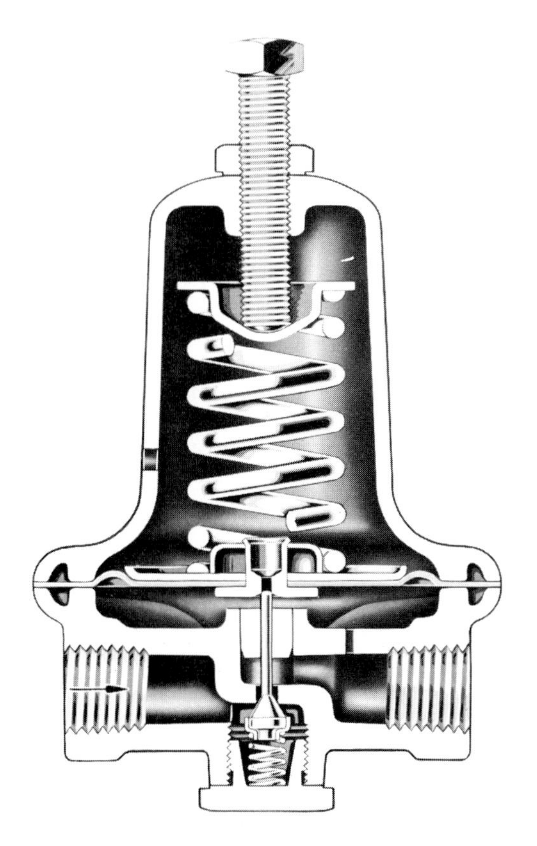

Figure 4.3. A self-contained, spring-loaded regulator (*Courtesy Fisher Controls*)

the outlet side of the unit exerts an opposing force against the lower side of the diaphragm. The valve plug will settle at some opening determined by the amount of pressure required to balance the force of the compressed spring. The outlet pressure is adjusted by varying the tension of the spring with the spring-compressing screw.

The field of pressure regulators is a broad and important one and cannot be completely explored here. However, knowing how to supply a regulated source of air at about 20 psi is necessary for the study of pneumatic control.

Developing a Pneumatic Controller

Although the regulator supplies a regulated source of air pressure, the problem of providing a convenient way of modulating, or varying, this supply according to the fluctuations that take place in the controlled variable of a system or process still exists. With this problem in mind, a pneumatic controller, capable of receiving and interpreting information from the measuring means and passing along corrective impulses to the final control element, must be developed.

Fixed and Variable Orifices

A fixed orifice in the controller is a simple arrangement that is designed to solve the problem. In the diagram in figure 4.4, the pressure

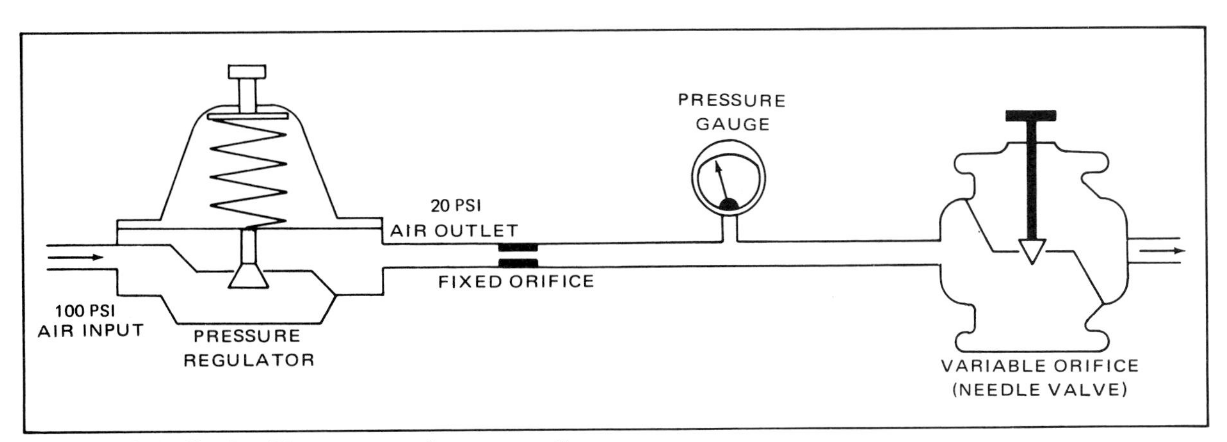

Figure 4.4. A fixed orifice as part of a pneumatic controller

on the left side of the fixed orifice is maintained at 20 psi by the regulator. If the variable orifice (in the form of a needle valve) is closed off tightly, the pressure to the right of the fixed orifice will also be 20 psi and will be so indicated by the pressure gauge.

Assuming that the variable orifice can be made somewhat larger than the fixed orifice—say, two or three times as large—it seems reasonable to suppose that air will escape from the large opening so fast that no appreciable pressure will build up between the two orifices, and the pressure gauge will indicate a value near zero. If a diaphragm actuator is substituted for the pressure gauge, changing the opening of the variable orifice will cause the actuator to follow the pressure changes. It seems obvious that unless the fixed orifice is made quite small, this system will waste a large volume of air, so some fairly small-sized orifice, about 0.01 inch in diameter, is used, and the development of the modulating means proceeds.

Nozzle and Flapper

The variable orifice shown in figure 4.4 is not readily adaptable to the needs of automatic control, so something other than that form must be designed. Using the same fixed orifice of 0.01 diameter, but substituting a 0.025-inch nozzle for the variable orifice, the system now approaches something practicable (fig. 4.5). With constant-pressure air fed to the fixed orifice, and the nozzle unobstructed, air will escape through the nozzle at a rate that prevents the buildup of any appreciable pressure between fixed orifice and nozzle.

If the pivoted flapper, which is a light strip of metal and is sometimes referred to as a baffle, is moved into position against the nozzle opening, pressure between orifice and nozzle will increase. The valve actuator will be operated, and the pressure gauge will indicate a pressure near the maximum available from the regulator. If the flapper is carefully retracted a very short distance from the nozzle, the pressure as indicated by the gauge will decline. Two interesting facts should be noted about the operation of the flapper and its effect on the controlling pressure:

1. Only a small force is required to position the flapper to produce a full range of pressure values for control. However, the force required is by no means a

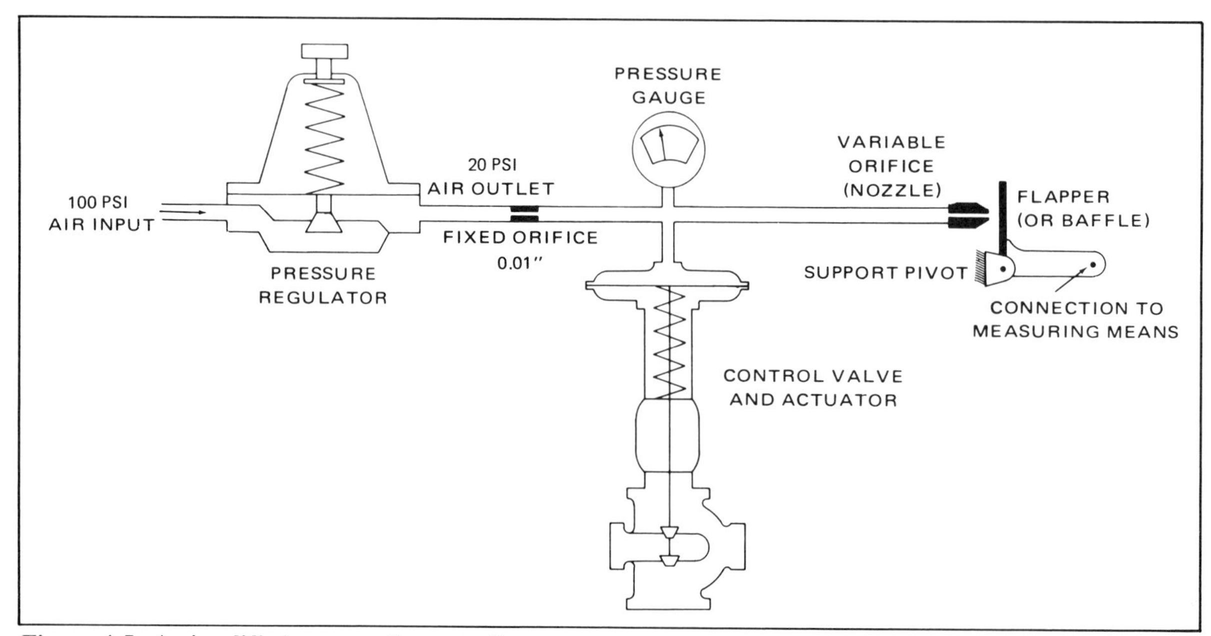

Figure 4.5. A simplified pneumatic controller

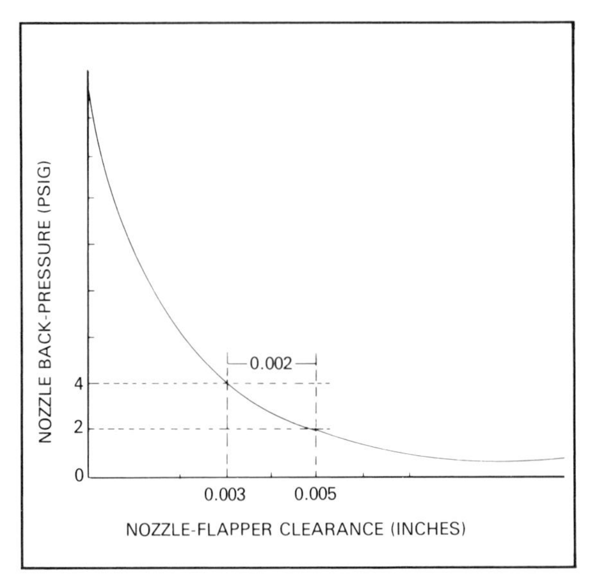

Figure 4.6. Effect of nozzle-flapper clearance on nozzle back-pressure

linear function, largely because of the opposing force of the jet of air as the flapper approaches the nozzle.

2. The amount of movement required of the flapper to produce the full range of values is incredibly small.

The idea that broad changes in control pressure can be effected by applying only a small positioning force is appealing, since good use can be made of the feeble forces available from the measuring means. On the other hand, the fact that only minute changes in the position of the flapper cause rather severe pressure changes presents problems that require extensive refinements to overcome.

It will be worth the effort to make a detailed study of the relations existing between nozzle back-pressure and nozzle-flapper clearance. The graph of these relations (fig. 4.6) shows that only about 0.01 inch of flapper movement produces enough change in nozzle back-pressure to drive the actuator from one extreme to the other.

Two-Position Pneumatic Controller

The new and unrefined controller is used in a liquid heating system (fig. 4.7). Note that a set-point device has been incorporated. It consists of a micrometer screw that moves the flapper pivot point along a line parallel to the nozzle. The flapper is connected to a Bourdon spring of the measuring means, and a rise in temperature causes the spring to flex in a way

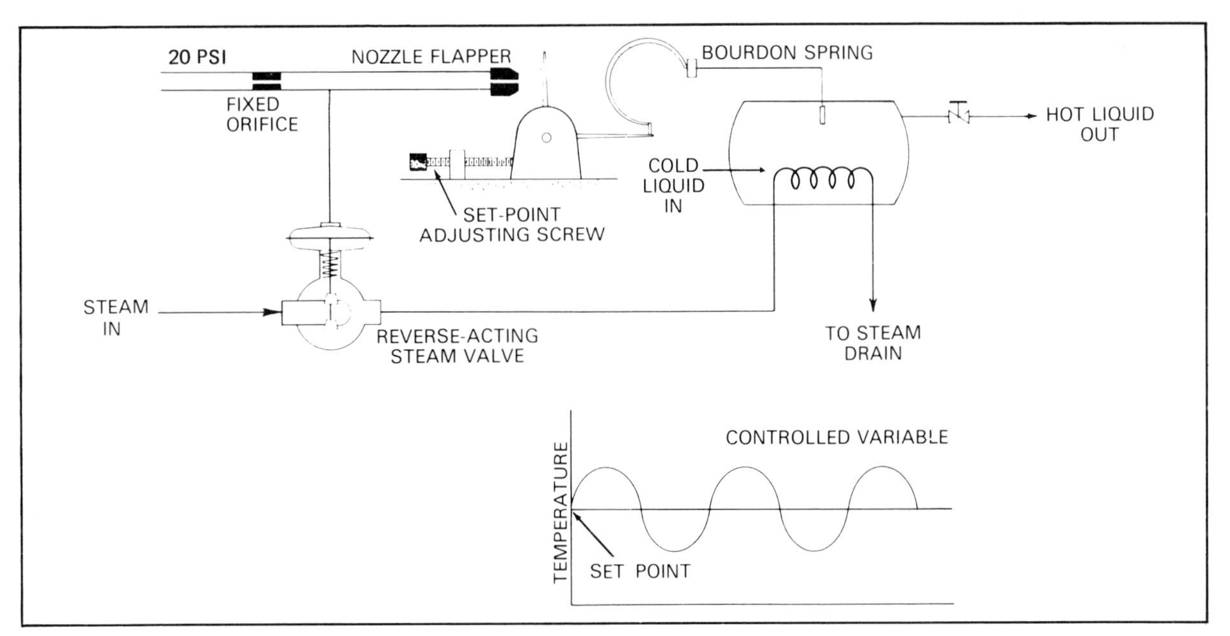

Figure 4.7. A two-position (on-off) controller in a liquid-heating system

that opens the gap between nozzle and flapper. The pressure to the valve declines; since this is a reverse-acting valve (to prevent overheating in event of air pressure failure), the plug begins to close slowly. The device is essentially an on and off controller, but it permits rather severe departures from the set point. The accompanying curve shows the changes in controlled variable.

This pneumatic two-position controller needs only a small amount of energy to trigger, but the sluggish action of the control valve is definitely not satisfactory. The cause is obvious: the fixed orifice passes air at such a low rate that the comparatively large volume of the actuator diaphragm case requires a considerable period of time to fill. Once it is filled, the slow bleed-off through the nozzle slows down its action in the reverse direction.

The answer to sluggish performance is a matter of supplying an air relay that can be actuated by a small volume of air and which, in turn, can control large-volume air impulses to drive the actuator. Such a device added to the present arrangement would result in making the system a good two-position controller.

The ultimate aim, of course, is not an on-off controller, since such controllers are not acceptable for many control applications. So, before beginning the design of an air relay, it will be a good idea to anticipate a sure problem. Looking again at figure 4.6, note the nonlinear relation between nozzle-flapper clearance and nozzle back-pressure.

The Air Relay

To avoid the problems that will arise from these effects, a nearly linear portion of the operating curve for nozzle-flapper clearance and nozzle back-pressure is selected. By selecting the range represented between 2 and 4 psi back-pressure, the linearity is acceptable, and any problem of flapper displacement caused by jet action at high back-pressures is eliminated. The immediate problem, then, is to design an air relay capable of using the 2 to 4 psi impulses of back-pressure to provide linear control of the 3 to 15 psi needed to operate the actuator.

A system that will work (fig. 4.8) has the usual fixed orifice and nozzle arrangement, but in place of the actuator diaphragm case, the nozzle back-pressure is allowed to actuate a small-volume bellows. Note that a push rod attached to the free end of the bellows operates a special form of ball plug valve. This special valve has two seats, and the small bellows is capable of driving the ball plug from one seat to the other.

The full 20 psi regulated air pressure is applied to the inlet of the special air relay valve, and the outlet is connected to the actuator diaphragm case. With the ball plug positioned between its two seats, some air will be applied to the actuator, and some will leak out around the push rod. Careful design of the valve can achieve a proportional relationship between nozzle back-pressure and the 3 to 15 psi pressure needed for operating the actuator. This relay is a crude but effective design. Other and better methods of building relays exist, but substituting this relay device into the control problem of figure 4.7 results in a fast-acting on-off controller.

Proportional Controller

Now that an air relay has been provided, attention is focused on the design of a proportional controller. The heart of the problem is to devise a means of using the almost microscopic flapper movement. Only about 0.002 inch of flapper travel is present to work with (fig. 4.6), as this slight movement produces a full range of pressure values from the air relay. Such small movement rules out the use of uncompensated mechanical linkages because of the lost motion and the backlash in such arrangements. A movement of 0.25 inch for the measuring means linkage is required. With this in mind, a system must be devised that will allow this 0.25-inch movement to result in a net change of 0.002 inch of the nozzle-flapper clearance.

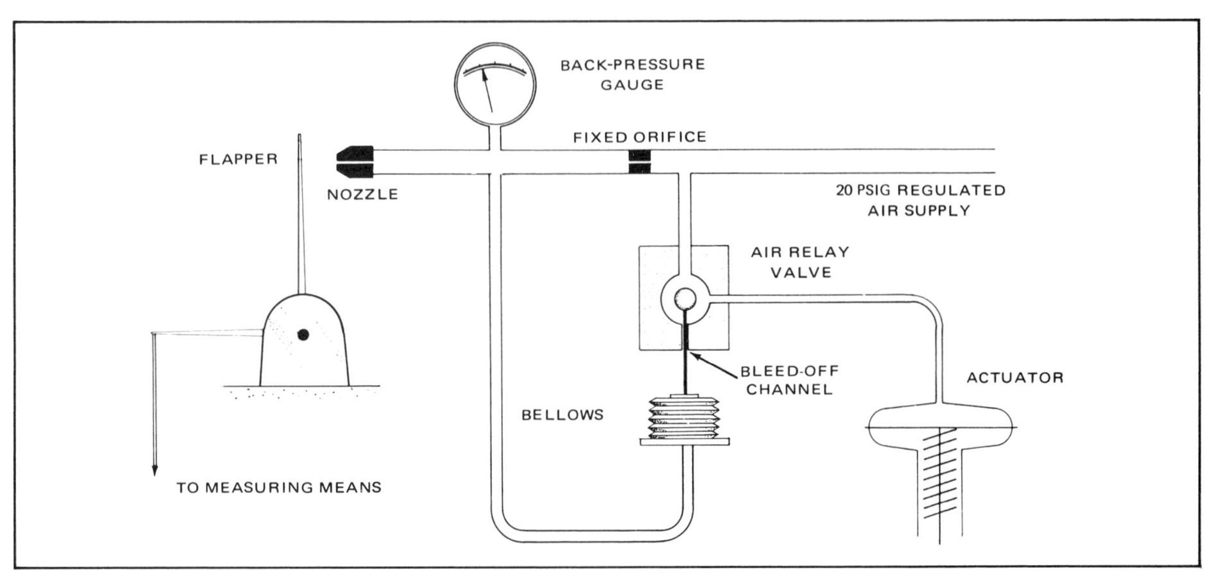

Figure 4.8. Using an air relay to provide linear control

The flapper might be actuated by two variables: (1) the action of the measuring means and (2) the related action of the air relay output pressure. The measuring means displaces the flapper. Then a bellows-spring arrangement, operated by the air-relay output pressure, drives the flapper back to near its previous position. This action provides a form of *negative feedback* and adds stability to the system.

At the outset, it is important to understand that all moving parts are restricted to motion in the plane of the diagram (fig. 4.9), as the levers and links are of such design that they prevent any component of motion perpendicular to the diagram. Rod *B* is a stiff

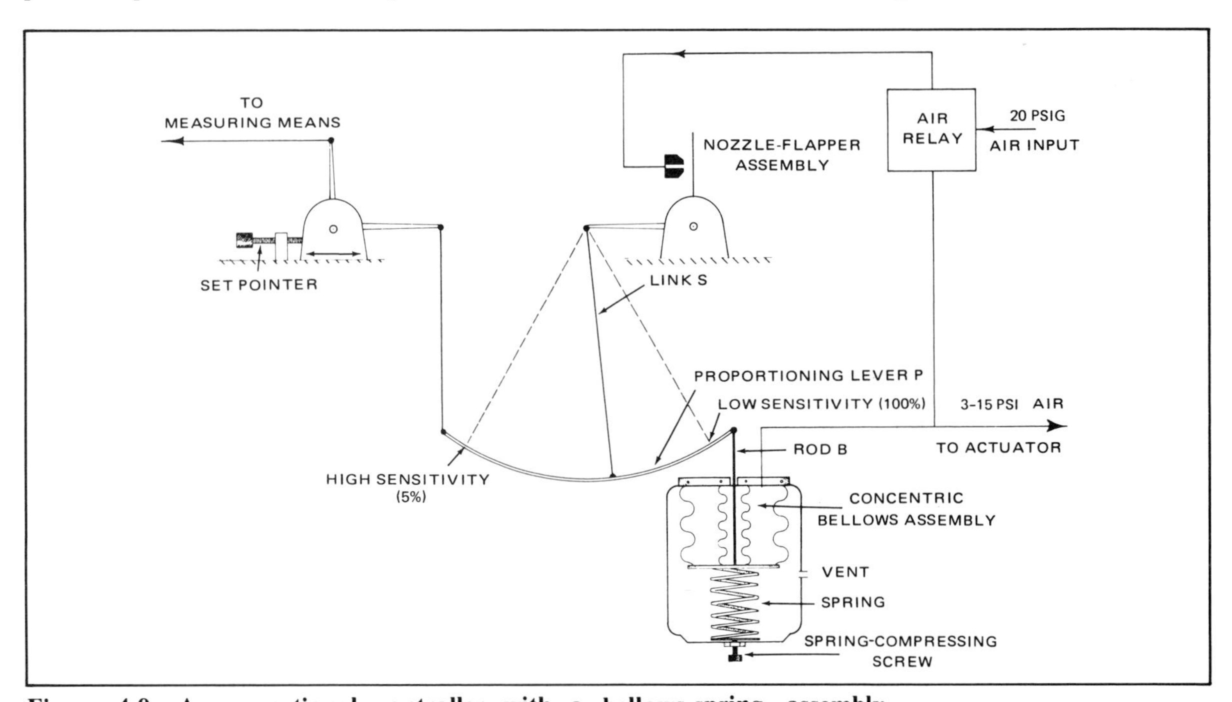

Figure 4.9. A proportional controller with a bellows-spring assembly

member capable of vertical motion only.

The proportioning lever *P* is a curved bar having a radius of curvature equal to the length of link *S*. Lever *P* is pivoted at the three points shown, but the location of the pivot point formed with link *S* is adjustable from one end of lever *P* to the other.

The bellows-spring assembly seems unnecessarily complicated, since the same proportional action can be obtained with a much simpler design, but anticipate a future need for this present design and accept its complexity for now.

The assembly consists of a pair of bellows arranged concentrically and fitted to common end plates, thus forming a single unit. The bellows assembly is fitted into a sturdy metal cylinder in such a way as to compress a special spring. A setscrew is provided for adjusting the tension of the spring. When assembled, the unit comprises two airtight compartments, one of which is the bellows and the other the metal cylinder surrounding the spring and bellows. The cylinder is vented. During the preliminary tests of the bellows-spring assembly, rod *B* moves a little more than 0.25 inch when the pressure applied to the bellows is varied from 3 to 15 psi.

When the regulated air pressure is applied to the control system, assuming that the flapper is midway between its limit of travel and thus positioned well away from the nozzle (that is, a midpoint reading of the measuring means), the nozzle back-pressure will be at a minimum. Under these conditions, the air relay (fig. 4.8) will rapidly transmit air to the actuator and the bellows (fig. 4.9). As the bellows expands, rod *B* pulls down on lever *P*, causing the flapper to approach the nozzle. The resulting back-pressure ultimately positions the ball-plug valve of the air relay. The system quickly reaches equilibrium with a nozzle-flapper clearance of about 0.004 inch as indicated by figure 4.6. It is found that the system will react quickly to any change in this clearance and will reposition the actuator to a new equilibrium position.

Link *S* can be positioned at any point along lever *P*, so the sort of action produced at several of these points can be studied. With link *S* attached to the left end of lever *P*, movement of the measuring means linkage produces a practically equal movement of the flapper, while movement of rod *B* has little or no effect on flapper position. It can be seen that the controller will be extremely sensitive to changes in the controlled variable under these conditions, because only 0.002-inch net flapper movement can cause a full range of output from the air relay. This effect produces an on-off controller action.

Now position the pivot point on lever *P* so that full movement of rod *B* causes 0.25 inch of flapper travel. Since rod *B* can move only a little more than 0.25 inch, the location of this pivot point will be very near the joint between lever *P* and rod *B*. In this position, link *S* is not greatly affected by movement of the measuring means linkage, although a slight movement of link *S* amounts to approximately 0.002 inch for a full range of travel by the measuring means linkage. For this position of link *S*, full range of the measuring means is required to produce full range of output from the air relay; so, for each position of the measuring means, there is a unique value of output pressure from the air relay. Thus, this arrangement has a 100% *throttling range*.

Somewhere along lever *P*, link *S* can be pivoted so that full 0.25-inch movement of the measuring means linkage moves the flapper 0.125 inch, and full movement of rod *B* moves the flapper 0.125 inch. For this position, movement of the measuring means linkage over only half of its total travel produces a full range of output from the air relay. Thus 50% throttling range is found in this system, since only half the range of the measuring means produces a full range of air relay output. For the other half range of the measuring means, the air relay output is either a maximum or a minimum.

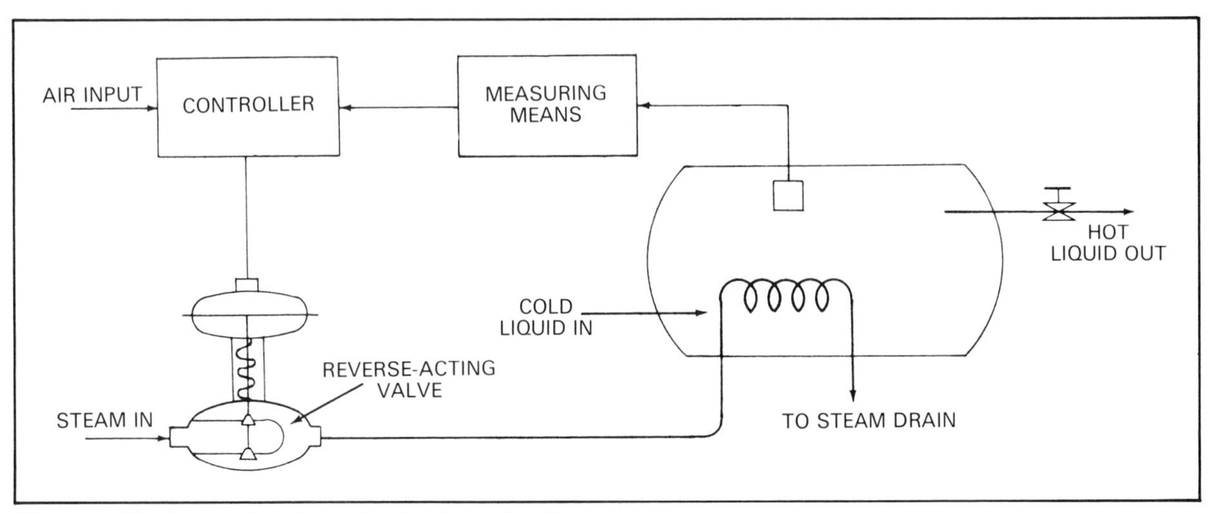

Figure 4.10. Proportional controller in a liquid-heating system

The Controller in Action

Suppose the controller is applied to a process in which a liquid is to be maintained at 150°C and, for normal conditions of load, the steam supply and energy requirements of the process are such that the steam control valve operates at 50% maximum flow. Suppose also that the measuring means indicates on a scale of 100°C to 200°C, and that this range of temperature will produce 0.25 inch of linear motion of the measuring means linkage. Link *S* will be set at full throttling range for the first part of the test (fig. 4.10).

With the set point of the controller calling for 150°C, the system is operating under normal load and is stable. What happens if the demand for hot liquid suddenly increases? This change upsets the energy balance, and the amount of steam normally flowing cannot maintain the temperature at the increased outflow of liquid. The temperature falls below 150°C, and this information goes through the measuring means and is transformed into flapper movement. The nozzle back-pressure will fall; the air relay output increases, forcing the reverse-acting steam valve to a wider opening and simultaneously actuating the controller bellows so that the flapper is repositioned. The system soon settles out at a new temperature somewhat below the desired value of 150°C and will remain stable at this new set of conditions as long as the load does not change.

The action of the controller really comes as no surprise, but it emphasizes a characteristic of all proportional controllers: in order to correct for a change in load on the process, a *proportional shift* in the value of the controlled variable must take place. If it is important to maintain the controlled variable at a set point, and it usually is an important consideration, development must continue. First, however, a shift in the position of link *S* will be investigated.

By moving link *S* to the 50% throttling range point on lever *P*, only half the full range of the measuring means will be required to throttle the steam valve from fully open to fully closed. With the same set of conditions, the controller is capable of maintaining the controlled variable within a band only half as broad as for 100% throttling range. Thus, the offset from the set point is only half as great, but there *is* an offset that should be eliminated. Further adjustment of link *S*, pushing it ever closer to its limit at the left side, does not seem to be the final answer.

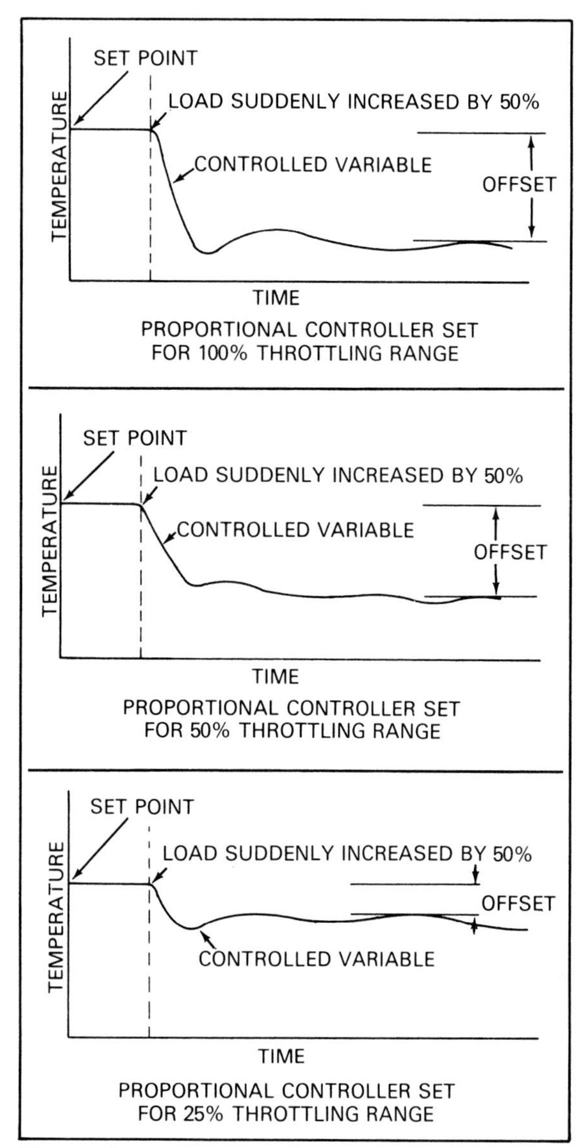

Figure 4.11. Performance of controller at different throttle range settings

When link *S* is at the 5% throttling range point, the controller begins to resemble an on-off device. See figure 4.11 for rough performance curves of the several settings of link *S*.

Mechanical Reset Adjustment

The controller needs an additional feature that will maintain the controlled variable at the set point regardless of wide load variations. A careful analysis of the possible changes that will correct the situation must be made. Obviously, in the hypothetical problem, the control valve is not passing enough steam to counteract the increased load. A means of increasing the air relay output pressure, and thus increase the valve opening, must be provided. Increased flapper-nozzle clearance will accomplish this. Also, if the spring-compressing screw is tightened so as to compress the spring, rod *B* and lever *P* in figure 4.9 can be forced back to the position they had at normal load conditions. As this change is made, the air relay output pressure immediately increases, the steam valve opens wider, and the temperature rises to the desired set point. Thus, an effective *manual reset* device has been discovered.

Automatic Reset Feature

The next problem is providing an automatic reset feature. What will be the effect of charging the cylindrical bellows-spring container with the output pressure of the air relay? The effect would be that of adding force in the direction of the spring's force. In fact, if equal pressure is applied to the bellows and the cylinder, the two resulting forces will be very nearly equal and opposite. The lack of equality is due to the slightly smaller area that the bellows pressure can act on to produce vertical motion. For purposes here this difference can be considered insignificant. With proper choice of the diameters of the inner and outer walls of the composite bellows, the inequality can easily be reduced to a factor of less than 1%.

Since air pressure within the cylinder can be used to oppose action of the bellows, the next step is to make the necessary alterations to admit the air relay output to the cylinder (fig. 4.12). Valves *X* and *Y* have been included to provide certain adjustments needed to make the new system acceptable.

Placing the new arrangement into the hypothetical control problem, with valves *X* and *Y* fully open, rod *B* remains virtually

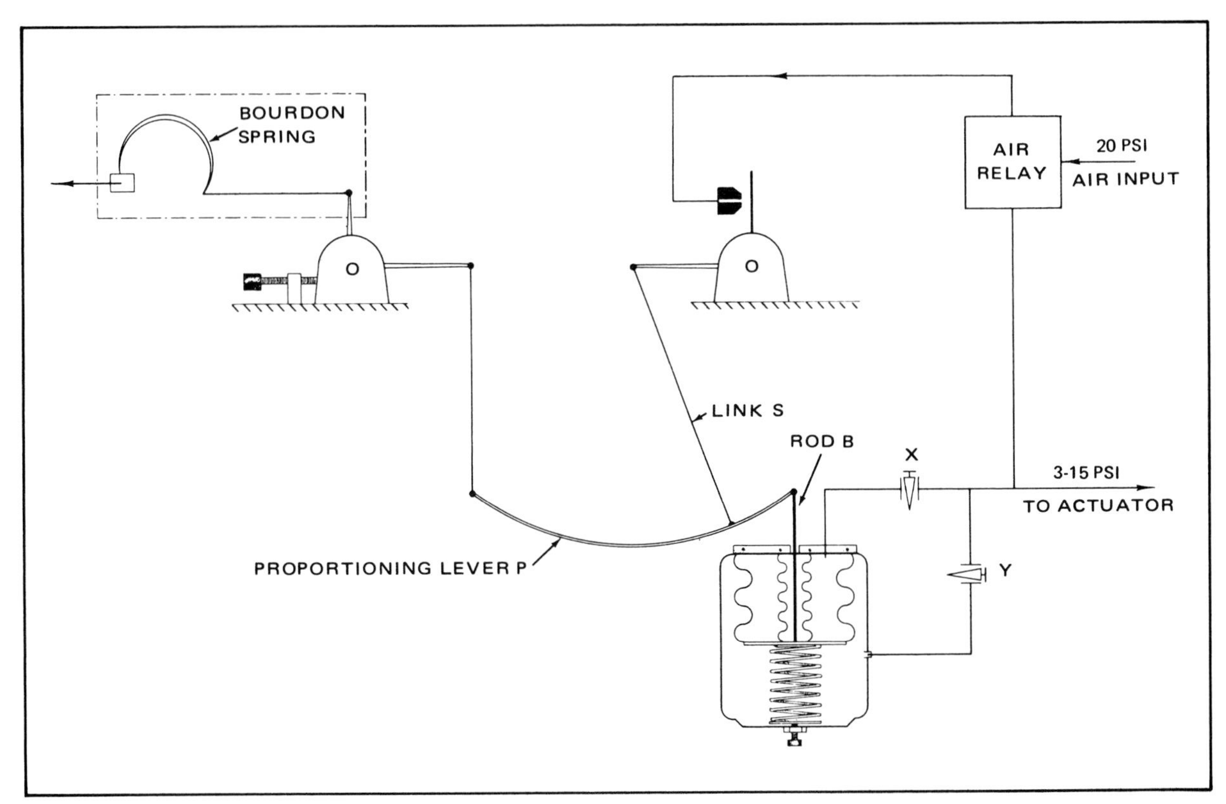

Figure 4.12. A proportional controller with automatic reset

tionary for all values of relay output pressure. Obviously, the simultaneous buildup (or decline) of equal pressure in bellows and cylinder results in a net force of zero. Operated in this manner, the controller will produce an on-off action; but, if valve *Y* is closed until the rate at which the cylinder becomes charged lags well behind the charging rate of the bellows, the effect will be heartening.

Having made a preliminary adjustment of valve *Y* in order to slow the charging rate of the cylinder, the control action on the liquid heating problem is tested. With the system running smoothly at normal load and temperature, suppose the load suddenly increases by 50%. The measuring means detects the temperature change and causes the flapper-nozzle clearance to broaden. The control valve opens wider, and the bellows of the controller expands rapidly, causing lever *P* to move in such a way as to narrow the nozzle-flapper clearance. The cylinder slowly charges with air and drives the bellows and rod *B* toward their midpoint of travel, causing the nozzle-flapper clearance to broaden. The reduced nozzle back-pressure ultimately results in wider opening of the steam valve, and this increased energy flow raises the temperature of the liquid. The system eventually settles and meets the new load requirements with liquid temperature maintained at the set point. The action caused by the slow charging of the cylinder tends to increase flapper-nozzle clearance and thus to keep the steam valve sufficiently open. At the same time, the heating effect of the steam is reflected in increased liquid temperature and in the efforts of the measuring means to reduce the flapper-nozzle clearance. The system can only settle at the set point. We now have a *proportional plus reset* controller.

Rate-of-Response Adjustment

Many processes are of such nature that, when the controlled variable deviates beyond certain narrow limits from the set point, their products are unacceptable and therefore rejected. Such processes must be provided with controllers that are capable of the most rapid corrective action so that the level of rejection is held to a minimum. The new controller, as judged by the performance curves of figure 4.13, does not appear to be able to restore equilibrium as quickly as desired.

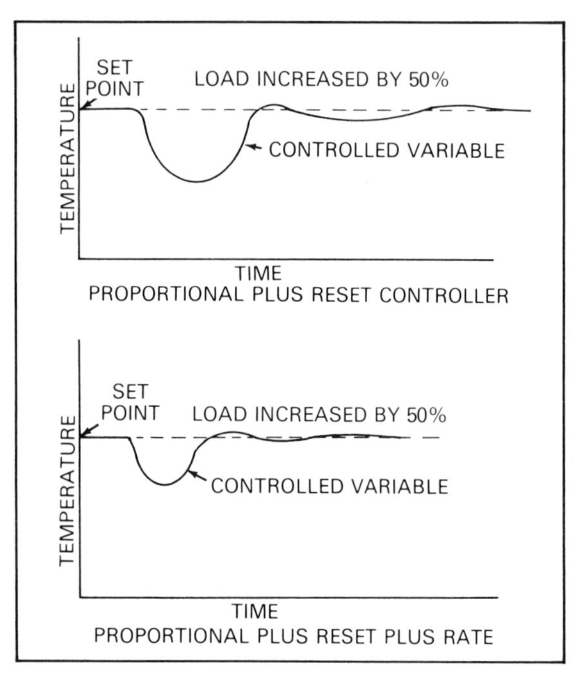

Figure 4.13. Performance curves for multi-mode controller action

Valve X can be used to control the rate at which the bellows receives its charge of air and the rate at which the charge can be exhausted. The opening of valve X is reduced until the rate of charge is only slightly greater than that of the cylinder. This new condition of charging rate is then applied to the hypothetical problem.

Again, with the process running smoothly at normal load, the demand for hot liquid is changed to require a 50% increase in load. The air-relay output increases rapidly, opening the steam valve wide. During the previous test, with valve X wide open, the bellows quickly expanded and narrowed the flapper-nozzle clearance, thus throttling the steam valve to a smaller opening. Now, however, the bellows is retarded in its action, so the steam valve stays at the wide position for a much longer time. Eventually, of course, the bellows begins to expand and slowly throttles the steam flow to a lower value. In the meantime, the added energy, made possible by retarding the action of the bellows, has done much toward restoring the controlled variable to the set point. Note the faster response by comparing the curves in figure 4.13. The device has gained further distinction and is now a *proportional plus reset plus rate controller.*

Summary of Controller Action

The experimental work on the pneumatic controller has produced most of the desirable features hoped for. Development has progressed from the initial sluggish, on-off device to a controller capable of four modes of operation—on-off (two-position), proportional band, proportional plus reset, and proportional plus reset plus rate. There are adjustments for set point, proportional band (from 0 to slightly over 100%), reset rate, and rate of response. The tests conducted to determine the adequacy of the controller centered around a single controlled variable temperature. It seems clear, however, that other variables—rate of flow, pressure, and liquid level, for example—could be controlled with equal facility, since it would be a simple matter to obtain the 0.125-inch linear travel necessary to fully actuate the controller.

The various modes of control that are possible with the controller, particularly when a combination is used to obtain maximum ability of the unit, are of such fascinating and important nature that it is worthwhile to discuss the roles of each mode in the hypothetical control problem. For instance, a straightforward proportional controller is not always satisfactory, since it cannot correct properly for sustained deviations of the controlled variable. On the other hand, the proportional characteristic is probably a desirable feature of the controller. Control effects derived from combining proportional, reset, and rate actions represent a remarkable intermingling of functions that produce desirable features.

In the hypothetical heating problem, each of the controller actions has its effect on the control valve.

Proportional action. The proportional action positions the valve in proportion to change in the controlled variable. If a 100% proportional band is used (that is, one in which full movement of the measuring means linkage produces exactly full movement of the valve plug), the motion of the valve plug is *directly* proportional to the deviation of the controlled variable. In equation form, change in controlled variable equals change in valve plug position. If a 50% proportional band is used, then there is no longer a *direct* proportion between deviation of the controlled variable and the motion of the valve plug, because now *half* the range of measuring means linkage can produce *full* operation of the valve. The ratio is 2 to 1 and can be stated in equation form as *change in controlled variable equals half the change in valve-plug position*. This process can be carried further, but the point is that the relation between controlled variable and valve-stem motion will always be a linear, or straight-line, function. The change in process input energy caused by proportional action is sustained only for such time as the controlled variable is displaced from the set point; thus, the energy change due to proportional action is only a temporary change when this action is followed by reset action.

Reset action. The reset action positions the valve at a rate proportional to the change in controlled variable. The input energy change caused by this action, unlike that in proportional action, will be a permanent one. Note that once the controlled variable is restored to the set point by reset action, the left end of the proportioning lever P is again in its normal position, that is, the position it had before the load change occurred (fig. 4.12). This action has eliminated the initial input energy change caused by proportional action.

Rate of change. The positioning of the valve by rate response is an action that is proportional to the rate of change in the controlled variable. When the load suddenly increases, for example, the controlled variable begins a rapid departure from the set point, its rate of change being a maximum at this time because as yet no correction measures are active in the process. The rapid rate of change of the controlled variable will cause the valve to open wide, because needle valve X will not permit a rapid charging of the controller bellows that would tend to narrow the nozzle-flapper clearance. Eventually, as the bellows of the controller begins to expand, the rate action will narrow nozzle-flapper clearance. Once the controlled variable has reached its maximum deviation, there is a moment at which its rate of change is zero. At this same moment, the valve-stem displacement caused by the rate response should also be zero, since it is proportional to the rate of change of the controlled variable. After its pause, the controlled variable begins changing again, seeking the set point. In this case, the rate of change is reversed; as a result, the component of valve-stem displacement due to rate response actually becomes less than that existing at the set point.

A set of curves (fig. 4.14) should help in visualizing the action of each mode of the controller following a quick change in load. Although proportional and rate features are active in resisting change and contribute to the system's stability, neither is responsible for any final correction; reset action has provided the change in energy input to compensate for load change. The shaded areas of figure 4.14 represent the input energy change caused by each mode of the controller.

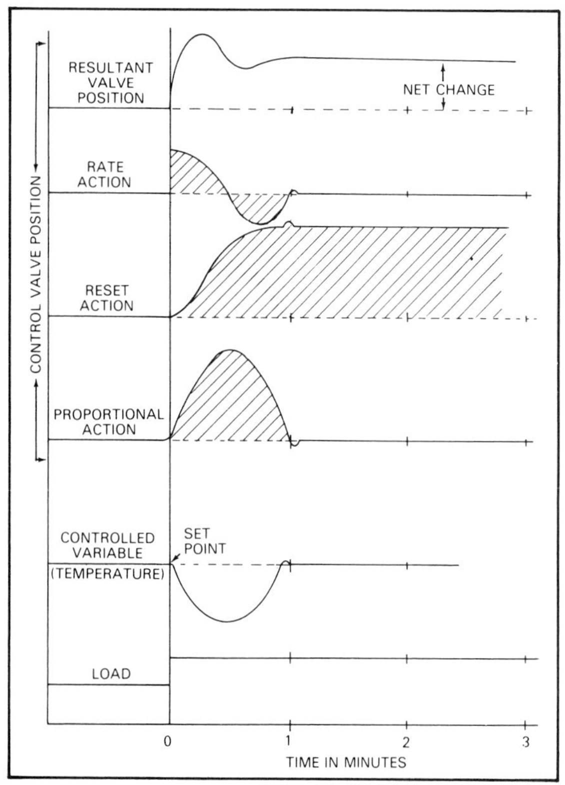

Figure 4.14. Response curves following quick change in load

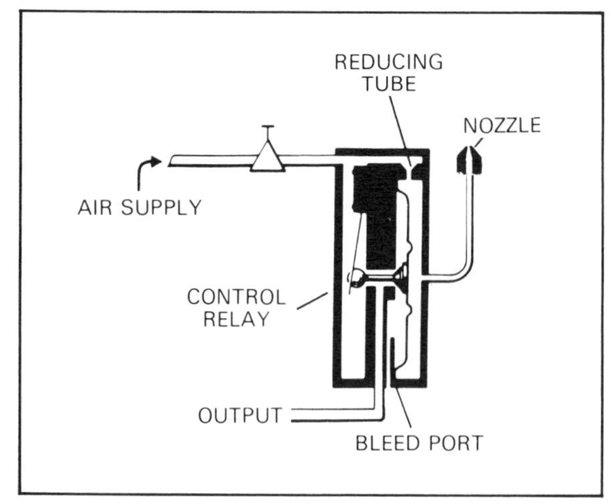

Figure 4.15. A continuous-bleed air relay (*Courtesy Foxboro*)

Commercial Pneumatic Controllers

Before considering additional components of pneumatic control, it will be helpful to study the design and other characteristics of pneumatic controllers that are available for commercial control applications. Examples will be limited to units possessing proportional plus reset plus rate capability.

Air Relays

Air relays almost invariably form an integral part of controllers, and they are generally one of two types: *continuous-bleed* and *nonbleed.* A continuous-bleed relay is one in which there is a constant loss of air from the relay unless the nozzle back-pressure is at either a maximum or a minimum. A nonbleed relay allows a loss of air only when it is necessary to bleed off the pressure applied to the valve actuator.

Continuous-bleed air relay. A continuous-bleed relay (fig. 4.15) is used extensively in instruments manufactured by Foxboro. Note that a similarity with the unit developed in this chapter exists, but the Foxboro relay is direct-acting—that is, output pressure is increased in proportion to nozzle back-pressure—and the one developed is reverse-acting.

Since the operation of this relay is similar to the reverse-acting relay discussed in the earlier part of the chapter, no additional effort will be made to explain its function; however, some of the physical characteristics of this commercial unit are interesting. The total movement of the valve stem is only 0.012 inch, and this motion can be caused by a change of only 0.75 psi pressure on the diaphragm. While operating within this small span of pressure, the output of the relay can be varied from 3 to 15 psi. The small pressure differential (0.75 psi) caused by nozzle back-pressure implies an incredibly small flapper movement; in fact, it amounts to only 0.001 inch.

Nonbleed air relay. In a nonbleed air relay (fig. 4.16), nozzle back-pressure is applied to the exterior surface of the large bellows. This pressure tends to compress the large bellows

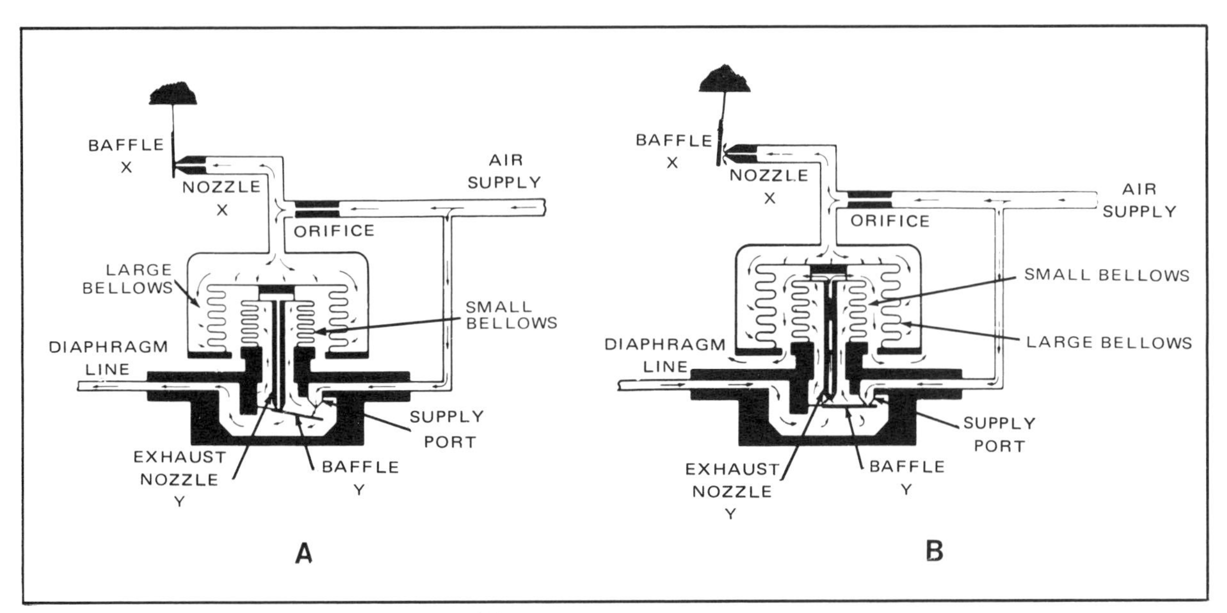

Figure 4.16. A nonbleed air relay *(Courtesy Honeywell)*

and the small interior bellows. An important relation exists between the sizes of the two bellows; it is such that approximately 5 psi pressure is needed within the small bellows to equal the force caused by only 1 psi pressure bearing on the exterior of the large bellows.

When there is minimum nozzle back-pressure, the 20 psi regulated air supply is closed off from the diaphragm line and principal components of the relay by the spring-loaded baffle *Y*. Once the nozzle back-pressure builds up, the supply port is opened, allowing air to pass simultaneously to the diaphragm line and the interior of the small bellows. Since a 5 to 1 ratio of pressure is needed to equalize forces of the two bellows, it can be seen that, once the pressure to the diaphragm line and small bellows becomes about five times that of the nozzle back-pressure, net force of the two bellows will be zero. The two-bellows assembly will return to normal, or neutral, position, and baffle *Y* will seal off the supply port. If there are no leaks, the system will remain in this charged condition until there is another change in nozzle back-pressure.

Assume that a lessening of the nozzle back-pressure takes place. The pressure caused by air trapped in the small bellows and valve-actuator diaphragm case will now be somewhat greater than five times that of the nozzle back-pressure. The small bellows will expand in an effort to reestablish an equilibrium of forces (fig. 4.16*B*). This action pulls exhaust nozzle *Y* away from baffle *Y*, allowing the trapped air to exhaust to the atmosphere until the pressure of the air trapped in the smaller bellows becomes just five times the nozzle back-pressure. At this point, baffle *Y* closes the two ports, and the system is in a stable, charged state.

Nonbleed relays waste little air and are reliable units. They are usually larger than bleed units of similar rating because of the need for large bellows. Large bellows are needed to avoid a serious dead spot that tends to occur at the equilibrium point. When the nozzle back-pressure has increased just enough to close off the exhaust port, bleed-off of the actuator will cease. If the nozzle back-pressure increases a slight additional amount, the dead spot arises because some finite force is required to overcome the effects of the spring action holding the flapper against the supply port. The increase in pressure required to overcome this force reflects a change in the controlled variable, but corrective action by the control valve will be delayed until this inherent force is overcome. Obviously, the use of large bellows will reduce the size of the dead spot, since the required force can be achieved with less pressure in a large bellows.

Air-O-Line Controller

Air-O-Line is a trade name applied to a series of controllers manufactured by the Process Control Division, Honeywell, Incorporated. The Air-O-Line controller (fig. 4.17) is basically capable of proportional plus reset action. It has two similar compound bellows assemblies acting in opposition. The inner space of each assembly is filled with a liquid, and a passage containing a needle valve is connected between the two assemblies. A rigid rod connects the free ends of the opposing small bellows, and a pin located on this rod is capable of imparting motion to the flapper.

To begin a description of operation, assume the system is in a stable condition when a change in controlled variable causes the flapper-nozzle clearance to increase. The loss of nozzle back-pressure causes the air relay to vent the valve actuator and the cavity surrounding the throttling bellows. Up to this time, the coil spring bearing against the large bellows on the right has been compressed by relay output pressure exerted on the exterior of the throttling bellows. But, with the decline of this pressure, the spring quickly drives the free ends of all bellows to the left. Very little liquid flows through the needle valve during this rapid motion. Motion will stop when the net force of the three springs is balanced by the effective force exerted on the throttling bellows. This action has tended to narrow nozzle-flapper clearance.

In the process of driving the free ends of the bellows to the left, the interior spring of the left bellows has been stretched, while that of the right bellows has been compressed, thus making the liquid pressure greater in the right bellows. Liquid will flow through the needle valve to the low-pressure side until equilibrium is established. As this occurs, the small bellows will tend to return to its initial position, thus actuating the flapper and causing a change in the air-relay output pressure to offset the load change.

The throttling range of the Air-O-Line controller must be augmented by a rate-action unit if rate response is required. Such a unit fits in the line between the air-relay output and the throttling bellows. The functioning of this unit is essentially the same as that accomplished by the needle valve used in the basic design, but it is refined to permit about 10 percent of the change in relay output pressure to be applied to the feedback bellows without any delay.

Foxboro Model 40 Controller

The Foxboro Model 40 controller is one of a family of pneumatic controllers available as single-mode, on-off units having proportional plus reset plus rate action (fig. 4.18). An enlarged view of the flapper-nozzle proportioning-lever complex appears in figure 4.19.

The keyhole-shaped dashed line represents a rotatable gear. The nozzle is mounted on a hexagonal hub of the gear, but the pigtail air connection is meant to convey flexibility and does not exist as the actual method of feeding air to the nozzle. The *striker arm and flapper form a single rigid unit* that pivots about a shaft located on the extremity of the gear. The gear can be rotated through 180 degrees, and this rotation is a means of adjusting the proportional band. It must be understood that rotation of the gear moves the hub, nozzle, and striker arm pivot.

A

Figure 4.17. An Air-O-Line pneumatic controller. *A*, controller in its normal housing with front panel removed; *B*, simplified schematic diagram of controller. *(Courtesy Honeywell)*

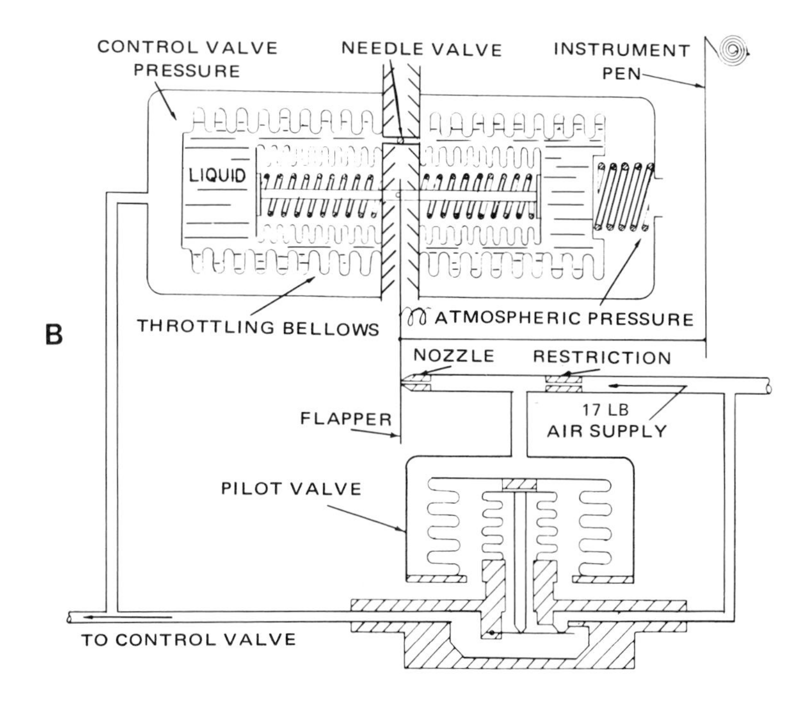

A

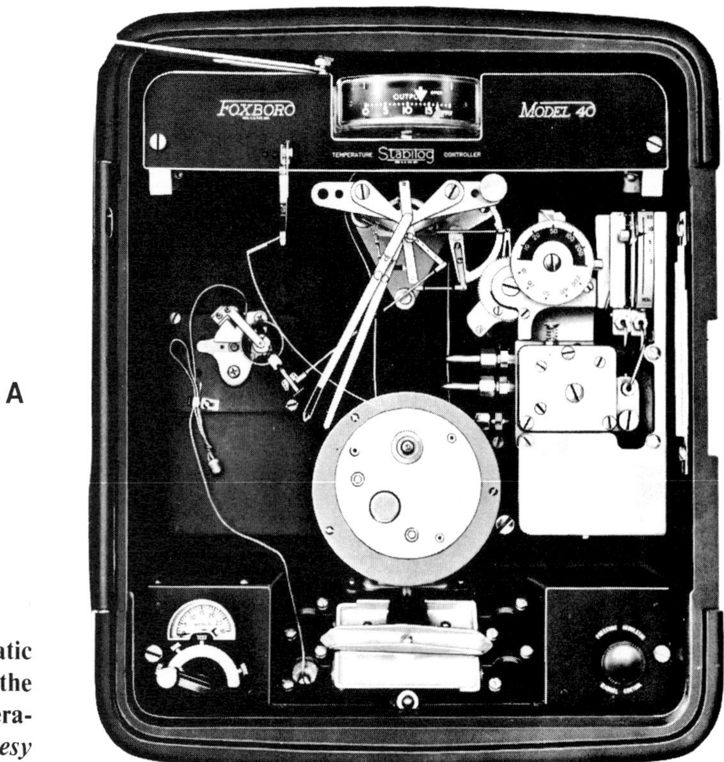

Figure 4.18. Foxboro Model 40 pneumatic controller. *A*, controller in its case with the cover and front panel removed; *B*, operational diagram of the controller. *(Courtesy Foxboro)*

B

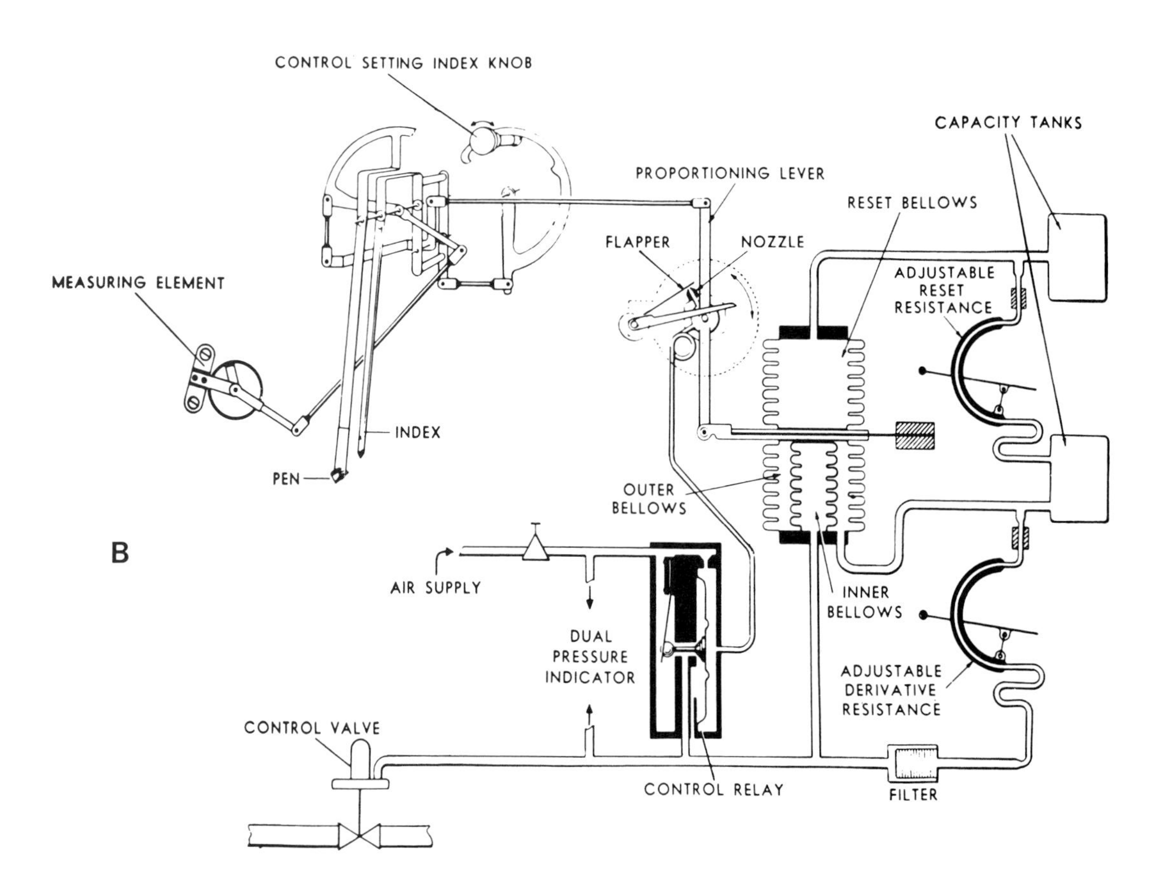

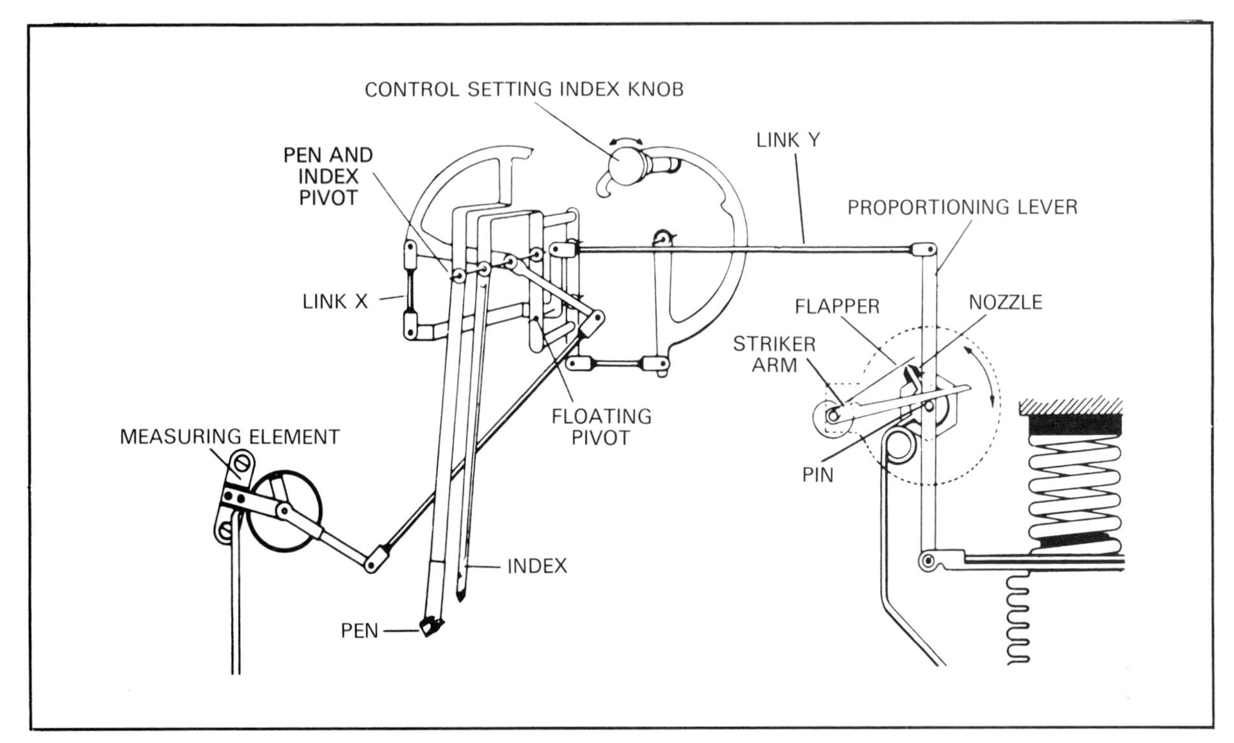

Figure 4.19. Flapper-nozzle assembly for Model 40 controller (*Courtesy Foxboro*)

The proportioning lever is pivoted at top and bottom only, and a pin is located at midpoint between the pivots. A spiral spring causes the striker arm to ride snugly against this pin, so it is clear that action of the proportioning lever can impart motion to the striker arm and flapper, thus changing flapper-nozzle clearance. However, if the striker arm lies in a horizontal position (that is, perpendicular to the proportioning lever), horizontal movement of the measuring means linkage (link *Y*) will have little or no effect on flapper-nozzle clearance, while movement of the bellows assembly will have maximum effect. This adjustment represents the position for maximum throttling range.

The nozzle and striker arm can be relocated to extreme positions by rotating the gear 90 degrees in either direction from the position of maximum band width. At these extremes, on-off action by the controller is obtained, because the striker arm will now be parallel to the proportioning lever, and only slight motion of link *Y* will have considerable effect on flapper-nozzle clearance. On the other hand, the vertical motion of the lever caused by bellows action will not be effective in bringing about change in flapper-nozzle clearance. Direct and reverse action of the controller is determined by the quadrants chosen for the proportional band. For example, with the striker arm in the horizontal position where maximum band width is obtained, rotating the gear in the clockwise direction moves link *Y* to the right, which causes the flapper-nozzle clearance to decrease. Had the gear been rotated counterclockwise, action of the measuring means would have caused the clearance to increase.

Features and functioning of additional parts of the controller are noteworthy. Referring to figure 4.18*B*, find the proportioning bellows, which is the compound assembly of two concentric bellows, and the associated adjustable derivative resistance. Note that the small bellows is connected to the relay output without intervening restriction, while the large outer bellows is retarded in its action by an adjustable restriction. This derivative resistance serves the same purpose as the needle valve in the basic controller developed, but it is a form of Bourdon tube whose flexure can be varied by the rate (or derivative) adjustment to produce a suitable range of flow rates. The capacity tanks add volume to the bellows section to which they are associated and assist in providing a more reliable and accurate response by the bellows.

The action of the controller is similar to that described for the basic controller developed earlier in the chapter, except that the Model 40 has the additional refinement embodied in the unimpeded air flow to and from the inner bellows. This bellows is of such size that it exerts a force only one-fifth that of the exterior unit for a given pressure application, but its presence adds stability to the system and keeps the valve actuator from responding violently when the controller is subjected to vibration or any other mechanical disturbance that might upset the nozzle-flapper relation momentarily (fig. 4.19).

Volume Booster Relays

Diaphragm motor valves are usually operated directly by the pneumatic output from controllers. An air relay is desirable for the purpose of speeding the response action of a valve. Occasional situations require that valve-actuator response be more rapid than is possible with the ordinary controller–air relay

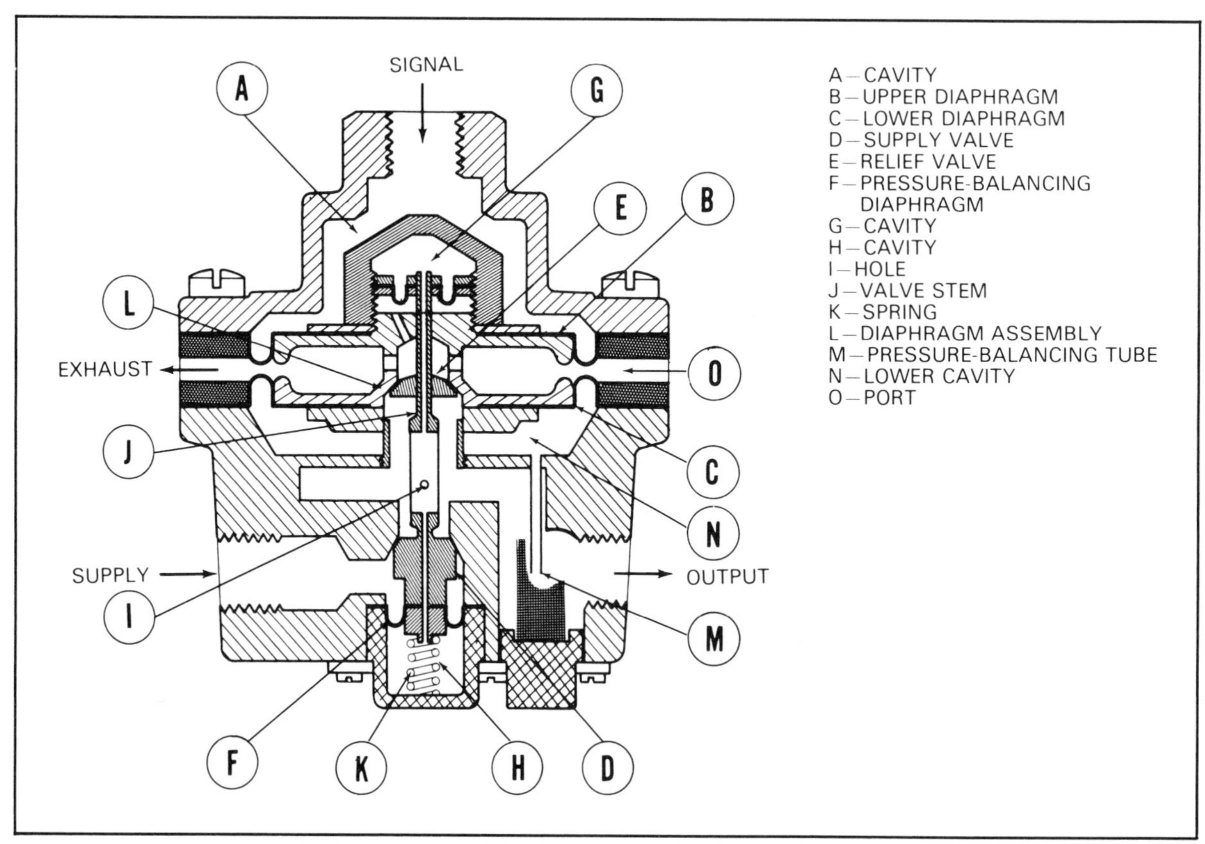

Figure 4.20. A booster relay, used to overcome response lag due to long connecting lines between controller and valve

combination. An example would be an installation having controller and valve separated by a considerable distance so that the combination of resistance and capacity of the connecting lines would slow valve action to an unacceptable degree. A *booster relay* installed at the valve location will overcome most of the response lag due to long connecting lines and insufficient output volume.

In the booster relay shown in figure 4.20, signal pressure is applied at the top of the relay and fills cavity *A* above the upper diaphragm *B*. The signal pressure is opposed by an equal pressure applied beneath the lower diaphragm *C*. This is the output pressure to the valve actuator. With equal pressures applied to the diaphragms, the supply valve *D* and relief valve *E* are seated as shown in the figure. Supply pressure tends to seat supply valve *D*, and this tendency is balanced by the same pressure applied to the top of the pressure-balancing diaphragm *F*. Output pressure is admitted to cavities *G* and *H* through a hole *I*, midway on the hollow valve stem *J*. Output pressure on top of supply valve *D* is balanced by output pressure beneath pressure-balancing diaphragm *F*. Forces are similarly balanced on the relief valve *E*. Thus, the valves are seated by the force of spring *K*.

An increase in signal pressure causes diaphragm assembly *L* to unseat supply valve *D* and increase the pressure in the output line. The air bleeds through the pressure-balancing tube *M* in the output line into the lower cavity *N*. The resulting pressure in the lower cavity will bring the diaphragm assembly *L* to its balanced position and seat the supply valve *D*.

A decrease in signal pressure will cause the pressure in the lower cavity *N* to raise the diaphragm assembly *L* and unseat the relief valve *E*. Excess pressure is vented through the relief valve *E* and port *O* until pressure balance is regained. The diaphragm assembly *L* then resumes its balanced position, seating relief valve *E*.

Valve Positioners

Volume booster relays are useful for improving the speed of response of valve actuators but are not apt to improve the accuracy of the valve positioning. Performing this task are valve positioners—devices that are capable of forcing valves with sticky stems and other adverse conditions to assume a precise position.

Controllers have been developed that are capable of responding to the most minute variations in the controlled variable. The immediate manifestation of this response is a change in the air relay output pressure, and this pressure change is to reposition the control valve plug. Should the controller be arbitrarily applied to just any control problem, its apparent inability to stabilize the process within the narrow band of values needed for satisfactory performance might be disappointing.

Close examination might reveal proper controller action associated with a stubborn control valve that refuses to respond to pneumatic nudges that are below a certain level of force. Several valve characteristics can upset the control pattern—the effects of flow through the valve, for example, and the friction between valve stem and packing material. Two forms of friction are involved here—*static* and *sliding*. The coefficient of static friction is usually significantly greater than that for sliding friction. In other words, a greater force is required to unstick the valve stem, but once it begins moving, this same force is capable of accelerating it. Such action can cause "overshooting" and instability, but the really serious problem is the dead spot caused by static friction. This might represent 5% or more of the total valve stem travel.

Returning to the hypothetical control problem, assume that a control valve having total stem travel of 1 inch is being used, and the objective is to maintain the controlled variable within 1% of the set point. This clearly requires positioning the valve plug within 0.01 inch of the optimum position. Now, since changing the diaphragm pressure from 3 to 15 psi (a net change of 12 psi) moves the valve stem 1 inch, a 1% change in pressure represents 0.12 psi. In order to meet the control requirements, the valve must respond to this rather small change in diaphragm pressure. If inherent forces in the valve are greater than the force tending to properly position the valve, the fineness of control needed obviously cannot be achieved. A ready solution to the problem would be the use of a larger diaphragm, since the force exerted by a diaphragm is proportional to the square of its diameter. However, the increased size of the unit and the greater volume of air needed for the operation lead to other methods.

Elementary Positioner

An analysis of the problem indicates a control valve that will not respond to diaphragm pressure changes of 0.12 psi. It seems logical to consider some means that will amplify these small pressure changes into a value that will exceed the force needed to overcome the static friction of the valve. A basic system capable of doing the job uses the original fixed orifice and nozzle, and diaphragm pressure is a simple function of nozzle backpressure. This is similar to the on-off controller discussed earlier and will suffer the same slow action because of the considerable time required for buildup and bleed-off of diaphragm pressure. Nevertheless, it will position the valve accurately.

The baffle plate (a form of flapper) in figure 4.21 is pivoted for rotary motion at the

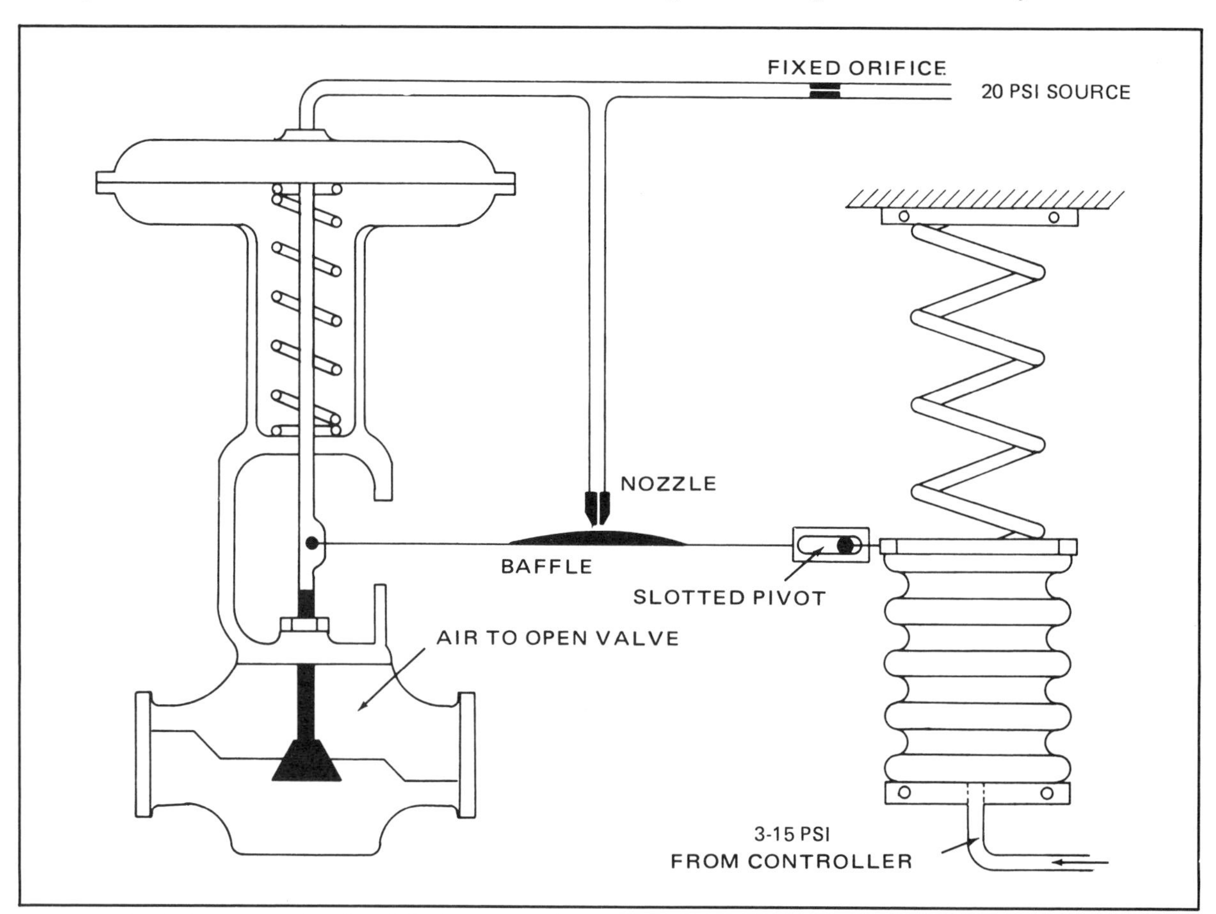

Figure 4.21. Schematic of a simplified valve positioner

actuator stem and at the bellows-spring assemby, but the pivot pin of the latter point is allowed to slide in an elongated slot. The purpose of the elongated slot is to compensate for the changing distance between pivot pins of the actuator stem and bellows-spring assembly as these move up and down. The curved portion of the baffle helps maintain a proper tangential relation with the nozzle. The baffle is resilient enough to withstand flexing action when it occasionally is driven tightly against the nozzle.

Assume that full travel of the stem causes nozzle-baffle clearance to change 0.5 inch and that a similar change is caused by bellows-spring action in response to a full range (3 to 15 psi) of pressure values from the controller. Now, if a change of 12 psi in controller output causes a 0.5-inch change in nozzle-baffle clearance, 0.12 psi will cause a change of 0.005 inch (0.5×0.01). Considering the sensitivity exhibited by a nozzle-flapper arrangement, this change in nozzle-baffle clearance will bring about a change in nozzle back-pressure so drastic that the diaphragm pressure will begin to either bleed off toward zero pressure or build up toward maximum, depending on direction of the clearance change.

Keeping in mind that an air-to-open, or reverse-acting, valve is used and that the controller's input *increases* for a *decrease* in controlled variable, look at the positioner's reaction to a fall of slightly less than 1% in the value of the controlled variable. Obviously, the response of the controller to this change will be an increase in output pressure of something less than 0.12 psi—say, 0.10 psi. The valve and actuator combination will not respond to such a change applied directly to the diaphragm, but this change will probably be enough to close off the nozzle tightly.

The pressure on the diaphragm will build up quickly as the fixed orifice will permit and if necessary will ultimately attain the full 20 psi to release the valve stem. Of course, in any properly operating valve and actuator, the response will be rather quick, and the correct relation between nozzle and baffle will be reestablished rapidly. The important point here is that a minute deviation of the controlled variable can be relayed and amplified into a very powerful force for positioning the valve.

Commercial Valve Positioner

The elementary positioner just described will work, but the addition of an air relay and refinements make it adaptable to a wide variety of valves and actuators—for example, a positioner that will adjust to the differences in valve stem stroke that might be encountered.

In the positioner shown in figure 4.22, a system of linkages and pivots allows adjustment for valve stem strokes ranging from 0.2 to 30 inches. The baffle is actuated by motion of a bar whose left end is attached to the free end of the bellows X and whose right end is moved by action of the valve stem (fig. 4.22*B*). Downward movement of the valve stem will pull the baffle away from the nozzle, while an increase in controller output will cause the bellows to expand and counteract this action.

The air relay used in this postioner is of unusual design. The 20 psi air supply is applied to the relay and enters the small cavity just above the cone-shaped, spring-loaded plug. The air supply also passes through the fixed orifice, which is a part of the relay, and then takes two paths—one to the nozzle and the other to an airtight chamber fitted over the upper bellows of the relay. A free passage for air exists between the interior portions of the upper and lower bellows, because considerable air space around each of the rods connects the upper and lower free ends of the bellows. Also, the air line leading to the valve actuator comes directly from the air space formed by the interior of these bellows.

Getting back to the hypothetical problem, assume that a *decrease* in controlled variable has just caused the controller to produce an *increased* output pressure. This increase, when applied to the main bellows of the positioner, narrows the nozzle-baffle clearance. The back-pressure in the relay rises, causing the upper bellows to compress, and this motion is transmitted through the two rods to the free end of the lower bellows. This action pulls the cone-shaped plug away from the seat, allowing a rapid inflow of air to the interior of the bellows and the actuator. At the same time, the baffle, which has one end attached to the free end of the lower bellows, is moved away from the nozzle, thus counteracting some of the effect produced by the increased back-pressure.

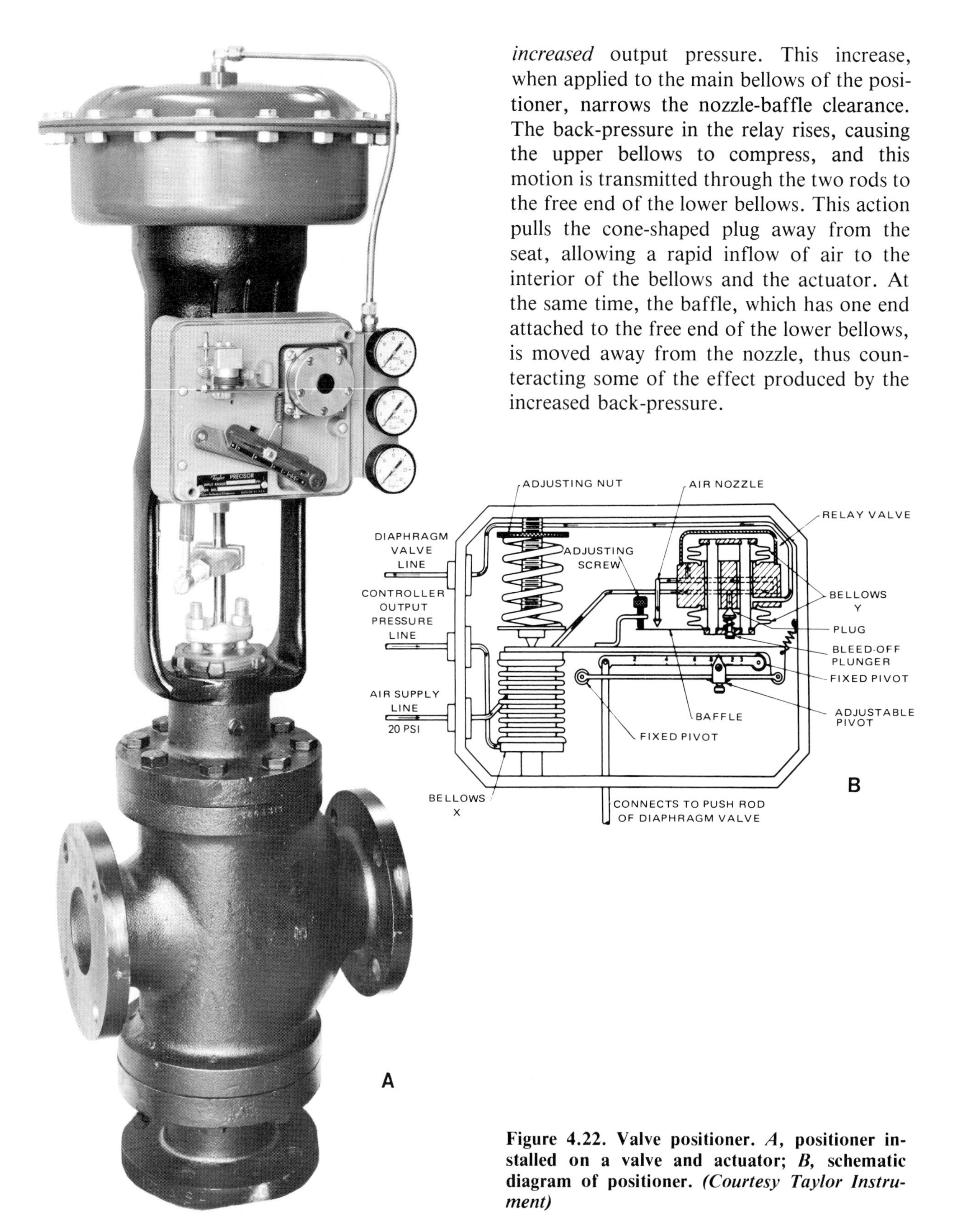

Figure 4.22. Valve positioner. *A*, positioner installed on a valve and actuator; *B*, schematic diagram of positioner. *(Courtesy Taylor Instrument)*

Should there now be a significant rise in the value of the controlled variable, the nozzle-baffle clearance in the positioner will increase, causing the nozzle back-pressure to fall. The upper bellows will expand, pulling the free end of the lower bellows upward. This action will unseat the bleed-off plunger and allow air from the actuator and the interior space of the bellows to bleed off until action of the valve stem and the system of linkages restores equilibrium.

In addition to the adjustable pivot used for adapting the positioner to various stroke requirements, an adjusting nut is used to vary the tension of the spring that opposes the main bellows. A light tension makes the positioner provide more action for small pressure variations. The adjusting screw that depresses the baffle is useful for making initial settings to adapt the positioner to a new process and valve combination; however, once this setting is made, further adjustment should not be necessary.

Positioner for Piston Pneumatic Actuator

A piston pneumatic actuator uses another type of positioner (fig. 4.23). In this positioner the air signal from a pneumatic controller is applied to a reversible bellows, which is attached to a beam that pivots about a fixed point. Special air relays are located on opposite sides of the pivot point, and the beam acts as a flapper in relation to nozzles that form part of the relays.

An increase in controller output pressure causes the bellows to expand, and this action tends to close off the nozzle of relay *X* and open wide the nozzle of relay *Y*. An increase in nozzle back-pressure in these relays causes them to pass more air into the cylinder, while reduced back-pressure will cause the relay to vent the cylinder until a correct proportion between nozzle back-pressure and cylinder pressure is achieved.

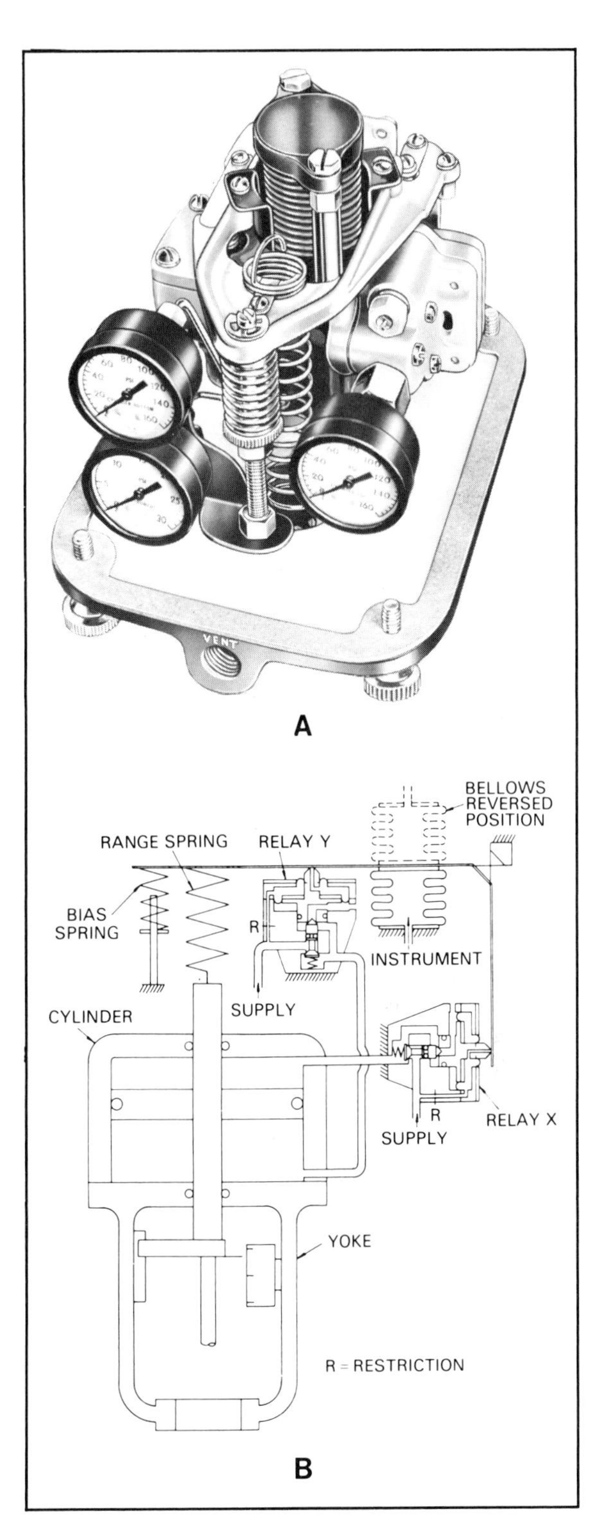

Figure 4.23. A valve positioner for a pneumatic piston actuator. *A*, positioner showing layout of components; *B*, schematic diagram of positioner and its connection to the actuator. (*Courtesy Fisher Controls*)

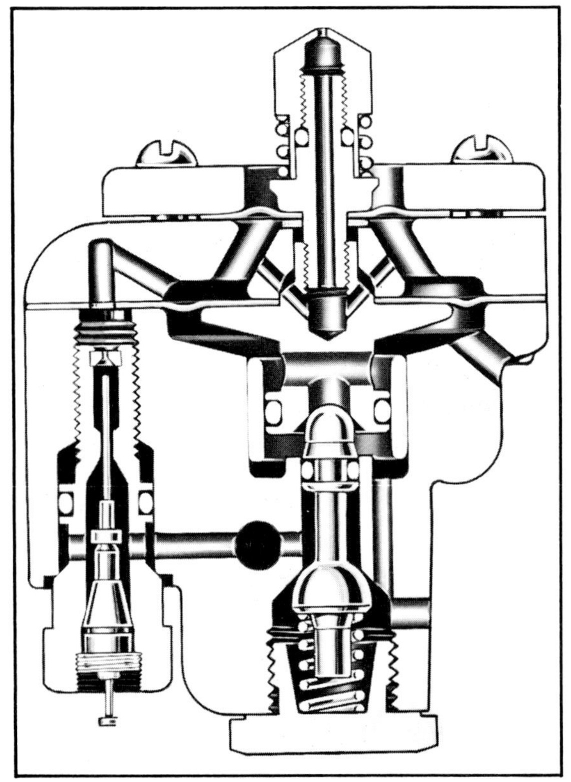

Figure 4.24. Relay used in the positioner shown in figure 4.23 ***(Courtesy Fisher Controls)***

Figure 4.24 is a sectional view of the type of air relay used in this positioner. Nozzle back-pressure is applied to the top of the diaphragm, while output pressure is applied to the small piston that acts as a seat for the exhaust valve. A balance of forces tends to develop between nozzle back-pressure on the large diaphragm and the output, or cylinder, pressure on the smaller area of the piston. This balance of forces represents a position of equilibrium.

Should the nozzle back-pressure decrease, the force of the output pressure acting on the pistonlike seat of the exhaust valve will drive the exhaust valve open, allowing air to escape out the exhaust vent. When the nozzle back-pressure increases, the force on the large diaphragm forces the pistonlike seat hard against the exhaust valve plug and drives the supply valve open. Air flows to the actuator cylinder and into the chamber containing the pistonlike exhaust valve seat. Soon, equilibrium between forces due to output pressure on the exhaust valve seat and nozzle back-pressure on the main diaphragm is established.

The study of additional commercial positioner units is not necessary since a positioner is a straightforward, one-purpose device whose principles of operation are far less involved than those of a controller. An understanding of the positioners described here will provide a suitable basis for further study.

Electronic Controls

It has been practical for some time to put together an assembly of electronic components that will duplicate any of the control effects that can be produced with pneumatics. The advent of solid state electronics gave electronic control systems a definite edge over pneumatics. The continuing development that has resulted in remarkable integrated circuits at low prices and high reliability has made electronic control an attractive way to go in instrumentation. In addition to high reliability and low cost, electronics has the advantage of almost instant transmission of signals from point to point in any system, however large, and easy adaptability to control systems in the form of microprocessors or computers.

The subject of electronic control can have vast dimensions. Even the largest texts and handbooks on the subject do not cover it in its entirety. Therefore, only some of the special circuits and components used in electronic control and various modes of control using electronics will be discussed here.

The student needs to have a good understanding of electricity and electronics in order to follow a discussion of electronic control systems. The discussion will not become deeply involved in electrical theory or mathematics, but an understanding of how resistance, capacity, and inductance affect electric circuits will be necessary for satisfactory comprehension of this section.

Analog Values in Current and Voltage

In the study of pneumatic control, an air signal of 3 to 15 psi was used as an analog representation of a controlled variable. The air signal was also used to operate valve actuators and other components. This one-to-five ratio (3 to 15) is carried over into electronic control, and the prevalent signal is a range of current values of 4 to 20 milliamperes. Another current range of 10 to 50 milliamps is less popular. In addition to current, ranges of voltage are used, the most popular being 1 to 5 volts, although ranges of 0 to 5 volts and 0 to 10 volts are also used. Typically, voltage ranges are used as input to recorders and indicators at a central location, while current ranges are found in control loops between central points and primary elements or transducers in the field or other more or less remote location.

The use of a current rather than a voltage signal to transmit from point to point is preferable because (1) the current signal is not affected by moderate but unintentional variations in circuit resistance and (2) signals are less subject to interference from power circuits, other communications signals, or other spurious electrical interference.

The Control Loop

The use of current signals to convey control information over a distance is probably preferable to using voltage signals. A control loop using signals of 4 to 20 milliamps is apt to work well for distances of 2 to 3 miles. Once the current signals reach a central control point, they are converted to a range of voltage signals, probably 1 to 5 volts. A 250-ohm resistor in the control loop provides an easy source for this range (fig. 4.25).

In its simplest form, a 4- to 20-milliamp control loop (fig. 4.25) consists of a regulated 24-volt DC power supply; a transducer or transmitter that accepts values of temperature, pressure, or other variable and converts these values to a range of 4 to 20 milliamps; a device for converting the 4- to 20-milliamp signal to 1 to 5 volts; and appropriate instruments that will respond to the 1- to 5-volt signal. In the loop shown, the instruments placed in parallel across the 250-ohm resistor must be of moderately high impedance (preferably about 1 megohm or more) to avoid undue loading of the 1- to 5-volt source.

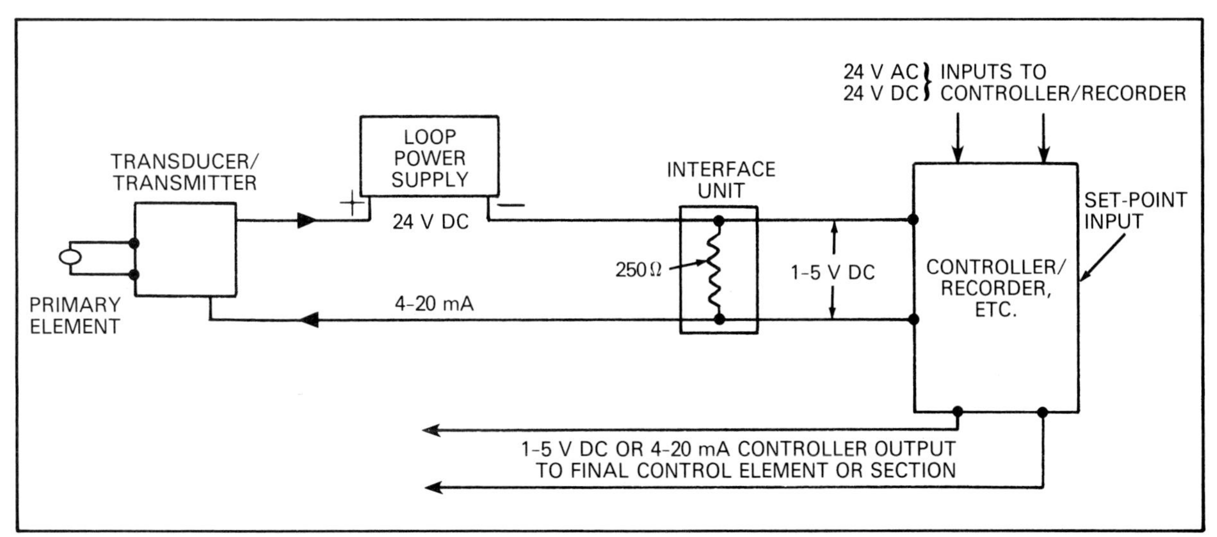

Figure 4.25. An example of control loops and associated elements

An electronic controller may easily incorporate the 250-ohm resistor as an integral part of its circuitry. On the other hand, some control systems include field interface units as separate modules ahead of the controller.

The Operational Amplifier

Operational amplifiers (op amps) existed for a considerable time as modules constructed from vacuum tubes, resistors, capacitors, and other components before appearing as integrated circuits (ICs). This remarkable device—the IC op amp—is of enormous importance to electronic instrumentation, because of the ease with which it can be adapted to perform a score or more of the functions needed in instrumentation.

The op amp is among the least complex of the large family of integrated circuits. However, its interior circuitry is complicated enough and should be disregarded when considering its application to instrumentation. It is enough to learn how to take advantage of the versatility of op amps, and this knowledge is acquired by studying the effects of applying signals (voltages) of various sorts to certain terminals and observing what happens at other terminals (fig. 4.26).

In figure 4.26*A*, terminals *1* and *5*, labeled *balance* (sometimes called *offset*) permit the application of an external voltage source to zero the output when equal voltages are applied to the inputs. Sometimes the tolerances of internal circuit elements of the op amp are such that offsetting voltages must be introduced to compensate for differences between the two channels. The terminal indicated as negative (−) in the drawing is called the *inverting input*, meaning that a positive voltage applied here will result in a negative voltage at the output. The other input terminal, marked positive (+), will cause an output having the same polarity as the input, so this is the *noninverting input*. The inverting input (*2*) is sometimes referred to as the summing point, and the noninverting input (*3*) as the reference junction.

Op amp characteristics and special functions. First, consider the ideal op amp. It will have the following characteristics:

1. Infinite gain
2. Infinite input impedance
3. Infinite bandwidth
4. Zero offset voltage
5. Zero output impedance

The inverting and noninverting inputs permit the op amp to function well as a differential amplifier, meaning that if the voltages

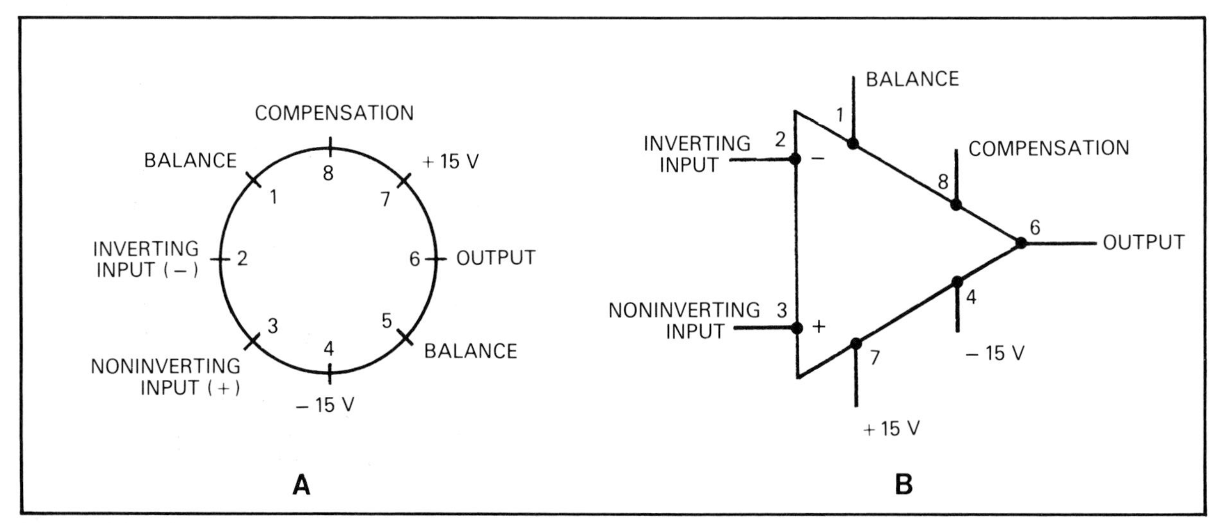

Figure 4.26. An op amp TO-5 package. *A*, typical pin configuration; *B*, symbolic layout.

applied at the inputs are equal (in absolute value and polarity), the output E_o will be zero. If the voltages applied at the inputs differ (in absolute value or polarity), output E_o will have some value other than zero.

The infinite gain characteristic assures that even an infinitesimal difference in input voltages will cause the op amp to produce an output voltage tending toward infinity but, in fact, limited to the value of the supply voltage. It is said that the op amp output will saturate to either positive or negative extreme with any difference, however small, in voltage values at the inputs.

The infinite input impedance characteristic implies that no current will flow between the input terminals. No op amp has ideal characteristics, but it is useful to assume ideal performance in many instances. A practical op amp does not have infinite input impedance, so current will, in fact, flow between its input terminals. A real op amp will not have infinite gain although its gain will be enormous, and saturation will occur with only small voltage difference between inputs. Output impedance of a practical op amp will be quite low, but never zero.

The op amp with negative feedback. The gain of an op amp can be reduced easily to a desired level by applying negative feedback between the output and the input (fig. 4.27).

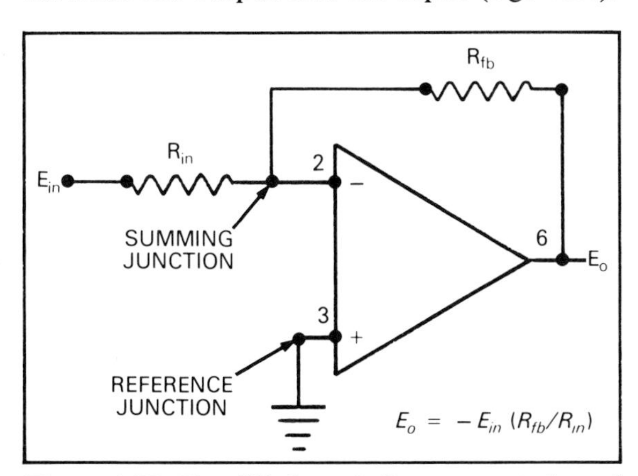

Figure 4.27. Simplified symbolic representation of an inverting feedback amplifier. Unless specifically needed in an illustration, terminals other than those for inputs and output are usually omitted.

Very simply, the gain for the arrangement shown will be R_{fb}/R_{in}, and the output voltage will be

$$E_o = -E_{in} \times R_{fb}/R_{in}$$

where

E_o = output voltage;
E_{in} = input voltage;
R_{fb} = feedback resistance;
R_{in} = input resistance.

The negative sign before E_{in} takes care of the fact that the polarity of the output will be reversed from that of the input. Clearly, if the resistances R_{in} and R_{fb} are of equal value, the gain, or amplification factor, will be unity (1).

For a stable condition of operation (nonsaturated state), the voltage measured at the summing junction of an op amp will equal the voltage at the reference junction. This fact allows circuits to be analyzed with greater ease. In those circuits where the reference junction is grounded, the summing junction is said to be at *virtual ground.*

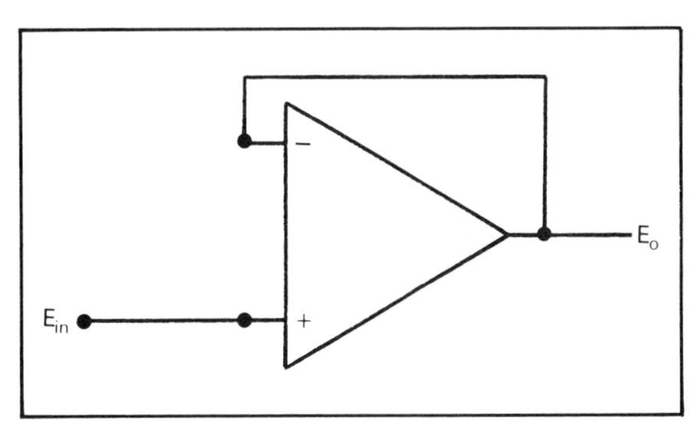

Figure 4.28. A noninverting, voltage-follower circuit. It has a gain of 1 but takes advantage of the high-impedance input and low-impedance output to match a high-impedance source to a low-impedance load. The circuit is noninverting.

A noninverting amplifier with unit gain (fig. 4.28) is useful for providing a match between a high impedance and a low one.

Capacitors and resistors can be used in the input and feedback circuits of op amps to form *integrator* and *differentiator* circuits (fig. 4.29). These circuits get their names from the mathematical functions that describe

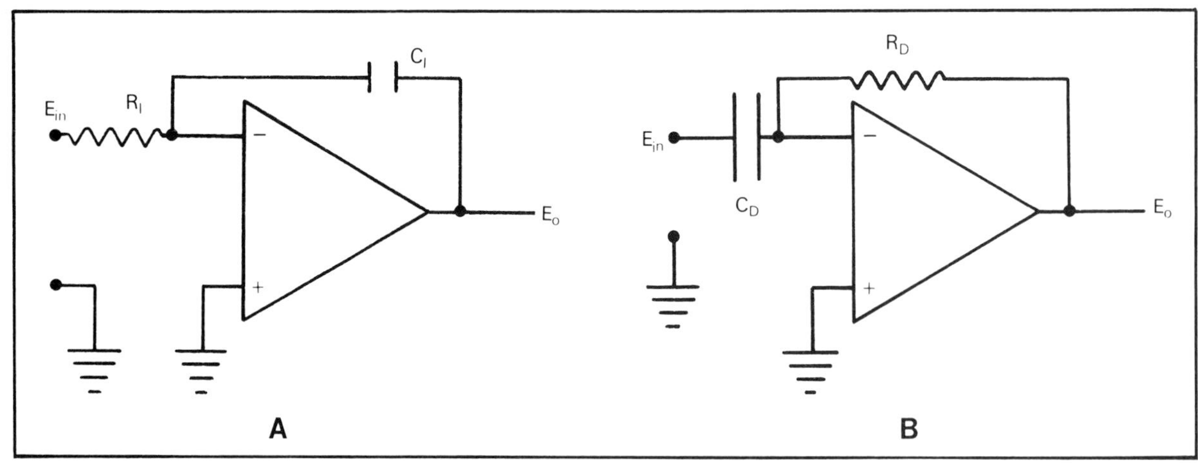

Figure 4.29. Two useful op amp circuits. *A*, an integrator circuit; *B*, a differentiator circuit. (Subscripts *I* and *D* stand for integral and derivative, respectively.)

their actions on an input signal. Note that the capacitor is in the feedback circuit in one instance (fig. 4.29*A*) and in the input circuit in the other (fig. 4.29*B*). This seemingly small change results in an astounding difference in the way the circuits act on the input signal, E_{in}. For example, in figure 4.29*A*, if a steady DC voltage is applied to the input, the op amp will quite quickly saturate, and the output, E_o, will be near the value of the supply voltage:

$$E_o = -E_{in}t/RC$$

where

E_{in} = a constant input voltage;
t = time in seconds;
R = input resistance in ohms;
C = feedback capacitance in farads.

Should a steady DC voltage be applied at the input of the circuit in figure 4.29*B*, the output E_o will quickly settle at zero value. The differentiator output responds only to a varying input at E_{in}. The equation for the circuit is

$$E_o = -RC(\Delta E_{in}/\Delta t)$$

where the quantity $\Delta E_{in}/\Delta t$ is the rate at which E_{in} is changing with respect to time t.

A differential amplifier. Figure 4.30 is a simplified version of a differential amplifier using an op amp. Equal inputs at E_{in1} and E_{in2} produce zero output at E_o. Any difference in voltage between E_{in1} and E_{in2} will produce an output whose value and polarity will depend on the inputs E_{in1} and E_{in2}. If E_{in1} is more positive than E_{in2}, the output will be negative, and vice versa. Also, the output will be determined according to the following equation:

$$E_o = (R_{fb1}/R_{in1})(E_{in2} - E_{in1}).$$

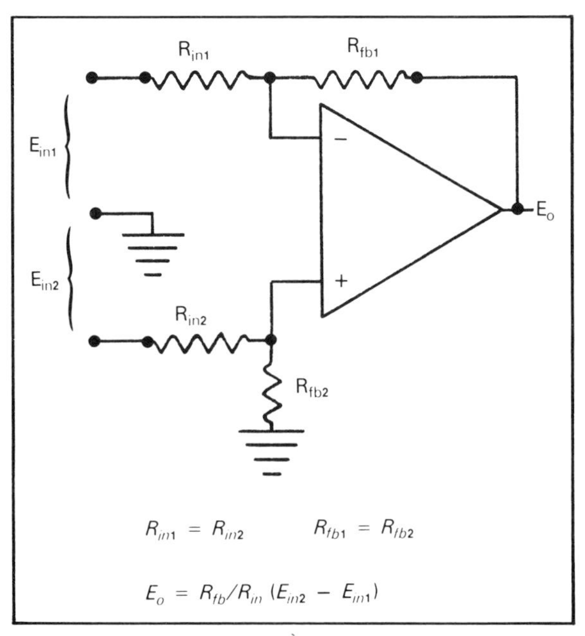

Figure 4.30. A differential amplifier circuit

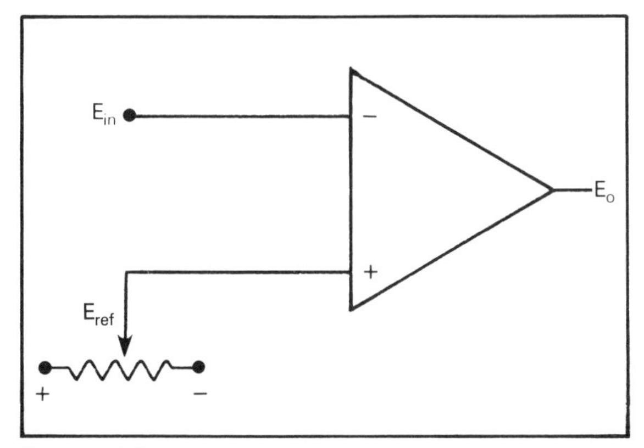

Figure 4.31. A comparator circuit

A form of differential amplifier is used as a *comparator,* a circuit that finds service in control functions. The op amp in this service has no feedback and is useful in on-off control loops (fig. 4.31). An input voltage, which can be the analog representation of a controlled variable, is applied to E_{in}. A reference voltage, E_{ref}, usually an analog representation of a set point for the controlled variable, is applied to the noninverting, or reference, input. Only a small deviation between E_{in} and E_{ref} will cause the op amp output to saturate toward positive or negative supply voltage, probably bringing about corrective action in the control system.

Other circuits for op amps. With the addition of relatively few external components, op amps can be made to perform many additional useful functions in instrumentation. The uses described above are generally limited to actions as amplifiers. Additional uses include voltage-to-current and current-to-voltage converters, square root extractors, integrators, differentiators, and various signal generators.

Power Supplies for Electronic Controls

Electronic control systems use a variety of voltages. Op amps typically require DC voltages of +15 and −15 volts (fig. 4.32). Close regulation of voltage at the value chosen is important. Usually transformers with two secondary windings, each driving a bridge rectifier, are used to obtain the DC voltages. Effective filters and regulators produce smooth current and close-tolerance voltage values.

In addition to the ±15-volt DC supply, a source of 24-volt DC (some systems may use 50-volt DC) will be needed to furnish control loop current. Also, a source of 24-volt AC, 60 hertz, may be required for chart drive motors and alarm circuits. Such a source may be as simple as a step-down transformer, because voltage regulation will not be critical.

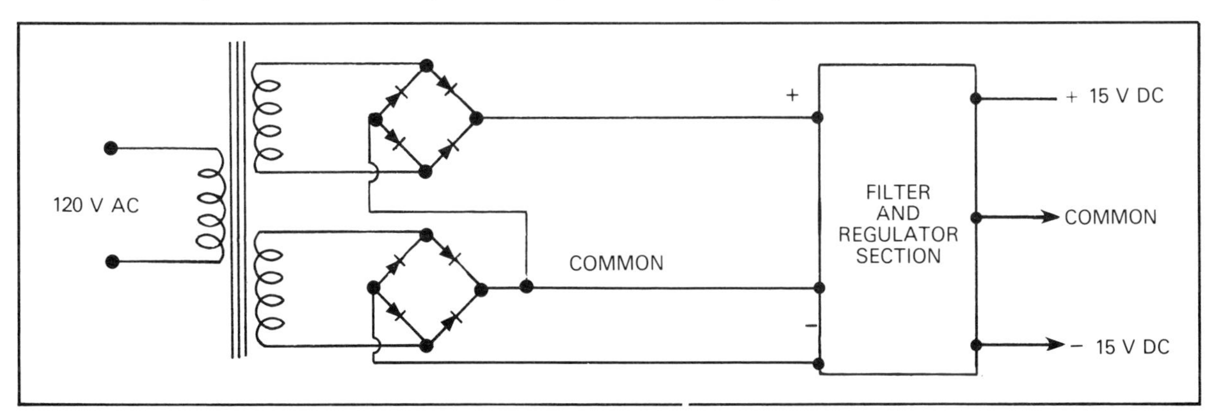

Figure 4.32. A simplified illustration of a power supply for electronic instrumentation. The ±15-volt DC output is a common choice, but other values are also used.

The 24-volt DC supply needed for control loop current should be reasonably well regulated, meaning that a variation of ± 1 volt would be acceptable. It is important to assure that voltage drops across various elements in the control loop, including line losses, do not threaten to approach the value of the power supply voltage. The loop power supply will in many instances provide operating voltage to transmitters in the control loop. Typically, a transmitter will cause a voltage drop of 12 to 15 volts. The recorders, indicators, or other receivers will need 5 volts. With 12 volts for the transmitter and 5 volts for the receiver, the remaining 7 volts limits the amount of line resistance in the loop to 7 V/20 mA = 350 ohms (fig. 4.25).

Converter Circuits

Signals between a control center and moderately distant points are commonly conveyed by a 4- to 20-milliamp current loop. At the control center, the current loop signals are usually converted to a 1- to 5-volt DC format by using a 250-ohm precision resistor in the loop. Clearly there will be some need to convert the 1- to 5-volt signals to corresponding 4- to 20-milliamp current values for transmission from control center to certain outside points. An op amp will assist in providing this conversion.

Voltage-to-current conversion. An op amp alone is capable of accurate control of current flow, but its current-carrying capability is limited to a few milliamperes. A power transistor, its bias controlled in turn by an op amp and field effect transistor (FET), can easily handle the current load (fig. 4.33).

In the simple voltage-to-current conversion shown in figure 4.33, points *A* and *B* are the summing and reference junctions, respectively. During normal operation, the voltages at these points will be equal. For an input of 1 volt at the reference junction, there must be a 1-volt drop across resistor R_1 in order to make the voltage at the summing junction equal

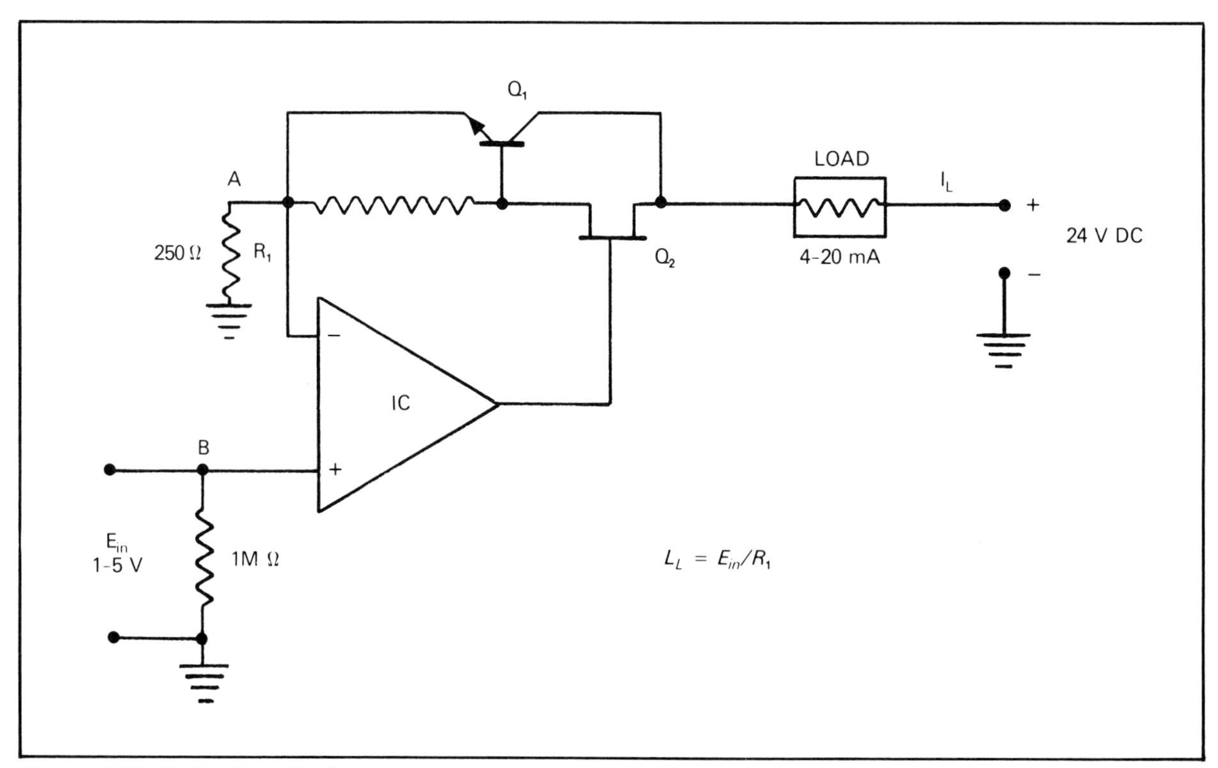

Figure 4.33. A simple voltage-to-current converter

that at the reference junction. A current flow of 4 milliamps through R_1 will accomplish this. Also, note that for an input of 5 volts the drop across R_1 can be made equal to this 5 volts by a current flow of 20 milliamps. All but an insignificant amount of this current flows through the load resistance, which is the current loop, so the arrangement provides an accurate conversion from 1 to 5 volts to 4 to 20 milliamps.

Other converter circuits and devices. Other circuits and devices provide means for going from values of pressure (psi) to proportional values of current or voltage. These are more complex devices than those needed to go from current to voltage and vice versa. Some of them will be described in the chapter on transducers, transmitters, and converters.

Summing and Scaling Circuits

Useful circuits for instrumentation include those that can add and scale quantities. An op amp with a few external components can be adapted to these circuits.

Summing circuits. In the summing circuit (fig. 4.34) the input resistances are equal to one another, and each has the same value as the feedback resistor. If E_{in1} has an input of +1 volt, and the other inputs are 0 volt, the output, E_o, will be −1 volt, the same as it would be for the usual unit-gain op amp circuit. When +1 volt is simultaneously applied to E_{in2}, in effect R_{in1} and R_{in2} have been paralleled. Equivalent to halving the input resistance, this effect will cause a gain of 2, thus producing an output of −2 volts. Should +1 volt now be applied at E_{in3}, the input resistance will be reduced, in effect, to one-third of the value of R_{fb}, producing a gain of 3. The output, E_o, will be −3 volts, or the negative sum of the three input voltages. This method of treating the summing action is not conventional but has good logic.

Assume +1 volt is placed at E_{in1}, +2 volts at E_{in2}, and +3 volts at E_{in3} in the circuit in

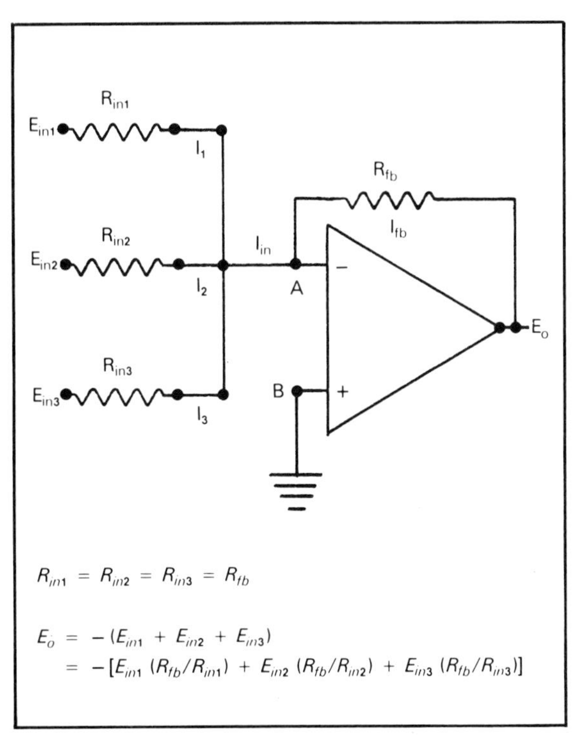

Figure 4.34. A summing circuit

figure 4.34. In this instance it is useful to take into account the current flow in the various circuit resistances, and this manner of diagnosing the circuit function is conventional. Remember that the summing point, *A*, is at virtual ground, so the relations among the resistances, voltages, and currents can be expressed as

$$I_{in} = E_{in1}/R_{in1} + E_{in2}/R_{in2} + E_{in3}/R_{in3},$$

but since the input and feedback resistances are equal to one another, the equation can be simplified as follows:

$$I_{in} = \frac{E_{in1} + E_{in2} + E_{in3}}{R_{in}}$$

$$= \frac{1 + 2 + 3}{R_{in}} = \frac{6\text{ V}}{R_{in}}$$

$$I_{in}R_{in} = 6\text{ V}.$$

Recall that for an inverting feedback amplifier, I_{in} equals I_{fb}. Since R_{fb} equals R_{in}, the voltage drop across R_{fb} is 6 volts, so the output, E_o, is −6 volts, the sum of the three

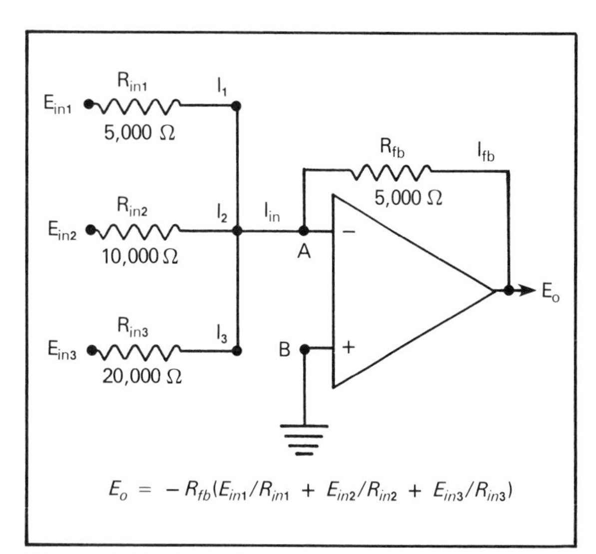

Figure 4.35. A scaling-adder circuit

input voltages, but with negative polarity. This arrangement contains input and feedback resistances having equal values.

Scaling-adder circuits. A scaling-adder circuit differs from the summing circuit in that each input resistor may differ in value from the others (fig. 4.35). Circuits having input resistances of various values present an interesting effect and find use in instrumentation.

In figure 4.35, numerical values have been assigned to better demonstrate the action of the arrangement. Note that each input—E_{in1}, E_{in2}, and E_{in3}—will produce an output component according to the value of its input resistance with respect to R_{fb}. E_{in1} will be amplified by a factor of 1 because $R_{fb}/R_{in1} = 1$. E_{in2} will be amplified by 0.5 (5,000/10,000), and E_{in3} will experience a gain of only 0.25.

Assume 5 volts is applied to each input. The output voltage will be the negative value of the sum of the scaled components, which are 5 volts, 2.5 volts, and 1.25 volts. E_o will be −8.75 volts.

Scaling adders can be used to accumulate numerical quantities of data from various sources. Assume there are three sources: 1, 2, and 3. They could be liquid levels in stock tanks of various sizes or flow rates in different sizes of piping systems. Each source produces signals of 1 to 5 volts between its zero and maximum value, but source 1 might produce quantities four times as great as source 3 for a given voltage signal. The scaling adder is an excellent method of weighting the data from each source so that the output, E_o, is an accurate account of total quantities.

Modes of Control

In pneumatic controls, controllers are fitted with mechanical adjustments that can vary the width of the proportional band, provide for returning a variable to its set point under conditions of varying loads, and adjust to the rate at which loads are changing. The proportional band adjustment is typically a means for varying the fulcrum position on a lever system. The adjustment for reset and rate functions are needle valves that control the charging and leakage rates for bellows assemblies.

Proportional control. A variety of electronic circuits may be used to provide the proportional mode of control. Typically the circuit will use either a potentiometer or a multiposition switch to select the proportional bandwidth. The potentiometer or selector switch is the equivalent of the shifting fulcrum in pneumatic control (fig. 4.36).

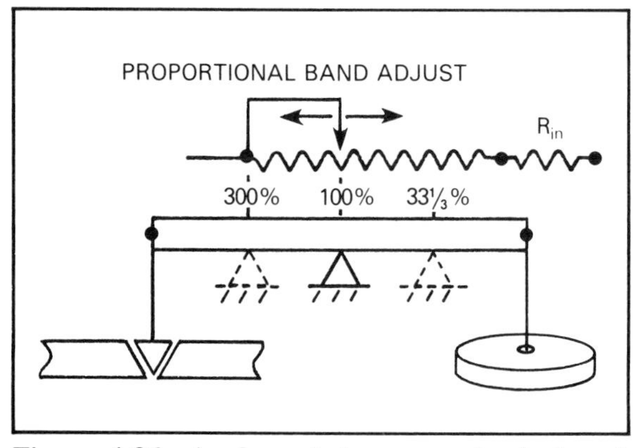

Figure 4.36. Analogy between mechanical and electronic proportional band adjustment. The shifting fulcrum has its counterpart in a variable resistance element.

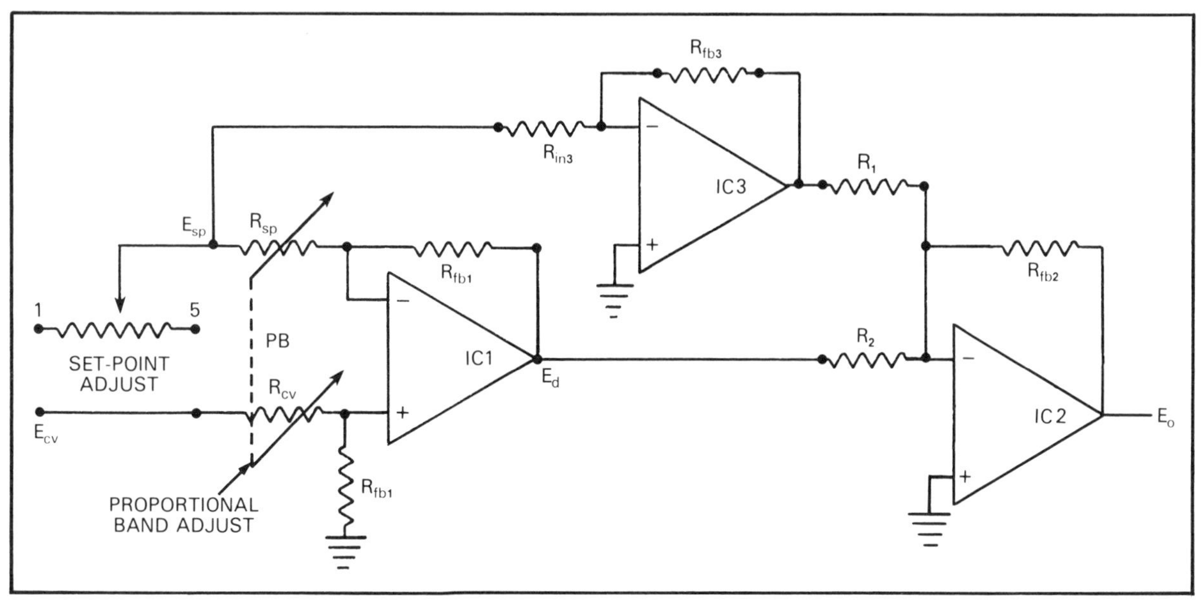

Figure 4.37. A proportional control circuit. The set-point adjust is a potentiometer capable of producing a range of 1 to 5 volts to apply to the inverting input of *IC1*. The proportional band adjust is a pair of variable resistance elements coupled to track with great precision.

A reasonably simple proportional control circuit will be used for discussion (fig. 4.37). It contains the minimum number of components to produce a theoretically functional circuit for proportional control.

A differential amplifier, *IC1,* gets one of its two inputs from a set-point control and the other input from the measuring means of the controlled variable output. The set-point control is capable of producing a 1- to 5-volt DC range, a range that represents the analog values of the controlled variable. The input from the controlled variable produces a like range of analog values. From knowledge of differential amplifiers, it is known that for equal inputs of voltage the output will be zero.

An adder circuit, *IC2,* follows the differential amplifier in the arrangement. The values to be added are the *inverted* set-point voltage and the output voltage from the differential amplifier. The voltage output of the differential amplifier is proportional to the *error voltage,* or *error signal,* that is the difference between the set-point voltage and the voltage that represents the controlled variable. Two variable resistance elements, R_{sp} and R_{cv}, are mechanically coupled to form the proportional band (*PB*) control. With each resistance element equal to R_{fb1}, the *PB* will be 100%, and the output from *IC1* will be $E_{cv} - E_{sp}$, reflecting unit gain for the amplifier. Adjusting the *PB* to 50% by making R_{sp} and R_{cv} half the value of R_{fb1} will cause *IC1* to have a gain of 2, and its output to be $2(E_{cv} - E_{sp})$.

Examine the action of *IC1*. With a +3-volt set point (E_{sp}) and a +3 volts from the controlled variable (E_{cv}), the system will be in equilibrium, so the output from *IC1* will be zero. With the proportional band set at 100%, assume that the voltage from the controlled variable measuring means falls to +2 volts. The output from *IC1* will be −1 volt. This comes from the equation for a differential amplifier:

$$E_d = R_{fb1}/R_{sp}(E_{cv} - E_{sp}).$$

Since R_{fb1} and R_{sp} are equal, the gain of the amplifier is unity, and with $E_{cv} - E_{sp} = -1$ volt, output E_d *will be* -1 volt. The equation will show that if E_{cv} becomes more positive than E_{sp}, the output of *IC1* will be positive, and so on.

Now look at what happens at the output amplifier, or adder *IC2*. With the set-point and controlled variable voltages equal, the output of *IC1* is zero, so it contributes nothing to the adder function of *IC2*. However, the set-point voltage has been inverted by *IC3* and applied to R_1 of *IC2*'s input. Since R_1 and R_{fb2} are equal, the inverted set-point voltage will have its polarity reversed by the inverting input of *IC2* and appear at E_o in its initial polarity and value of $+3$ volts. Whenever the controlled variable and set-point voltages differ, there will be an output at *IC1,* and this output is applied to *IC2* through resistor R_2. The action of *IC2* will cause the values of the set-point voltage and the *IC1* output voltage to be combined to form the *IC2* output. The output of *IC2* is the signal that ultimately positions the final control element. The output from *IC2,* E_o, can be summarized by the following equation.

$$E_o = (R_{fb1}/R_{cv})E_d + E_{sp}.$$

The quantity (R_{fb1}/R_{cv}) takes into account the proportional band setting. It is in fact the amplification factor for *IC1.*

Reset mode. The reset mode functions to return the controlled variable to the set point. An electronic controller may use an op amp connected as an integrator to provide this mode. For this discussion an integrator stage is added to the proportional control circuit of figure 4.37. In addition to the integrator, an inverter stage is added to change polarities to an acceptable form. The proportional plus reset circuitry is shown in figure 4.38.

Recall that if a fixed input is applied to an integrator for an extended time, its output will saturate to an appropriate positive or negative value. The speed with which this action occurs depends on the time constant of *RC,* that is, the product of reset resistor R_I and capacitor

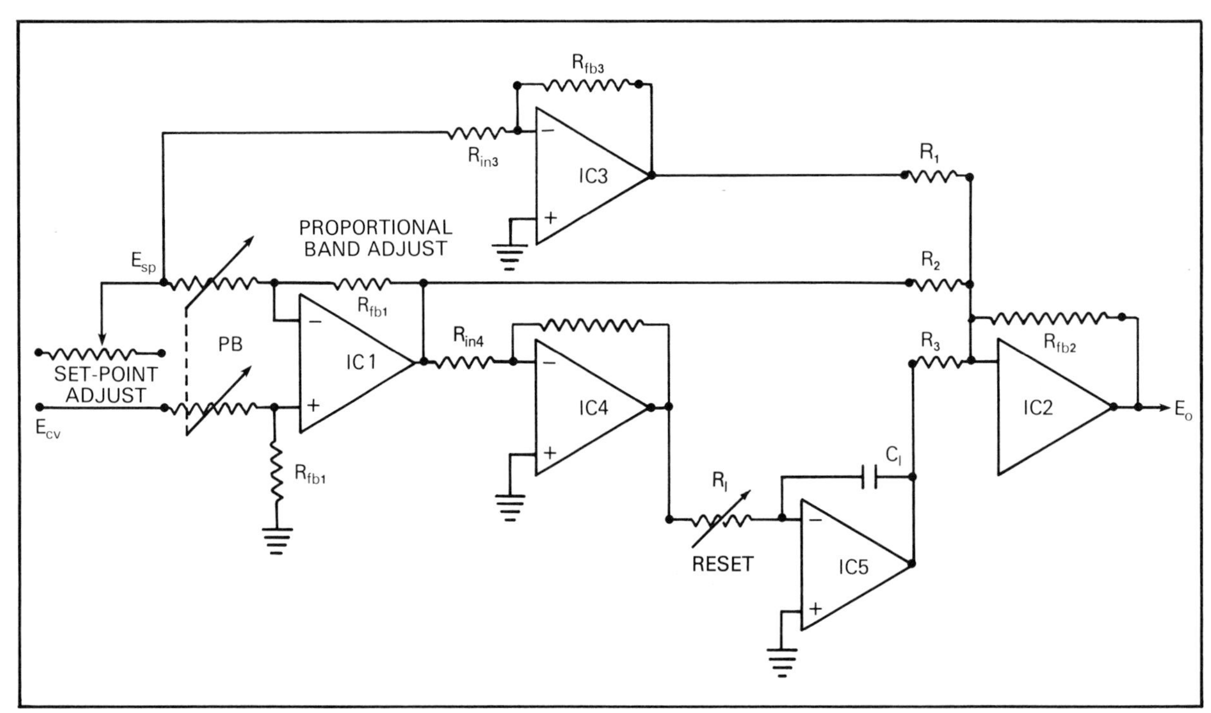

Figure 4.38. Op amps *IC4* and *IC5* have been added to the proportional control circuit to provide the reset mode.

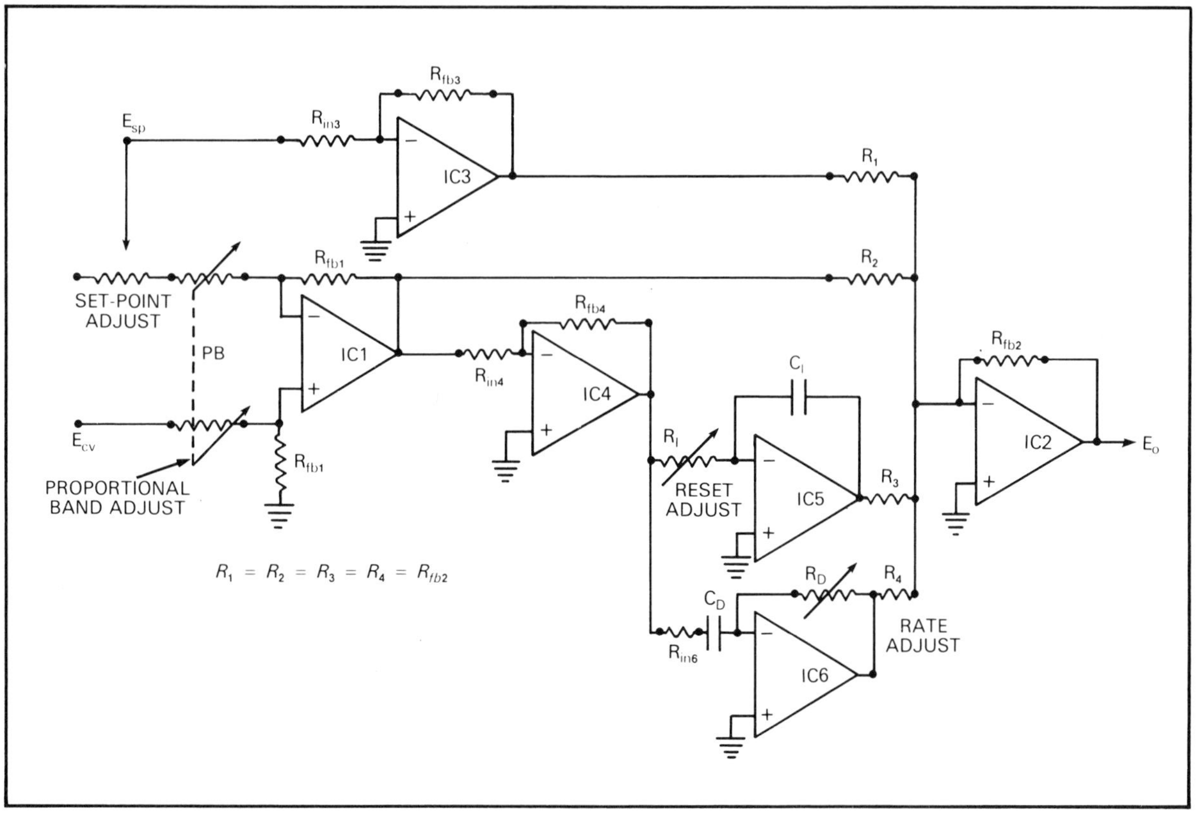

Figure 4.39. One example of the circuitry for three modes of a controller

C_I. The time constant is usually expressed in seconds.

Any error voltage in the system (fig. 4.38) will pass from the output of *IC1,* through inverter *IC4,* and to the input of *IC5* by way of the reset control, R_I. With this application of input voltage, *IC5* will have an increasing output that is applied to the adder stage, *IC2.* This voltage from *IC5* will be added to that already present from the set-point input (at R_1) and voltage produced by the proportional mode action (at R_2). Until such time as the controlled variable is returned to the set point, the integrator stage will continue producing a voltage of appropriate polarity to drive the final control element toward a position of stability. Once the controlled variable is returned to the set point, the reset action will cease, because the input to *IC5* will be zero.

The *reset rate* is the reciprocal of the time it takes for reset action (which is the voltage output of *IC5*) to equal the response of the proportional mode of the controller (which is the output voltage of *IC1*). Reset rate is expressed in *repeats per minute.* Assume the proportional mode response (output of *IC1*) is −1 volt, and the reset mode matches this −1 volt in 30 seconds (half a minute). The reciprocal of this time is 2, so the reset rate is said to be 2 repeats per minute, meaning that in 1 minute the −1 volt will be repeated twice to produce an output of −2 volts. However, in reality the controlled variable will probably be well on its way toward the set point before the reset action reaches −2 volts.

The rate mode. The rate mode can be provided by an op amp differentiator circuit. Figure 4.39 is similar to figure 4.38 with the differentiator stage, *IC6,* added to furnish the third mode of a three-mode controller circuitry.

As pointed out earlier, the rate mode is incorporated in a controller to respond to the rapidity with which a controlled variable changes value. Proportional plus reset controllers are usually sufficient to maintain a system within required bounds of control. However, some systems have problems because of rapid and sizable changes in controlled variables, and the system is of a nature that these changes must be corrected as rapidly as possible. It is this sort of problem that the rate mode can help solve.

The differentiator, unlike the integrator circuit, will not respond to a steady voltage applied to its input components, R_{in6} and C_D. Should the controlled variable begin a deviation from the set point, this action will cause an error voltage to be applied at R_{in6}. While this error voltage is changing, *IC6* will have an output, and this output will be in proportion to the *time rate* at which the change is taking place. Once the rate of change stops, that is, when the error voltage becomes a steady value, the output from *IC6* becomes zero and remains at zero until the controlled variable begins changing again. Once the error voltage reaches its peak and begins receding toward zero, the rate mode comes into play again. This time its action will be the reverse of what it was when the error voltage was increasing. It will tend to retard action of the final control element to affect the controlled variable, to prevent serious overshooting of the set point. In this way the rate mode will help limit the deviations of the controlled variable and settle it at the set point more rapidly than would be likely with just proportional and reset modes.

The action of the rate mode is measured in *minutes* and is called *rate time*. To grasp an understanding of its meaning, imagine a controlled variable changing value such that its analog rate of change is V volt per minute. The proportional action of the controller will respond to this change and at some point in time, t_p minutes, will have caused a response in the final control element of $X\%$. Now, add rate mode to proportional mode and check the time required to achieve the same $X\%$ change in the final control element. Call this time period t_{pr}. It should be smaller than t_p, and the difference, $t_p - t_{pr}$, expressed in minutes is the *rate time*. Rate time is adjusted by varying the resistance of R_D, which effectively changes the amplification factor, or gain, of the *IC6* stage.

In this discussion of how modes of control may be accomplished with electronic circuits, *theoretically* functional circuits were used as a means of teaching circuit actions. Real electronic controllers contain many refinements, varying from one manufacturer to another (fig. 4.40).

Deviations of the controlled variable used to describe actions of the circuits in the sections on modes of control were much greater than those apt to be encountered in a real system. Working with whole numbers, however, has merit when the important need is to gain an understanding of how a system functions.

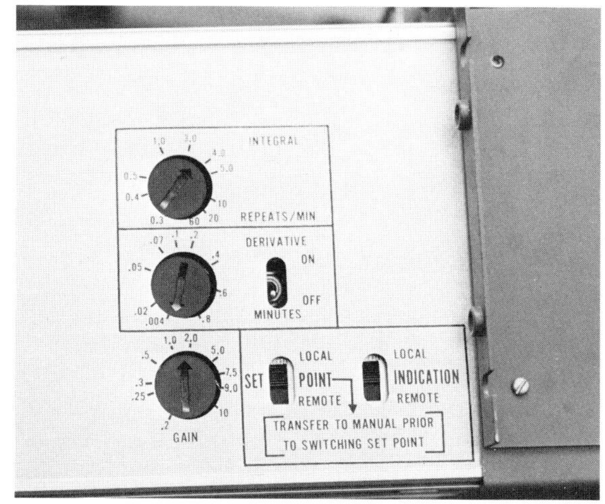

Figure 4.40. The adjustment area on an electronic controller. The manufacturer of this unit uses *integral, derivative,* and *gain* in lieu of *reset, rate,* and *proportional band,* respectively. The set-point and indication switches provide for remote setting and observation of the controlled variable. The automatic/manual set-point switch is located on the front end of this controller (not visible). (*Courtesy ACCO Bristol Division*)

Conclusion

The purpose of this chapter has been to provide a basic understanding of automatic control concepts. The treatment has been elementary when compared with the considerations that must be taken into account by design engineers. Nevertheless, the non-mathematical apporoach will make it easier for those interested in the subject to pursue it to whatever level their capability and need carries them.

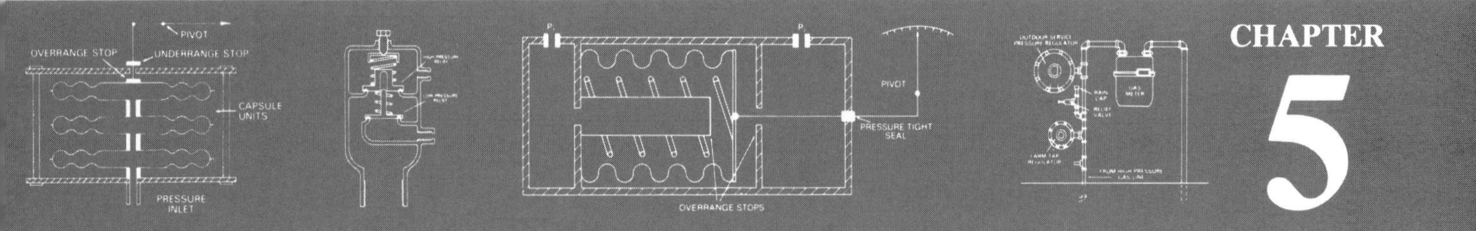

CHAPTER 5

Pressure Measurement and Control

The measurement of pressure is perhaps the most important single function in automatic control, not because of the value of pressure measurement itself, but because its measurement is used as a means of indicating and controlling other variables in the system. Although the *control* of pressure does not rival the overall importance of temperature control, it serves important needs.

The primary concern here is with the measurement and control of pressure in pipes, vessels, and other components of a system; however, many of the principles also apply to the measurement of other variables such as temperature and rate of flow. Although most of the methods and instruments relate to control applications, a few instruments serve only as standards for checking other measuring means.

Units of Pressure Measurement

A large number of means are used to measure the value of pressure. The word *measure* is used in a broad sense because in some instances pressure is not literally measured—its effect is used to actuate a measuring means that might not contain an indicator. For example, the free end of a Bourdon tube may be attached directly to the flapper of a pneumatic controller and the controlled pressure applied to flex the tube. In this case the Bourdon tube is the *primary element* and, in accord with the definition, it measures the *controlled variable* although no graduated scale need be present.

Scales

Pressure is measured as *force per unit area.* The natural environment contains a reasonably uniform atmospheric pressure of about 14.7 psi. Since it is always present and has about this value, one is not consciously aware of it. As a consequence, one scale of measurement ignores the presence of this atmospheric pressure and arbitrarily establishes a zero value for normal atmospheric pressure. Such a pressure scale is called *gauge.* Most pressure gauges indicate gauge pressure, meaning the force per unit area in excess of that brought about by the ambient air pressure. Gauge pressure values below zero are commonly referred to as *values of vacuum.*

Another pressure scale is called *absolute.* An absolute scale has no negative values. It begins at zero pressure (a pure, unattainable vacuum) and progresses into positive values. An absolute scale indicates about 14.7 psi for atmospheric pressure, while a gauge pressure scale will indicate zero for this pressure.

Absolute and gauge pressure scales find important application in instrumentation and control. It is important, however, to be certain which scale is in use.

Range of Values

Another consideration of interest in pressure measurement is the range of values to be measured. The methods used in measuring differ considerably according to the range of pressures involved. The range of absolute pressures that would encompass almost all controlled systems and processes could safely be established as those values ranging from 1 micrometre of mercury to 10^4 psi. One micrometre of mercury is equal to about 2×10^{-5} psi. Gauge pressure below atmospheric should be shown by negative values (a vacuum of -7 psig, for example).

An interesting aspect of pressure measurement is that the low end of the range of values is stated in terms of a linear measure of distance—*micrometre,* a millionth of a metre. The upper limit is stated in true units of pressure—*pounds per square inch.* The term *millimetre of mercury* is accepted nomenclature for expressing pressure. Such terms stem from the use of a mercury column for measuring pressure, and they merely express the height of the column above a reference level. The use of millimetres, or inches, to express large values of pressure is not the usual practice. In the sense used here, large and small values might be reckoned to those above and below standard atmospheric pressure, respectively. Low pressures are commonly referred to in terms of millimetres of mercury, and *very low* pressures are expressed in micrometres of mercury. It should be clear that use of a mercury column for direct measurement of pressure in the micrometre range is not feasible, but the concept of inferring this magnitude of column height is a very useful and realistic one. A special form of column gauge, called a McLeod gauge, is capable of measuring pressures as low as 1 micrometre of mercury. The practice of measuring pressure in terms of millimetres of mercury is apt to continue for some time.

Another term that is used in measuring pressure is *head.* Head is defined as the pressure resulting from gravitational forces on liquids. It is measured in linear distance and is thus a deviation from the use of true pressure units, just as is the case with the use of millimetres of mercury. When one mentions a *hydrostatic head* (or just *head*) of 300 feet, for example, the implication is that a pressure of 130 psi exists:

$$300 \times 0.433 = 129.9 \text{ psi.}$$

This calculation assumes that water is the liquid providing the head, and, since water exerts a pressure of 0.433 psi for each vertical foot of height, converting head into pressure is a simple matter.

In a pressure spectrum ranging from zero to values exceeding 7,000 psi (fig. 5.1), typical subranges can be associated with appropriate elements used in measuring the range. A liberal overlapping of subranges exists. Intermediate subranges use *direct* means for measurement (those depending directly on pressure for indication), while the extreme subranges use inferential means for measurement (those indicating a result of some by-product of pressure, such as change in electrical resistance).

Mechanical Pressure Elements

Mechanical pressure elements comprise those that may be considered as the direct measuring means, in which the existence of a fluid pressure causes mechanical motion of some part of the measuring means. Such devices are used extensively in the intermediate ranges of

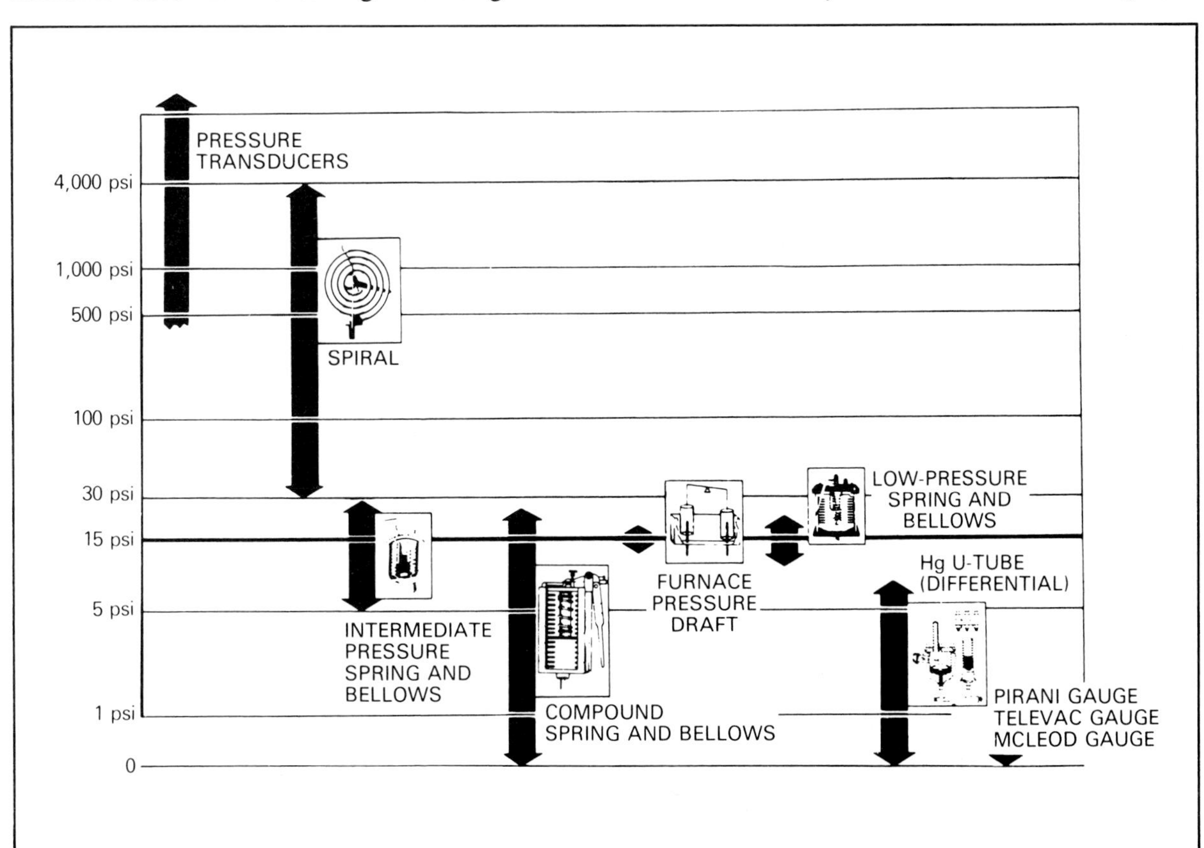

Figure 5.1. Ranges of pressure appropriate to various types of measuring devices (*Courtesy Honeywell*)

pressure, typically from a value of less than 3 ounces per square inch to well into the thousands of psi range. This is not to imply strict limitations on mechanical devices, for they can be made capable of measuring values near zero pressure. The variety of mechanical pressure elements is almost unlimited. Some of the more useful ones are Bourdon springs (or tubes), diaphragms, bellows, liquid manometers, and bell-type gauges.

Bourdon Tube Elements

The design of the Bourdon tube, also called Bourdon spring, involves the use of mathematical and physical concepts that have been developed through the combination of established laws and empirical (or cut-and-try) findings. A Bourdon tube obeys Hooke's law; that is, within elastic limits, its free end will experience a movement that is proportional to the fluid pressure applied. In this respect, fluid pressure applied uniformly to the exterior surface of the tube tends to cause motion opposite to that caused by pressures applied to the interior of the tube.

Materials used for forming Bourdon springs are divided into groups that characterize the type of metal and the hardening method used to achieve qualities related to corrosion resistance, ease of forming, and strength.

The Bourdon tube is used in three forms: the simple C-tube, the spiral, and the helical. Each has an important place in measurement, and each possesses an advantage that makes it suited to a particular application. The advantage exhibited by a given tube in a given situation is frequently related to the free-end movement desired and the space configuration—that is, the shape and size of the space into which the Bourdon tube is to be fitted.

C-tube. The C-tube is by far the most commonly used Bourdon element. It is easier to manufacture in uniform quality, and large quantities can be made to occupy space that is both shallow enough and small enough in breadth to be commensurate with legible dials. The free-end motion, sometimes called tip travel, for a given pressure application is smaller than for spiral or helical units, because the tip travel is directly proportional to the total angle subtended by the tube. Most C-tubes subtend an angle less than 345 degrees, so it is clear that a helical tube of several turns is capable of exhibiting more motion than a C-tube. Amplifying the small motion of the C-tube free end through an arrangement of levers or sector-pinion gears is common practice (fig. 5.2). The expanded motion is used to position an indicator hand on a gauge. There are applications where the wear of gears and linkages, and the friction they induce, make this arrangement unacceptable.

The C-tube finds extensive use in pressure controllers. For this service, motion between the C-tube and the sensing device of the controller—the flapper and its associated mechanism in a pneumatic controller, for example—is not usually amplified. C-tubes used in this service are chosen for their stability and ruggedness, as they must be capable of flexing about the set point for long indefinite periods of time, yet always be capable of adjusting to pressure changes with microscopic accuracy.

Spiral and helical tubes. Spiral and helical Bourdon tubes (fig. 5.2) have a common advantage over the simple C-tube. Each provides greater tip travel for a given pressure application. The indicator hand and the pen arm are driven directly by the motion of the free end of the tube. Gauges and recording instruments using this form of drive achieve sensitivity with less inherent friction and lost motion than that caused by adding linkages and gears to a C-tube. However, some controllers and recorders incorporate a system of linkages between the Bourdon spiral or helical tube and the pen arm or indicator. Such an installation permits a minimum amount of effort and material in changing the recorder or controller from use on one variable to another; for example, to change a controller from the control of pressure to the control of

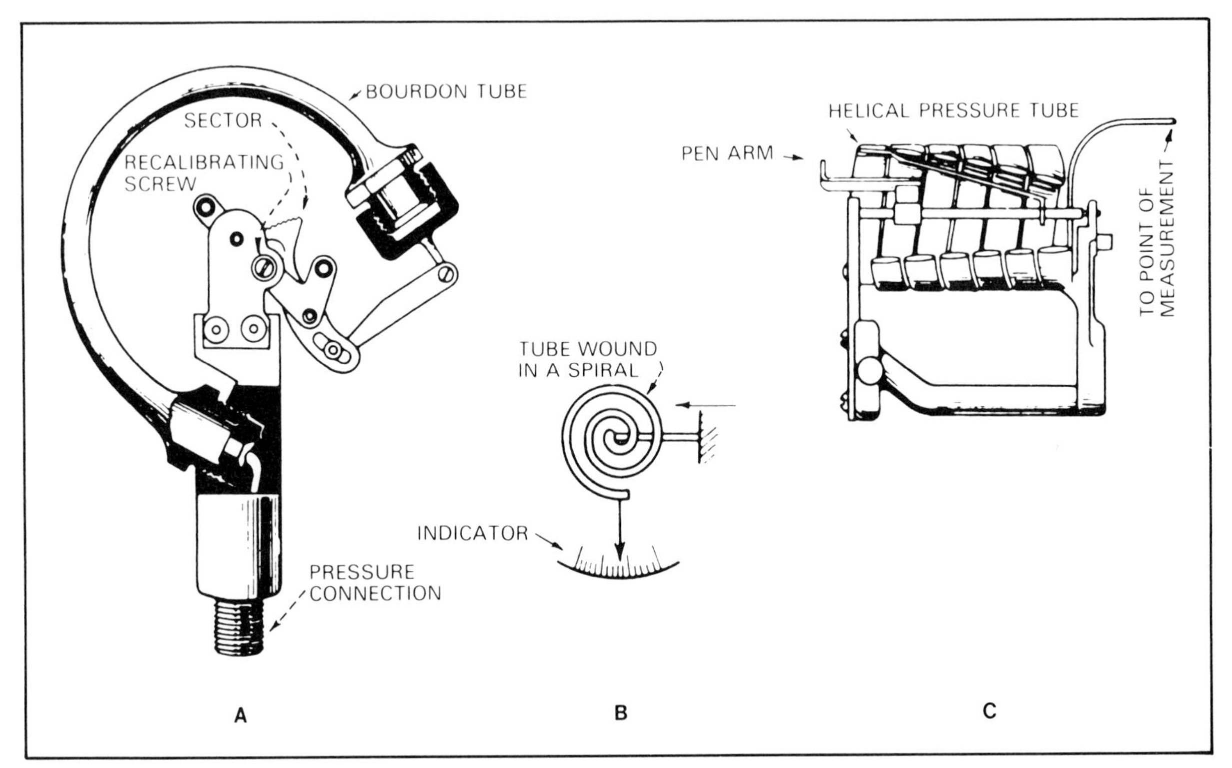

Figure 5.2. Three basic types of Bourdon tubes. *A*, simple C-tube; *B*, spiral; *C*, helical.

temperature, only the Bourdon tube need be substituted, rather than the entire assembly that includes the pen arm or indicator.

Spiral and helical tubes are as popular for controller use as the simple C-tube. They are particularly well-adapted for use in controllers that incorporate a recording or indicating mechanism because of their ability to provide sufficient unamplified motion to drive a recording pen or an indicator hand.

Bourdon tube gauge. Any of the types of Bourdon tubes can be adapted to the usual concept of a differential pressure gauge, although their use for this purpose is not common because of the relative lack of sensitivity of these devices for measuring the extremely small pressure differences that are so commonly encountered in practice. A differential pressure gauge is one that provides the application of pressure from two sources and a device that measures the pressure difference of the two sources. A common Bourdon tube gauge indicates the difference in fluid pressure existing on its exterior and interior surfaces. In a basic arrangement (fig. 5.3), the pressure difference between two sources, P_1 and P_2, would be measured.

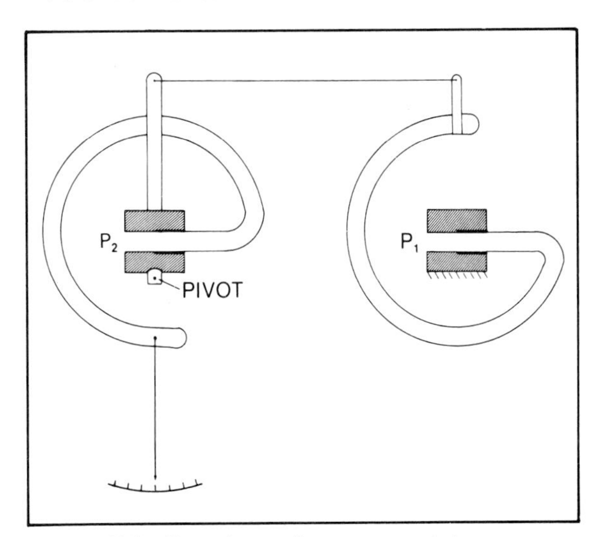

Figure 5.3. Bourdon tubes arranged to measure differential pressure

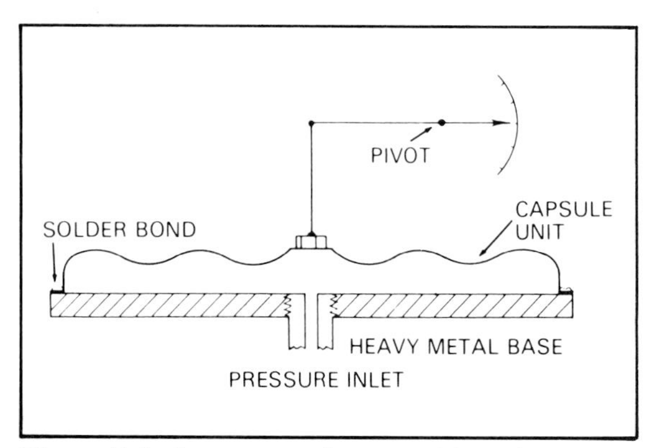

Figure 5.4. A pressure element with a single-shell metallic diaphragm

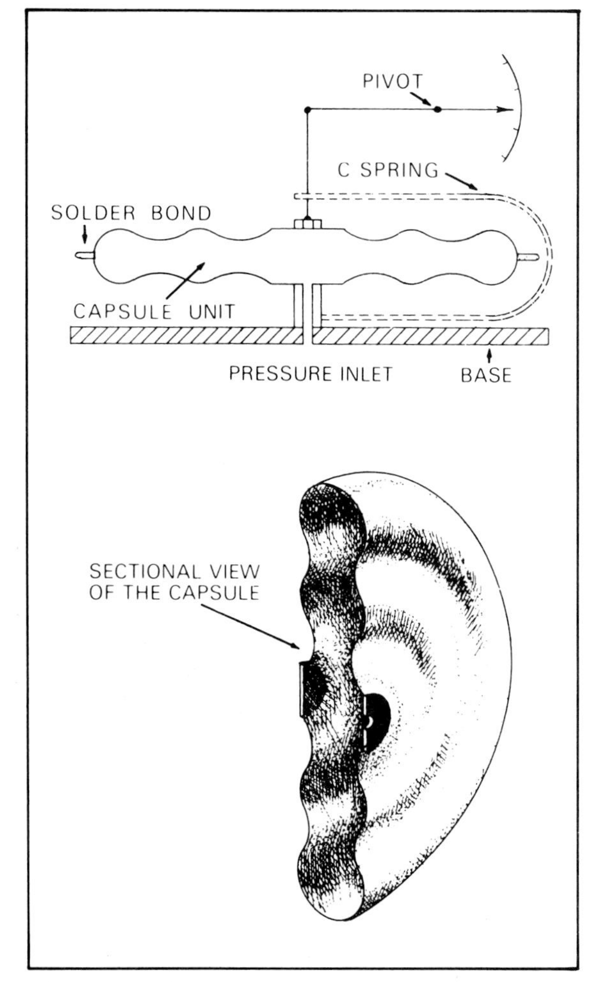

Figure 5.5. A pressure element with a capsule of two metallic diaphragms

Diaphragm Pressure Elements

Two broad types of diaphragm pressure elements are the metallic and the non-metallic, or slack, diaphragms. Both are characterized by sensitivity to small pressure changes and find extensive use for differential pressure measurement in the low-pressure and low-vacuum ranges.

Metallic diaphragm elements. A metallic diaphragm element may take several forms (figs. 5.4, 5.5, and 5.6). The elementary form (fig. 5.4) consists of a single diaphragm shell soldered to a heavy base. Pressure is applied to the cavity formed by the base and diaphragm shell. A diaphragm shell is a relatively broad, shallow, cuplike form that is made of thin metal chosen for qualities that relate to elasticity, ease of forming and heat treatment, and resistance to corrosion. The variety of metals used is large and includes brass, phosphor bronze, beryllium copper, and stainless steel.

A common practice is to press corrugations into the shell during the forming process. This is usually done by mechanical or hydraulic presses. Corrugations vastly improve the performance of a diaphragm element; in fact, it is possible to deflect a properly corrugated element three or four times as much as a flat

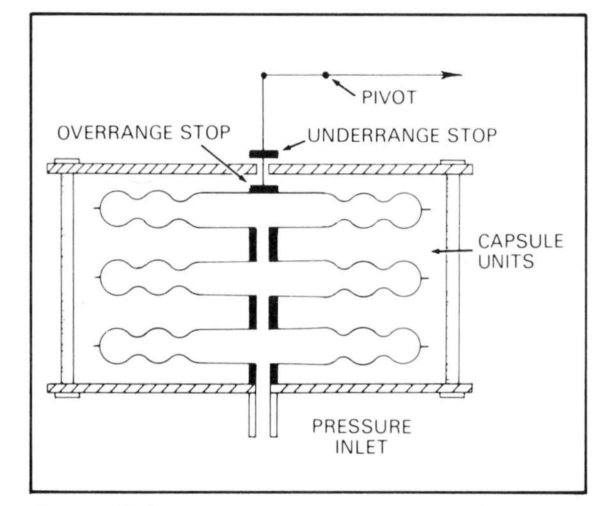

Figure 5.6. A compound element with three diaphragm capsules

sheet and still retain a proportional relation between applied pressure and deflection.

A popular form of element (fig. 5.5) consists of two diaphragm shells soldered, or welded, together to form a *capsule*. Such an arrangement provides twice the amount of motion as the single-diaphragm type. The capsule finds popular use in *aneroid* barometers, which are liquidless gauges used to observe atmospheric pressure. Should the capsule unit in figure 5.5 be evacuated and sealed off, the device would function as a crude absolute pressure gauge. Any barometer, of course, is a special form of absolute pressure gauge. In commercial forms such as that shown in figure 5.5, a C-spring (shown dotted in the figure) resists the action caused by evacuating the capsule and provides a more linear relation between pressure and deflection. Springs are not used in most other applications of *metallic* elements, because these elements are usually able to provide their own suitable pressure-deflection characteristics.

Greater sensitivity can be achieved for a diaphragm instrument by adding capsules in such a way that the sum of the deflections of all units is used to actuate the pointer or any other component being used. Total deflection of such a compound device is directly proportional to the number of capsules. Figure 5.6 shows a compound metallic diaphragm gauge using three capsules.

Nonmetallic diaphragms. Nonmetallic, or slack, diaphragm elements find frequent use for measurements of very low pressure, where the pressure on one side of the diaphragm differs only by a small amount from that existing on the other side.

Nonmetallic diaphragms are made of a variety of materials, such as leather, impregnated silk, and a number of plastic substances having trade names such as Teflon, Neoprene, and Koroseal. They are formed from circular sheets of the material and sometimes contain one or more corrugations.

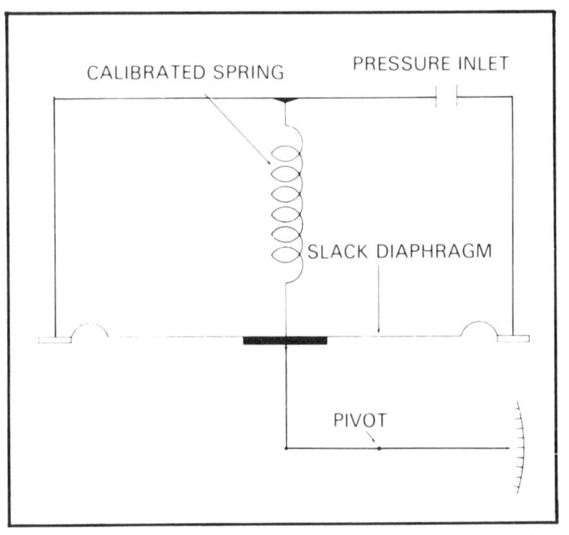

Figure 5.7. A nonmetallic diaphragm with calibrated spring

A simple form of a slack diaphragm gauge (fig. 5.7) has a single corrugated ring in the diaphragm material and the presence of a coiled spring. Unlike a metallic element, the nonmetallic diaphragm ordinarily does not have suitable pressure-deflection characteristics, so it must be provided with a calibrated spring whose force constant provides one of the important factors in determining the applied pressure. The other important factor, of course, is that determined by the area of the diaphragm. Most slack diaphragms are so flexible that the pressure constant associated with them is very low.

Use of diaphragm elements. Diaphragm elements lend themselves readily to applications requiring accurate response to small pressure changes. Gauges using diaphragms can be made so sensitive that a full-scale deflection can be obtained with a pressure change of only a few psi. Draft gauges, which measure the pressure of air within the furnace of a boiler, are commonly made with slack diaphragm elements. Liquid-level gauges, which must be very sensitive to minute changes in pressure in order to be accurate, often use metallic diaphragms (fig. 5.8). The pressure-sensing unit is located as

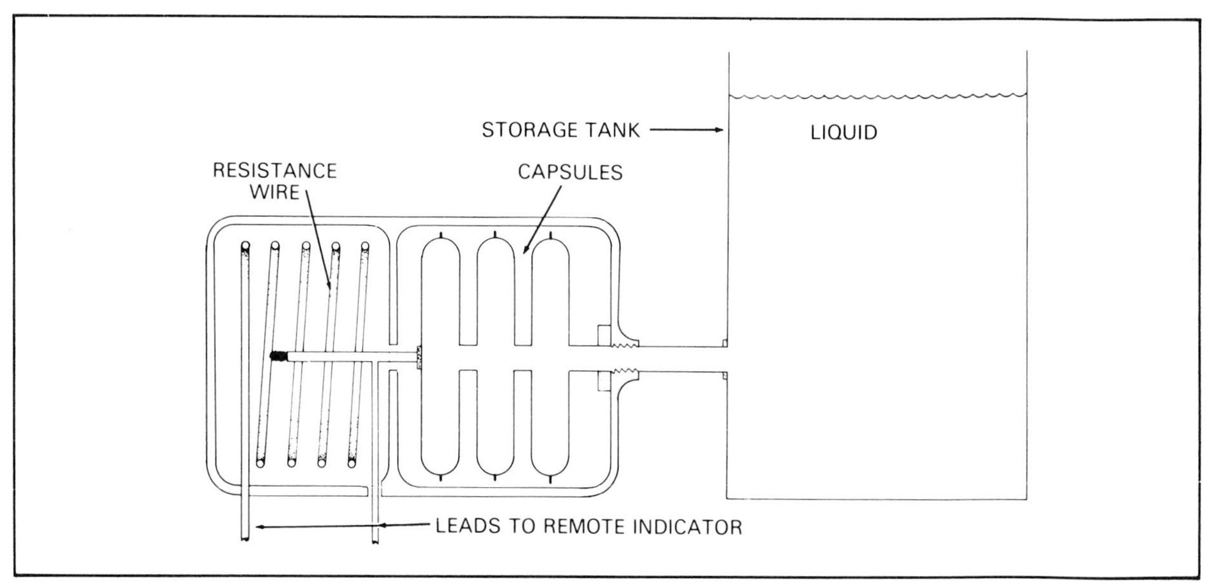

Figure 5.8. One method of using a metallic diaphragm element to measure liquid level

near the lowest level of the liquid as is practicable in most installations. In some instances, it might be located somewhat above or below the bottom of the tank or vessel whose liquid level is being monitored, but the unit must be located not higher than the minimum level that is to be measured. The arrangement shown in figure 5.8 is provided with a form of rheostat, or a variable resistance. The contracting and expanding action of the capsules causes the movable contact of the rheostat to move across the resistance wire, varying the resistance from one value to another as the liquid level in the tank or vessel is interpreted in terms of the value of resistance existing between the two leads.

Diaphragm capsules can also be arranged to make an effective differential pressure gauge (fig. 5.9). This instrument has more versatility than is immediately evident. Connecting P_1

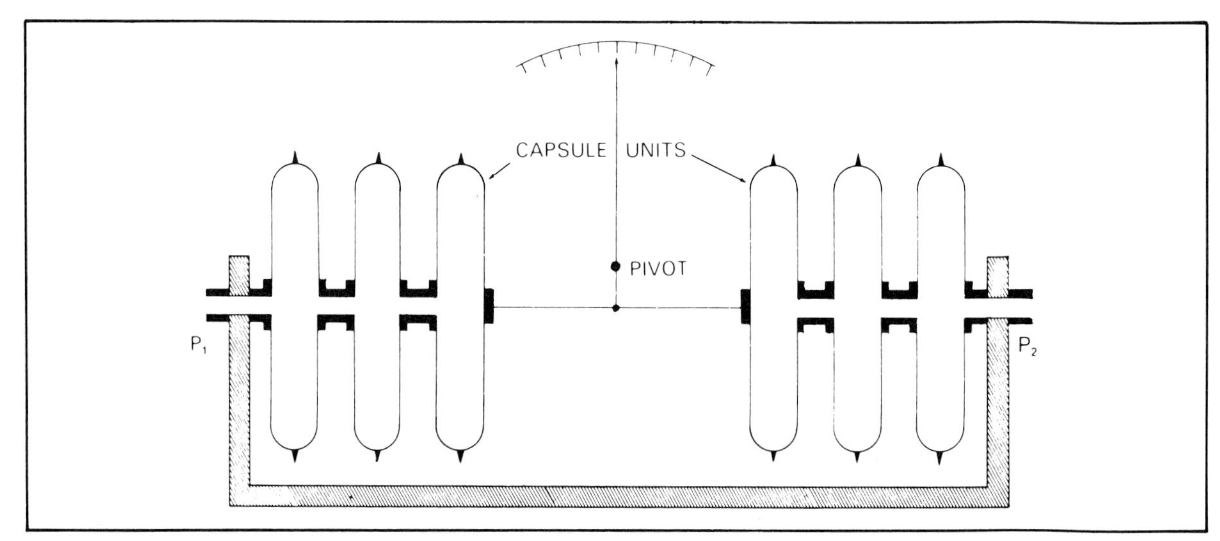

Figure 5.9. An arrangement of diaphragm capsules for measuring differential pressure

and P_2 to different pressure sources results in a moderately sensitive means of measuring the differential pressure. For this service, the gauge dial might be calibrated with values ranging on either side of the zero mark, which is determined when the inputs to P_1 and P_2 are equal. Now, consider use of the gauge with P_1 vented to the atmosphere and P_2 used as the input for pressure sources to be measured. The device will be useful and quite sensitive for measuring values above and below atmospheric pressure. The equilibrium point on the scale can be marked as *zero*, values to the left of zero as millimetres of mercury *vacuum*, and values to the right of zero marked as pounds per square inch *gauge*. Other units of measurement could be used, or the point of equilibrium could be marked as being the value of the existing pressure (atmospheric), in which case the gauge would read in absolute values of pressure.

The gauge here is used with P_1 open to the atmosphere. It should be clear that the vented capsules perform as a spring in this instance, and that the sensitivity of the instrument is less than it would be if the capsule assembly attached to P_2 was free of this restraining force. On the other hand, consider the effect of sealing off P_1 so that the *quantity* of air trapped within the associated capsules is held constant. When the pressure into P_2 increases, the expansion of the capsule assembly associated with P_2 will not only have to overcome the springlike action of the sealed capsules, but also the additional force necessary to compress the air trapped in the capsules. A related action will take place if the pressure at P_2 expands the sealed capsules. Upon expansion, the increased volume within the capsules lowers the internal pressure, while the pressure on the external surfaces still equals the atmospheric pressure; thus, the net force due to this pressure differential tends to hold the capsule assembly at the equilibrium position. Obviously, sealing off one set of capsules has further reduced the sensitivity of the instrument but not its versatility.

Bellows Pressure-Sensing Elements

A bellows element is similar to a diaphragm element composed of a series of metallic capsules, although some characteristics distinguish bellows elements.

Materials. The metallic materials used in the manufacture of a bellows closely parallel those used for a metallic diaphragm, and the selection of a particular metal depends largely on the corrosive conditions to which the unit will be exposed. Stainless steel, beryllium copper, and metals with trade names like Monel and Everdur are in common use, each being noted for its resistance to most forms of corrosive action.

Design. Bellows are made in a variety of sizes, ranging from ¾ inch to 12 inches in diameter and from ¼ inch to 30 inches in axial lengths. Bellows have a wide application in services that are not closely related to pressure measurement; for example, they are frequently used to convert hydraulic pressure to linear motion for the operation of mechanical devices. The bellows that are used for pressure measurement are limited in variety and size.

A bellows unit is usually manufactured from a thin seamless tube of the selected material. The tubing is subjected to a machine that exerts pressures of sufficient magnitude to stretch and form the metal into the *convolutions*, which is the name given to the corrugated effect seen in a bellows. Base plates and other fittings are soldered, brazed, or welded to the ends of the bellows, although end plates are attached during the forming process in a few instances. Sometimes the process of installing the end plates makes the end convolution *inactive*; that is, it will produce no linear motion in response to pressure application. Those convolutions that *do* respond to pressure are called *active* convolutions.

The amount of linear motion that will be caused by applying a given pressure to a bellows unit is directly proportional to the number of active convolutions; it is proportional to the square of the radius of the bellows and is related in some way to several constants that are determined by such factors as thickness of the metal and its elastic characteristics. Bellows as a class are considerably "stiffer" than diaphragm elements; the stiffness accounts for the fact that they are generally less sensitive and have a broader range of capability for pressure measurement. Obviously, sensitivity can be improved by using lighter materials and a large number of convolutions.

Uses. Bellows are used extensively in pneumatic controllers, air relays, and valve positioners. In each of these applications, the bellows performs a function of *measuring,* although this is usually not its only function. In air relays, valves are closed and opened; in the valve positioner and controller, flappers are positioned; but, in each of these actions, the bellows is *measuring* some value of air pressure in order to position a contrivance.

Although similar in many respects to a cascaded group of diaphragm capsules, bellows are frequently used in conjunction with an opposing spring. (This arrangement is not common in the use of capsules, although springs are used with slack diaphragms.) A spring used to oppose the action of a bellows will lower its sensitivity but increase its range-of-values capability. The life of a bellows, measured in cycles of expansion and contraction, is determined to large extent by the amount of pressure applied and the length of stroke permitted.

As an example, assume that a certain bellows, 1½ inches in outside diameter and having 14 active convolutions, can be expanded not more than 0.4 inch without exceeding safe limits of elasticity. Furthermore, the maximum pressure to be applied is 150 psi. Now, if this bellows is operated through its full range, alternately applying and removing 150 psi pressure, its life will be extremely short—on the order of hundreds of cycles. If the stroke is limited by some means, such as a spring or positive overrange stop, the bellows will endure 6,000 or 7,000 cycles before failure. Reducing the cycling pressure and stroke each to 50% of maximum will extend life of the unit to something in excess of 50,000 cycles. From these considerations, it is clear that a bellows of any given size can be made to endure for a larger number of cycles by incorporating an arrangement that allows the bellows to work against a spring. The advantage lies in the fact that a smaller physical unit can be used for a given pressure, while a disadvantage is the loss of sensitivity.

The use of springs to oppose the action of a bellows does not always signify an effort to prolong the life of the unit. Springs are frequently used as a means of providing a calibrating agent. This was the purpose of the spring used in the early stages of the development of a pneumatic controller. A differential pressure gauge can use a spring-opposed bellows (fig. 5.10). By evacuating the interior of the bellows through P_1, the gauge can be used as an absolute pressure gauge within the range of values permitted by the overrange stops. The life of a bellows is greatly increased if used with a spring and overrange stops.

Liquid Manometers

The use of liquid columns as a pressure-measuring means is usually thought of as a laboratory method, but several forms of devices are used in measuring and control applications that use the force available from liquids because of their buoyant effect, their height relative to some reference level, or both of these.

Manometer is a name given to pressure gauges capable of responding to small pressure changes. The word comes from the Greek terms *mano,* meaning thin or rare (as a gas or vapor), and *meter,* meaning to measure. Thus *manometer* probably designated a gas or vapor gauge at one time, but today it is usually reserved for use in speaking of mercury or other liquid-type gauges for measuring low differential pressures.

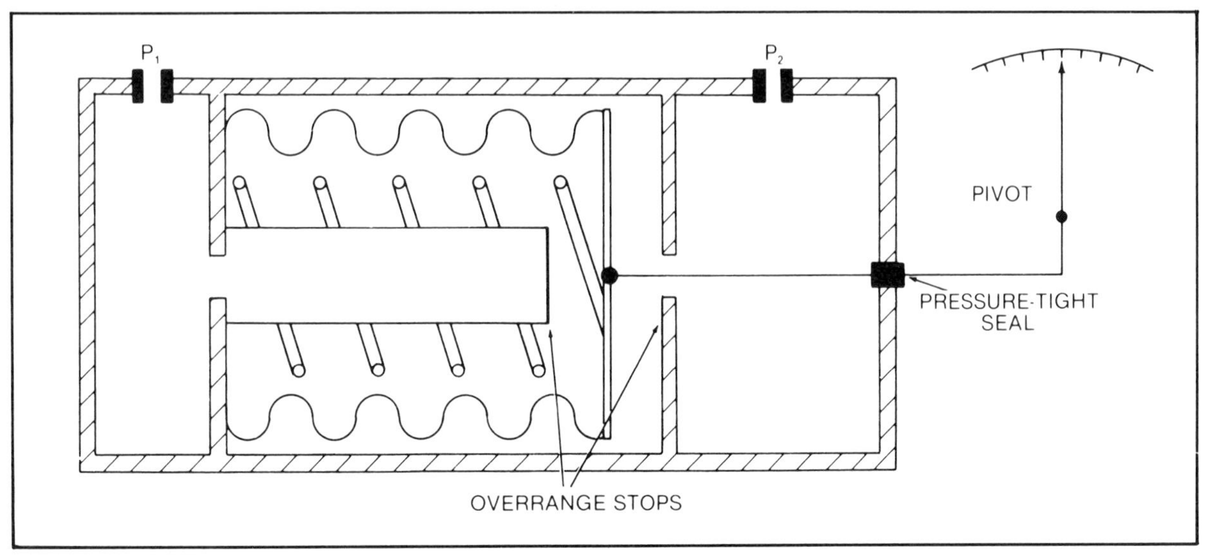

Figure 5.10. Differential pressure gauge using a spring-opposed bellows

Bell-Type Gauges

A bell is another type gauge used to measure differential pressure (fig. 5.11). The gauge depicted has a relatively heavy, cast-iron, bell-like device floating in a pool of mercury, and the cavity formed between the bell interior and the surface of the mercury is tightly sealed by the mercury-bell contact area. If the pressure existing in cavity V_1 is equal to that in V_2, as would be the case if P_1 and P_2 were

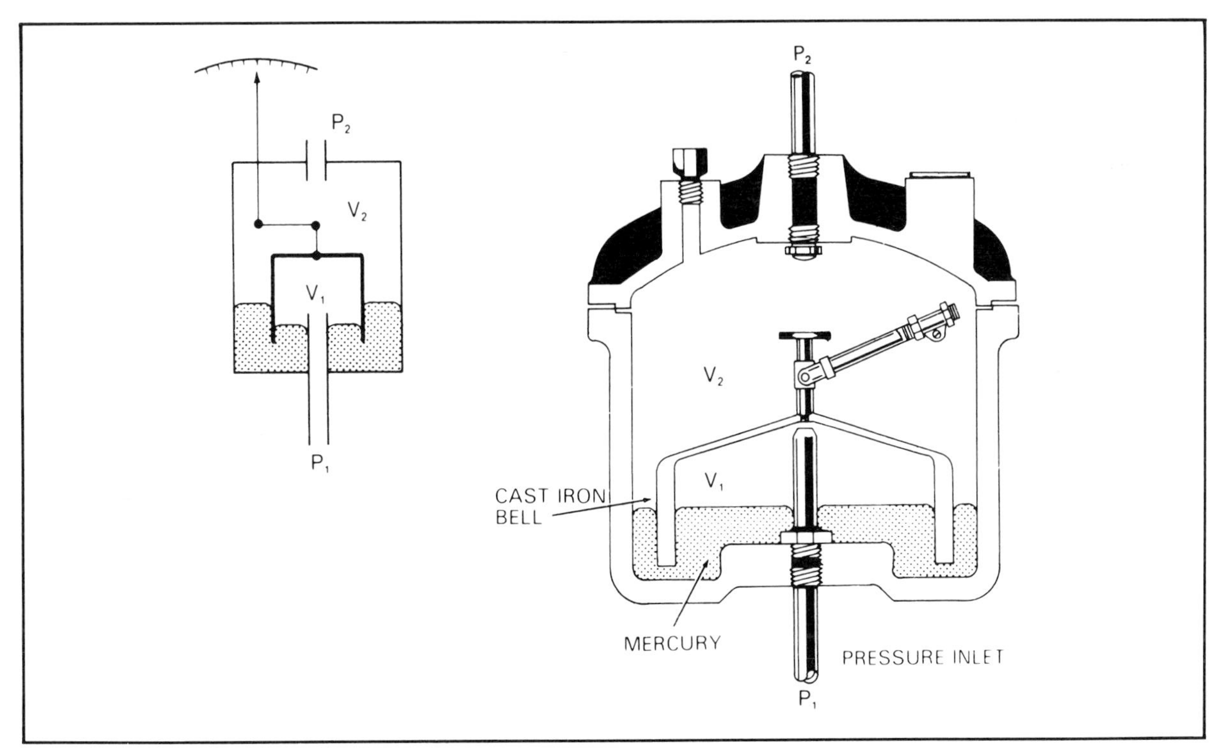

Figure 5.11. A mercury-bell differential pressure gauge

open, the vertical position of the bell will be largely determined by the weight of the bell pulling it down and the buoyant effect of the mercury pushing it up. An equation that expresses an exact balance of forces might be written as follows:

$$W + (P_2 \times A_E) = F + (P_1 \times A_I)$$

where

W = weight of bell in pounds;
P_1 = pressure in V_1;
P_2 = pressure in V_2;
A_E = effective exterior area of bell;
A_I = effective interior area of bell;
F = buoyant force.

The weight of the bell can be considered as a constant. The areas A_E and A_I are the *effective* areas on which the pressures can act to produce vertical motion and are equal to $\pi R^2{}_E$ and $\pi R^2{}_I$ where R_E and R_I are the exterior and interior radii of the bell, respectively, at the surface of the mercury. Obviously, the bell will settle to some point that makes the two sides of this equation equal.

Assume that P_2 is open to the atmosphere and therefore a constant and that A_E and A_I are constants—as they would be for a straight-sided bell—then the left side of the equation is a constant. Now what happens when the pressure in V_1 is increased to something higher than atmospheric? The balance of forces in the equation is upset by the variable term $P_1 \times A_I$. Since the left side of the equation is a constant, the buoyant force term must adjust to the new conditions and restore the balance. As the bell is lifted out of the mercury by the upward force ($P_1 \times A_I$), less and less mercury is displaced until eventually the weight of mercury displaced is lessened by exactly the same amount as the force change brought about by increasing P_1 above atmospheric pressure. The bell will again be in equilibrium.

In order to make practical use of this bell-type instrument, a lever is attached to the bell by means of a pin and yoke. The other end of the lever drives a shaft that passes through a gastight seal in the instrument housing. The sensitivity of the instrument is determined by the bell diameter and its mass. Sensitivity is commendable. Accuracy and stability are good, although the gauge might suffer a slight dead-spot error due to the output shaft sticking in the gastight seal. The heavy construction of the case and cover implies that the unit can withstand very great static pressures.

Protective Devices for Pressure Instruments

Measuring devices frequently need protection from environmental factors that might damage or destroy them. A protective component must effectively isolate the instrument from ruinous conditions while allowing the transmission of pressure information between source and instrument.

Pulsation Dampeners

The output of reciprocating pumps or compressors is a typical example of a severely pulsating pressure. Pressure gauges or other instruments attached directly to these outputs will experience unnecessary wear and abuse. Many simple attachments can prevent these pulsations from reaching the instrument, and each of them accomplishes this by retarding the *response rate* of some part of the pressure line leading to the instrument. A capillary tube, for example, cannot transmit the rapid pressure changes in a pulsating system, and, if a small length of capillary is placed ahead of the instrument, pressure reaching the latter will be of average value. The use of a capillary section to accomplish dampening is not popular because it is easily clogged by foreign matter. It also has the disadvantage of being nonadjustable, although this is not a serious shortcoming if the installation and conditions related to the pulsation are of a permanent nature. The adjustable needle valve dampener (fig. 5.12*A*) contains a felt filter section to aid

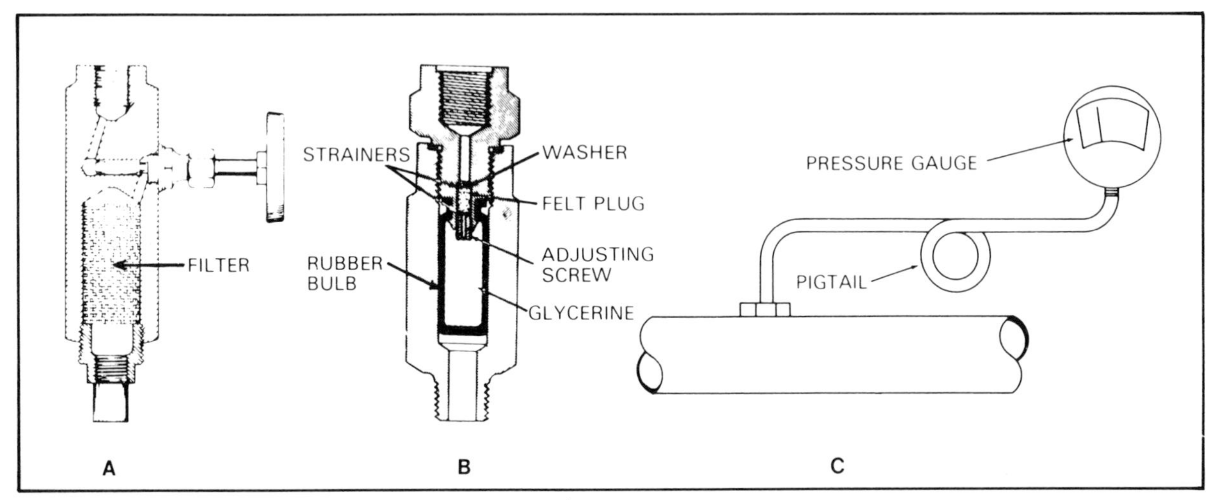

Figure 5.12. Protective devices for instruments. *A* and *B*, pulsation dampeners; *C*, isolation seal.

in preventing the introduction of contaminants that might clog the valve. Another form (fig. 5.12*B*) uses a rubber bulb and provides a closed system for meter, small passage, and filters, thus assuring freedom from contaminants. The rubber bulb is filled with glycerine or other relatively inert liquid, and process fluid acts on the exterior surface of this bulb. Pressure of the process is transmitted through the rubber wall to the glycerine, which then actuates the measuring device.

Isolation Seals

Sometimes certain properties of a fluid will adversely affect the pressure-measuring instrument. Some process fluids are maintained at temperatures that would be injurious to many gauges or other instruments, while other processes may involve fluids so viscous that they clog Bourdon tubes and other elements. In each of these cases, it is advisable to prevent direct contact between process fluid and measuring instrument.

A simple and effective means of protecting a gauge from the high temperature of some process fluids is using a pigtail, or complete turn, in the tubing leading to the gauge (fig. 5.12*C*). Steam and other vapors tend to lose temperature and condense in the pigtail. The liquid formed by the condensation usually will settle in the lower half of the pigtail when the system is shut down, thus being ready to act as a cushion when the system is again in operation.

High Vacuum Measurement Gauges

What is *high vacuum*? Of course, it implies a low, or very low, absolute pressure, but since other methods are used to measure ultra-high vacuum, establishing a range of values for the two categories is desirable, however roughly they may be defined and regardless of the amount of overlapping between one and the other. The high-pressure limit of a high vacuum is vague indeed, but hardly anyone considers an absolute pressure greater than 20 millimetres of mercury as a high vacuum. However, most authorities regard pressure values less than 1 millimetre of mercury as high vacuum. The range of pressure values from 1 micrometre of mercury to 1 millimetre of mercury will be called the *high-vacuum range,* and those values below 1 micrometre, the *ultra-high vacuum range.*

Most of the devices already discussed can be adapted to measure pressure values to as low as 10 millimetres. The Bourdon tube gauge, least sensitive of these, is commercially available for measuring down to this value, while the bellows and diaphragm gauges are useful for measuring values as low as 0.1 millimetre of mercury. The instruments that have practical worth for measuring pressures below this value are the McLeod gauge, Pirani gauge, and the thermocouple vacuum gauge. Ionization gauges are used almost exclusively to measure ultra-high vacuum.

The McLeod Gauge

The McLeod gauge is a form of column instrument capable of measuring pressures down to 1 micrometre of mercury column. This instrument finds virtually no use in automatic control application, because it does not adapt to remote transmission of its measurement. However, it is used extensively as a standard of measurement to check the accuracy of other instruments, although accuracy is not one of its strong points; errors greater than 1% are common. It is used as a standard because it is relatively simple and inexpensive for a gauge capable of such low-pressure application.

The McLeod gauge can take several forms. In one form of the gauge (fig. 5.13), the mercury of the system is maintained near the initial level mark, while the unknown pressure is applied to the top connection. In time, the capillaries and other glass tubing will assume the unknown pressure. The necessary amount of air pressure is then applied to the mercury reservoir, forcing mercury to rise in the measuring system until it reaches the top index mark. Mercury in the left tube will naturally seek the same level as that of the center tube but will be resisted by the compressed residual air or other gas trapped above the mercury. This residual gas is used as the measure of the unknown pressure.

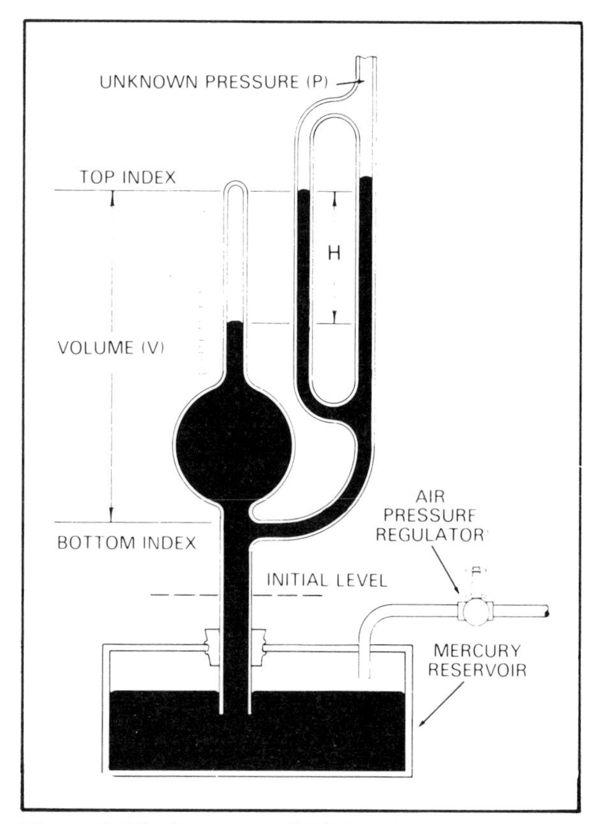

Figure 5.13. One type of McLeod gauge

The ability of the McLeod gauge to indicate and distinguish between pressure values in micrometre increments is due to the relatively enormous volume difference between the left and center columns. This big difference in volumes is due to the bulb section located below the left capillary, and the difference between the two volumes provides for an expanded scale that makes minute pressure changes readily visible.

The left and center capillaries shown in figure 5.13 are of equal diameter. The volume per unit length of the capillaries multiplied by the density of mercury is furnished by the manufacturer as one of two constants used in calculating pressure with the instrument. The other constant factor used is the volume (V), also furnished by the manufacturer.

Applying this instrument to an actual measurement, the low-pressure source is attached to the unknown pressure connection, while the mercury is maintained at the initial level mark. The pressure within all the tubing soon becomes equal to that of the unknown pressure (P). Now the mercury is slowly forced up into the measuring system. Note that once it reaches the level marked by the bottom index line, it seals off the left member of the assembly, thus trapping a known volume (V) of residual gas. The mercury is forced higher until it finally stabilizes at the top index mark of the center tube. In the left tube, the level has lagged behind, because of the force exerted by compressed residual gas.

A relation exists between the pressure above the mercury in the sealed left column and the height of mercury in the center column (H). Since the two columns are of equal and uniform diameter, the pressure can be calculated by using the relation between the height of a column of mercury and the pressure it produces. In order to develop the equation for calculating the unknown pressure (P), a different route will be taken to determine the pressure in the left column (p).

$$p = F/L^3 \times H,$$

where

p = pressure in left column;
F/L^3 = force density of mercury; and
H = height of mercury in the center column.

Boyle's law states that pressure changes are proportional to volume changes; this knowledge is used to determine pressure P. Letting v and V be the volumes of the compressed and uncompressed gas respectively, and p and P the pressures of the two volumes,

$$PV = pv;$$

but, since

$$p = F/L^3 \times H,$$

and

$$v = H \times L^2$$

where L^2 is the cross-sectional area of the capillaries,

then,

$$PV = H^2 \times F/L^3 \times L^2$$
$$= A \times H^2$$

where A is a constant furnished by the manufacturer, equal to $F/L^3 \times L^2$.

Then,

$$P = \frac{A \times H^2}{V}$$

where V is a constant furnished by the manufacturer.

Should the above calculations be carried out using force density of mercury in pounds per cubic inch, area in square inches, volume in cubic inches, and height in inches, the answer will be in pounds per square inch. The constants supplied by the manufacturer are apt to be in appropriate metric units that will cause the answer to be in micrometres of mercury.

Note that pressure P is proportional to the square of the height H. For this reason, this form of McLeod gauge is usually referred to as having a *square law* scale.

Pirani Gauge

The Pirani gauge measures *inferential pressure* and is *not* concerned with the forces that are involved with pressure, but with certain electrical effects that are observed in an environment of rarefied air or other gas. These observations can be related with pressure values.

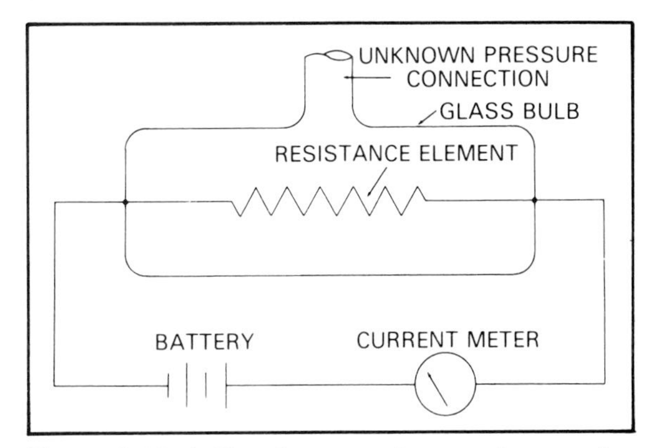

Figure 5.14. A Pirani gauge using a resistance element

The Pirani gauge shown in figure 5.14 is a simple electric circuit of battery, ammeter, and a pressure-controlled glass bulb containing a quantity of resistance wire whose resistance is a function of temperature (its length and diameter being considered constant). Current flowing in the resistance element causes it to heat up, thus changing its resistance. Some of the heat, which is energy being dissipated at the same rate the battery supplies it, is carried away by the supporting electrical leads, some by radiation, some by gas thermal conduction, and an insignificant amount of convection.

Only *gas thermal conduction* is an important function of pressure. Heat loss by gas thermal conduction reaches a high and fairly stable value for pressures greater than about 10 millimetres of mercury. For pressures above about 1 millimetre of mercury, this form of loss declines with pressure, becoming approximately linear at pressures below about 0.1 millimetre of mercury.

With a given current flowing in the resistance element and a given low gas pressure existing in the bulb, the temperature and resistance of the element will reach stable values. If the pressure within the bulb is reduced to values within the range for which gas thermal conduction and pressure are definitely related, heat loss from the element will decrease and its temperature will increase to a higher value, thus increasing its resistance and lowering the current flow.

A practical Pirani gauge (fig. 5.15) has two similar resistance elements—one of them evacuated and sealed off, the other fitted with the unknown pressure connection. These

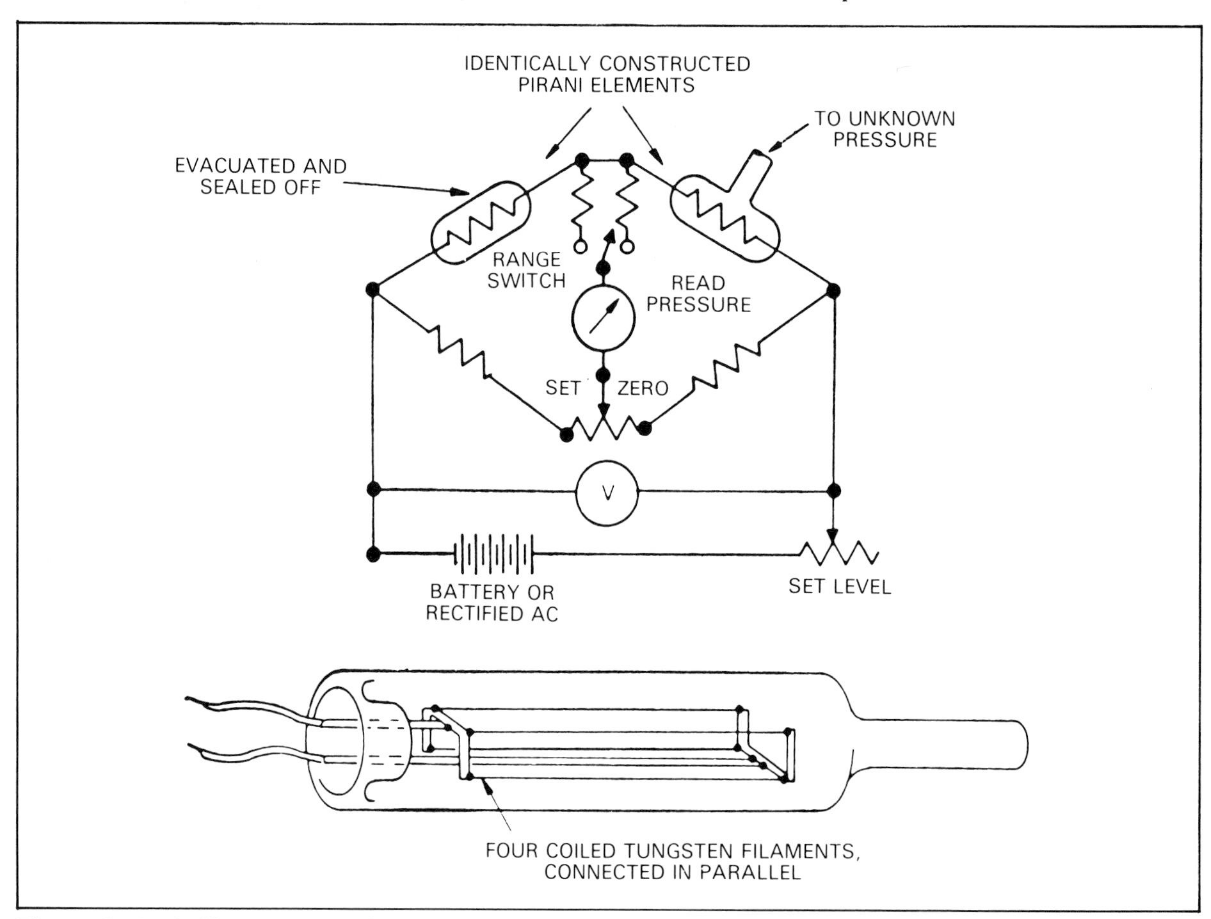

Figure 5.15. A Wheatstone bridge circuit for a Pirani gauge

elements form arms of a Wheatstone bridge circuit. The connections are such that when the unknown pressure is equal to that in the evacuated element, current flow and resistance values of the two elements are equal. Consequently, no current flows through the pressure-indicator meter. Once the unknown pressure rises in value, the ability of its associated resistance element to dissipate heat will increase, lowering its resistance and upsetting the balance of current flow. This will be reflected as a pressure reading on the meter.

The use of two resistance elements in the way shown provides for compensation of voltage variations. Since the pressure indication is a measure of current caused by unequal resistance values, the system is unaffected by the sort of voltage variations that might be expected in practice. The range switch provides the gauge with two ranges, from 0 to 20 micrometres of mercury, and from a few micrometres to about 1 millimetre of mercury.

Pirani or other hot-wire pressure gauges suffer several forms of error. The most important form is related to the changing rate at which the resistance element loses heat by radiation. New, shiny elements radiate at low rates; but, with age and use, the surface becomes oxidized or otherwise darkened, and the rate of heat loss from radiation increases. The effect of this is to cause "high" readings, perhaps as much as 20 micrometres of mercury. The Pirani gauge is popular, not because it is a precision instrument, but because of its simplicity, low cost, facility for remote indication, ruggedness, and application to a very important pressure range.

The Thermocouple Vacuum Gauge

The thermocouple vacuum gauge is similar to the Pirani instrument in that both depend on heat loss by gas thermal conduction for their operation.

In a simple form of thermocouple gauge

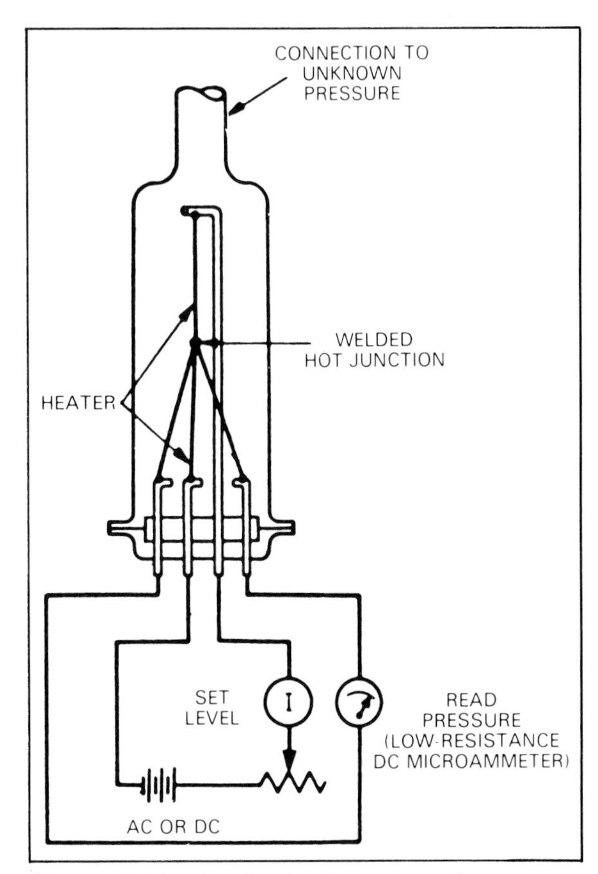

Figure 5.16. A simple thermocouple vacuum gauge

(fig. 5.16), the thermocouple element is welded to the midpoint of a heater filament, and this joint represents the hot junction of the thermocouple. Its cold junction is one of the connections to the base terminals. Heating of the hot junction and filament is caused by current flow in the latter, and rate at which this heat is dissipated at low pressure is a function of pressure, just as for the Pirani gauge. The thermocouple, which produces a voltage proportional to the differences in temperature between its hot junction and cold junction, is capable of actuating a sensitive current-measuring device—a microammeter, for example. The microammeter can be calibrated to read in pressure units.

Thermocouple gauges can be made to have usable ranges of from a few to several hundred micrometres of mercury, although their accuracy is no better than that of the Pirani gauge. For the type of unit described here, manufacturers provide a standard calibration curve; then, in order to compensate for production variations among units, a particular and appropriate heater (or filament) current is specified to bring individual units as near as practicable to the standard.

Ionization Gauges

The measurement of ultra-high vacuum—pressures below about 1 micrometre of mercury—is a difficult process to carry out with accuracy. The use of ionization gauges is almost exclusive in this range. An ionization gauge is one in which molecules of the residual gas are bombarded with electrons or other particles. This bombardment causes some of the molecules to lose electrons, thus forming positive ions. The ions and loose electrons are then used in current flow measurements, which are correlated to the quantity and type of residual gas.

The sensing element of one type of ionization gauge (fig. 5.17) resembles an ordinary triode vacuum tube, but the voltages used on the electrodes are very different, as can be seen from the highly positive grid. In order to prevent burnout or rapid deterioration of the incandescent heater, or cathode, current is not applied to the heater until the bulb is evacuated to a very low pressure, of the order of a few micrometres of mercury at most.

With the heater energized to incandescence, electrons are emitted and then are attracted to the positively charged grid. Some of them are captured by the grid, but many of them pass between its broad spirals and travel on to the plate, which is negatively charged with respect to the grid. The electrons are then driven back toward the grid, and many of them strike molecules of residual gas in their travels, dislodging electrons and forming positive ions that are attracted to the plate. The amount of current flow in the plate circuit is proportional to the number of ions formed; if the electron emission from the cathode is closely controlled, the number of ions formed depends on the quantity of residual gas.

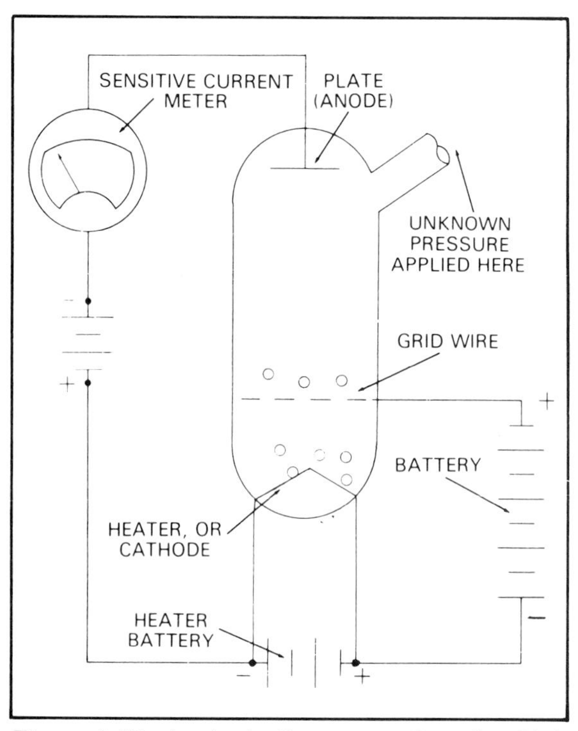

Figure 5.17. An ionization gauge for ultra-high vacuum measurement

Measurement of plate current flow cannot be accomplished as simply as the figure implies, since the value of this current is apt to be a fraction of a microampere. Elaborate forms of meters using solid-state amplification are needed to accurately display the feeble current flow.

Several other ionization-type pressure gauges are used to measure ultra-high vacuum. One type uses a cold cathode and very high plate, or anode, voltage. Another form uses a radioactive substance that emits alpha particles to bombard and ionize the residual gas. The principle of operation is basically the same in each instance, but each form has some advantage that makes it suitable for a particular application.

Although the principle of ionization gauges is based on simple logic, the problem encountered in the design, calibration, and use of a worthy instrument, including the accessory equipment needed to maintain and measure proper currents and voltages, is by no means resolved by simple expedients. Its complex nature prohibits an extensive treatment here.

Purposes of Pressure Control

Pressure has many sources and is controlled for varied purposes. Some important sources of pressure are natural—for example, the pressure existing at a wellhead in an oil field or the hydrostatic pressure of a liquid caused by its relative height. Of course, pressures are created artificially by compressors, other pumps, or boilers for one use or another. To be useful, however, pressure must be *controlled.*

Sometimes pressure is controlled for safety purposes. For example, steam boilers are equipped with relief valves to prevent excessive pressure that might destroy the boilers and cause personal injury. Petroleum stock tanks are fitted with relief valves to allow release of vapor pressures that could seriously distort or even rupture the tanks.

Frequently, pressure is regulated to certain specific values because pressure-operated devices, such as pneumatic controllers and valve actuators, operate most efficiently and effectively when the pressure supplied to them is within design limits.

Often, in controlled processes or systems, pressure must be accurately controlled in order to produce a desired effect. For example, the decomposition of raw material supplied to a process might be determined very narrowly by the combined effect of temperature and pressure to which it is exposed.

Thus, pressure must be controlled for safety, for efficiency of operation, and for the physical and chemical processing of materials.

Pressure Control Devices

The control of pressure can be achieved in a number of ways, but each is involved rather closely with controlling the *flow of something.* If pressure is produced by artificial means, as with a boiler or air compressor, it can be controlled by regulating the rate of pressure production. Thus, the pressure in a boiler is controlled by adjusting the flow of fuel oil to its furnace, and the pressure of a compressed air system might be controlled by an on-off pressure-operated switch that starts and stops a motor-driven compressor.

The pressure in some pipe systems is often controlled by establishing a hydrostatic head in an open or vented vessel (fig. 5.18). With inflow and outflow adjusted to maintain a fixed liquid level in the tank, a particular pressure will exist at the pressure tap. If the inflow is increased, the liquid level and pressure will rise, while the reverse is true if the inflow is reduced. The liquid level and

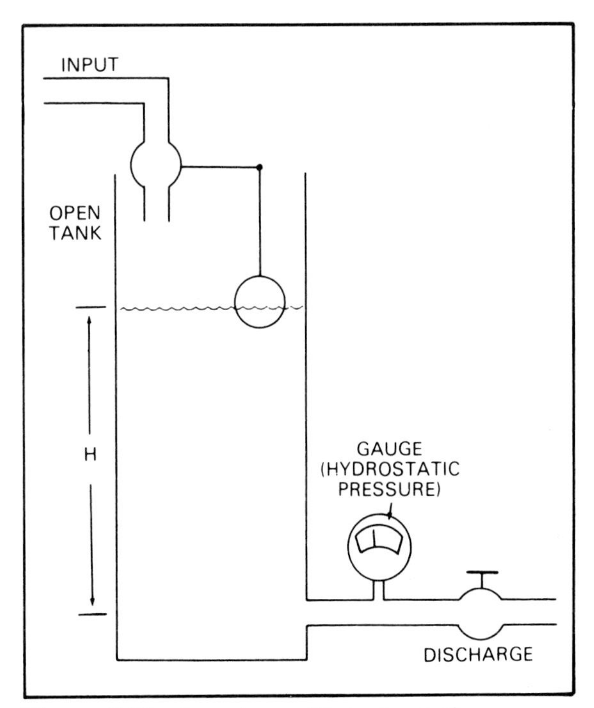

Figure 5.18. A system with hydrostatic pressure control

pressure can also be manipulated by varying the rate of outflow. These methods of control achieve their effect by varying the height of liquid and thus its hydrostatic pressure. Such methods are of practical value.

When the pressure in a line must be greater than the hydrostatic head can provide, a closed vessel (fig. 5.19) can be used. Now, by regulating the gas or air pressure above the liquid in the vessel, the outflow pressure can be made equal to the hydrostatic pressure plus the pressure existing on the surface of the liquid.

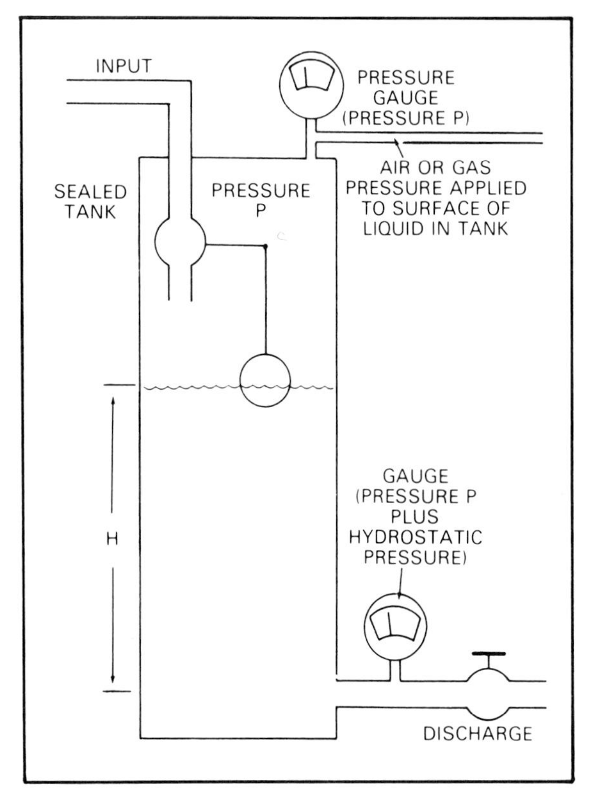

Figure 5.19. A system with air or gas pressure control

The bulk of concern with pressure control is a matter of regulating *reduced* pressures, for in most cases, the situation will be one in which a control valve or pressure regulator marks a dividing point between a certain pressure source and a lower pressure whose value must be controlled. In common practice, the high-pressure side of control is called *upstream* and the low-pressure side, *downstream.*

Although the great majority of pressure control arrangements are directed to the regulation of the low-pressure side, there are important exceptions. Safety valves for boilers and vapor relief valves for petroleum stock tanks are obvious examples, since their usual task is to prevent the buildup of dangerous pressures on the *high-pressure* side, while they also act as a seal against loss of steam or valuable vapors at pressures below the danger point. Another exception is the regulator that maintains fixed pressure in an oil and gas separator by releasing fluid into a line on the low-pressure side. Controllers such as these are commonly called *back-pressure regulators.*

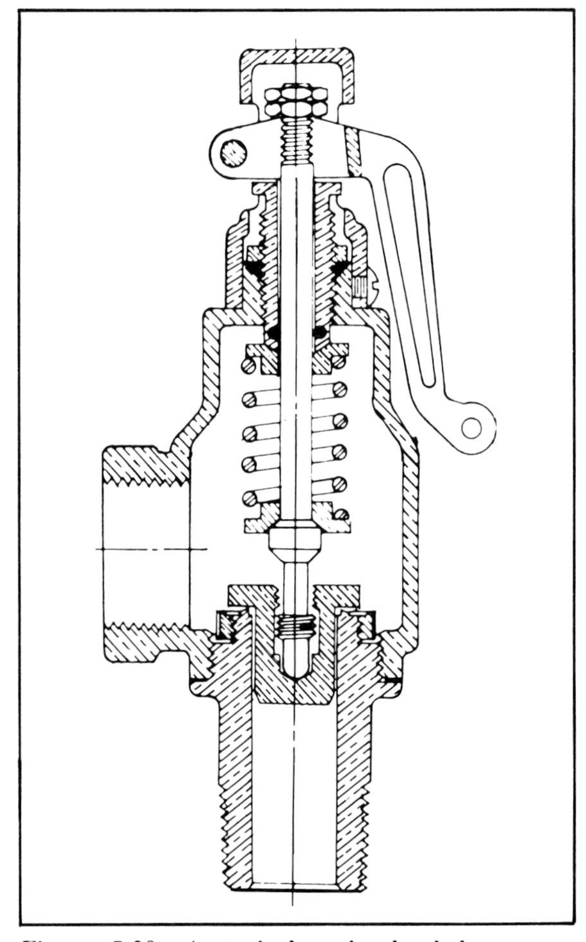

Figure 5.20. A typical spring-loaded pressure relief valve

Pressure Relief Valves

Pressure relief valves are sometimes appropriately called safety valves, because their sole purpose is to prevent damage to equipment and possible injury to personnel. The most common form of relief valve is one designed to release pressures above a set value, which can be varied by adjusting the loading on the valve stem. A form of spring-loaded relief valve (fig. 5.20) is commonly used to protect against overpressure in hot-water and compressed air systems. Another form of pressure relief valve (fig. 5.21) is capable of protecting against both overpressures and underpressures. An example of its use is on a petroleum stock tank whose construction is of a nature that will not permit wide excursions in pressure. An oil tank should be kept under some slight pressure in order to prevent the loss of volatile compounds through vaporization, so a relief valve capable of venting the tank before serious conditions of pressure arise is desirable, rather than an unrestricted venting policy. A tank that has endured very high temperatures during the day may cool off sufficiently at night to cause enough condensation to reduce the pressure to a dangerously low level. Unless this partial vacuum is relieved, the tank may collapse.

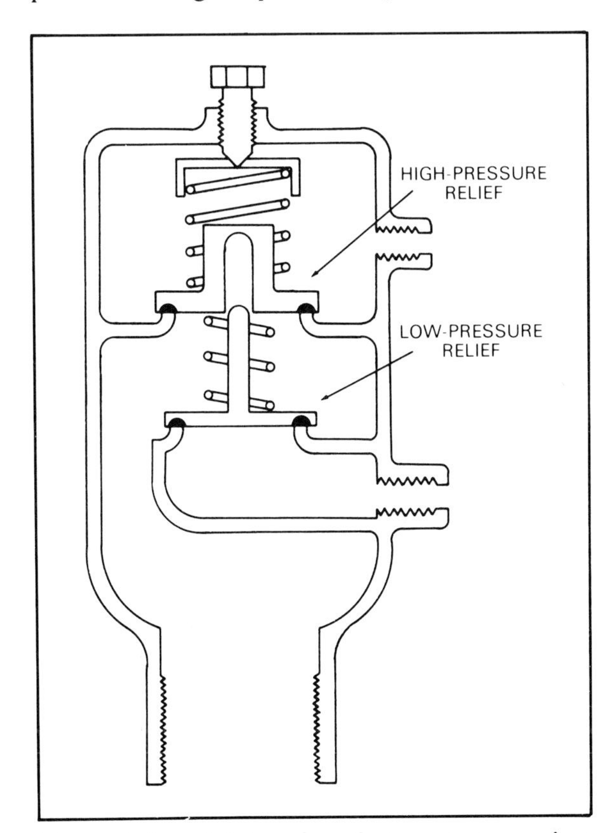

Figure 5.21. Relief valve that protects against overpressures and underpressures

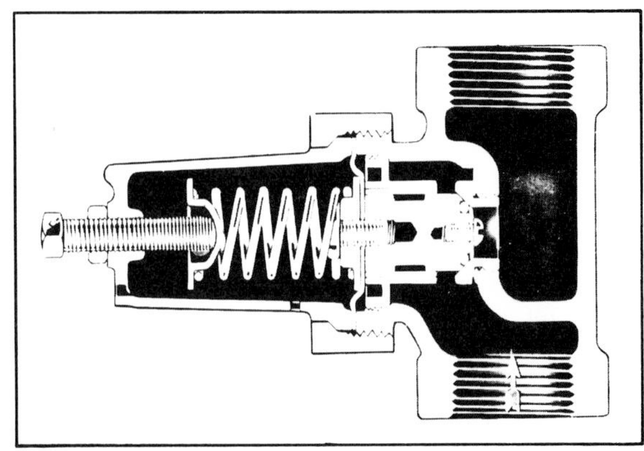

Figure 5.22. A pressure relief valve (*Courtesy Fisher Controls*)

A third form of pressure relief valve (fig. 5.22) serves to protect a second-stage regulator, meter, and any additional attachments in event of failure in a first-stage regulator (fig. 5.23). A typical installation is

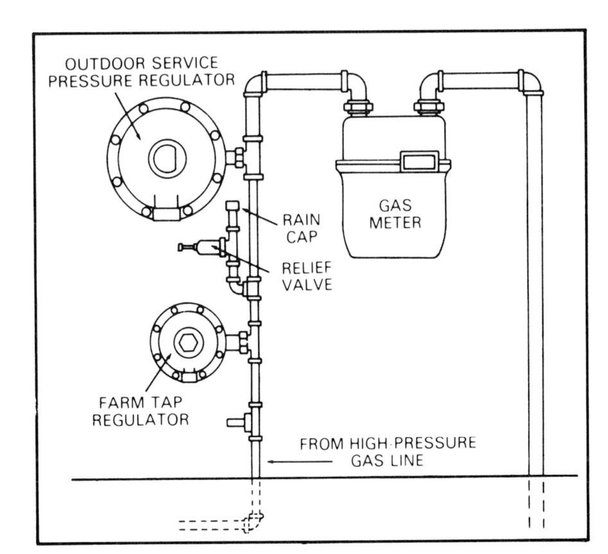

Figure 5.23. Location of instruments in a rural gas line (*Courtesy Fisher Controls*)

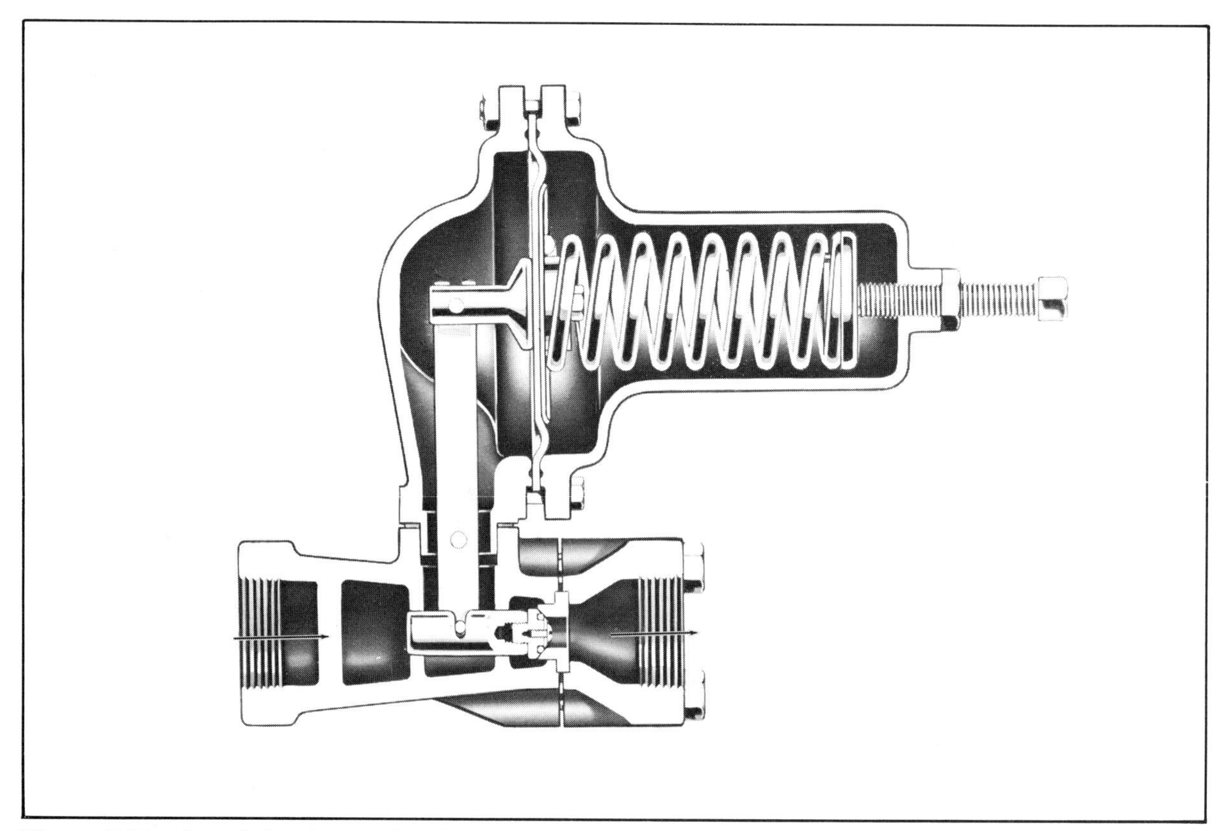

Figure 5.24. A relief valve used with a pump (*Courtesy Fisher Controls*)

in a rural gas line between the first and second stages of pressure regulation. It is easy to imagine the magnitude of the hazard that would be created by allowing gas to enter the piping and appliances of a home at 700 psi. Urban gas lines, of course, do not carry this great pressure.

Another type of relief valve is one designed to operate between the suction and discharge ports of a pump (fig. 5.24). Pressure applied to the inlet side of the valve encounters the normally closed plug and the diaphragm whose motion actuates a pivoted lever attached to the valve plug. Motion of the diaphragm caused by pump discharge pressure is opposed by the adjustable spring tension. Valves in this sort of service prevent the buildup of dangerously high pressures on the discharge side of pumps in the event that a valve in the line is inadvertently closed or some other condition prevents sufficient release of the pump discharge.

Pressure Regulators

The terms *pressure regulator* and *pressure controller* are not synonymous. Two very general characteristics distinguish one form of device from the other:

1. A regulator uses energy from the controlled medium to actuate the controlling means, while a controller probably uses energy of another source.
2. A regulator may combine functions of some of its elements so as to have a single element acting as a component of both measuring and controlling means, while such means of a controller are distinctive sections.

Regulators, particularly the more simple forms, are not usually considered capable of the fine degree of control that characterizes pressure controllers and are usually found in service that requires only reasonably good regulation and a limited variation of flow rates. Pressure controllers are generally associated with a need for close control, pressure recording facilities, transmission of measurement, or all of these. Controllers are frequently closely attended, while regulators are often in isolated locations that mean long periods of service without attention.

Simple regulators consist of three elements—measuring element, loading element, and a variable-area port—while more extensive designs may include pilot relays. Measuring elements are usually diaphragms acting against the load element, which can be a spring or a weight. The variable-area port may be any form of plug and seat but is usually a disc type with soft material—Neoprene or rubber—for positive shutoff. The simple units do not have large volume capacity since the variable port, when fully opened, is still quite small—from 0.125 to 0.625 inch in diameter.

As a fuel, natural gas can be used best at very low gauge pressures, but efficient transmission usually requires high pressures. Regulators, therefore, find wide use in providing the several stages of pressure reduction necessary between high-pressure pipeline and low-pressure fuel use. The gas pressures used for ordinary domestic service are quite low, being about 5 or 6 ounces per square inch.

Intermediate-pressure regulators. Regulators are usually constructed to serve a reasonably narrow range of pressure values, allowing the manufacture of a regulator that will provide close regulation within the design pressure values. Using a single regulator to produce an output pressure of 0.75 to 7 psi from a source of 700 psi will probably result in poor performance.

Thus, an intermediate-pressure regulator is used. Designed for pipe sizes of ¾ to 1½ inch, this regulator is capable of handling input pressures of up to 425 psi, and its output pressure may range from 5 to 100 psi, depending to a large degree on the choice of load spring and the value of the input pressure. The particular unit shown in figure 5.25 is called a *farm tap regulator* because of its frequent use in providing gas service to farm dwellings from relatively high-pressure lines. Its location in the system is shown in figure 5.23.

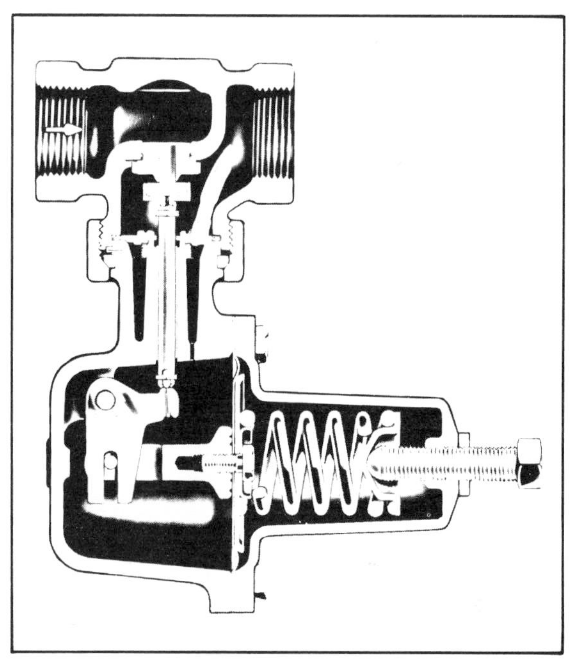

Figure 5.25. An intermediate-pressure regulator (*Courtesy Fisher Controls*)

The operation of this regulator is simple and similar to the unit studied in connection with pneumatic control. In addition, certain properties apply to most regulators of this type. Assume that the regulator is set to deliver 14.5 psi and that the setting was made with pressure on the system but with no flow. Once flow begins, a sharp drop in outlet pressure—perhaps to 12 or 13 psi—will occur, but as flow increases to higher values, outlet pressure will rise and possibly exceed the set value by a small amount. The pitot tube that extends into the outlet stream is responsible for the favorable part of this action, because it has a venturi effect for any appreciable flow. This tends to reduce pressure on the lower side of the diaphragm, causing the valve to open wider and raise the pressure slightly. The pressure drop, noted to be most significant just as flow begins, is caused by (1) an increase in effective area of the diaphragm as the valve opens and (2) the waning force of the spring as the valve opens. Coupling these facts together to achieve a lower balancing force and consequently a lower outlet pressure holds the valve in equilibrium.

Orifice sizes for this regulator range in diameter from 0.125 to 0.5 inch in increments of 0.125 inch. Five color-coded load springs provide the 5- to 100-psi range of outlet pressures. This regulator has no internal relief valve, and if arranged as shown in figure 5.23, a follow-up relief valve is indicated to protect against excessive pressure at the outlet. A low-pressure regulator includes an internal relief valve as an integral design feature.

Low-pressure regulators. The operation of a low-pressure gas regulator (fig. 5.26) is similar to the farm tap unit shown in figure 5.23 but incorporates an internal relief valve. The center of the diaphragm is connected to the linkage-shaft-pivot assembly that operates the disc valve plug. This center connection forms a relief valve that is normally held closed by the tension of the small inner spring. Under usual conditions, outlet pressure pushes up on the diaphragm and tends to close off flow through the regulator. However, if the disc valve fails to restrict flow sufficiently, because of improper seating, for example, pressure will continue to build up on the lower side of the diaphragm. Eventually the pressure on the diaphragm will cause the

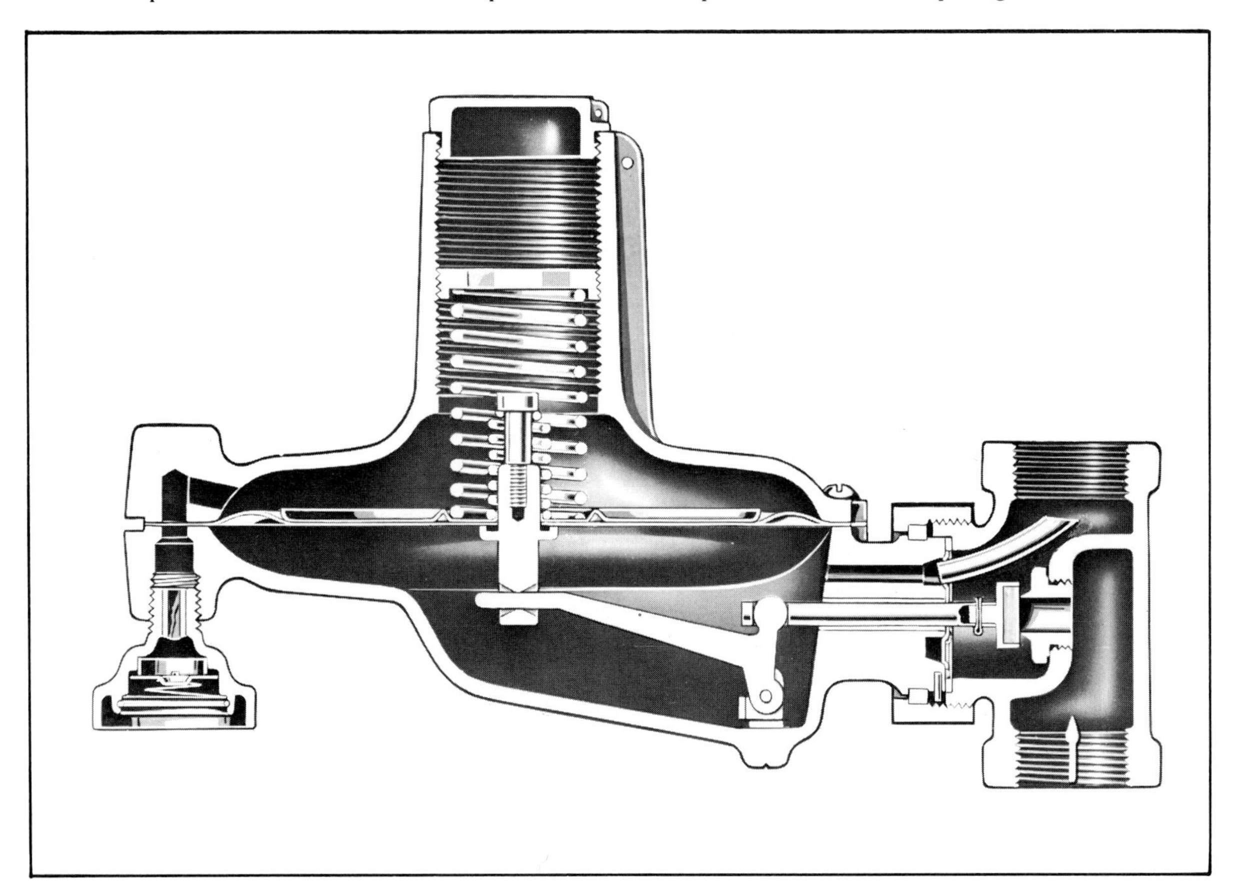

Figure 5.26. A low-pressure gas regulator (*Courtesy Fisher Controls*)

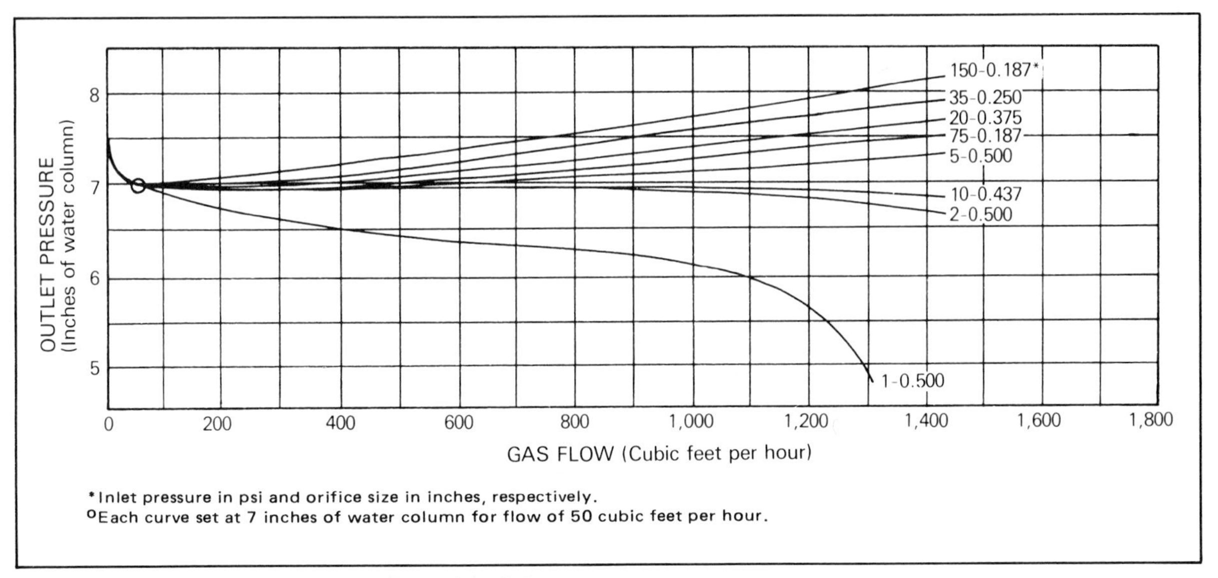

Figure 5.27. Performance curves for a 1-inch low-pressure regulator. Data are for flowing gas of 0.6 specific gravity, 14.7 psia pressure, and temperature of 60°F. (*Courtesy Fisher Controls*)

inner spring to be compressed, releasing gas into the chamber above the diaphragm and out the screened, bug-proof vent. The relief valve will open when the outlet pressure exceeds the desired setting by about 0.5 psi.

This low-pressure regulator is available in pipe sizes of 0.75 to 1.25 inch. At least six orifices are available in sizes from 0.125 to 0.5 inch, and inlet pressures range from about 0.3 to 150 psi. Standard springs and orifices permit a range of outlet pressures from 0.1 to 1 psi, and a special model with high-pressure diaphragm assembly will permit outlet pressures to about 7 psi.

The performance curves for the low-pressure regulator (fig. 5.27) are a good indication of its flow volume and regulation capability. Each curve is associated with a particular inlet pressure and orifice size; for example, 150–0.187 represents an inlet pressure of 150 psig and an orifice diameter of 0.187, or 3⁄16, inch. A point of interest is that the load spring in each instance was set to deliver an outlet pressure of 7 inches of water column for a flow of 50 cubic feet of gas per hour. Note the sharp rise to 7½ inches of water column at flow rates below this value. The performance of this regulator from shutoff to maximum flow is remarkably good.

High-pressure regulators. Regulators of large capacity and simple operation are also available, but they differ in minor aspects from the simpler types with limited capacity. Some of these large units are weight-loaded, and most of them use valve plugs and seats that resemble those found in control valves.

The regulators described so far possess diaphragms that are subjected to spring loading and atmospheric pressure on one side and downstream, or controlled, pressure on the other side. A disadvantage of such an arrangement is that, as the desired outlet pressure increases, the diaphragm must be made heavier, smaller, or both, in order to cope with the overall force that increases as the square of the diaphragm radius for a given pressure. Making the diaphragm heavier or making it smaller adds to its stiffness and lessens its sensitivity. A method that provides for a small difference of pressure on the two sides of the diaphragm will permit the use of

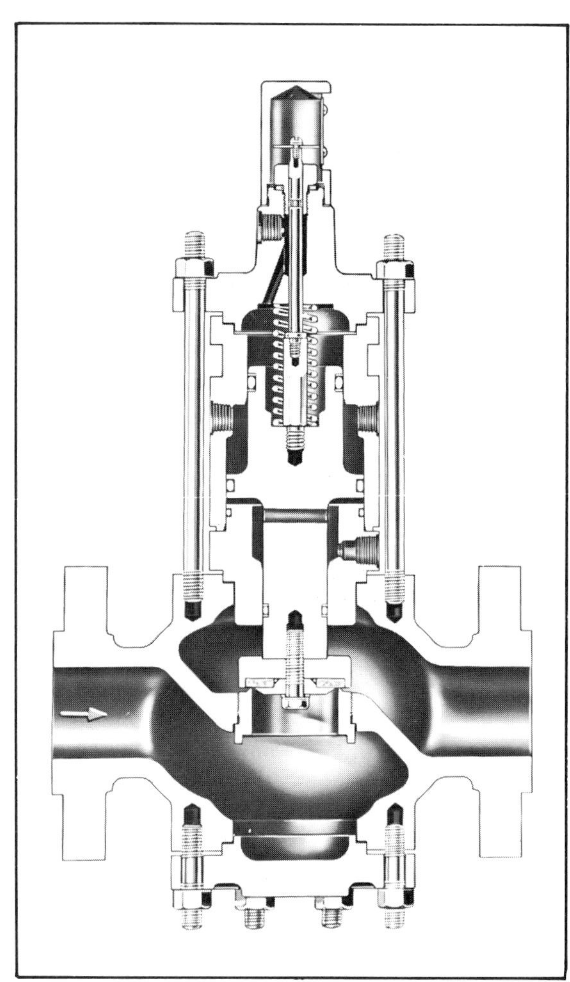

Figure 5.28. A high-pressure regulator (*Courtesy Fisher Controls*)

broad and light diaphragms having good sensitivity.

One type of high-pressure regulator is pilot-operated and piston-actuated (fig. 5.28). It differs in several important respects from the types described so far. Although a high-pressure regulator, it uses a single-seated valve. The force required to position the valve plug in the presence of high pressures is obtained from the use of an external pilot valve and a piston actuator. Piston actuators can be made very small and yet develop great force through the use of high pressures. On the other hand, the qualities that make diaphragms suitable positioning agents—broad and flexible for sensitivity and sufficient linear travel—cause them to be unsatisfactory for use where the differential pressures are quite large. This regulator, available in five pipe sizes from 1 to 6 inches, will accept inlet pressures to 1,000 psi and produce outlet pressures of 40 to 450 psi. The 6-inch size is capable of passing as much as 16 million cubic feet of gas per hour!

The regulator operates with a piston actuator and an external pilot valve as diagrammed in figure 5.29. The piston shown is a compound unit of several diameters and fitted with O-rings to provide good seals. Fluid pressure acts on two surfaces to drive the piston down and on one surface to drive it up. Inlet pressure on the valve plug also tends to force the piston up.

The upstream pressure is introduced to the top of the piston *D*, which, being larger in area than the port area *F*, balances the forces on the valve plug. The difference in areas between the piston *D* and port *F* also ensures tight shutoff of the regulator. The downstream pressure is brought in on top of the piston *E*, and the loading pressure is applied to the bottom of the piston *E* by the pilot. Assume that the outlet pressure is below

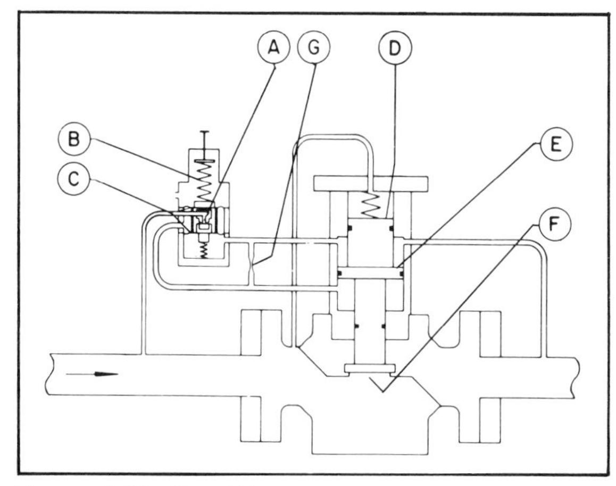

Figure 5.29. Simplified drawing showing sections of a high-pressure regulator and its external pilot valve. *A*, pilot inner valve; *B*, pilot valve spring; *C*, pilot diaphragm assembly; *D* and *E*, elements of the compound piston; *F*, port; *G*, orifice. (*Courtesy Fisher Controls*)

the setting of the pilot valve spring *B*. The bottom side of the pilot diaphragm assembly *C* has a lower pressure than the setting of spring *B*. Thus, the spring force will cause the pilot diaphragm assembly *C* to move downward, opening valve *A*, which supplies additional loading pressure to the bottom of the piston *E*. This creates additional pressure under piston *E*, forcing it upward and opening the main valve *F*. When the downstream gas demand has been satisfied, the reduced pressure tends to increase. This pressure increase acts on the bottom of the pilot diaphragm head assembly *C*, pushing it up and closing pilot inner valve *A*. With the pilot inner valve in the closed position, the loading pressure under the piston is reduced by bleeding downstream through orifice *G*.

Back-Pressure Regulators

So far the discussion of pressure control has involved the control of downstream pressure. Equally important is the problem of controlling upstream pressure by regulating the rate at which fluid is allowed to flow downstream. The control of upstream pressure is more commonly called *back-pressure control,* and controlling back-pressure involves maintaining a fixed pressure in a vessel of some sort—a fractionating tower or an oil and gas separator, for example.

Regulators not too different from those used for downstream pressure are used for back-pressure control. One regulator that is commonly found in this service is a weight-loaded, double-ported type (fig. 5.30).

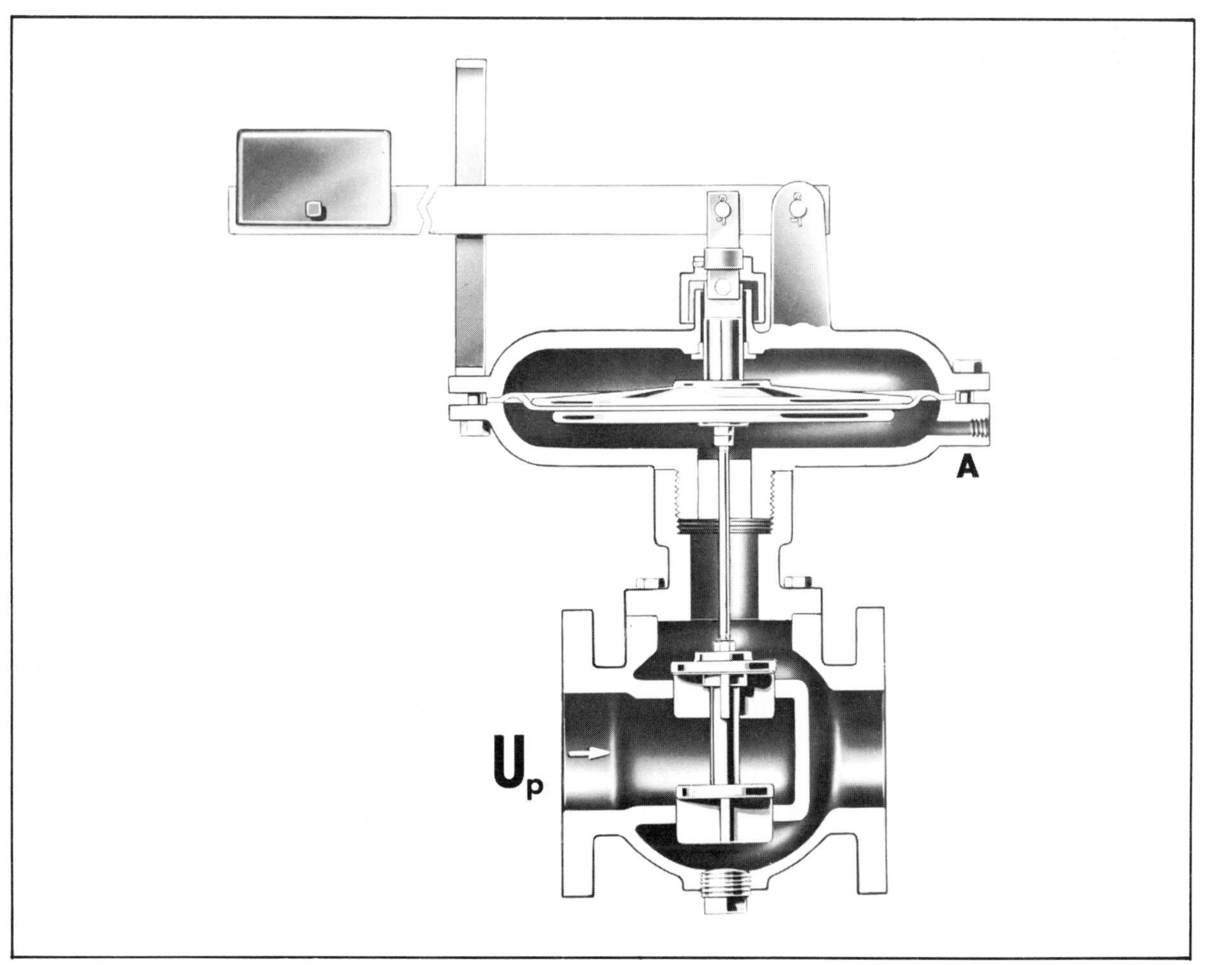

Figure 5.30. A weight-loaded back-pressure regulator

Weight loading has been losing popularity in recent years in favor of spring and pressure loading, probably because of these disadvantages:

1. Weights have considerable mass to be moved, and once in motion their momentum tends to make them overshoot. This action sometimes causes serious oscillations in the controlled variable.
2. Weight loading requires more space than other forms and also requires that the regulator be positioned accurately so that weights and levers move without pivotal binding.
3. Weights and levers are apt to be readily accessible to tampering, thus ruling out their use in some critical applications.

Weight loading, however, does have a few advantages:

1. Its design and manufacture are simple.
2. It can be adjusted easily for an almost unlimited range of loading forces.
3. Some designs provide facility for effectively reversing action of the weights on the valve plug.

The operation of the weight-loaded regulator shown in figure 5.30 is as simple as any considered thus far, but its double-ported valve and the form of plug used clearly indicate its ability to handle rather large volumes of fluid. The large lever carrying the weight is supported and pivoted at one point, while another pivot point is attached to a shaft that passes through the upper cover of the diaphragm case and is connected to diaphragm and valve stem.

When used for back-pressure control, this regulator has upstream pressure U_p applied to the lower diaphragm case through fitting *A*. This requires the installation of a short piece of small-diameter pipe or tubing. Some similar back-pressure regulators have built-in channels that introduce upstream pressure into the diaphragm case without external piping. When the back-pressure attains a value of such magnitude that its force on the diaphragm exceeds the opposing force of the weight loading, the valve plug will open. If the source of the back-pressure is capable of producing sustained flow, as from an oil and gas separator stage, the plug will soon settle at some reasonably stable opening, and the back-pressure will remain quite close to the desired value.

Pressure Controllers

In many ways regulators resemble some pneumatic-actuated valves; for example, regulators and control valves make similar use of spring loading and diaphragms. Another slightly different form of pressure control is the controller, which uses the conventional types of control valves and offers certain advantages over the regulator.

A characteristic that has come to be associated with regulators is that they use energy from the controlled medium to accomplish certain bits of work necessary to regulate pressure. This idea works fine for some applications, but many instances require other means to achieve the control needed. Consider the task of controlling the pressure in a pipe system that is carrying a fluid that is very viscous or contains even a small proportion of solid particles. Obviously, such fluids are not suitable for passing through small orifices or for actuating diagrams and pistons. Other valid reasons for choosing pressure *controllers* over pressure *regulators* are that controllers (1) allow rapid and easy change of pressure control set point; (2) are adaptable for remote indication, recording, and setting of pressure values; (3) offer a ready choice of on-off, proportional band, and reset and rate capabilities; and (4) provide superior performance for use in very high-pressure applications.

Commercial pneumatic controllers are capable of providing excellent pressure control when used with a control valve appropriately fitted to the control problem. A simple and effective pneumatic controller finds wide use in many applications ranging from on-off action with adjustable differential to proportional action with band adjustment from about 0 to 50%. Bourdon tubes of special design are used for pressures up to 7,500 pounds per square inch gauge, while bellows elements are available for low-pressure control work.

Proportional band controller. One form of commercial pressure controller is the proportional band controller that is used to control downstream pressure (figs. 5.31 and 5.32). Similar to a simple pneumatic controller, this device has no air relay; however, the valve actuator operates fast enough for most situations. Because the controller is located on the valve assembly, the volume and retarding action of lengthy pneumatic lines is eliminated. The flapper is attached to the free end of the Bourdon tube and rides against an adjustable fulcrum pin, which is the proportional band adjustment. The lower part of the flapper is positioned over the nozzle, and a light spring holds the flapper snugly against the proportional band adjusting pin. The tension of this

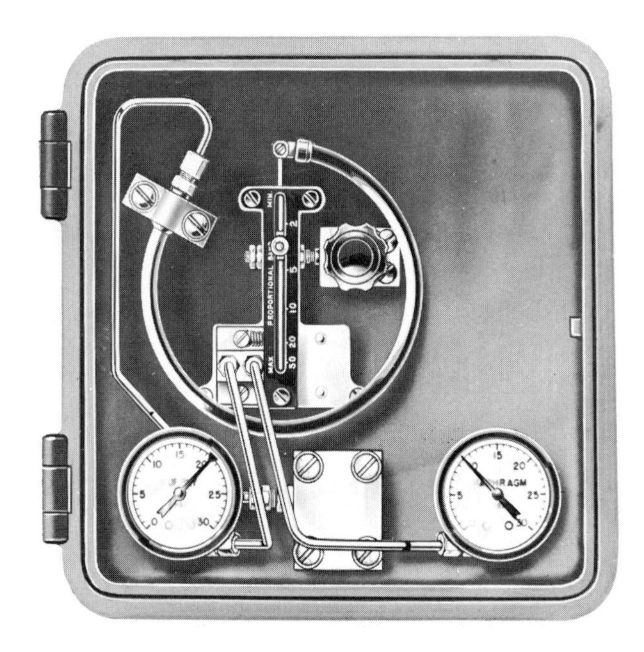

Figure 5.31. A proportional band pressure controller (*Courtesy Fisher Controls*)

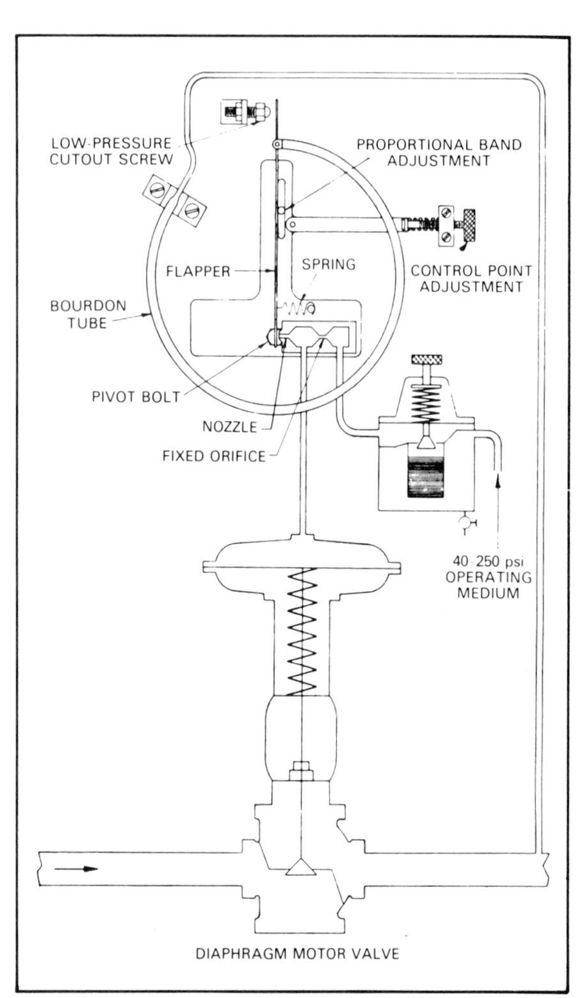

Figure 5.32. Diagram of a proportional band pressure controller

spring prevents any backlash that might arise from a loose fit between flapper and Bourdon tube. The control point adjustment moves the base plate assembly, including the flapper. Since the upper end of the flapper is fixed to the Bourdon tube, which does not move during this adjustment, the flapper is forced to pivot about the proportional band adjustment and thus change flapper-nozzle clearance. This change in clearance effectively alters the control point setting.

Snap-action controller. A variation of the proportional band controller is the snap-action controller, which uses a small permanent magnet in place of the tension spring

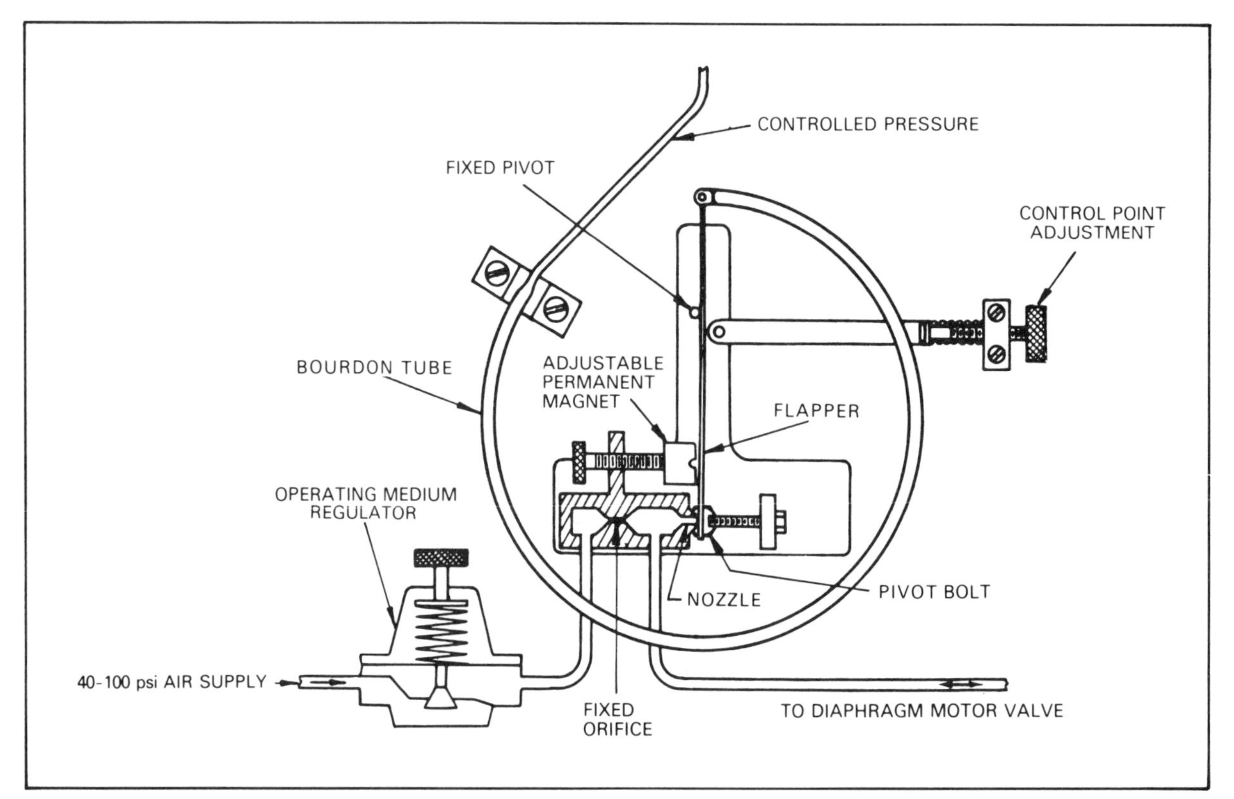

Figure 5.33. Snap-action operation is obtained by installing a magnet. (*Courtesy Fisher Controls*)

(fig. 5.33). The position of the magnet with respect to the flapper can be varied by an adjusting screw. The action of this form of the controller is strictly on-off, and the fulcrum pin that was the proportional band adjustment becomes a fixed pivot. The operation is such that as the flapper moves toward the nozzle and the face of the magnet, a critical point is reached at which the magnet quickly snaps the spring-steel flapper tightly against the nozzle. When conditions change and the Bourdon tube, which tends to position the flapper away from the nozzle, attains a certain value, the attractive force between flapper and magnet will be overcome, and the flapper will be quickly snapped away from the nozzle.

One use of the snap-action controller is to control flow from a pressure-flowing oilwell (fig. 5.34). Fluid pressure is built up in the sealed casing by gas escaping from petroleum in the well, and this pressure forces petroleum up and out of the tubing that extends down into the liquid. The purpose of the snap-action controller is to allow liquid outflow only for a certain range of casing pressure values. Note that gas pressure from the casing is filtered, reduced in pressure, and used in lieu of an air supply for the controller and actuator. Casing pressure is also applied to the Bourdon tube, and below a certain value of casing pressure the control valve is tightly closed. As pressure mounts, the snap action of the controller opens the valve wide, allowing well fluid to flow to the oil and gas separator. Casing pressure will decline as the fluid is expelled out the tubing, and liquid will probably fall below the lower end of the tubing, in which case the casing pressure will drop rapidly because of gas passing up the tubing. The controller will shut off flow, and the cycle will be repeated.

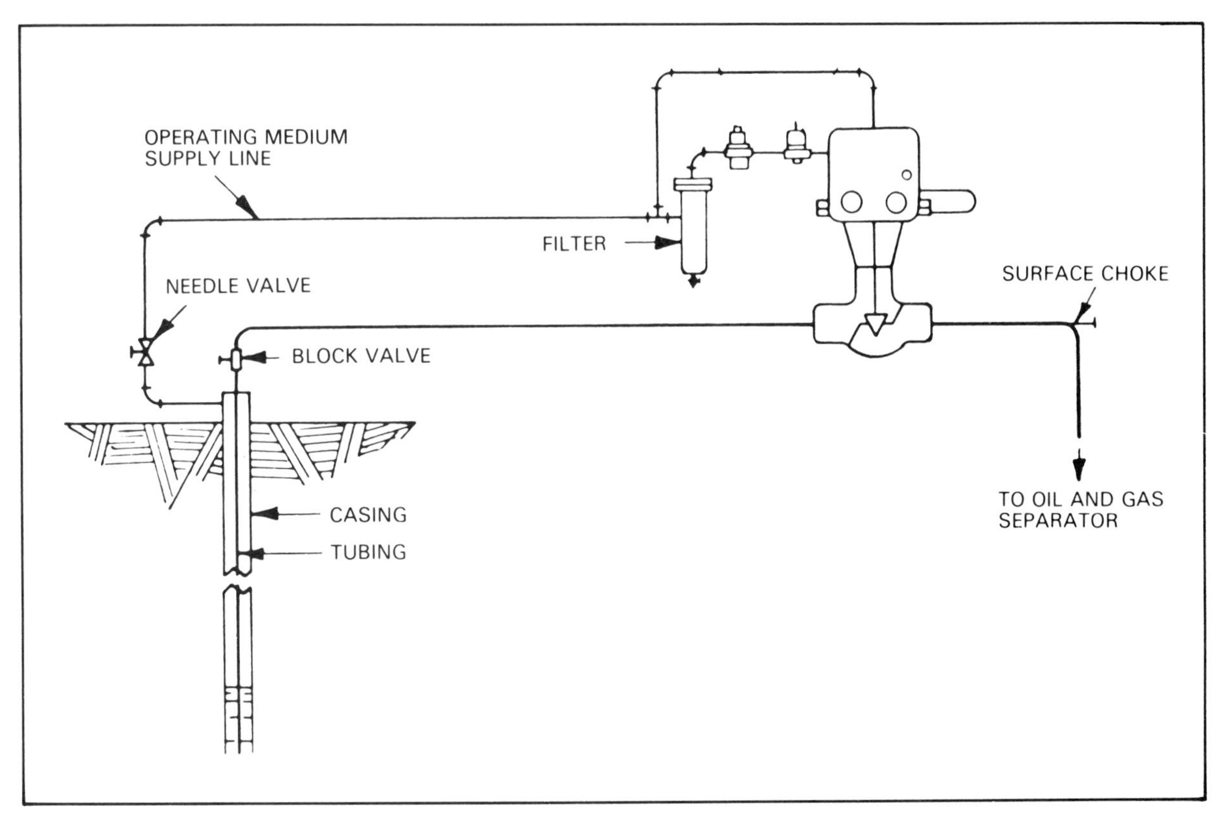

Figure 5.34. A snap-action controller used to control flow from a pressure-flowing well (*Courtesy Fisher Controls*)

The control system just described poses an interesting question. What is the controlled variable? Although not immediately obvious, it is the casing pressure and not the flow of petroleum to the separators. The latter is the control agent. More obvious is the fact that a pressure *regulator* could not accomplish this *stop-cocking* control, as this form of service is called.

Snap-action, on-off control is also used in some high-pressure, high-capacity gas line installations. Controller action is a function of downstream pressure that is maintained between maximum and minimum values determined by the differential setting of the controller. The adjacent downstream pressure in such a system is apt to be of almost sawtooth form, the value rising and falling in steep straight lines. If large volume and great distance are characteristics of the line, the abruptness of the pressure variations will be well dampened at great distances downstream. The advantages of on-off control as opposed to throttling action is unmistakable when used in gas lines that contain critical amounts of water vapor. A throttling valve will normally represent a flow point that divides two considerably different pressures. As the high-pressure gas passes through the throttling valve it expands rapidly and loses temperature. For certain conditions the loss in temperature is sufficient to cause formation of ice in the throttling valve. Since the on-off action permits either no-flow or full-flow, the tendency for ice to form at the valve is minimal. The disadvantage of on-off action in any high-pressure service is the severe shock to which valve and line are subjected when the momentum of the flowing gas must be absorbed so rapidly.

In these applications of the snap-action controller, the fluid pressure available from the controlled medium could have been used as a matter of convenience. A suitable filter and pressure reducer are inexpensive means of providing the actuating energy when compressed air sources are not conveniently available.

Differential pressure controller. One type of differential pressure controller is a pneumatic controller that contains a measuring means capable of responding to the difference of two pressure sources, neither of which must necessarily be atmospheric pressure. This type of controller can be used to maintain a differential pressure value, for example, in a fractionating column whose purpose is the separation of volatiles from petroleum (fig. 5.35). In this column a particular and constant rate of flow of vapor from bottom to top must be maintained, and a good way to measure flow rate is to observe the pressure difference existing between the upper and lower sections of the column. The rate at which vapors rise in the column will be closely related to the amount of heat energy introduced into the bottom section, and this can be appropriately controlled by a narrow-band proportional controller. For proper separation of volatiles, the vapor flow rate and, therefore, the column differential pressure must be accurately controlled, although control of other variables can also be used to achieve separation.

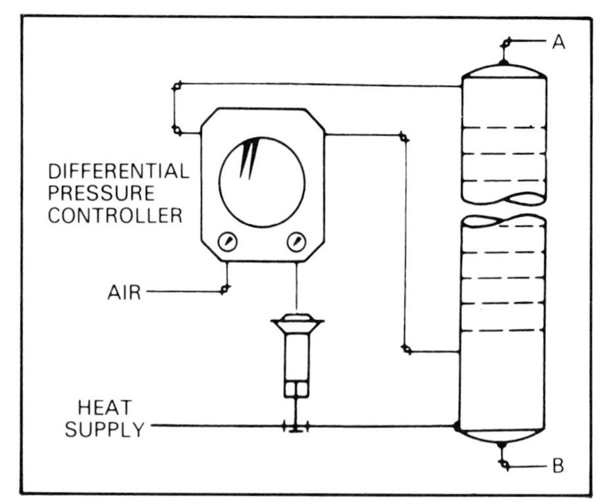

Figure 5.35. One use of a differential pressure controller (*Courtesy Taylor Instrument*)

Again, a puzzling question is raised: "What is really being controlled in this instance?" The differential pressure in the column is being controlled because this is the only variable to which the controller will respond, but differential pressure is merely used to indicate vapor flow. Vapor flow is caused by the heat supplied to the column, and the heat supply is what must ultimately be controlled.

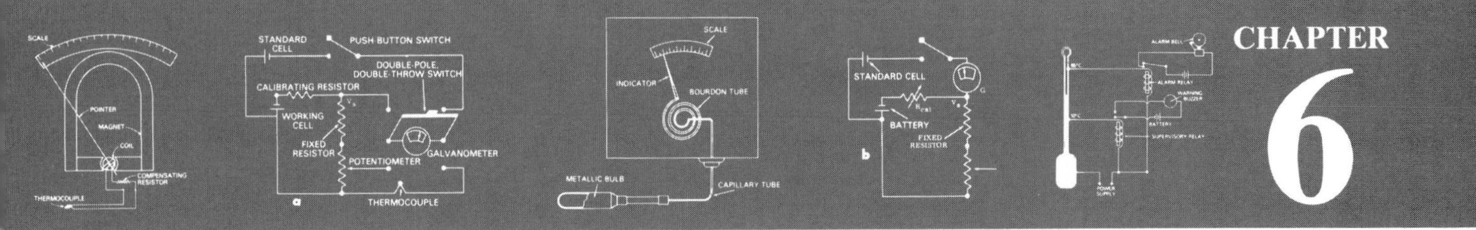

Temperature Measurement and Control

Temperature is the most important variable encountered in automatic control. It is also the only important variable whose quantitative value cannot be readily determined by direct means. Temperature has a profound effect on almost any process. Its effect on personal comfort alone shows that it can bring about some spectacular events. Temperature frequently acts with other variables to produce effects that are interrelated. The physical laws that establish dependency between temperature and pressure and between temperature and volume are widely known. Less obvious is the relation between humidity and temperature, or the ability of air to contain moisture at different temperature levels.

Because temperature has such a pronounced and predictable effect on the matter with which it is associated, its measurement by inferential means is a relatively simple task, and the variety of effects achieved and used are considerable. Common means of temperature measurement include the expansion and contraction of metals (bimetal thermometers), the changes in volume and pressure of liquids and gases (filled-system thermometers), the change in electrical resistance and the thermocouple effect (electrical measuring devices), and the correlation of radiated energy or color and brightness with temperature (pyrometers) (fig. 6.1).

Each of these manifestations can be used in a variety of ways to indicate temperature. For example, the expansion of metal caused by temperature change can be used in two ways: (1) by correlating the linear expansion of a metal rod with a given temperature range or (2) by combining two metal strips with dissimilar expansion coefficients and noting free-end movement with change in temperature.

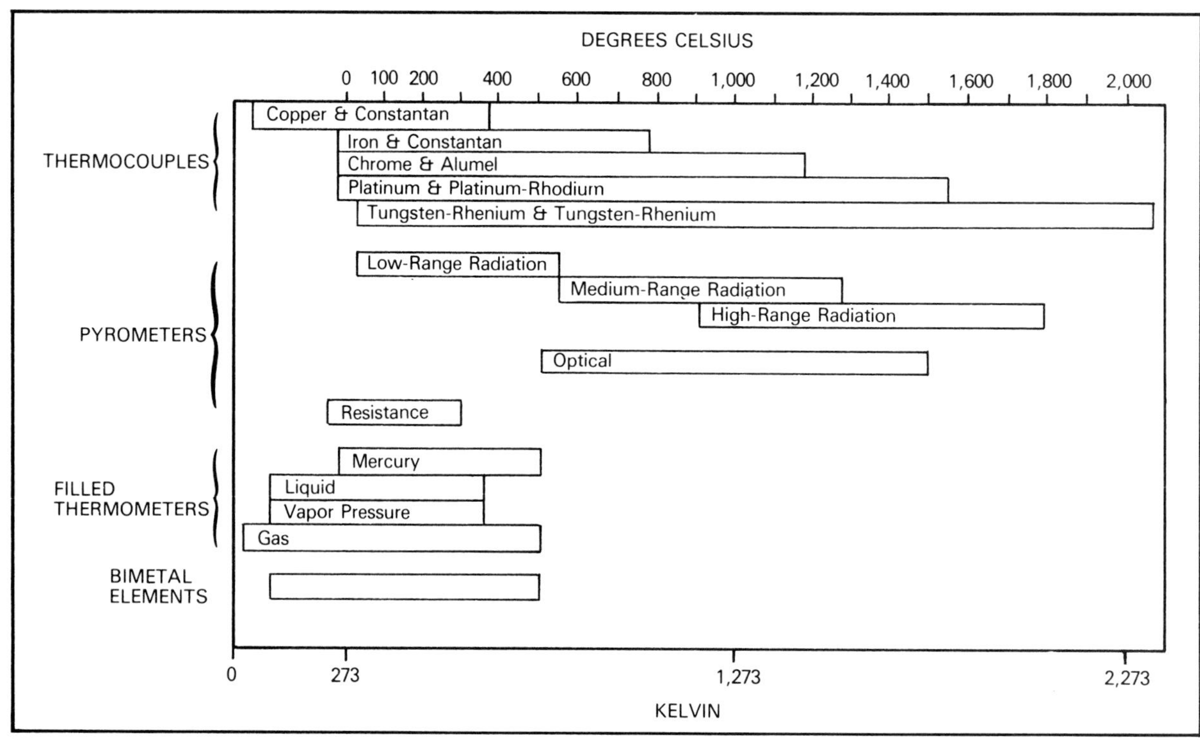

Figure 6.1 Types of temperature-measuring devices and their ranges

Temperature Scales

Temperature is the measure of molecular activity of a substance. The activity is a random motion of molecules, an irregular darting about of minute quantities of mass. Mass in motion possesses kinetic energy. As heat is added to a substance, the amplitude of motion of the molecules increases. This increased activity, representing the increased heat stored in the substance, is measured as temperature.

If all heat were removed from a substance, it is logical to suppose that the molecular activity in the substance would cease. If the substance has no heat, its temperature must have absolutely no value. This fact implies the existence of an *absolute zero temperature* and is a situation akin to the absolute pressure concept, for in each instance a theoretically attainable zero value is postulated from which there is only one way to go—*up,* or toward positive values. This theory is the basis for the absolute temperature scale—the kelvin (K). Also important to the scientific world is the Celsius scale (°C), formerly known as the centigrade scale.

The Fahrenheit scale (°F) and its absolute temperature conversion, the Rankine scale (°R), enjoy domestic and commercial use in English-speaking nations but are not commonly used in scientific work. (Note that the degree symbol ° is not written with K.) Figure 6.2 shows how the Celsius and Fahrenheit scales compare and how they relate to the absolute scales, kelvin and Rankine.

Celsius Scale

The Celsius scale, devised by Swedish astronomer Anders Celsius, is based on the freezing and boiling temperatures of water, with 0°C chosen for the freezing point and 100°C for the boiling point. These values were established a long time ago and represented at that time two criteria whose values could most easily be reproduced with acceptable accuracy.

Kelvin Scale

The kelvin scale, also known as the international thermodynamic scale, possesses at least one convenient relation to the Celisus scale: a 1-degree change in Celsius is equivalent to a change of 1 kelvin. The Celsius scale has its zero at 273.16 K.

Kelvin, named for its developer Lord Kelvin (William Thomson), is more representative as an *interval* of temperature than it is as an everyday scale of temperature. For example, a change from 40°C to 20°C results in a new temperature of 20°C. It is also a change of 20 K. However, a change from 303.15 K to 253.15 K equals 50 K, but not 50°C. The new Celsius temperature would be −20°C.

A considerable amount of scientific work performed prior to Lord Kelvin's time pointed to the existence of an absolute zero temperature somewhere near −273°C. Jacques Charles, a French scientist, carried out important experiments concerning the behavior of gases when subjected to various conditions of temperature and pressure. He noted, for example, that if a quantity of gas trapped in a tube was maintained at a constant pressure while the temperature was varied, the volume occupied by the gas also varied. In fact, he found that for constant pressure conditions the volume of the gas increased uniformly as

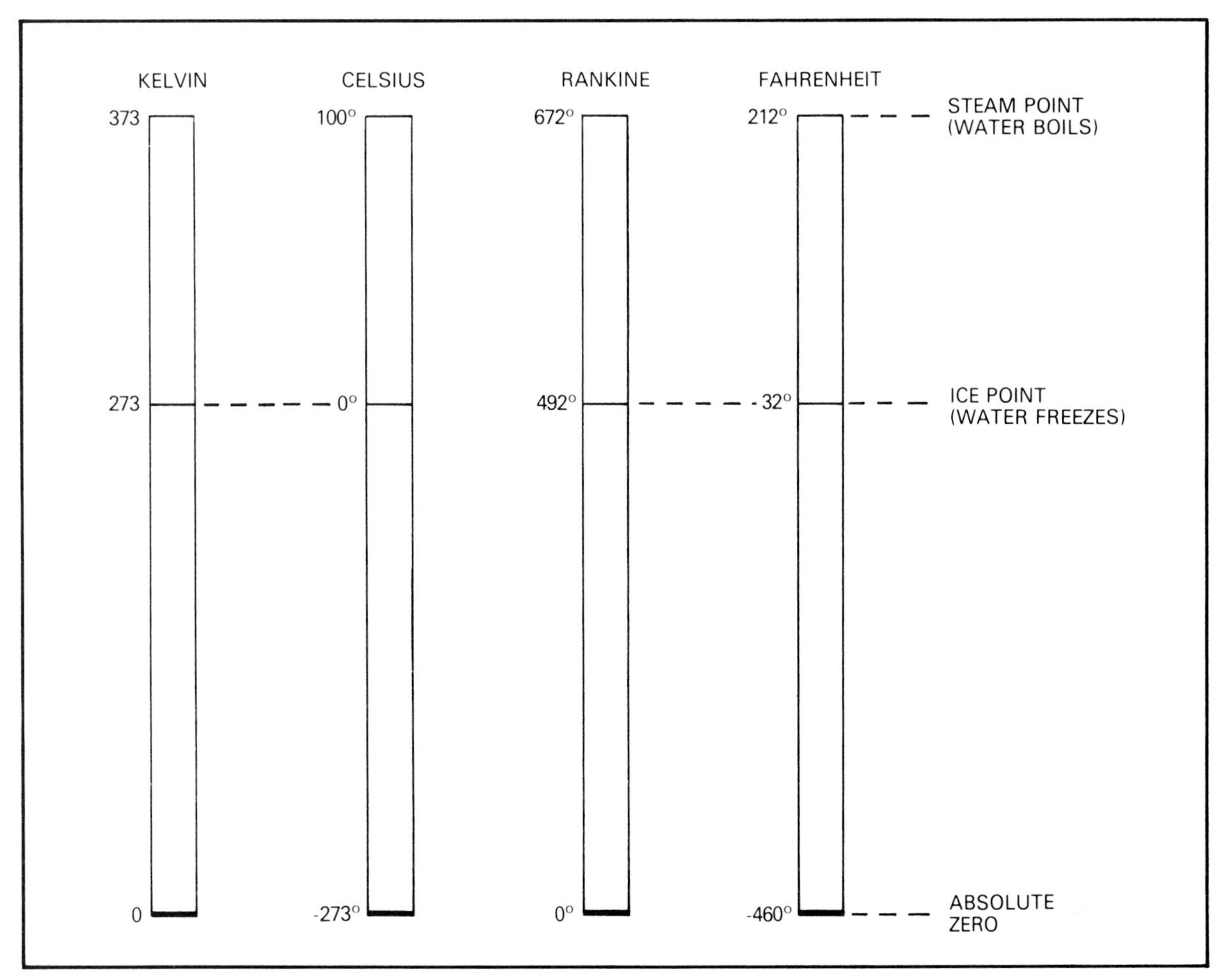

Figure 6.2. Comparison of four temperature scales

the temperature was increased from 0°C to 100°C, and at 100°C the volume was about $1\frac{100}{273}$ times that at 0°C. Since this was a uniform growth in volume and amounted to about $\frac{1}{273}$ of the volume at 0°C for each degree increase, it seemed logical to devise another temperature scale that enabled calculations based on the idea that the volume of gas is *directly* proportional to its temperature as measured on this new scale.

Obviously, a scale similar to the scale used by Charles, but having its zero mark at 273 degrees below the present zero, would accomplish the purpose. Ideas like this probably led Lord Kelvin to search deeply into the nature of heat and energy and to devise the absolute thermodynamic temperature scale as it is known today.

Liquid-in-Glass Thermometers

The clinical, or fever, thermometer, one of the very first pieces of apparatus the newborn baby becomes acquainted with, is probably the best-known liquid-in-glass thermometer. This and other liquid-in-glass thermometers operate on the principle that liquids expand and contract with changes in temperature; that is, they experience a volumetric change. It is this change in volume that facilitates the measurement of temperature with a glass thermometer.

Elements of Glass Thermometers

A liquid-filled glass thermometer usually consists of a glass bulb, a thin column in a glass stem, a liquid fill, and an indicating scale (fig. 6.3). Some glass thermometers also have contraction and expansion chambers and protective devices. The glass used is a careful mixture of silicon, borax, calcium, alumina, lead, and possibly other substances. The coefficient of expansion of such glass is only about 0.12 of that for mercury, one of the most common liquids used in glass thermometers.

Bulbs and columns. Various techniques of glass blowing and forming are used to provide a column having a bore that is only a small fraction of a millimetre in diameter. The special external shape of the stem acts as a lens that magnifies the image of the mercury or liquid in the column. The indicating scale is often acid-etched directly on the glass stem. The bulb of a thermometer can be formed by sealing and heating one end of a capillary tube. Careful blowing into the open end shapes the bulb to the desired form.

More expensive thermometers use a bulb prepared from a different glass composition and then fused to the column. Bulbs commonly have about one thousand times the volume of the column. Large-volume bulbs

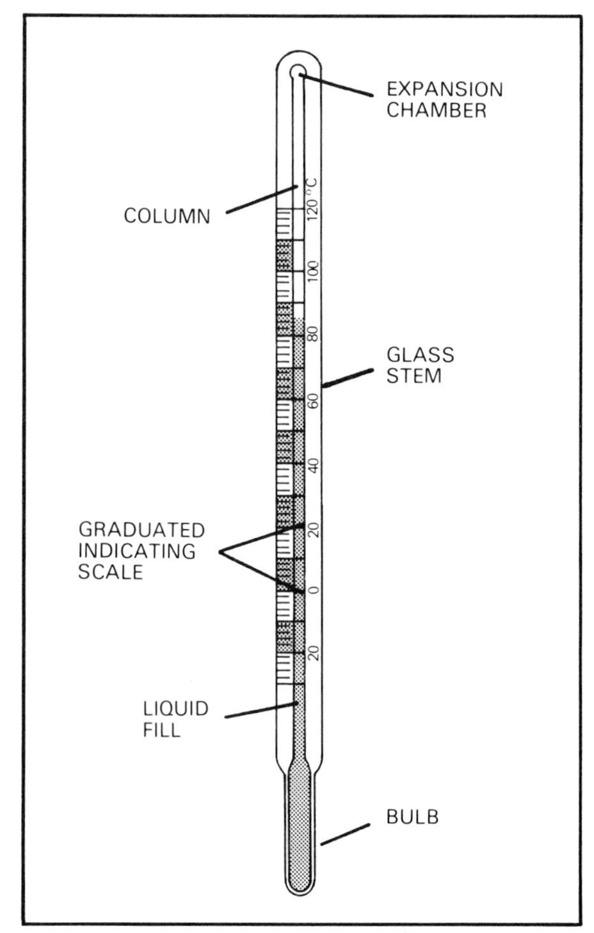

Figure 6.3. Simple liquid-in-glass thermometer

are used on thermometers having a narrow range of temperature, while small-volume bulbs are found on wide-range thermometers. The bulbs and columns of liquid-filled thermometers are filled by subjecting the units to a high vacuum, then allowing the desired liquid to be drawn into the column and bulb.

Fills. Liquids such as mercury, alcohol, toluene, and pentane are used as fills in glass thermometers. Mercury-filled thermometers can be made for measuring temperatures between −30°F and 925°F. The low-temperature measurement capability can be extended to approximately −75°F by use of a mercury-thallium filling. High-temperature mercury thermometers, capable of measurements to 1200°F, could be achieved by using an inert gas filling above the mercury—argon, nitrogen, or carbon dioxide, for example. However, this is *not* practical in a glass thermometer, although quartz is suitable for such temperature.

Pentane and alcohol are suitable fills for many low-temperature applications. Pentanes have the chemical formula C_5H_{12} and are paraffin-type hydrocarbons that boil at temperatures not much above ordinary room temperature—less than 100°F, for example. Pentane thermometers are available in ranges extending from −330°F to near 30°F. Alcohol can be used for measuring temperatures as low as −150°F and is commonly used in inexpensive thermometers with ranges suitable for casual weather observation. Liquids other than mercury are usually colored with a suitable dye to make them easily visible in the thermometer column.

Contraction and expansion chambers. Some thermometers also have a chamber just above the mercury bulb (fig. 6.4). The 32°F mark, sometimes called the ice point, is just below the enlarged cavity, while the column above the cavity forms a uniform scale ranging from 400°F to 600°F. The enlarged chamber separating the 32°F to 400°F marks has a volume equal to that occupied by 368 degrees of the thin-bore column. Because of its action, the enlarged cavity in the column is called a contraction chamber rather than an expansion chamber. Clearly the scale between 32°F and 400°F is contracted into a linear spread equal to the top-to-bottom length of the contraction chamber, while the range of 400°F to 600°F is spread over a distance that allows accurate reading of temperature. The 32°F mark in this instance provides a means of making a periodic check on the accuracy of the instrument, or at least a check of the relation between the volume of the bulb and the quantity of mercury it contains. Deviation from the ice point during this check would indicate that age had affected the accuracy of the instrument, probably by changing the effective volume of the bulb.

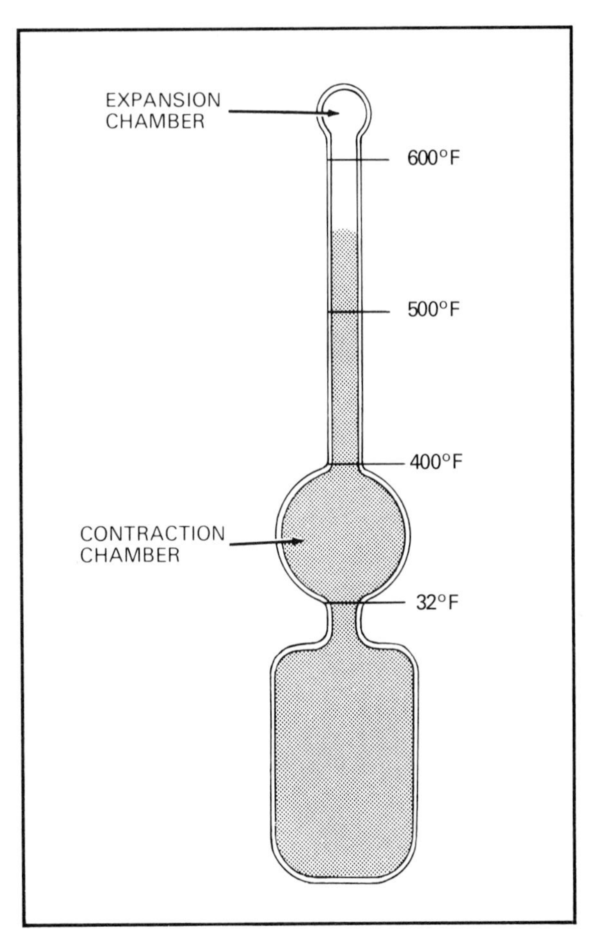

Figure 6.4. Thermometer with contraction and expansion chambers

Protective devices. Most glass thermometers are provided with some sort of protection against breakage, although such provision usually has an adverse effect on the response of the instrument. A pencil thermometer can be installed in a well, known as a *thermowell,* that is fitted securely into a pipe (fig. 6.5*A*). The pipe carries the fluid whose temperature is to be monitored. In this installation, the thermowell not only protects the thermometer against damage but also prevents the escape of fluid from the pipe. If the fluid is corrosive, care is exercised in the choice of material for the well. Other factors such as operating pressure and the need for rapid response by the thermometer also influence the design of these thermowells. Thermowells are sometimes filled with mercury, oil, or other stable liquids that permit a rapid transfer of heat from well wall to thermometer bulb. For those installations that will permit its use, a simple guard (fig. 6.5*B*) will protect the bulb against physical attack while allowing a free flow of fluid about the bulb. The response of the thermometer is not seriously affected by the guard.

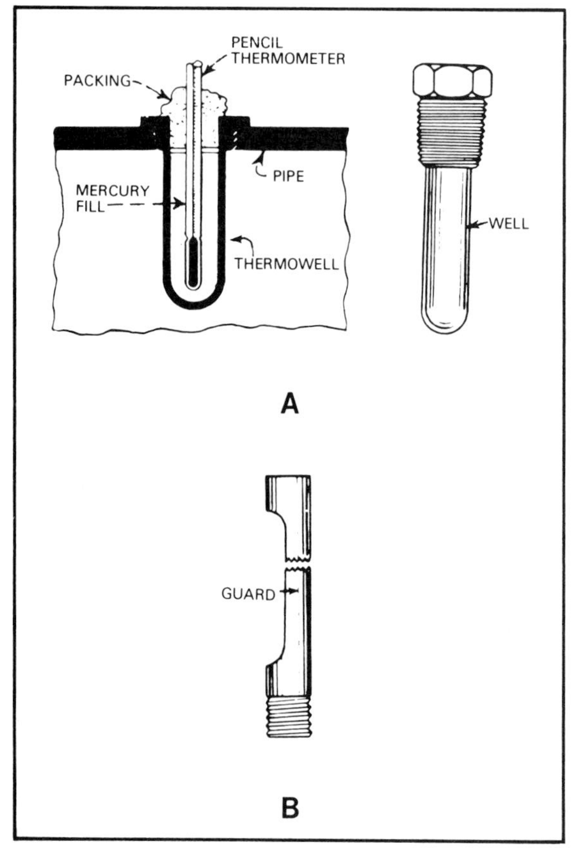

Figure 6.5. Protective devices used with glass thermometers. *A*, thermowell; *B*, guard.

Uses of Glass Thermometers

Very broadly, glass thermometers are used for measuring the temperature of liquids, vapors, and gases. They find wide and important use for manually controlled boilers, as medical and laboratory instruments, and as household tools but have limited application for automatic control systems. Except for a very small number of applications, glass-thermometer elements are limited to uses that are satisfied by visual monitoring of temperature. Glass thermometers are relatively immune to attack by corrosive fluids and are therefore of great value in experimental work involving strong acids and alkalis. Their degree of inertness and ease of cleaning are favorable points in avoiding contamination of the fluids.

Often, the construction of a unit is dictated by its use. For example, the two mercury-filled glass thermometers shown in figure 6.6 are basically similar but mounted differently. Thermometer *A* is a rugged instrument intended for permanent installation in a rough environment. Its mercury-filled bulb is protected against physical force and other attack by a thermowell, which forms a part of the mechanical structure of the thermometer. The threaded part of the well screws into a vessel or pipe. The thermometer column is protected by a heavy-duty glass and metal housing. The right-angle bend in the column permits the scale part of this thermometer to stand upright when installed in the side of a vessel. The scale is formed by a metal plate containing etched or embossed numerals and division marks. The straight pencil thermometer (*B*) is

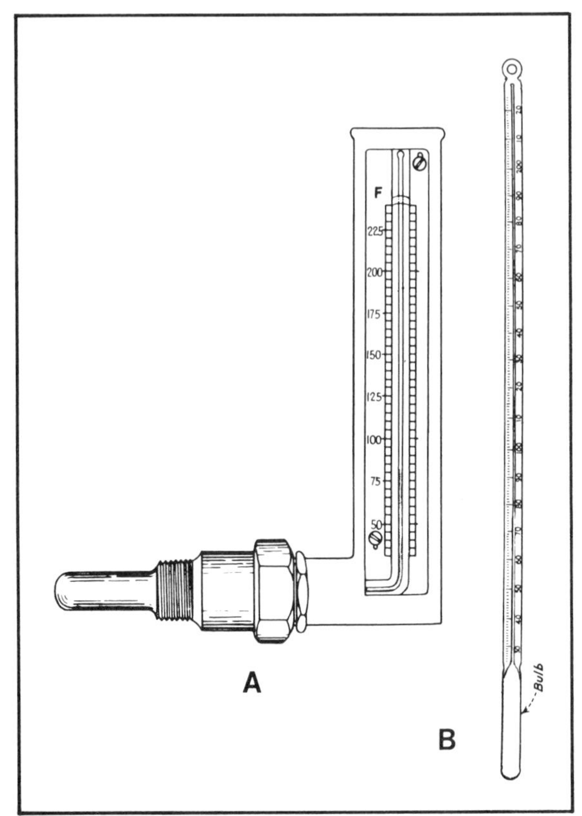

Figure 6.6. Two types of mercury-filled glass thermometers. *A,* mounted in thermowell for rugged use; *B,* fragile, pencil type for portable use.

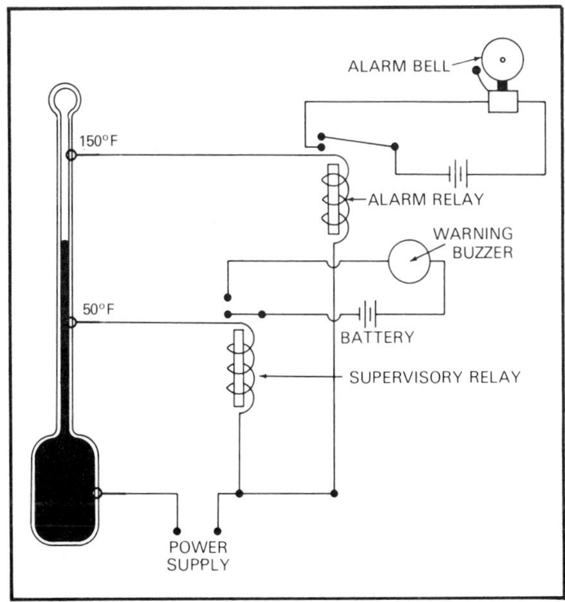

Figure 6.7. A mercury thermometer as a component of a temperature alarm system

a portable instrument of obviously fragile nature.

A form of mercury thermometer can be used as a component in a temperature alarm system (fig. 6.7). Used in this way, the mercury thermometer is probably more correctly termed a *thermostatic alarm element.* This device might be installed in remote locations and used to warn of fire or other unsatisfactory temperature conditions. The type shown contains two circuits, each of them using the mercury column to complete a part of its circuit. Electrical contact with the mercury is made by metal pins that pass through the glass of the bulb and column. The supervisory circuit is normally completed by the mercury that rises from the bulb to some point above 50°F. With current flowing in this circuit, the supervisory relay operates to open the warning buzzer circuit.

Should the power supply fail, the thermostatic element get broken, or the monitored temperature fall below 50°F, the supervisory relay will let the circuit to the buzzer be completed, and a warning will be sounded. In the event of fire or a dangerously high temperature, the mercury rises to the 150°F mark and causes the alarm bell to sound.

Thermostatic elements of this kind are available in an almost unlimited variety of ranges and for a multitude of purposes, although bimetal types of sensitive elements fitted with mercury switches are beginning to overshadow the use of thermometer-type elements for this purpose.

Filled-System Thermometers

Filled-system thermometers are simple in design and relatively inexpensive to manufacture. They may be simple direct-reading devices whose sensitive bulbs or other sensing elements form a rigid compact entity, or they may consist of compensating elements that offset effects of a varying ambient temperature. Filled-system thermometers are used in process control to provide remote indication and recording of temperatures.

Types of Filled Systems

Thermometers of the filled-system type are classified into four groups:

Class I—liquid-filled thermometers, except mercury units;
Class II—vapor-pressure thermometers;
Class III—gas-pressure thermometers; and
Class V—mercury-filled thermometers.

An earlier classification system listed mercury thermometers as Class IV, but the present designation is one adopted and recognized by the Scientific Apparatus Manufacturers Association. It should be noted that bimetal and electrical systems are excluded from this classification. Furthermore, the definitions accompanying the classifications exclude liquid-filled and mercury-filled *glass* thermometers.

Each of the classes is subdivided by affixing an alphabetical symbol to the Roman numeral. These symbols usually designate the sort of compensation used to bring about corrective effects to offset environmental conditions, or, in the case of vapor-pressure units, they may signify whether the sensitive bulb is to be used for measuring temperature above or below the ambient temperature.

Liquid- and mercury-filled thermometers. Classes I and V are *completely* filled systems that operate on the principle of volumetric expansion. This similarity between mercury-filled and other liquid-filled thermometers permits treating them as a single class although some significant differences exist. A Class I system may be filled with toluene, alcohol, pentane, or silicone fluids, while a Class V system uses mercury. A plain Class I liquid-filled system is an uncompensated instrument and has very limited use. Class IA is fully compensated, and Class IB has case compensation only. All Class V mercury systems are compensated, either fully (Class VA) or case compensated (Class VB).

Vapor pressure thermometers. Class II, or vapor-pressure, thermometers are *partly* filled systems that operate on a pressure principle. All Class II systems use a volatile filling such as methyl chloride, butane, propane, ether, or hexane. The choice of liquid will depend somewhat on the temperature range to be measured and the fact that the boiling point of the liquid must be lower than the lowest temperature to be measured. Methyl chloride is suitable for very low temperature measurements, for example, while ethyl alcohol can be used for high temperatures.

Vapor-pressure systems are subdivided into four classes: IIA, IIB, II C, and IID. Those in Class IIA are designed for measuring temperatures that are always higher than ambient temperature, while Class IIB is a low-temperature system, or one used for measuring temperatures whose values are always below ambient. Classes IIC and IID are capable of measuring temperatures at, above, and below ambient.

Gas-filled systems. Class III systems contain gas under pressure. The type of gas and its pressure are determined largely by the temperature range to be measured. Gas-filled systems are divided into subclasses IIIA and IIIB. Class IIIA is fully compensated, while IIIB has case compensation only. Gas systems are capable of measuring extremely low temperatures as well as reasonably high values. Nitrogen and helium are commonly used as gas fills.

Elements and Uses of Filled Systems

Filled systems consist of (1) a metallic bulb containing a fluid—liquid, gas, vapor, or mercury—whose volume or pressure responds to temperature changes; (2) a capillary tube that provides a means of transmission between the bulb and the indicating device; (3) an indicating device using a spiral Bourdon tube to drive a pointer or recording pen; (fig. 6.8) and sometimes (4) compensating elements to offset the effects of a varying ambient temperature.

Filled systems are trouble-free and accurate enough for most work, although the failure of any part of the sealed system necessitates the replacement of the Bourdon tube, capillary tube, bulb, and any compensating elements. Filled systems are designed as remote reading or recording devices, although systems having distances between indicator and bulb of more than 100 feet are apt to prove unsatisfactory for one reason or another. They exert enough force to operate not only indicators and records, but control apparatus as well.

Of the four types of filled systems, gas-filled thermometers have the greatest measurement capabilities (table 6.1). In addition to measuring extremely low temperatures and reasonably high temperatures, gas systems lead in response speed, followed by vapor-pressure, liquid, and mercury systems in that order. All filled systems are quite fast in responding to temperature change. The response of systems is determined by the filling fluid, capillary length, bulb size, and other characteristics, as well as the medium in which the bulb is located. If the capillary tubes are

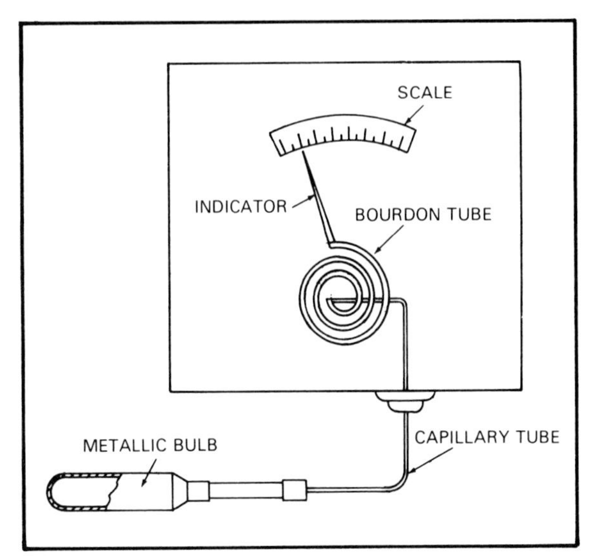

Figure 6.8. Filled-system thermometer with no compensating elements

TABLE 6.1
RELATIVE MERITS OF FILLED SYSTEMS

System	*Temp. Measuring Capability* *Minimum*	*Maximum*	*Advantages*	*Disadvantages*
Liquid, Class I	−300°F	700°F	Has linear scale, small bulb.	Requires case and capillary compensation; elevation error occurs.
Vapor Pressure, Class II	−300°F	660°F	Is low-cost; has fast response; does not require compensation.	Has nonlinear scale; elevation error occurs; erratic response is possible at ambient temperature.
Gas, Class III	−435°F	1,400°F	Has low- and high-temperature capabilities; has linear scale; withstands overranging.	Has large bulb.
Mercury, Class V	−60°F	1,200°F	Has linear scale, small bulb.	Requires case and capillary compensation; elevation error occurs; does not have low-temperature capability.

not excessively long, any of the filled systems will prove equal to other forms of measurement in speed.

Fills and bulbs. Class I liquid-filled systems are filled with toluene, alcohol, pentane, or silicone fluids. Class V systems have mercury fills. Class III gas-filled systems commonly use nitrogen and helium as fills. A basic requirement of all Class II vapor-pressure systems is that the volatile filling (methyl chloride, butane, propane, ether, or hexane) exists as a liquid at the coolest part of the system and as a vapor at the hottest part. During its normal use, a Class IIA system will be filled with a liquid that has a low vapor pressure at ambient temperature. It is possible for liquid and vapor to exist together at one end of the system, as can be seen in figure 6.9.

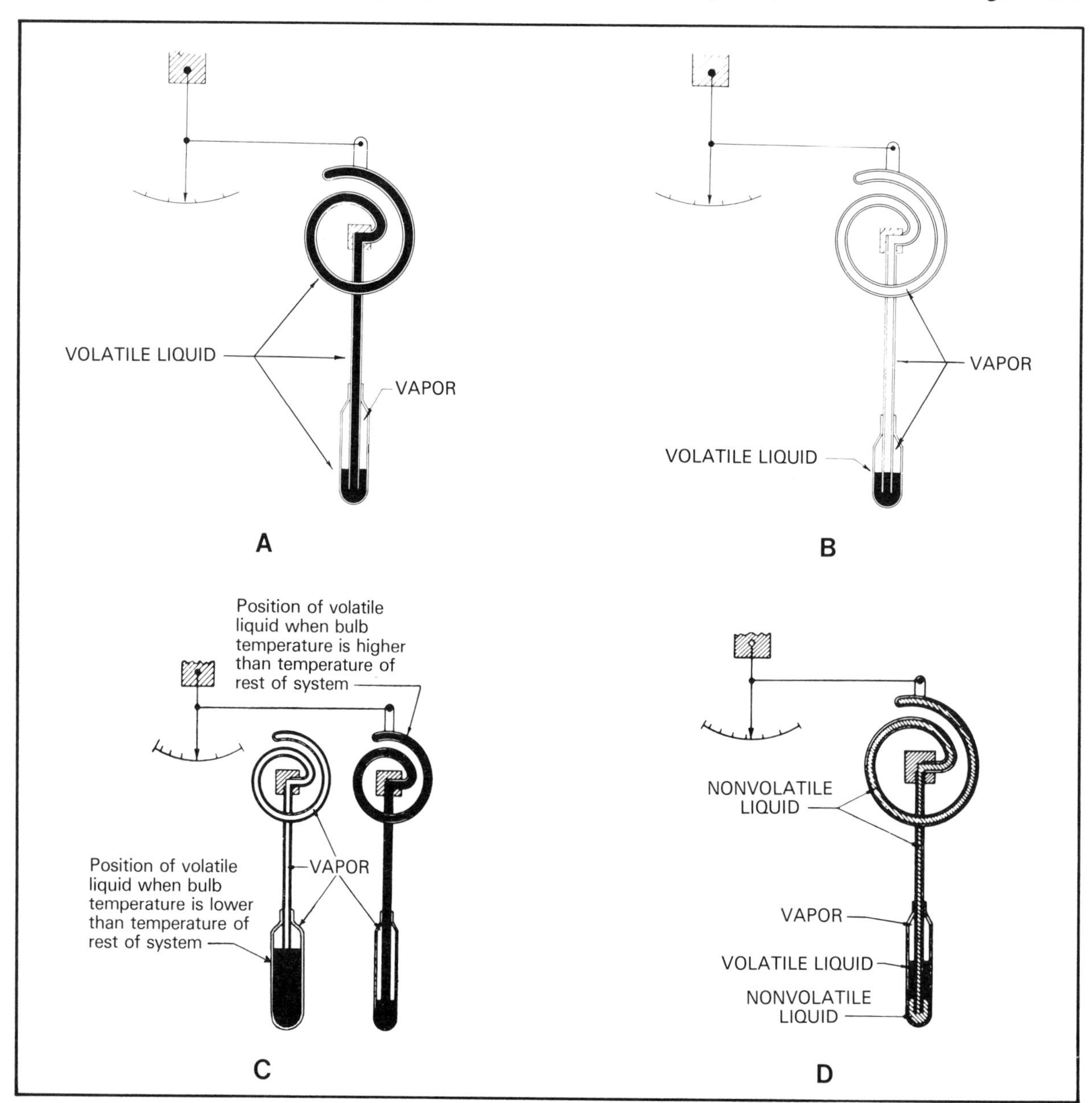

Figure 6.9. Types of vapor-pressure systems. ***A,*** **Class IIA;** ***B,*** **Class IIB;** ***C,*** **Class IIC;** ***D,*** **Class IID.**

Class IIC systems (fig. 6.9*C*) are designed to measure above and below ambient temperature and are similar to Class IIA (fig. 6.9*A*) but have somewhat larger bulbs. In measuring low temperatures, all the liquid will be in the bulb, so it must be large enough to hold all of it and still have a vapor space. This class is unsatisfactory for service where the bulb temperature makes regular excursions across the ambient value because erratic action and delay occur.

For measuring temperature that normally crosses the ambient value, a Class IID system should be used (fig. 6.9*D*). This system is not erratic at or near ambient temperature because the Bourdon and capillary tubes are always filled with stable fluid that has no tendency to vaporize or solidify over the range of operation. Action of the Bourdon tube is strictly a function of the vapor pressure in the bulb.

Capillary tubes. Capillary tubes are usually made of steel alloy and must be chosen for their inert character when used with the fluid in the system. The bulb and Bourdon tube must also be selected for their compatibility with the system fluid. Pressures in filled systems are apt to approach 2,500 psi, so bulbs and tubes are built to withstand this pressure without distortion. The very small internal dimensions account for this ability to some extent. Capillaries vary in inside diameter; 0.02 inch is a typical value.

Bimetal elements. Some filled systems use bimetal elements as compensating components. Bimetal elements utilize two alloys with widely different thermal expansion coefficients. Two metal strips are used in various configurations to provide certain corrective factors in filled-system thermometers. (See figure 6.10 and the discussion on bimetal thermometers.)

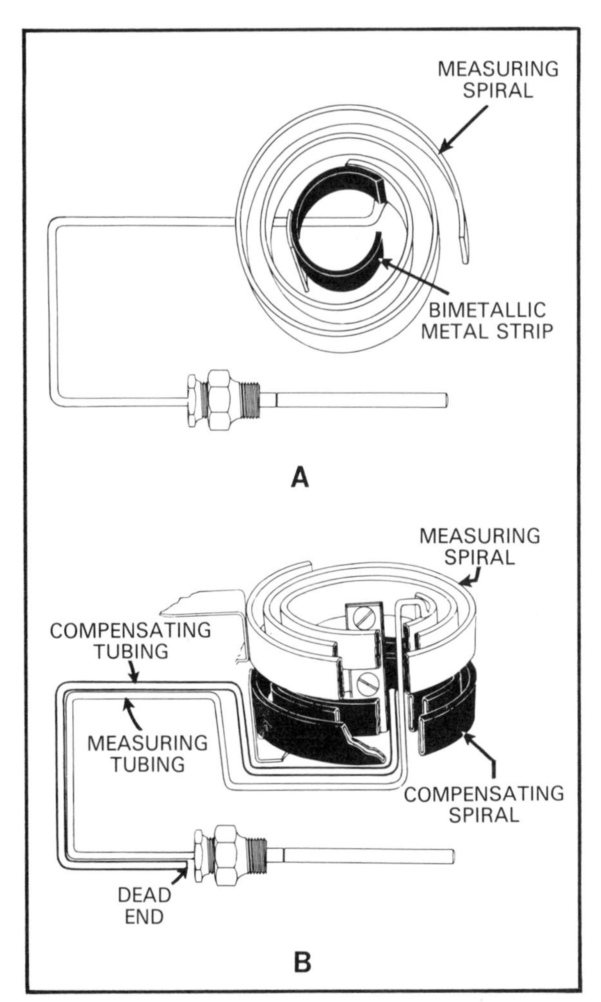

Figure 6.10. Bimetal elements used in filled systems as compensating components. *A*, for case compensation; *B*, for full compensation.

Errors in Filled Systems

Capillary tubes used in filled systems introduce several interesting factors into temperature measurement, principally error and time lag, although no error is caused in some classes of vapor-pressure systems, and the capillary can be made large enough to avoid serious time lag in these systems. Time lag caused by the capillary is far less important than the error caused by wide-ranging ambient temperatures. Two other forms of error that can occur in filled systems are barometric errors and bulb-elevation errors.

Ambient temperature changes. The error in a filled system caused by ambient temperature variations along the path of the capillary arises from the fact that liquid within the capillary experiences a volumetric change with temperature. Now, if the length of the capillary is such that the volume of its contents becomes a significant part of the total system volume, an unacceptable error can occur with only slight variations in ambient temperature.

To change generalities about length of capillary and error into more specific information, consider some actual figures. Capillaries typically have an inside diameter of 0.02 inch, which is quite small, but a capillary tube 100 feet long would have a volume of 0.377 cubic inch. Such a volume could easily represent more than half the total volume of many systems, since volumes of 0.06 to 0.6 cubic inch are commonly used in Class I systems.

Bulb volumes in liquid- or mercury-filled systems have a direct effect on the *temperature range span,* that is, the change in temperature required to cause full-scale motion of the indicator or recording pen. Large bulbs have small spans, and vice versa. For example, a 0.6 cubic inch bulb might have a span of only 25°F while one with 0.06 cubic inch could have a span of 290°F. A rule of thumb concerning error introduced by ambient temperature changes along the capillary tube is

$$E = 0.00004 \times S \times L \times \Delta T$$

where

E = error in °F;
0.00004 = an empirical factor;
S = span of °F;
L = length of capillary in feet;
ΔT = ambient temperature variation in °F.

It should be realized that this equation, only an approximation at best, applies to the typical liquids (other than mercury) that might be used in the filled system.

The desirability of having the bulb of a system operate either always above or always below ambient temperature arises from the rather severe and variable error that accompanies the transition of the bulb temperature across the ambient value. Assume in a Class IIA system that the temperature of the capillary and Bourdon tubes becomes higher than the bulb temperature. The liquid in the tubes becomes unstable, and vapor bubbles form and break up the liquid column, producing pressure surges that cause erratic response of the indicator. This action continues until all the liquid has been vaporized or driven back into the bulb. The same erratic action will occur each time the bulb temperature approaches near or passes through the ambient temperature.

Barometric errors. The vapor pressure and gas-filled systems (Classes II and III) are subject to errors caused by variations of air pressure, although design factors usually preclude errors in excess of 0.35% of the full range of the instrument. Liquid- and mercury-filled systems have no significant barometric error.

Bulb-elevation errors. Bulb-elevation errors can affect the accuracy of Classes I, II, and V, while gas-filled systems are virtually free of this error. Elevation error is caused by the difference in elevation between bulb and indicator and is a direct result of the hydrostatic pressure caused by the relative elevation of one end of the system with respect to the other end. Such error is proportional to the density of the fluid *and* the difference in elevation of bulb and Bourdon tube, and it is inversely proportional to the range span.

The error in Class I systems is really quite small and that in Class V will be several times as great for a given set of conditions, but still small. For example, a mercury system with a range span of 100°F and a bulb-elevation difference of 20 feet will have an error less than 1°F. Vapor-pressure systems in which liquid fills the Bourdon and capillary tubes are subject to the greatest error from this cause, and errors twice that of a mercury system will be encountered for similar installations and ranges.

Compensation of Filled Systems

As pointed out, the accuracy of a filled system, except a vapor-pressure system, is affected by the changes in temperature that occur along the path of the capillary tube and in the enclosure that contains the Bourdon tube. If the ambient temperature at these locations does not vary appreciably, the system can be calibrated at the mean ambient temperature, and no additional measures or compensation is necessary to assure accuracy. However, if excursions of ambient temperatures are of such nature that an unacceptable error might occur, several means for offsetting the error are available. Of these, two forms of error compensation are basic: case compensation and full compensation.

Case compensation. Case compensation takes care of the ambient temperature changes that occur at the indicator. Therefore, case compensation is suitable for use only where short runs of capillary tube are used. A knowledge of bimetal elements, described in the section that follows, is necessary to understand how the compensation takes place.

In the unit shown in figure 6.10*A*, a curved bimetal element is anchored securely at its mid-section, while the so-called fixed end of the Bourdon tube is bonded to one end of the bimetal element. If the temperature in the case containing the Bourdon tube rises above that for which the system is calibrated, fluid in the Bourdon tube will expand, causing an erroneous high reading whose magnitude will be proportional to the ambient temperature rise, the internal volume of the Bourdon tube, and the kind of fluid used. However, if the bimetal compensating element is selected carefully, the flexing of this element with changing temperature will reposition the Bourdon tube in a way necessary to offset the erroneous reading. This sort of case compensation is used with equal success on mercury-, liquid-, or gas-filled systems.

Full, or capillary, compensation. One type of fully compensated filled system uses two Bourdon tubes and two capillary tubes (fig. 6.10*B*). The compensating Bourdon tube here acts in a manner similar to the bimetal compensator; that is, it repositions the fixed end of the measuring Bourdon tube in order to offset errors arising from ambient temperature conditions. The compensating capillary tube is of the same length and filled with the same fluid as the measuring tube. It follows the same path between bulb and indicator case, but no bulb is associated with the compensating capillary tube. Therefore, response of the compensating system is restricted to the same ambient temperatures that affect the measuring system.

Another form of capillary compensation used with mercury-filled systems, usually in association with a bimetal case compensator, consists of placing an invar wire within the capillary tube. Invar is an iron-nickel alloy that has an extremely small coefficient of thermal expansion. Compensation is achieved by the fact that the combined volume of mercury and invar wire within the tube experiences a change with temperature about equal to that experienced by the capillary tube so that no change in pressure is produced at the Bourdon spring.

No compensation. Vapor-pressure systems need no case or capillary compensation, and it might be worthwhile to consider briefly the characteristic that accounts for this desirable feature. The vapor pressure of a liquid is a function of temperature, and as the temperature of a closed system increases, the pressure and the quantity of liquid that is vaporized also increase. If the ambient temperature associated with the capillary increases somewhat, the vapor within the capillary naturally tends to expand and increase the system pressure. Every such effort, however, is thwarted by the fact that some vapor immediately condenses and restores pressure in the system to a value appropriate to the *temperature of the liquid in the bulb.*

Bimetal Thermometers

The principle used in bimetal thermometers has been known for more than a century, but the development of alloys with widely differing thermal expansion coefficients provided the means for making bimetal elements extremely useful and practical components of temperature-sensitive devices.

Principle of Operation

A bimetal thermometer utilizes two strips of metal (fig. 6.11), one having an extremely low coefficient of thermal expansion and the other a rather high coefficient. A nickel-iron alloy containing about 36% nickel, called *invar,* has a thermal expansion coefficient that is very small, while certain other alloys containing nickel, chromium, and iron have comparatively large coefficients. The two strips might be held together by welding, brazing, or riveting.

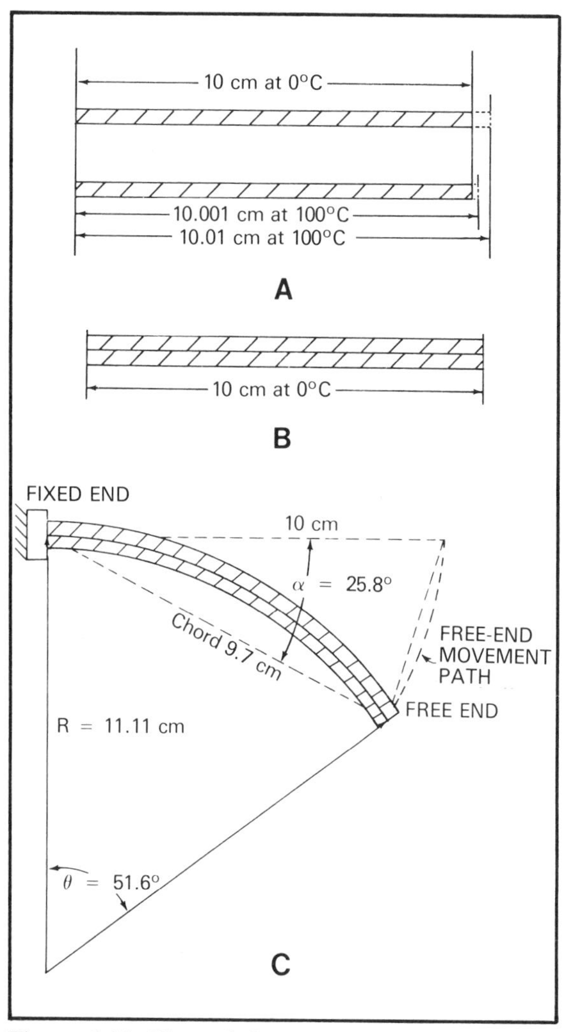

Figure 6.11. Bimetal thermometer elements

For some particular value of temperature, the bimetal element will be flat and straight, but because of the dissimilar thermal expansion coefficients of the two metals forming the strip, variations of temperature away from this particular value will cause the strip to bend into an arc. Since each of the individual strips expands according to its characteristics, and since one strip must expand to a greater length than the other, the formation of an arc takes place. The arc formed by the outer strip will be longer than the arc formed by the inner strip.

To study the effects of temperature change, assume the temperature of the bimetal element shown in figure 6.11*B* has been increased from 0°C to 100°C. Assume also that at 0°C each strip of the element has a standard length of 10 cm, each is 0.1 mm thick, and one strip has an expansion coefficient of 10^{-5} per degree Celsius while the other (the lower strip) has a coefficient of 10^{-6} per degree Celsius.

These coefficients are factors in determining how much the linear dimensions of the strips change with temperature. The change in length is a triple product whose factors are (1) the standard length of the strip (10 cm for this case), (2) the change in temperature (100°C), and (3) the thermal expansion coefficients. Thus, with the conditions chosen, the changes in length for the upper strip will be

$$10 \text{ cm} \times 100°\text{C} \times 10^{-5}/°\text{C} = 10^{-2} \text{ cm},$$

and for the lower strip,

$$10 \text{ cm} \times 100°\text{C} \times 10^{-6}/°\text{C} = 10^{-3}.$$

The amount of linear change is clearly quite small, but since these strips are bonded together and must bend into an arc to satisfy the linear changes experienced by the individual strips, the free-end movement will be quite impressive.

The curved form of the bimetal element after being heated to 100°C is shown in figure 6.11*C*. The values shown for the angles, chord length, and radius (*R*) can be arrived at without too much difficulty by the mathematically inclined reader, and the free-end movement will be found to be about 4.4 cm.

Forms of Bimetal Elements

With this fundamental understanding of bimetal elements, it is time to note some relations that might not be obvious. For example, in an arrangement that has one end of the element fixed and the other free to move (fig. 6.11*C*), free-end movement will be (1) proportional to the change in temperature, (2) proportional to the square of the length, and (3) inversely proportional to the thickness of the element. From these facts, it can be seen that a long, thin element is desirable for sensitivity.

To obtain such an element in a compact form, spiral and helix shapes (fig. 6.12) are used, just as the Bourdon tubes are shaped for pressure measurement.

Spiral elements. Spiral elements are popular household temperature control devices and are used as sensitive elements in automatic chokes of automobiles. They are not used extensively in industrial controls or industrial thermometers. The spiral element has the disadvantage of being broad and flat and thus not suited for placement in thin stems that are considered desirable for most purposes.

Helical elements. Helical elements can be made into comparatively thin forms that can be fitted into small-diameter tubes of the type most desired for fast-response thermometers (fig. 6.13). The multiple-helix element consists of two or more coaxially wound coils formed from a single length of bimetal material. The multiple-helix element has two main advantages over the single-helix:

1. For a given amount of deflection, the multiple-helix element forms a shorter unit and therefore requires less immersion depth for effective measurement.
2. A multiple-helix element can be designed to have little or no axial travel while responding to temperature change, while a single-helix element experiences a significant amount of thrust while twisting or untwisting. Such axial movement may cause contact between the pointer and dial or glass cover.

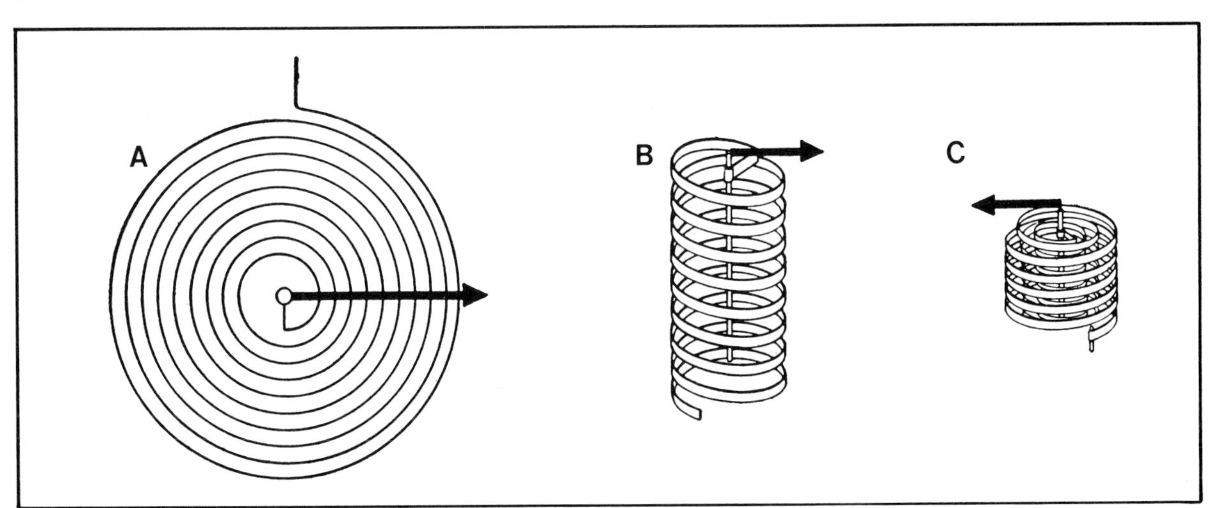

Figure 6.12. Bimetal elements in wound forms. *A*, spiral; *B*, single-helix; *C*, multiple-helix.

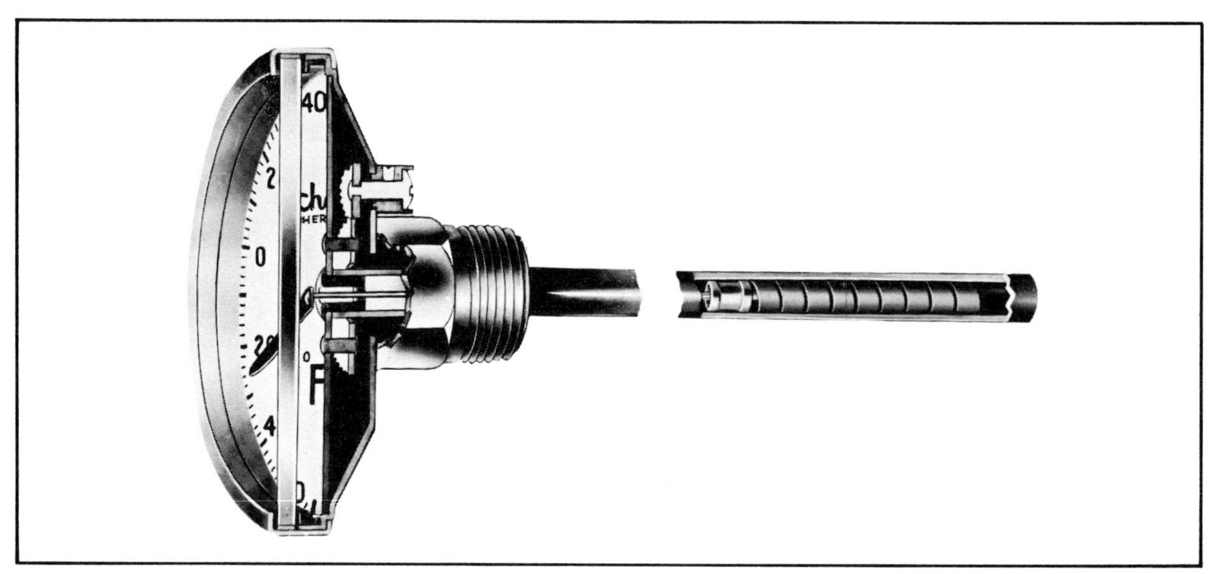

Figure 6.13. Thermometer with helical bimetal element (*Courtesy Rochester Manufacturing*)

Characteristics and Uses

Most of the thermometers using bimetal elements have uniform scales, but no bimetal element is uniform in its motion over a very wide range of temperatures, although some are linear for as much as 400°F change. Uniform-scale, accurate thermometers with ranges covering a span of 250°F are common. The maximum limits of temperature measurement with bimetals seem to be from −300°F to 1,000°F, but at very low temperatures the rate of deflection with temperature change falls off seriously because the coefficients of expansion of the two metals in the element tend to approach a common value. After prolonged exposure to high temperatures—those approaching 1,000°F—bimetal elements seem to undergo a change in characteristics that causes them to read low for high values and high for low values of temperature.

Because of their acceptable accuracy and great ruggedness, bimetal thermometers are used for about the same purposes as glass thermometers, except for medical use and a few laboratory applications. In addition to being rugged, they are capable of withstanding severe overranging without damage and are easier to read than liquid-in-glass thermometers.

As pointed out in the previous section, bimetal elements are used to provide certain corrective factors in filled-system thermometers.

Electrical Measuring Devices

Electrical systems of temperature measurement overcome many of the disadvantages of the other systems. Electrical systems are by no means free of disadvantages, but they are capable of measurements up to 5,000°F, can be made extremely small, and possess unrivaled accuracy. The minimum temperature that can be measured by electrical means is about −320°F, somewhat inferior to a gas-filled system in this respect.

Two principal limitations of the other systems of temperature measurement are (1) the low maximum value of temperature capability (about 1,200°F to 1,400°F) and (2) the difficulty of providing remote transmission of temperature values. It is true that all the systems using Bourdon tubes and some others can provide pneumatic transmission but are limited in their range. Other shortcomings, some of them serious for particular applications, are that the sensing elements have a large thermal mass and that their large physical size precludes their use in restricted spaces.

The Relations of Temperature to Electrical Resistance

Temperature changes in a substance are apt to have a spectacular influence on many of its physical characteristics—its linear dimensions if it is a solid or liquid, its volume if a gas, and its vapor pressure if a liquid. The electrical characteristics are usually no exception, and several useful relations exist between the temperature and the electrical properties of most substances.

Thermoresistive materials. The electrical resistance of most elemental metals—silver, copper, and platinum are three such metals of importance—is a nearly linear function of temperature. The oxides of some thermoresistive metals—including those of copper, iron, magnesium, nickel, tin, and zinc—have a temperature-resistance characteristic that is not only far more pronounced than the pure metals, but is *negative*—that is, resistance decreases as temperature increases. These powdered oxides are carefully prepared by compressing them into particular shapes and then heat-treating them to form a ceramiclike body. Electrical leads may be embedded during preparation, or contacts can be plated to the finished body. These temperature-sensitive elements are called *thermistors* and are now important for measuring temperatures below about 500°F, although their reliability and accuracy are well below those of ordinary metallic resistance elements.

The relation between the temperature and the electrical resistance of a conductor is conveniently dealt with in terms of coefficients that describe the change in resistance with change in temperature. Platinum, for example, has a resistance-temperature coefficient of 0.00392 Ω/Ω/°C over the range of 0°C to 100°C. This means that for any particular temperature in this range, each ohm of resistance a platinum wire contains will become 1.00392 ohm for each degree Celsius increase, or 0.99608 ohm for each degree decrease.

An example in the use of the coefficient for platinum will do much for a good understanding. Assume that a coil of platinum wire has exactly 100 ohms resistance at 0°C and that in boiling ethyl alcohol the resistance increases to 130.58 ohms. The resistance increase is 130.58 minus 100, or 30.58 ohms. Each degree of temperature increase from the reference value of 0°C causes an increase in resistance of 0.00392 ohm for each ohm of resistance as measured at the reference temperature. Thus, the new temperature—that of boiling ethyl alcohol—can be calculated using the equation,

$$T = \frac{R_t - R_0}{R_0 \times a}$$

where

T = new temperature in °C;
R_t = resistance at temperature T in ohms;
R_0 = resistance at 0°C in ohms;
a = resistance temperature coefficient of platinum.

Substituting,

$$T = \frac{130.58 - 100}{100 \times 0.00392} = 78°\text{C}.$$

Measuring resistance. Having noted the small change in resistance that accompanies a temperature variation of the platinum wire and the three significant figures allotted to expressing the resistance-temperature coefficient, the thought occurs that measuring resistance with the required accuracy must pose a problem. Coupled with this thought is a suspicion that current flow through the resistance element must be kept to a very low value in order to avoid heating the element and thus creating an error.

One instrument useful for measuring resistance accurately and with a minimum amount of current flow through the circuit is the Wheatstone bridge. The Wheatstone bridge is an electric circuit consisting of four resistors, a galvanometer, and a voltage source connected in the manner shown in figure 6.14*A*. The source of voltage is usually thought of as a battery, but permanent industrial installations frequently use step-down transformers and rectifiers to obtain the necessary voltage from regular alternating-current power lines. The galvanometer, *G*, is a low-current meter having its zero at the center of its scale, so that pointer deflections in either direction are conveniently made. Three of the resistors—R_1, R_2, and R_3—are precision units and at least one of them is variable. The fourth resistor, R_x, is the unknown unit and represents the resistance element of the temperature measuring device.

Briefly, operation of the bridge is as follows. The three precision resistors are adjusted so that zero current flows in the galvanometer. Note that if R_1 and R_2 are made equal, and R_3 is adjusted to the same value as R_x, no current will flow in the meter, because the voltage drop from *X* to *Y* will be exactly equal to that from *X* to *Z*. In order to produce zero current through the meter, it is necessary to have the ratio of resistance so that

$$R_1/R_2 = R_x/R_3.$$

From this,

$$R_x = (R_1/R_2) \times R_3.$$

Where the highest accuracy and stability are demanded, modifications of the bridge can be made in order to eliminate the error that can be caused by (1) changes in contact resistance of the variable resistors and (2) changes in resistance of the electric leads between the bridge proper and the resistance-temperature elements.

To eliminate error caused by changes in contact resistance of the variable resistors, three fixed resistors and a single variable resistor, S_1, can be used (fig. 6.14*B*). Since the moving contact of the variable resistor is carrying no current for a balanced condition

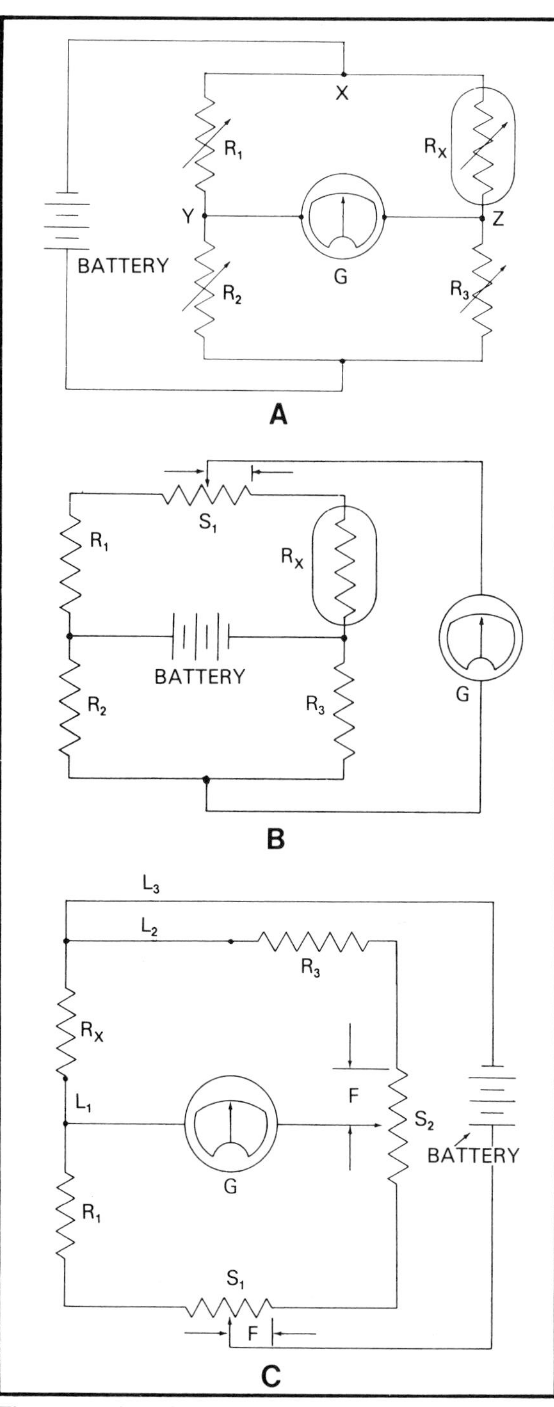

Figure 6.14. Bridge circuits for resistance devices used in electrical measurement of temperature. *A*, Wheatstone bridge; *B* and *C*, bridge modifications.

(meter reading zero), its contact resistance is not important.

Changes in resistance of the electric leads are taken care of by another circuit arrangement as shown in figure 6.14*C*. The automatic compensation for lead resistance is achieved by the use of *three* leads from the bridge proper to the sensing element. S_1 and S_2 are variable resistors that share a common drive shaft; that is, they are driven in unison. L_1 and L_2 are the identical leads between the sensing element and the bridge proper; if they follow the same path, their resistances will be equal at all times. Note that the third lead from the battery is connected so that the resistance of L_1 is added to the R_x leg of the circuit while that of L_2 is added to the R_3 leg. Letting R_L be the resistance of the matched leads, L_1 and L_2, and F the *fractional* part of the total resistance of S_1 and S_2 (that is, F is a factor that has values of 0 to 1.0), the equation of the circuit is

$$\frac{R_x + R_L}{R_3 + FS_2 + R_L} = \frac{R_1 + S_1 - FS_1}{FS_1 + S_2 - FS_2}.$$

If this equation is to hold for all values of R_L, then the right side of the equation must be equal to 1 at all times. Thus,

$$\frac{R_x + R_L}{R_3 + FS_2 + R_L} = 1,$$

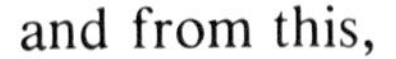

and from this,

$$R_x = R_3 + FS_2.$$

Thermocouples

Thermocouples are electric devices useful in measuring a large range of temperatures from perhaps −300°F to 3,000°F. As was true with resistance elements, the problem of accurately measuring minute changes in an electrical quantity arises. The voltage output of the thermocouple must be measured, and changes of 100°F are apt to produce only 2 or 3 millivolts of change in output voltage.

Several thermoelectric effects are associated with thermocouples, but only those that cause a current to flow when two junctions of a thermocouple are at different temperatures are of interest here. Junctions are the connections that join two dissimilar metals and may be made by twisting, soldering, or brazing together wires made of the metals. Butt welding (fig. 6.15) is popular because it provides a joint of minimum mass and yet has good physical strength.

Thermocouple materials. Materials used for forming thermocouples and the means of joining them together depend largely on the temperature range to which the thermocouple will be subjected. Copper and constantan are two metals often used for thermocouples

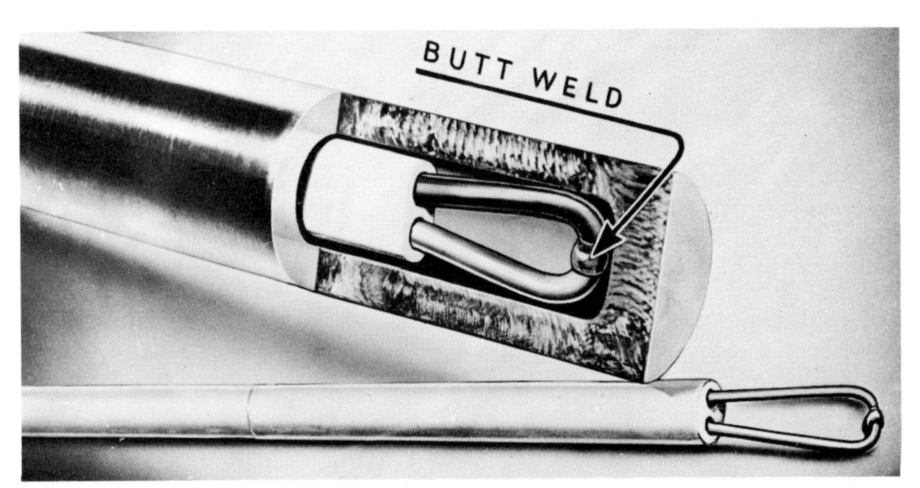

Figure 6.15. A butt-welded thermocouple (*Courtesy Honeywell*)

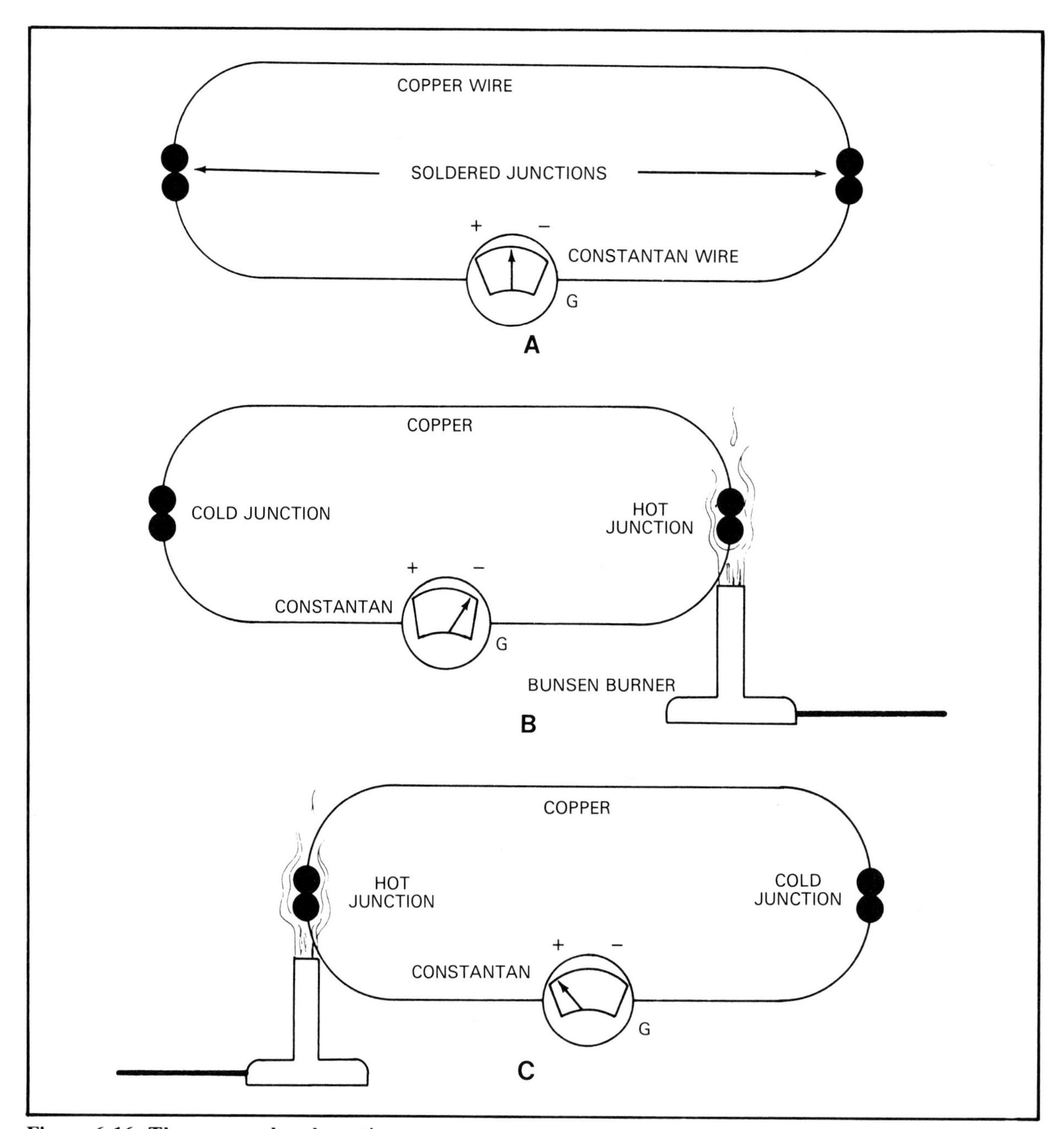

Figure 6.16. Thermocouple schematics

(fig. 6.16). In figure 6.16*A* the junctions are shown at identical temperatures. No current flows when the junctions are at equal temperatures, but if one junction is heated (fig. 6.16*B*), current will flow, indicating the existence of a potential difference between the two junctions. In the schematics the junctions are identified as *hot* and *cold,* but in practice they are better identified as *measuring* and *reference* junctions in order to avoid confusion.

When one junction is heated to a temperature somewhat higher than the other, current will flow in a certain direction. If the other junction is then heated to a higher temperature, current flow will reverse (fig. 6.16*C*). The positive and negative signs associated with the galvanometer signify the terminals of the meter. When the positive and negative leads of a current source agree with the polarity of the terminals, the pointer deflects to the right, and vice versa.

The copper and constantan thermocouple is popular for measuring temperatures in the range of −300°F to 600°F. This combination of dissimilar metals produces a relatively high output voltage but suffers from erratic performance and deterioration at temperatures above 600°F. A thermocouple made of platinum and platinum-rhodium is suitable for temperatures as high as 3,000°F, but it must be protected against corrosion by a gas-tight protecting tube (usually a ceramic tube) for temperatures above 1,800°F. Iron and constantan combine favorably for a range of temperatures of −300° F to 1,500°F because of relatively low cost, resistance to oxidation for all but higher temperatures, and excellent voltage output.

Measurement of thermocouple output. The output voltage of most thermocouples is feeble. For example, if a reference temperature of 32°F is used, a copper-constantan thermocouple will have an output of approximately 9.52 millivolts at 400°F, while an iron-constantan unit will produce 11.03 millivolts. Despite this low output, sensitive d'Arsonval millivoltmeters are capable of reading such values directly. A temperature-measuring device with this type circuit arrangement is sometimes called a *millivoltmeter pyrometer* (fig. 6.17), and it finds wide use for measuring high temperatures where high accuracy is not required. The reference junction frequently is located with the millivoltmeter, and no special effort is made to maintain it at a fixed temperature. Compensation is sometimes provided by special resistors or by bimetal elements acting in conjunction with the spiral hair spring of the meter.

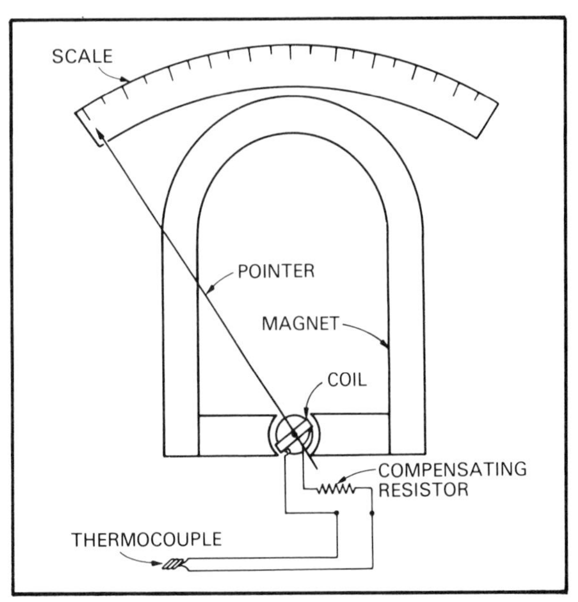

Figure 6.17. Schematic of millivoltmeter with thermocouple (*Courtesy Honeywell*)

As a means for measuring temperature, thermocouples of proper design and calibration are as accurate as the output voltage measurements, and great precision can be obtained through the use of standard cells and potentiometers. A standard cell is any special electric cell capable of producing a particular voltage whose value is stable under a given set of reproducible conditions. The *Weston cell* (fig. 6.18) is the best known of several standard cells and is found in two forms—saturated and unsaturated. The saturated Weston cell is

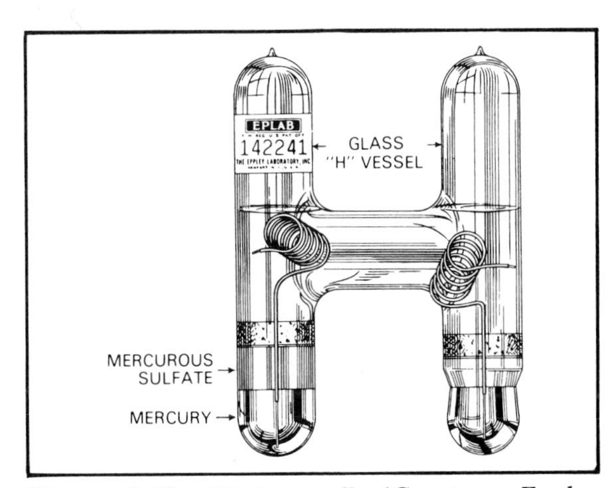

Figure 6.18. Weston cell (*Courtesy Eppley Laboratory*)

sensitive to temperature, and its output at 20°C is 1.018636 volts. The unsaturated Weston cell is not sensitive to temperature but suffers a significant voltage drop with age. It has an output when new of about 1.019 volts and declines about 100 microvolts per year. The saturated cell is used by standards laboratories, mostly because of its extremely small change in output with age, while the unsaturated cell is used commercially. Neither of the cells will endure more than a few microamperes of current drain without damage.

Although standard cells continue to enjoy great popularity, developments in solid state physics have made practical small power supplies that are capable of accepting an input of 102 to 132 volts of 50 to 60 hertz and delivering a stable direct current output of about

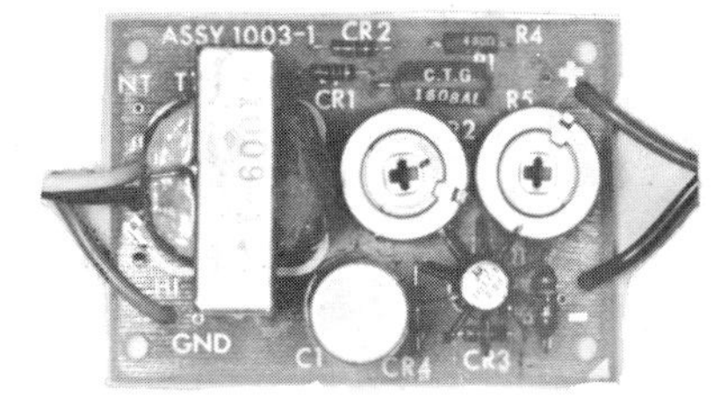

Figure 6.19. Regulated power supply (*Courtesy Corning Glass Works*)

1 volt (fig. 6.19). Their output voltage is adjustable between the limits of 0.900 volts and 1.070 volts. Their output voltage is so stable that a change of 20°C in temperature causes a voltage change of less than 2 millivolts, and a line voltage change of 15 volts causes a change in output voltage of only 0.5 millivolt. Additional good points of such standard voltage sources include (1) the ability to deliver appreciable current (up to about 10 mA) without being overloaded or inducing error, (2) potential years of service without attention, and (3) the ability to endure severe environments—high or low temperatures and high relative humidity.

The success of most power supply type standard voltage sources is due to the employment of *Zener diodes.* A Zener diode is

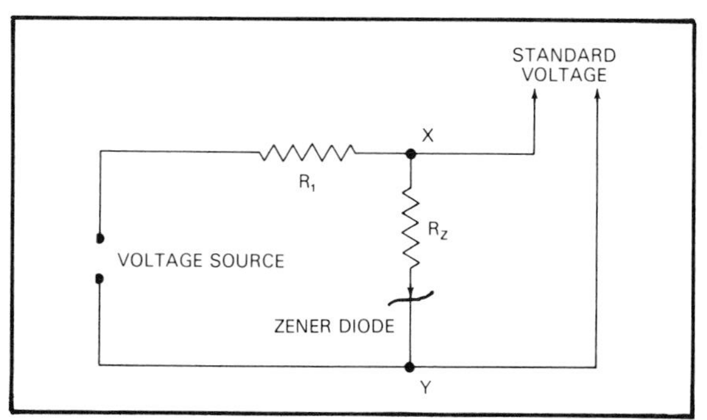

Figure 6.20. Circuit diagram of a Zener diode

remarkable for its ability to maintain a reasonably constant voltage across its terminals under certain conditions. The simple circuit diagram in figure 6.20 shows a possible way to achieve a stable source of voltage. A particular value of voltage will be developed between points X and Y, that is, across the terminals of the Zener diode. The resistor R_z is not a separate component but represents the internal resistance of the diode. Even a very small change in the value of voltage across X and Y will be counteracted by a relatively drastic change in current flow through the diode. This change in flow will cause a different voltage drop across resistor R_1, the net result being that a stable voltage will be maintained.

Potentiometers provide an excellent means for using standard cells for calibrating the voltage from another source, because they enable one to make comparisons between the standard and the unknown voltage without requiring any more than a feeble flow of current from the standard cell. Potentiometers may take many forms, but they are basically resistors with sliding contacts that move from one end of the resistor to the other. In many instances the potentiometer is called a *slide wire* because it is formed by a circle of resistance wire around a drum. The drum rotates while metal fingers ride on the resistance wire.

The circuit in figure 6.21*A* is useful for explaining the use of the potentiometer and a standard cell for accurate measurement of

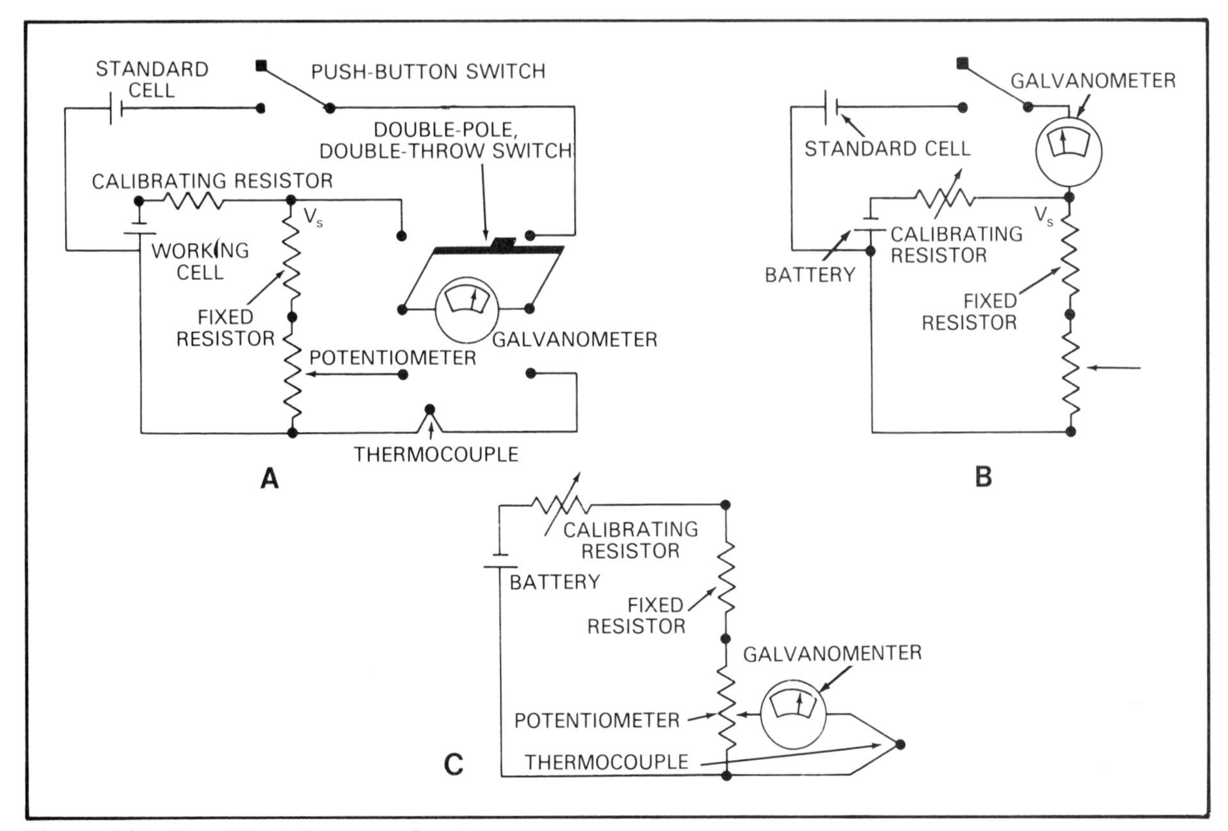

Figure 6.21. Simplified diagram of a thermocouple circuit. ***A*, with a potentiometer; *B*, during calibration phase; *C*, in the measuring phase.**

voltage output from a thermocouple. During the calibration phase (fig. 6.21*B*), the working cell delivers current at all times through the series combination of the fixed resistor, potentiometer, and the calibrating resistor; consequently, a voltage drop will occur across each of these resistors. If a 1.5-volt cell is used as the working battery, the calibrating resistor must have a value of resistance that enables easy adjustment for reducing the voltage to that of the standard cell at point V_S, which is the connection between fixed resistor and calibrating resistor. The voltage drop across the potentiometer should be quite small, that is, not very much greater than the maximum voltage output expected from the thermocouple. This means that the bulk of the voltage drop will occur across the fixed resistor. It is essential that the fixed resistor and the potentiometer be stable and accurate, because any error caused by them cannot be detected by the routine calibrating procedure.

A primary requirement for accuracy is to have a known voltage at point V_S (fig. 6.21*B*), for then the resistance values of the potentiometer and fixed resistor can be used to calculate the thermocouple output. The voltage wanted at point V_S is that of the standard cell, 1.019 volts. A voltage produced by the working battery is always present at point V_S, and to check the value of this voltage relative to the standard cell, the push-button switch is pressed momentarily. This action effectively parallels the standard cell with a voltage source equal to the working battery voltage minus the voltage drop across the calibrating resistor. If there is no deflection of the galvanometer needle, this is a good indication that the two voltages are equal and that calibration is satisfactory. On the other hand, if the needle deflects one way or the other, current is flowing in the standard cell circuit, and the calibrating resistor must be manipulated until a no-deflection condition indicates that voltage at point V_S is equal to that of the standard cell. The double-pole, double-throw switch is then returned to the thermocouple position.

When in the measuring phase (fig. 6.21*C*), the potentiometer has a voltage drop across it due to the flow of current from the working battery. As noted earlier, however, this voltage drop is not apt to be very large, probably no more than 15 to 20 millivolts depending on the temperature range for which the instrument is designed. One lead of the thermocouple is connected to the lower end of the potentiometer, while the other connects to one terminal of the galvanometer whose other terminal is connected to the sliding contact of the potentiometer. The small voltage produced by the thermocouple and the small voltage determined by the position of the slider of the potentiometer are effectively placed in parallel, with the galvanometer forming one of the connecting links. If the two voltages do not agree in value and polarity, a current flow will occur, causing deflection of the galvanometer pointer. Adjustment of the slider one way or the other will probably produce a balanced condition. The potentiometer is calibrated either in degrees of temperature or in millivolts. If calibrated in millivolts, tables that express temperature and output voltage as related functions of thermocouples can be used to determine temperature. Temperature of the reference junction must be accounted for properly.

The error caused by variations in temperature of the reference junction can be severe if the measured temperature is not very much higher or very much lower than the reference junction temperature. If the reference temperature is assumed to have a set value, then each degree of temperature that the reference junction deviates from this set value will produce a 1-degree error in the measured temperature. If the reference temperature is 0°C and measurements are being made near 100°C, then a variation of 5°C at the reference junction will cause a rather significant error of 5°C unless the variation is accounted for. On the other hand, if measurements in the vicinity of 1,000°C are being made, 5°C variation of the reference junction temperature will cause an identical error of about 5°C, which is not likely to be significant.

Compensation for reference junction temperature. The need to maintain a precise fixed value of reference junction temperature or to compensate for its changes ultimately depends upon the accuracy required for the measurements being carried out. This accuracy, in turn, depends upon (1) the deviations of the reference junction temperature from the calibrated value—that is, the reference junction temperature at which the system is supposed to produce correct measurements—and (2) the difference in temperature between that of the reference junction and the measured temperature.

For accurate temperature measurements, particularly those carried out in a laboratory or for an important experiment, the temperature of melting ice (0°C) is frequently used as the reference standard. Such a temperature is easy to maintain by simply immersing the insulated reference junction in a container filled with cracked ice and water. Such a system, of course, is not practical for many permanent installations, so other ideas are used to reduce or eliminate error caused by reference junction temperature variations.

Most of the common methods merely *compensate* for changes in reference junction temperature rather than attempt to maintain a fixed value of temperature. Some forms of millivoltmeter pyrometers incorporate a compensating device in the form of a bimetal spiral that acts in conjunction with the spiral hair spring of the millivoltmeter, thus varying the tension on the meter pointer. The bimetal spiral either adds to or detracts from the overall tension on the pointer, depending on

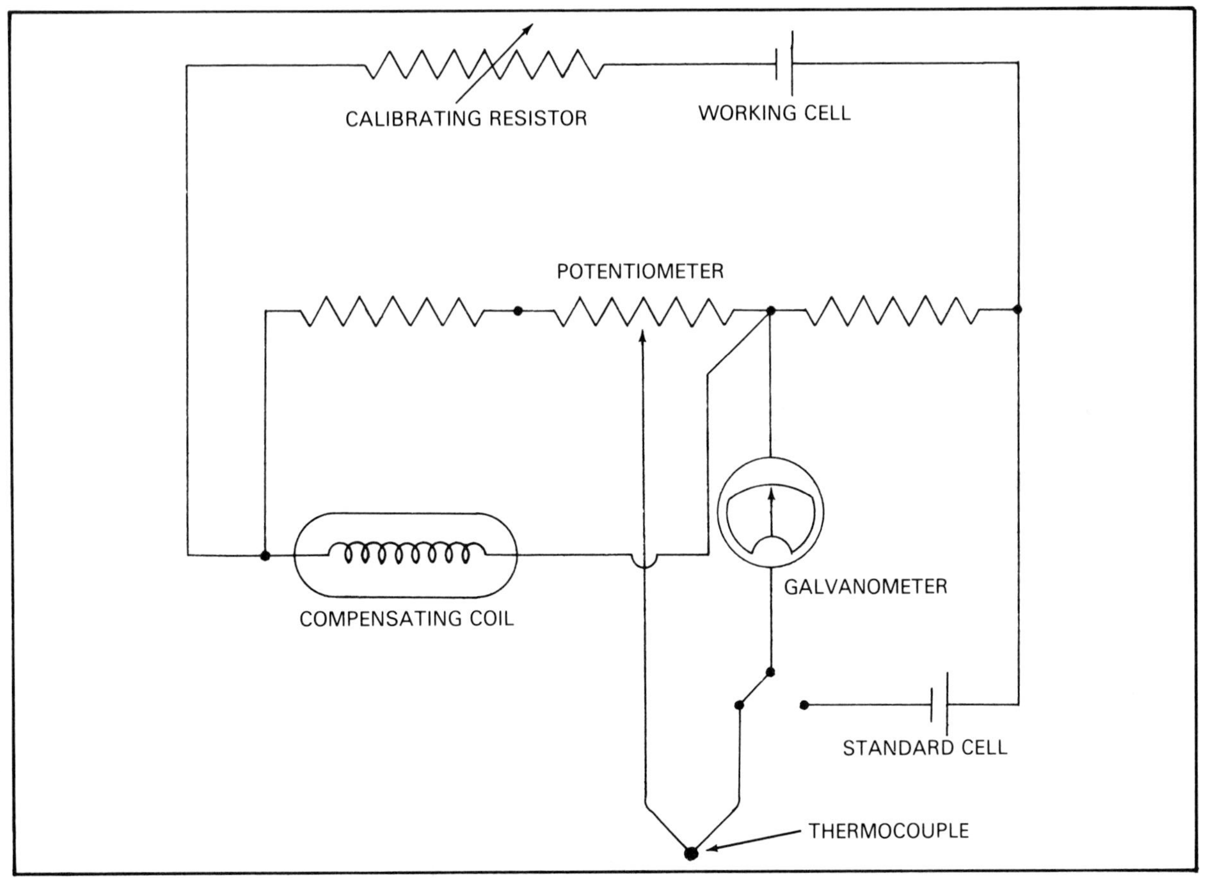

Figure 6.22. Thermocouple circuit with a wire coil that compensates for reference junction temperature changes

whether the reference junction temperature is above or below the correct value and whether the measured temperature is above or below the reference value.

Using a compensating coil of fine wire near the reference junction is another way of achieving compensation (fig. 6.22). The temperature-resistance characteristic of this coil is such that the upper terminal of the galvanometer is, in effect, made more or less negative as the temperature of the compensator coil and reference junction deviates above or below the reference value. This method of compensation will give excellent results if the relevant characteristics of the coil are carefully chosen to match the thermocouple and other parts of the circuit.

Pyrometers

The systems and devices described thus far enable the measurement of temperatures that extend from near absolute zero to as high as about 3,200°F. The thermometers, sensing bulbs, resistance elements, and thermocouples used in these systems have a common requirement—they have to be directly exposed to the temperature being measured. The upper limit of approximately 3,200°F is primarily a result of the inability of most materials to remain stable at such temperatures. In fact, very few materials remain solid at temperatures above 3,000°F. In order to carry out measurements in this upper range of temperature that would destroy most measuring elements subjected directly to its effects, another side effect that can be related with temperature must be found. A form of inferential measurement must be devised—one that allows the equipment to remain at a safe temperature while inferring the true temperature of the process being monitored.

It is a familiar fact that the heat of a hot substance, particularly very hot solids, can be detected at some distance by the sensation of warmth conveyed by radiation. It is also quite well known that when metal is heated to high temperatures, it will begin to emit light in the visible spectrum, beginning with dull red and progressing through brighter reds to an almost white incandescence. Less known is the fact that heating sand, clay, glass, and other solids will produce similar effects. With this knowledge available, radiation and optical pyrometers were developed, needing only the refinements required to correlate radiated energy and the color and brightness of incandescence with temperature.

The fundamental theories and laws that account for the successful operation of radiation and optical pyrometers are the results of extensive experimentation and physical and mathematical analysis by a large number of noted scientists. An unqualified understanding of the theories and laws involved requires considerable knowledge of physical optics, atomic physics, and a mathematical background appropriate to these subjects. However, in order to appreciate the usefulness of these high-temperature devices, it is only necessary to accept the fact that there are relations between the temperature of a substance and the energy it radiates and between its temperature and its color and brightness.

Optical Pyrometers

The optical pyrometer (fig. 6.23) uses the principle of comparing the color and brightness of an object with those of a controlled tungsten filament to determine temperature. The objective lens forms an image of the hot source in the plane of the filament of an incandescent bulb. An ocular lens in the eyepiece enables the observer to focus on the filament and the hot source image, while the red filter allows the observer to see only the red light emitted by the filament and the hot source, so the comparison to be made is in the relative brightness of hot source and filament.

The human eye possesses a remarkable ability to detect slight differences in brilliance between two adjacent sources of light if the two are of the same color, and the red filter provides for a nearly pure single color. The brightness of the filament is varied by controlling its current flow, which is adjusted by the rheostat until the observer can no longer distinguish the outline of the filament against the background of the image formed by the hot source. Temperature of the source is then determined by referring to a calibration curve that shows the relation between the temperature of the filament and the current flow through it.

Since the accuracy of the pyrometer described is dependent to considerable degree upon accuracy of current flow measurements, a potentiometer method is sometimes used to determine current flow, as shown in figure 6.23*A*. Temperature of the filament is controlled by rheostat, *R*, and current is supplied by battery, *B*. The potentiometer, *P*, is a precision unit having a finely calibrated dial. Calibration could conceivably be in temperature, current, or resistance, since all are interrelated. Associated with the potentiometer is a standard cell and galvanometer. A balanced condition of the galvanometer is obtained concurrently with the correct temperature of the filament, providing extremely accurate determination of the current flow.

The life and stability of tungsten filaments used in pyrometers are adversely affected when filaments are operated at temperatures higher than 3,200°F. In order to extend the usable range to and above 5,200°F, *absorbing screens* are placed between the hot source and

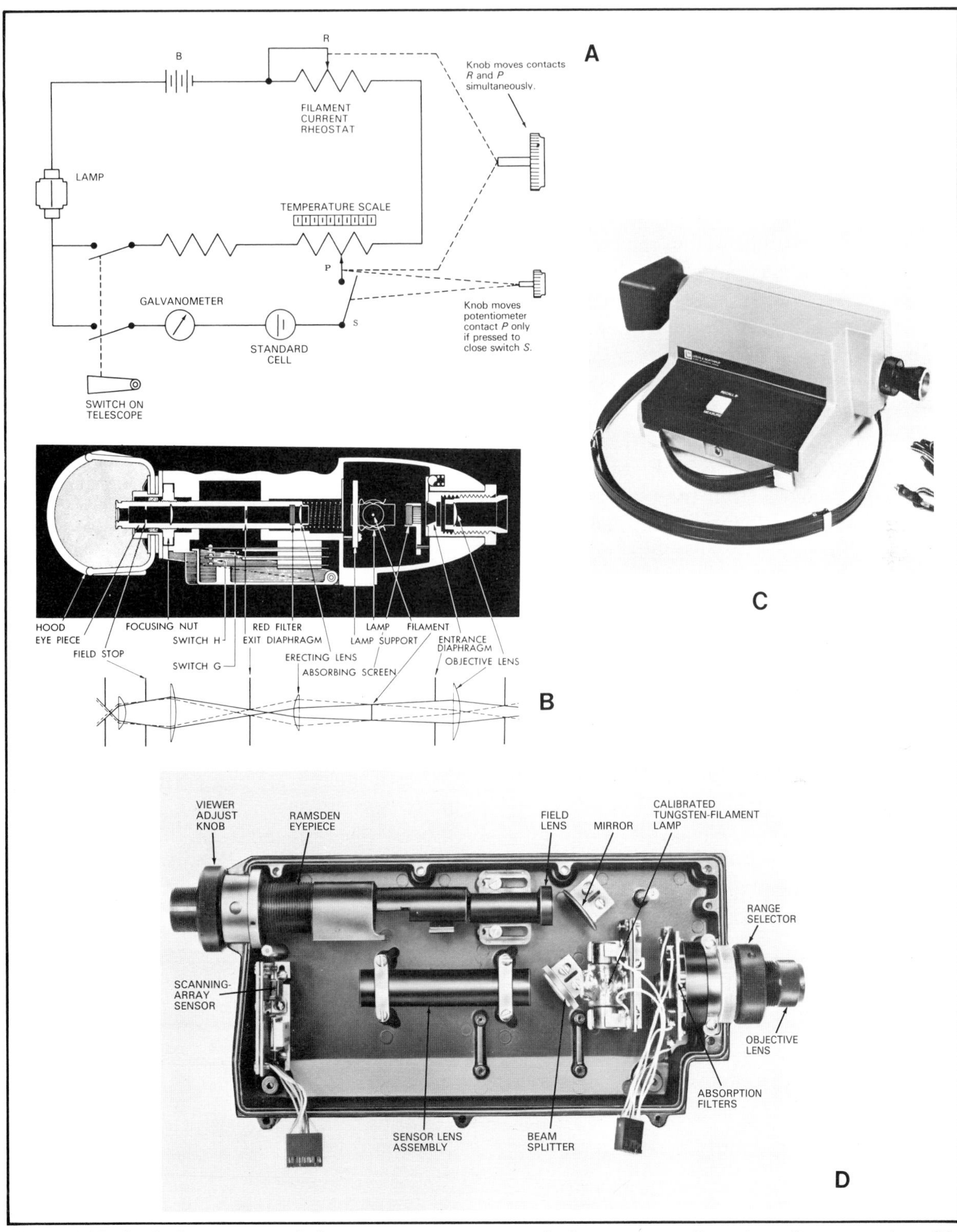

Figure 6.23. Optical pyrometers infer temperature from color and brightness. *A,* simple schematic diagram of basic idea; *B,* cross-sectional view and ray diagram of a commercial unit; *C,* automatic optical pyrometer; *D,* automatic pyrometer with eyeshield and electronics removed to show optical components. *(Courtesy Leeds and Northrup Instruments)*

the tungsten filament. Such screens have the effect of reducing the brilliance of the light from the source, so that filament temperature can be maintained at a safe low level.

Industrial uses of the optical pyrometer include gauging the temperature of molten metals and furnace interiors, as in steel mills. The hand-held and manually operated unit (fig. 6.23*B*) has served well in the past and is still in common use. More modern versions, using advanced solid state electronics, are almost automatic in operation (fig. 6.23*C*).

The operator sights the target through the optical system in the usual manner, then presses a *measure* switch that energizes the electronics and the filament of the tungsten lamp. Looking at figure 6.23*D*, note that the target image passes from the objective lens, across the tungsten-filament lamp, and on to a beam splitter. The beam splitter sends the combined image of filament and target to the upper lens system and the eyepiece. This combined image is also passed through the sensor lens assembly to the scanning-array sensor.

The scanning-array sensor scans across the dual image of filament and target and signals the microcomputer part of the pyrometer so that adjustments to the tungsten-filament current can be made. The current will be adjusted to the value needed to match the target and filament. The microcomputer converts the current value to an analog of temperature. A switch provides a choice of temperature scale, either Fahrenheit or Celsius, and the temperature is displayed on a digital readout of light-emitting diodes (LEDs). The display appears below the image of target and tungsten filament. The last temperature reading is retained in memory and may be displayed by pressing a *recall* button.

Although the optical pyrometer discussed here generally finds use as a portable instrument, the principles embodied in its operation could be applied to provide a fixed installation fully capable of unattended automatic operation.

Radiation Pyrometers

While the optical pyrometer infers the temperature of a hot substance by measuring the brightness of a particular color emitted, the radiation pyrometer infers temperature by measuring a portion of the energy radiated (fig. 6.24).

One of the laws dealing with radiated energy is called the Stefan-Boltzmann law. It states that energy radiated by a body is proportional to the *fourth power* of the body's absolute temperature. Other factors involving radiated energy are the surface area of the body and its *emissivity*—a factor that expresses the ratio between the actual energy radiated from the body and the energy that would have been radiated had the body been a perfect radiator. The Stefan-Boltzmann law

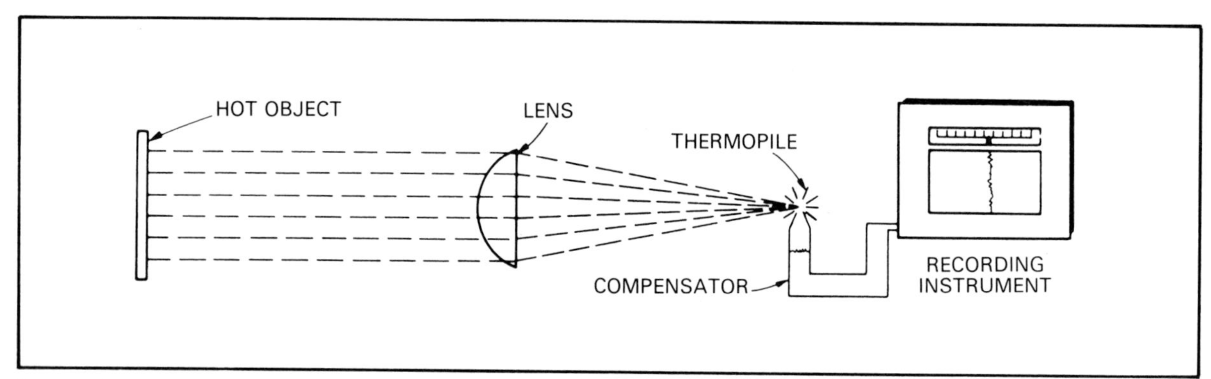

Figure 6.24. A fundamental circuit and optical arrangement of a radiation pyrometer containing a thermopile. (*Courtesy Honeywell*)

guarantees that radiated energy is a function of temperature. The measurement of a portion of the radiated energy and the eventual conversion of this finding into satisfactory expression for temperature is so filled with imponderables that the law is of little practical assistance.

Several methods are used for collecting and measuring a little of the total energy radiated by a hot source. Most of the methods used in radiation pyrometers result in producing an electric effect—a change of voltage from thermocouples, for example, or a change in resistance of a thermistor or resistance element. Only the type containing thermocouples will be considered here.

You probably have witnessed the effect of using a lens to focus sunlight on a piece of wood or paper. As the lens is adjusted to form a very small, bright spot of light, smoke is emitted almost immediately from the wood or paper. In this instance, radiant energy from the sun is collected from a relatively broad expanse—about the area of one side of the lens—and concentrated into the fine and brilliant spot. Since only a small amount of the energy is lost in being concentrated, the heating effect available within the small area of the bright spot is enormous. This same principle is used in the radiation pyrometer, and the concentration and utilization of radiant energy is so efficient that temperatures as low as 125°F can be measured.

In order to obtain high sensitivity, a *thermopile* is used in the thermocouple-type pyrometer. A thermopile (fig. 6.25) is a compact collection of several thermocouples connected in series. Radiant energy is focused on

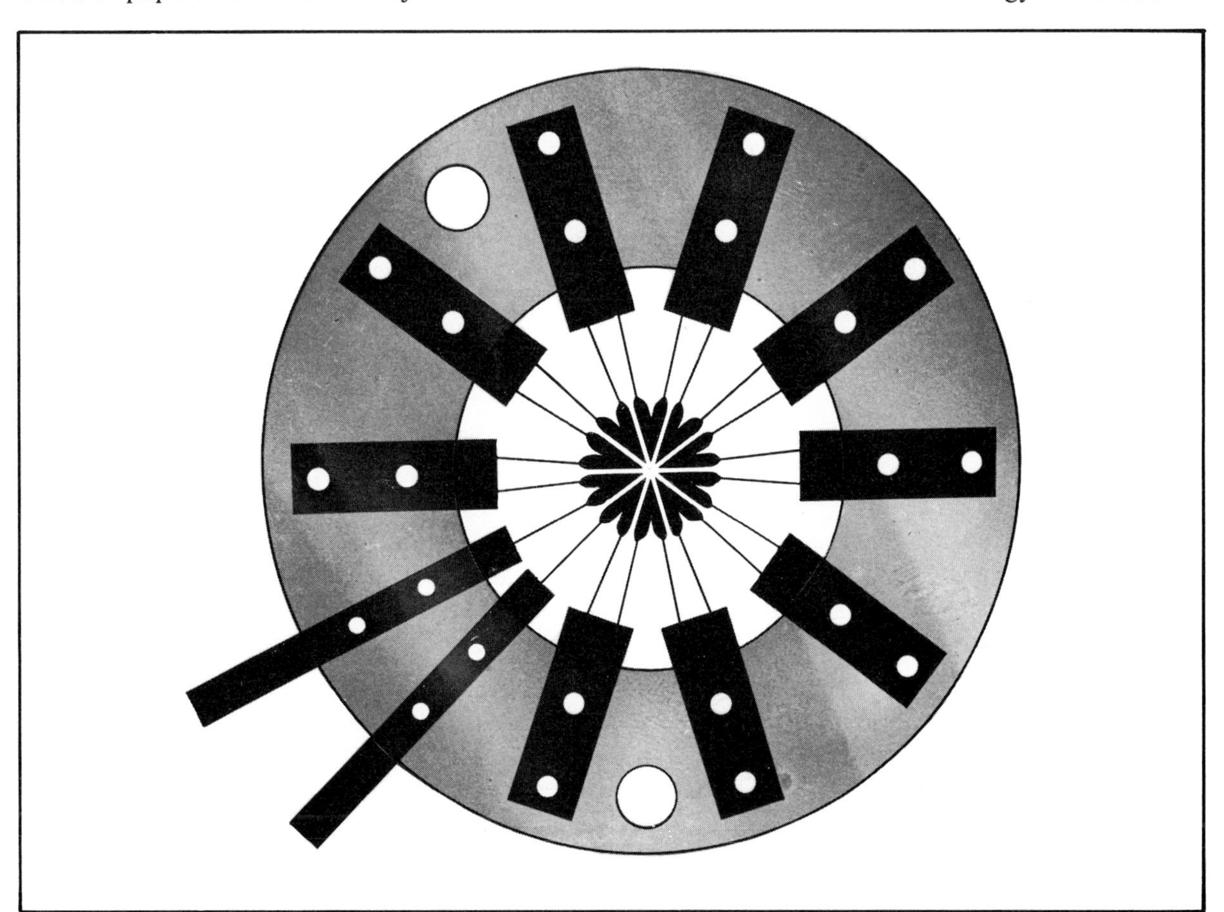

Figure 6.25. A thermopile composed of ten thermocouples in series (*Courtesy Honeywell*)

Figure 6.26. A phantom view of a radiation pyrometer using a thermopile ***(Courtesy Honewell)***

the group of thermocouples that occupies an area equivalent to a circle of less than 5 mm in diameter. The reference junctions are located at points of attachment between thermocouples and the series-connecting strips. In a pyrometer using a thermopile (fig. 6.26), the objective lens focuses an image of the target or body whose temperature is to be determined on the thermopile. Opposite the objective lens is a sighting window that permits viewing the target and accurately aiming the pyrometer. The line of sight passes through the center of the thermopile assembly.

The response of the pyrometer described is rapid; 98% of a temperature change is accounted for within two seconds. Radiation pyrometers find application where other primary elements cannot be used—for example, for measuring the temperature of moving targets or furnace interiors. They possess certain advantages over optical pyrometers, but they are inferior to the optical type in accuracy and versatility: they must be calibrated for the particular range and type of process they are to be used with. Because of the electrical output, radiation pyrometers are easily fitted to needs requiring automatic control, such as the regulation of furnace temperature.

The primary calibration of a radiation pyrometer involves an accurate source of radiant energy that provides an adequate field of view for the pyrometer, temperature stability, and a uniform emissivity over the entire range of calibration. Primary calibration is hardly ever conducted by other than the manufacturer. Secondary calibration is carried out by using the "working" pyrometer in conjunction with one having a certified primary calibration.

It is of interest to note that the distance between pyrometer and target has no effect on the measurement, *provided the optical system field of view is completely filled by the target.* This requirement means, roughly, that the target diameter must be no less than 5% of the distance between target and instrument.

Temperature Control

Temperature is not only the most important variable in industrial and commercial processes; it is also important in everyday life. Numerous items in the home and in automobiles, for example, require some degree of temperature control. For the purposes of this text, two broadly classified forms of temperature control will be considered: (1) controlling the flow of a fluid, as natural gas to the burner of a hot-water heater, and (2) controlling the flow of electricity through a heating element. The discussion of temperature control is simplified by the mere need to control flow as a function of temperature.

Safety Controls

Two forms of relief valves are designed for use as safety controls in liquid systems. One form contains a fusible plug that is designed to melt at some specified temperature (fig. 6.27*A*). Melting of the plug releases fluid from the vessel or pipe to which the relief valve is attached. Note also that a pressure relief valve is incorporated with the unit. The other form uses a bellows unit and sensitive bulb filled with an expansion fluid (fig. 6.27*B*). Excessive temperature causes the bellows to extend and force open the soft disk plug. Once the temperature returns to a safe level, the valve closes, and this fact represents an advantage over the fusible plug valve that, once blown open by excessive temperature, continues to release fluid until properly attended to.

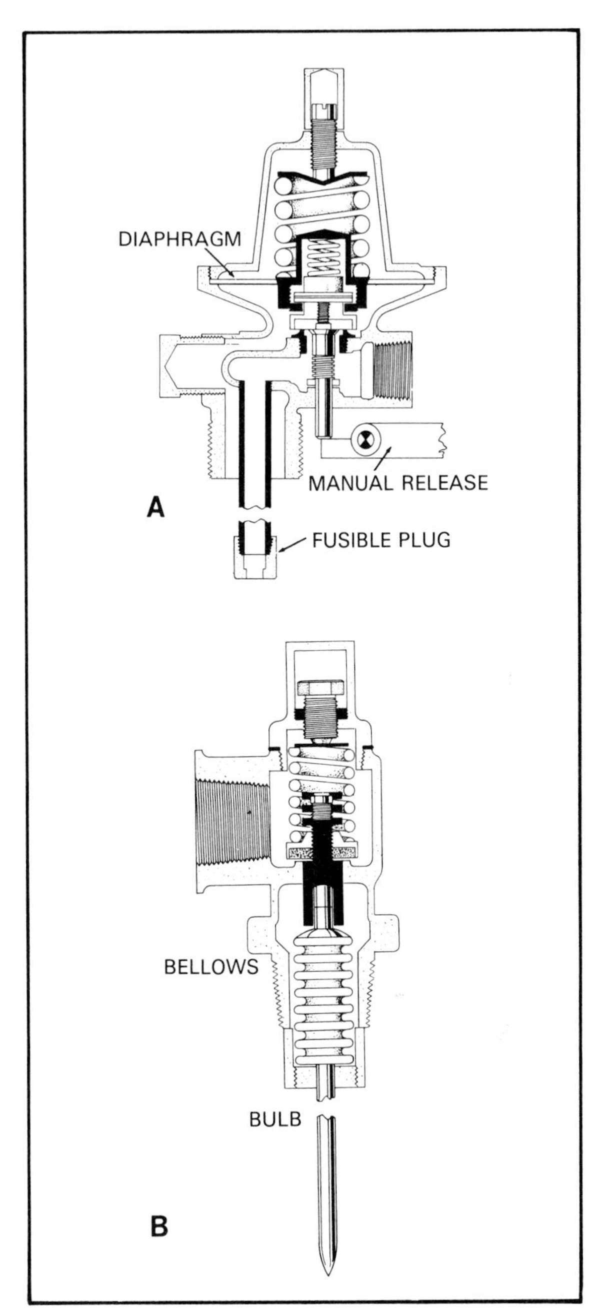

Figure 6.27. Temperature safety control valves. *A*, with fusible plug; *B*, with bellows unit and sensitive bulb.

Relief valves such as these find use on hot-water systems that are supplied by water sources of unpredictable pressure values. Ordinarily, it would be logical to feel that a pressure relief valve would also take care of excessive temperature conditions because excessive temperature will cause an increase in pressure. Close examination shows that for erratic source pressures, this logic is not valid. Suppose, for example, that a system receives water from a source whose pressure varies from 10 to 40 psi. A pressure relief valve would have to be set to some value above 40 psi in order to avoid unnecessary loss of fluid during periods of high source pressure.

On the other hand, if excessive temperature occurs during periods of minimum pressure, fluid will be pumped back into the source, unless a *temperature-sensitive* relief device is provided. Eventually the vessel will be boiled dry and possibly damaged severely.

Regulated Mixing Valves

The idea of mixing a hot liquid with varying amounts of a cooler liquid in order to obtain temperature control is popular for some uses. One common example is a combination hot-water heating system and domestic hot-water supply. The heating system typically uses water heated to about 190°F, a temperature that would be extremely dangerous for the household hot-water supply because of the severe scalding effect of water that hot. The household hot-water supply begins in a set of coils immersed in the 190°F water of the heating system tank or boiler and is then mixed with water from the regular cold-water piping system. A common form of self-operated mixing valve suitable for this type service is a three-way, piston-type proportioning unit (fig. 6.28). This valve allows the same rate of total flow, regardless of its piston position, but proportions the flow from each inlet in accordance with temperature. The operating mechanism consists of a relatively long temperature-sensitive expansion element that positions the piston. Fluid is forced to flow in a helical path about the element in order to provide rapid and uniform response to temperature changes. It is logical to suppose that other forms of three-way proportioning valves could be used with pneumatic controllers and diaphragm actuators to control temperature, but such systems are not common. The methods used to control temperature of fluids in industrial processes usually do not involve the direct mixing of two fluids, but rather the transfer of heat from one fluid to another through an intervening medium such as the metal wall of a vessel or pipe. Mixing valves are self-operated temperature regulators that control temperature by mixing two liquids.

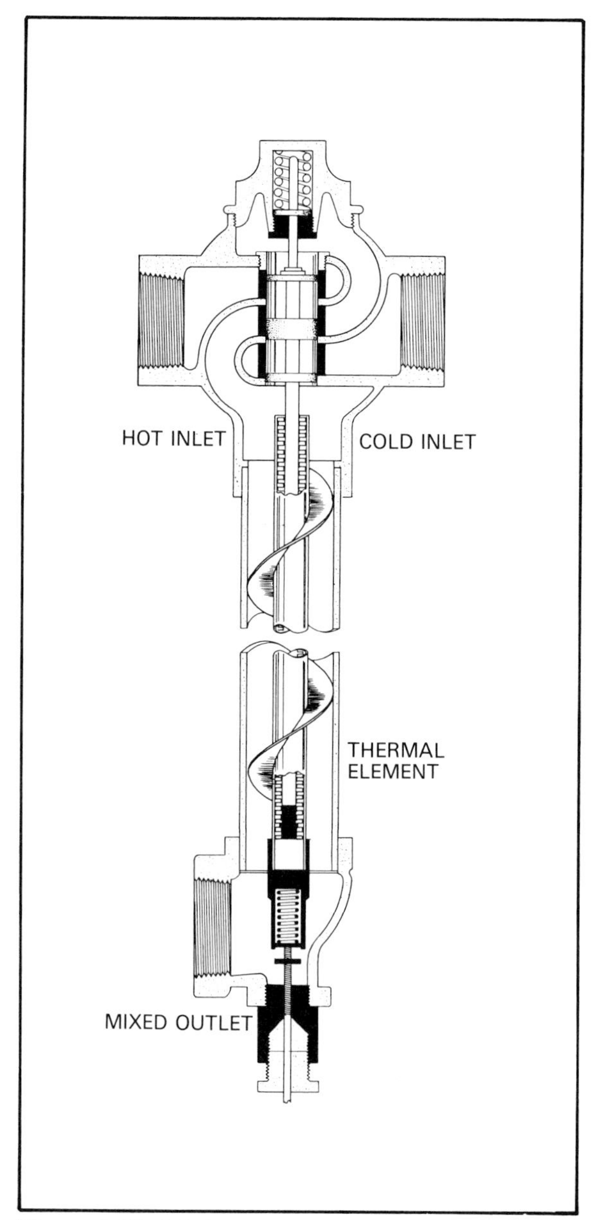

Figure 6.28. Mixing valve

Flow Regulators

Temperature regulators control the flow of a control agent, such as steam, hot water, or natural gas. A common form of temperature regulator is one used extensively with systems employing steam as a control agent

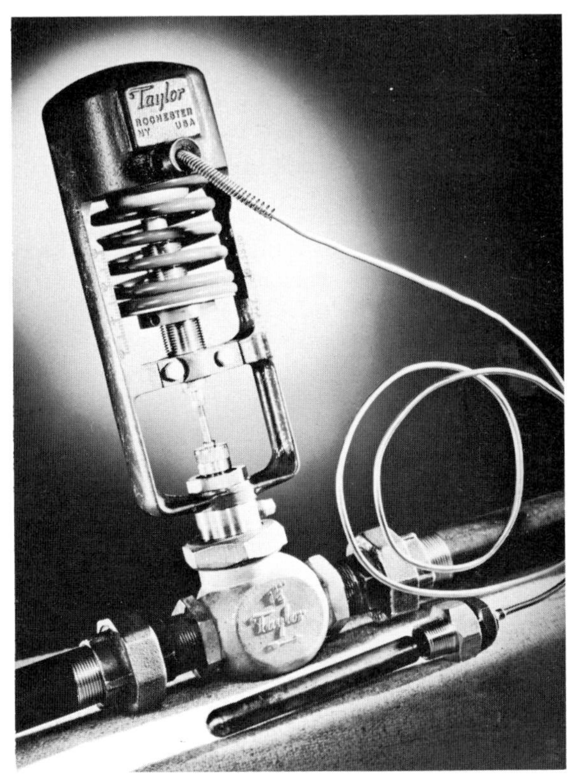

Figure 6.29. Temperature regulator using steam as a control agent

(fig. 6.29). In this type of regulator, the actuator is a vapor-pressure system consisting of a sensitive bulb, capillary tubing, bellows unit, and compression spring. The valve is a double-ported poppet globe valve designed to give smooth operation and throttling action. The unit is normally open and is therefore not a *fail-safe* device. Should the capillary tube,

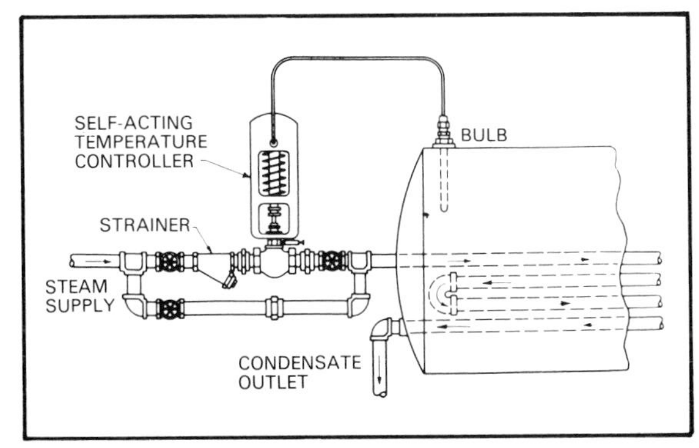

Figure 6.30. A possible piping installation using a steam-controlled temperature regulator)

bulb, or bellows fail due to leakage, this regulator provides no safety shutoff of the control agent.

Normal operation is as follows. The valve is installed in the steam line (fig. 6.30), and the sensitive bulb is fitted into the vessel whose temperature is to be controlled. The compression spring holds the double-ported valve open until the force due to pressure in the bellows unit overcomes the force of the extended spring. Pressure in the bellows, of course, is proportional to the temperature of the volatile fluid in the sensitive bulb. As pressure in the bellows causes the fluid to expand, the spring compresses and the valve plug moves toward the closed position. For steady loads, the valve plug will take up some position that represents a balance between energy input and energy output of the tank.

Special Controller Elements

Self-operated devices, including the mixing valves, are satisfactory means for maintaining desired temperatures for most purposes, but some installations require more precise control, remote indicators or recorders, or particular valve actions not readily available from self-operated regulators—reset and rate actions, for example.

Pneumatic controllers are adaptable to the control of most variables. In most applications pneumatic controllers use some form of Bourdon spring to actuate the linkage that varies nozzle-flapper clearance. In order to control temperature, any of the filled systems of measurement could conceivably be used to drive the Bourdon spring. The choice will depend on the range of temperatures involved, speed of response required, and distance between various points of the control loop.

One example of using precise control of temperature to regulate flow rate is in a fractionating tower capable of separating a feed supply into two fractions—that is, two hydrocarbon products. When operated properly, this system (diagrammed in fig. 6.31) is able to accept a steady input at the feed supply and produce the two fractions with acceptable purity, although additional stages may be required to achieve even greater purity. Several important variables are controlled to maintain equilibrium, including raw material feed rate, heat supply rate, and rate of withdrawal of the two fractions. A particular percentage of concentration of the upper fraction—the

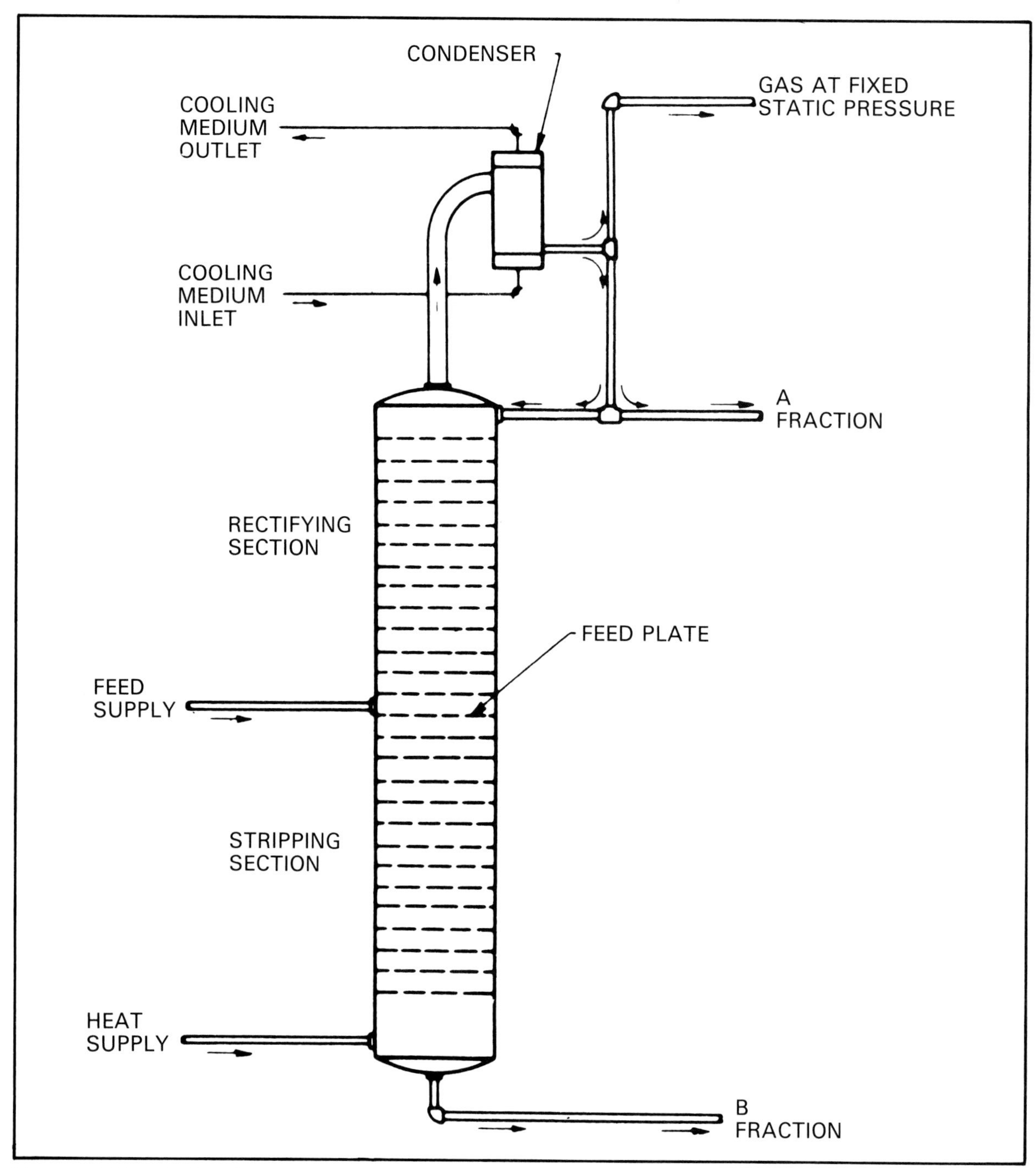

Figure 6.31. A fractionating tower requiring precise control of temperature (*Courtesy Taylor Instrument*)

most volatile of the two—is desired at all times, so this concentration is the variable whose control will affect the other variables. Obviously, if the rate of withdrawal of overhead fraction is too great, its content will include too much of the bottom fraction, and vice versa, so the rate of withdrawal of the upper fraction must be maintained at whatever rate is needed to assure the proper concentration.

Two variables, pressure and temperature, have a determining effect on the separation of the feed material into the two fractions, and a measure of the temperature at the upper fraction outlet indicates quite accurately the concentration of upper fraction. If the temperature can be controlled effectively, then the proper concentration of upper fraction can be controlled. The temperature controller (fig. 6.32) maintains the desired percentage of upper fraction by *regulating* the *rate at which the upper fraction is withdrawn* from the process. Of interest is that pressure control could have achieved the same end since the two variables, pressure and temperature, are interdependent in this process.

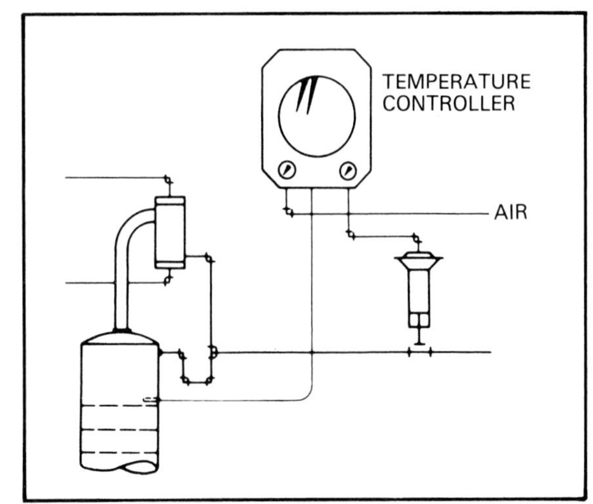

Figure 6.32. Temperature controller installation for fractionating tower (*Courtesy Taylor Instrument*)

Another example of temperature control is a temperature-controlled electroplating bath used to tin-plate continuous sheets of steel (fig. 6.33). The plating bath in this instance

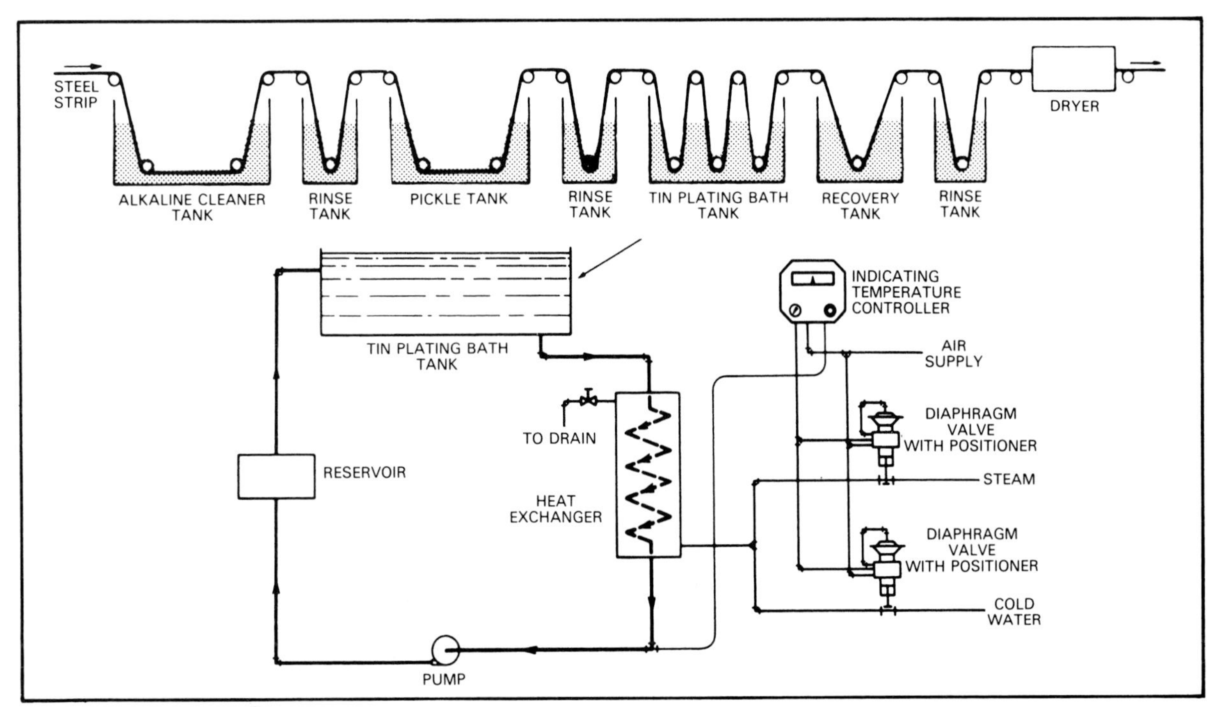

Figure 6.33. Temperature control system for an electroplating process (*Courtesy Taylor Instrument*)

requires a particular temperature for maximum efficiency; but for prolonged operation, internal heat caused by the heavy flow of electric current in the liquid increases the temperature excessively, and cooling is required. On the other hand, during the initial period of plating the temperature is too low, and heating must be used.

The heat exchanger shown in the figure is a copper coil surrounded by a jacket containing the cooling or heating liquid. The plating bath is circulated continuously and at a high flow rate through the plating tank and heat exchanger by a centrifugal pump. A temperature-sensitive bulb is connected near the lower end of the heat exchanger, and its capillary tube is connected to the Bourdon tube of the pneumatic controller. Output of the controller leads to two diaphragm valves with valve positioners. One of the two valves is direct-acting and controls the cold-water supply to the heat exchanger, while the other valve is reverse-acting and is in the steam line.

When the system is started up, the plating solution is apt to be cold, so the pneumatic controller will have a maximum control pressure output. This will close the cold-water line and open the steam valve wide, allowing maximum heating of the plating solution. As the temperature rises, the controller will begin to close off the steam valve and open the cold-water valve until equilibrium is finally established. Although two valves and a pneumatic controller are used in this hypothetical installation, a three-way proportioning valve probably would be capable of maintaining temperature with sufficient accuracy; however, for purposes of providing a permanent record of temperature, the pneumatic controller with temperature recorder has a definite advantage.

Summary

The measurement of temperature has been a matter of interest to people for a long time. Temperature was measured with crude devices long before anyone had a proper understanding of what it was. Today various scales and devices are used to measure and record the extensive range of temperature values.

Major types of measuring devices are liquid-in-glass thermometers, filled-system thermometers, bimetal thermometers, electrical measuring devices, and pyrometers. Of these, liquid-in-glass thermometers are commonly found in domestic and laboratory use, while filled-system thermometers are used extensively in industrial processes. Bimetal thermometers are rugged as well as accurate within a given range, and some bimetal elements are used to provide corrective factors in filled systems. Electrical systems are used for measuring a wide range of temperatures accurately, and pyrometers are designed for measuring extremely high temperatures.

Temperature, like pressure, is a part of the natural environment. Unlike pressure, which is comparatively stable, the natural fluctuations of temperature are of such magnitude that special controls are needed for protection against its extremes and for making it functional.

Much similarity exists between temperature measurement and control and pressure measurement and control. In both, Bourdon tubes and other pressure-sensitive elements are used extensively to actuate indicators, recording pens, and controller elements. In later chapters, particularly those concerning the measurement and control of liquid level and flow, this similarity becomes even more apparent.

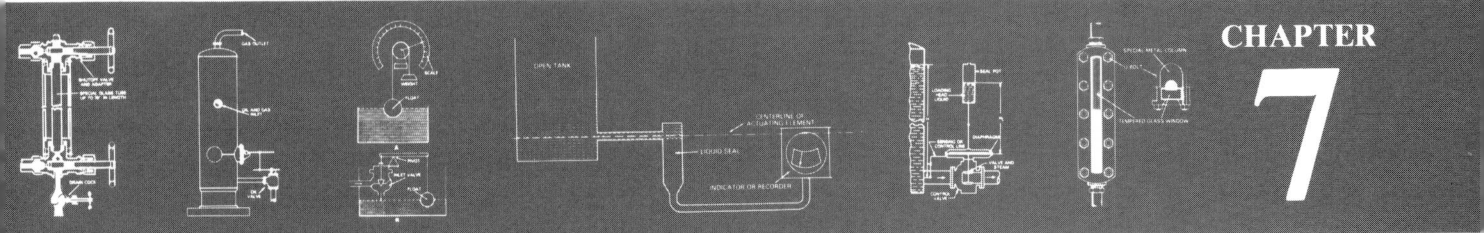

CHAPTER 7

Liquid-Level Measurement and Control

Of all the variables that are of interest in measurement and control, liquid level is certainly the most straightforward. It is the variable most easily measured directly, and it is unique in its simplicity of dimension. Whereas pressure is force per unit area, flow is volume per unit time, and temperature is the measure of the nebulous activity of molecules, liquid level is merely a measure of length. Despite its simple character, liquid level lends itself readily to numerous means of inferential measurement.

The measurement and control of liquid level does not rival either the incidence or the importance of that associated with temperature for industrial processes, but it shares with flow measurement a position of immense importance in certain applications that closely relate to the delivery or sale of liquid products. It is particularly important in the petroleum industry, where the determination of liquid level in many cases has a direct bearing on the amount of money that changes hands between customer and producer. In such instances the accuracy of liquid-level measurement and control must be of highest practicable quality.

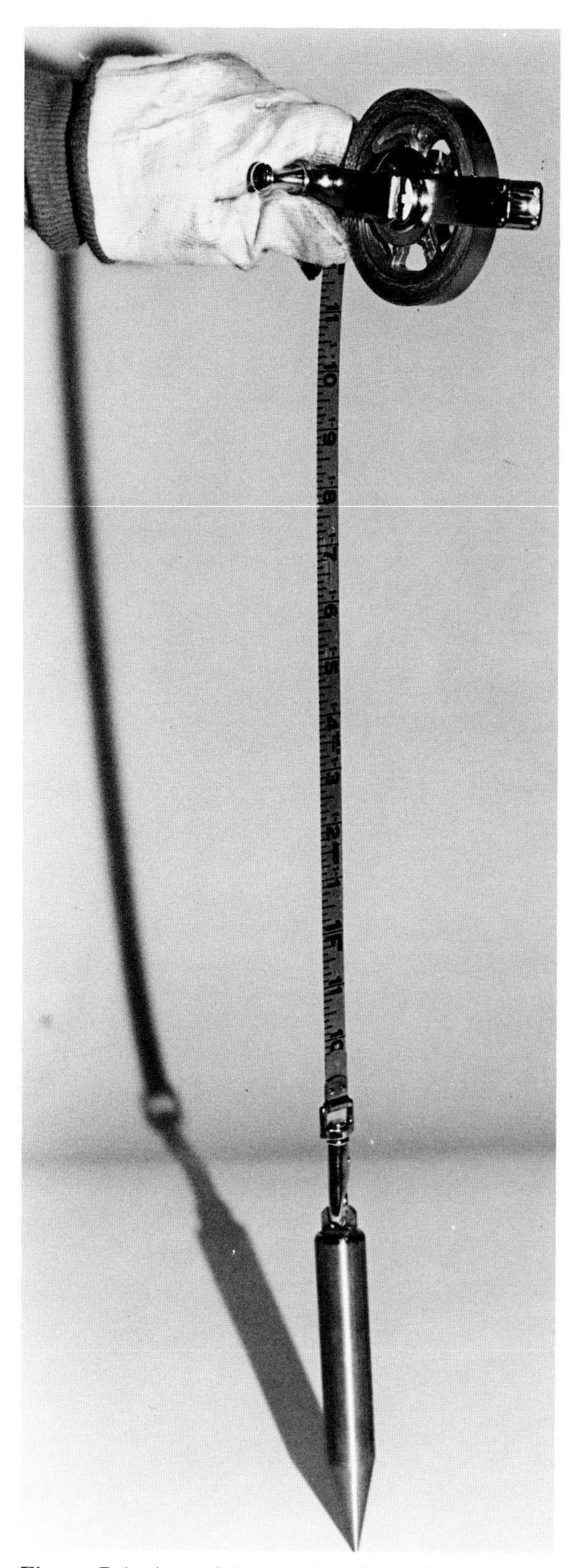

Figure 7.1. A steel tape and bob used to measure liquid level in stock tanks

Direct Measuring Devices

The earliest form of liquid-level measurement was probably the use of sticks or poles to test the depths of streams. This system of measurement is still used far more widely than one might suppose. *Dipsticks* used to check the oil level in automobile engines and transmissions is just such a primitive, though satisfactory, system. *Gauge rods,* or sounding rods, such as those used at gasoline service stations and for the fuel-oil supply of certain domestic heating systems are still used for measuring the quantity of liquid in buried tanks.

Flexible lines fitted with end-weights and called *chains,* or lead lines, have been used for centuries by seafaring men to gauge the depth of water under their ships. Steel tapes having weights like plumb bobs and stored conveniently on a reel are still used extensively for measuring liquid level in fuel-oil bunkers and petroleum storage tanks. Figure 7.1 shows such a liquid-level gauge used for measuring or, more correctly, *gauging* the height of oil in a petroleum stock tank. Gauges of this type may be 18 feet or more in length. Their accuracy is quite good at a standard temperature of 60°F.

Gauge cocks can be used as a means for checking liquid level in a tank or other reservoir. They are merely finger- or lever-operated petcocks, which are normally closed. If liquid flows out when the cock is opened momentarily, the level of liquid must be at least as high as that of the gauge cock. The Model T Ford automobile of many years ago used two gauge cocks as a means of checking engine oil level, but today such cocks are found on engines only as secondary devices or as rough checks of the proper operation of other liquid-level indicators.

Another form of simple liquid-level measurement is the *sight glass,* also called a gauge glass. Sight glasses serve extremely well for many applications and are of two basic types (fig. 7.2). The type shown on the left is suitable for gauging an open tank, or any

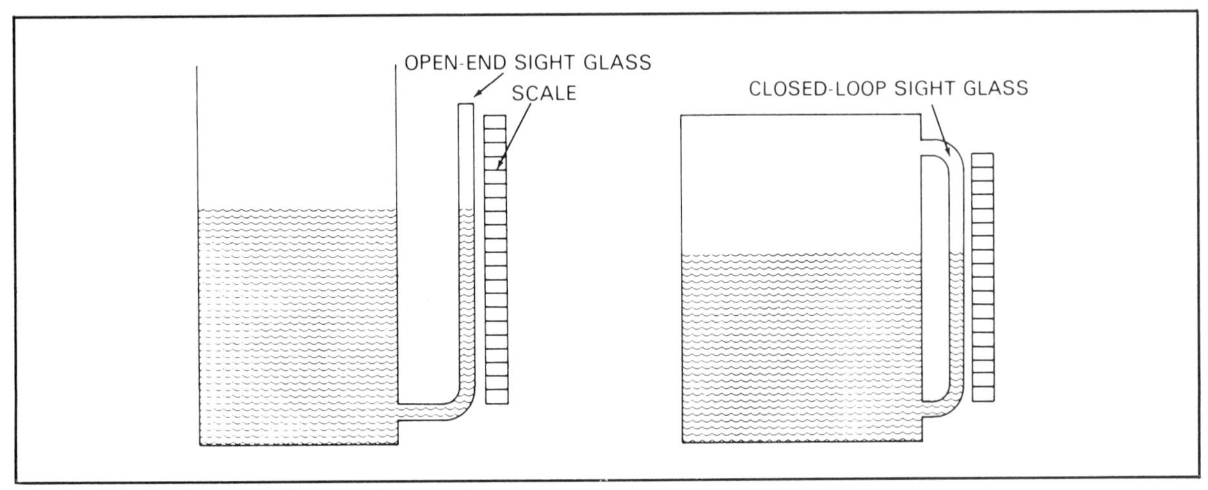

Figure 7.2. Basic types of sight glasses: *A*, open or vented vessel; *B*, pressurized vessel.

vessel with a free vent to the atmosphere; it is seldom used today, although it is occasionally seen as a lubricator on old types of stationary engines. The other type shown represents the form found on pressurized vessels, although its use is also desirable for nonpressurized vessels in most instances because it forms a closed system and therefore cannot allow a loss of liquid. Gauge glasses provide a visual indication of liquid level that is continuous and are therefore superior to dipsticks, tapes, and rods in this respect.

One form of sight glass (fig. 7.3) consists of a glass tube fitted between two special valves. Adapters can be used for fitting the gauge to the vessel whose liquid level is to be monitored. The purpose of the metal balls is to block off fluid flow should the sight tube rupture. Ordinarily the balls allow free passage of fluid, but should the velocity of flow reach an appreciable value, as it would in the case of a ruptured tube, the balls would be driven tightly against their seats and the loss of fluid stopped. Tubular glass for this sort of gauge is available in lengths up to about 6 feet and for pressure up to 600 psi, although safe operating pressure decreases with increasing length. Tubular glass gauges are used extensively on low-temperature vessels and low-pressure steam boilers.

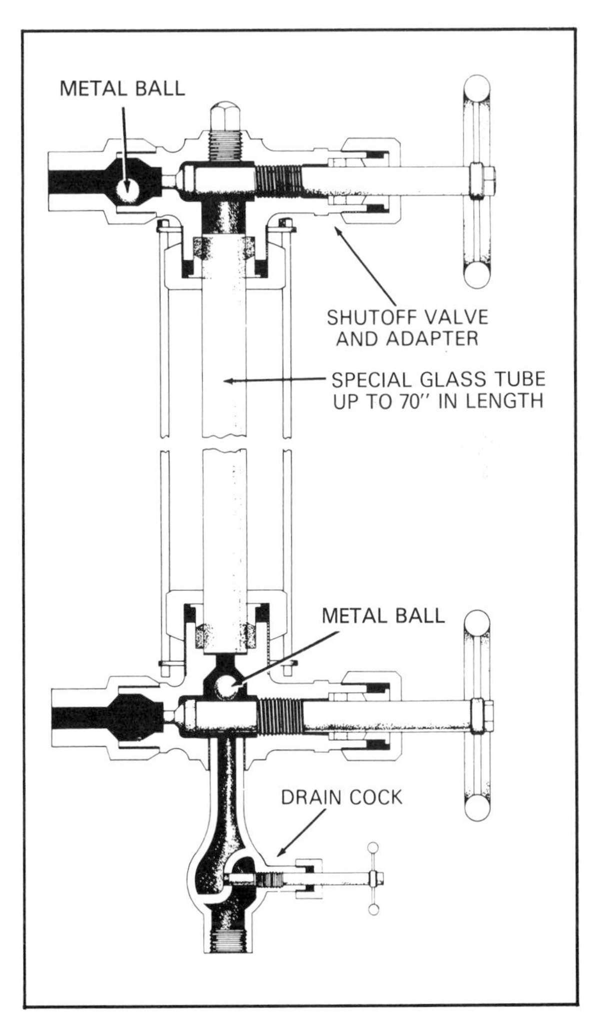

Figure 7.3. Sight glass for low-pressure, low-temperature systems

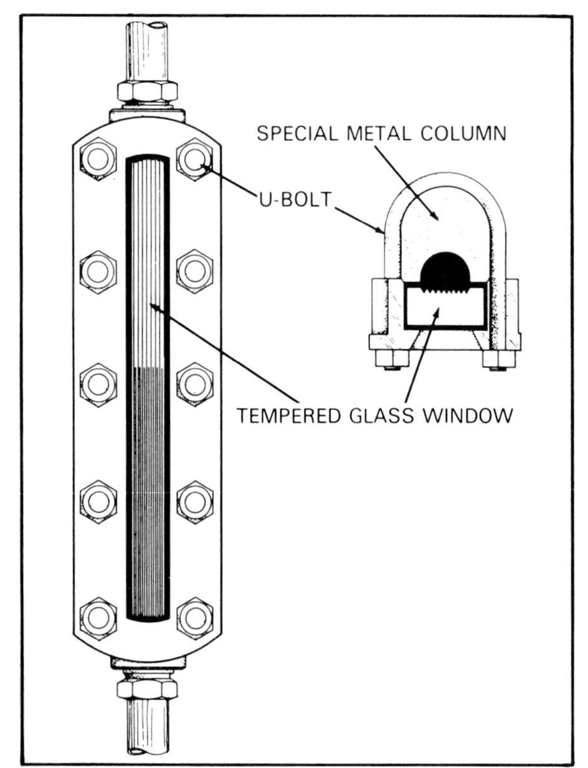

Figure 7.4. Sight glass for high-pressure, high-temperature systems

For boilers of high pressure and temperature, the form of gauge shown in figure 7.4 is usually used. It is made of heavy brass and heavy flat glass windows. Sometimes the windows have grooves cut in one side, as shown in the figure, and this feature provides a very clear demarcation at the liquid level. It is a desirable feature when the liquid is clear and colorless, like water in a boiler. The distinct line of demarcation arises from the fact that light entering the glass plate above the level of the liquid strikes the surface of the grooved side of the plate and is reflected in a way that produces considerable light. On the other hand, light passing through the glass plate below the upper level of the water or other colorless liquid finds water in contact with the grooved surfaces, and since the water has a higher index of refraction than the vapor or steam above it, the light is reflected in a way that diminishes the amount of light appearing at the outer surface of the sight glass.

Instruments Using Buoyancy

Many forms of buoyancy-type instruments are available, but each uses the principle of a buoyant element that floats on the surface of the liquid and changes position as the liquid level varies.

Direct Operation

One form of such a liquid-level indicator is capable of indicating from zero to maximum level (fig. 7.5*A*). Many other float-type indicators and controllers have float elements

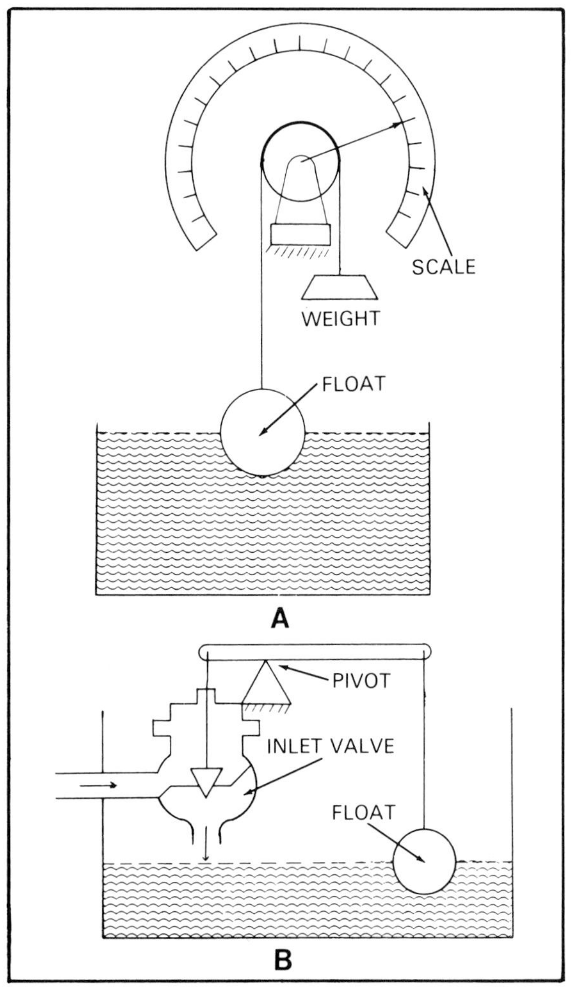

Figure 7.5. Buoyancy-type instruments with (*A*) full-range indicator and (*B*) restricted-range controller.

whose vertical motion is restricted to a given range (fig. 7.5*B*) by a ball float controlling an inlet valve. Motion of the float is restricted to perhaps a fourth of the overall possible liquid-level change by the limits of travel of the valve stem. This simple form of level control is in widespread use for controlling water level in livestock troughs and in water closets.

A commercial form of float-operated liquid-level controller (fig. 7.6) is designed to be mounted in a flanged hole of a pressurized vessel or mounted on a float chamber. Overall vertical travel of the float is limited by the length of the float rod and the angle of arc through which it is free to swing. Motion of the float turns a shaft that protrudes from the flanged housing through a stuffing box. The linkage attached to the rotary shaft can be arranged either to close or to open the valve

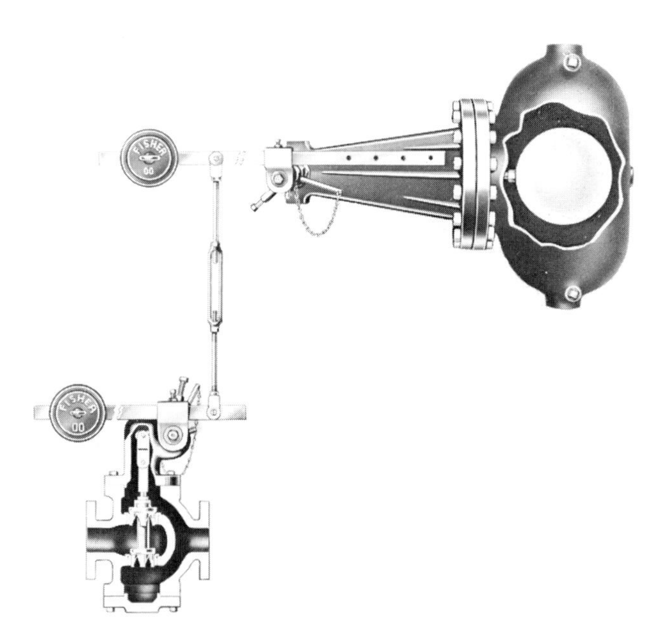

Figure 7.6. A float-operated liquid-level controller mounted on a float chamber (*Courtesy Fisher Controls*)

with a rise in liquid level. Liquid-level controls of this sort can be used to control dump valves on oil and gas separators (fig. 7.7).

When the float element is placed in a float chamber mounted external to the main vessel,

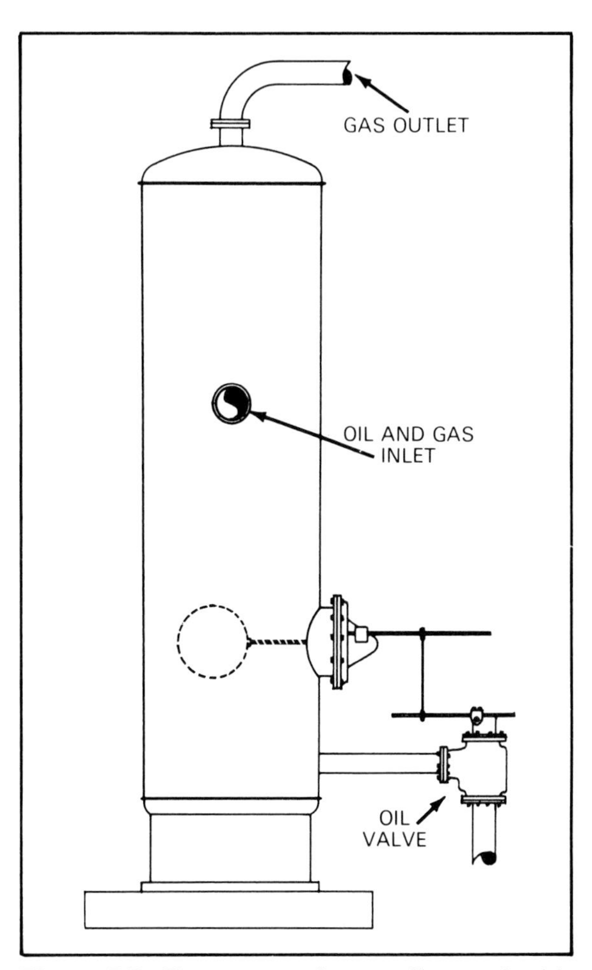

Figure 7.7. Float-operated controller used with dump valve on oil and gas separators (*Courtesy National Tank*)

the arrangement has the advantages of providing easy servicing and being free from effects of turbulence that may exist in the main vessel.

Pilot-Valve Operation

In buoyancy-type control devices, the force needed to operate the control valves is provided by the buoyant effect of the float, and consequently the suitability of such a control system is seriously limited for some applications by the frictional and other forces present in the valve and the rotary-shaft stuffing box. The use of a pilot valve removes from the float mechanism that part of its load caused

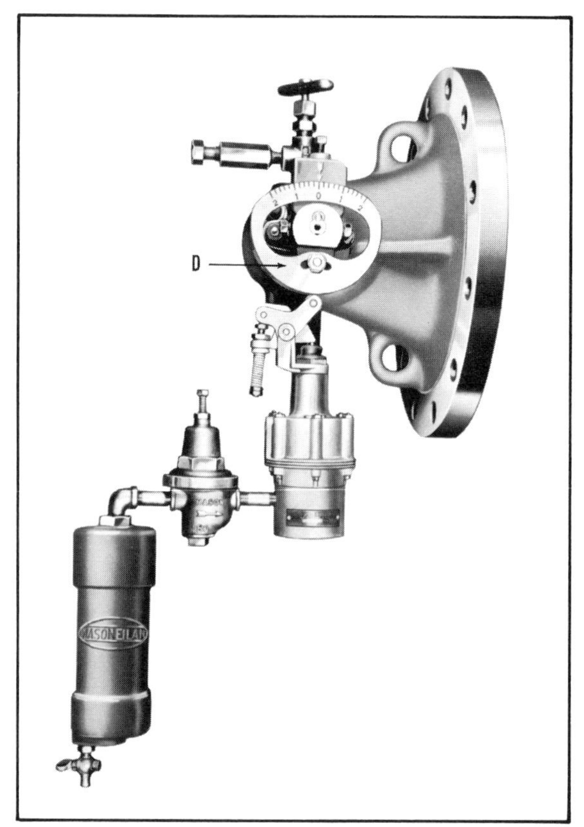

Figure 7.8. A pilot-operated controller (*Courtesy Mason-Neilan*)

by valve operation. Figure 7.8 shows a pilot-type controller that is flange-mounted.

Operation of the pilot is as follows (fig. 7.9). Air or other suitable gas under pressure is applied at *A*. An adjustable cam is attached to the rotary shaft and is therefore driven by the motion of the float element; adjustment of the cam is made at *D* (fig. 7.8). Roller *B* of the pilot-valve assembly rides against the cam and is linked to a pedal that actuates pilot stem *C*. Downward movement of *C* acting through steel spring *E* closes exhaust port *M* and opens inlet port *L*, allowing supply air to flow into the supporting cavity and out *H* to the control-valve diaphragm. As pressure on the diaphragm and in the lower cavity of the pilot valve increases, bellows *F* and *K*, whose exteriors are under the same pressure because of the equalizing port *G*, are compressed against spring *E*. Eventually spring *E* will be compressed until, at equilibrium, inlet valve *L* is closed, and pressure on the diaphragm actuator is maintained at a fixed value. Should the cam move in the opposite direction, the exhaust port will open, allowing pressure on the actuator to bleed off until equilibrium is again established.

Pilot-operated liquid-level controllers are capable of giving excellent control and can be made to withstand considerable pressure.

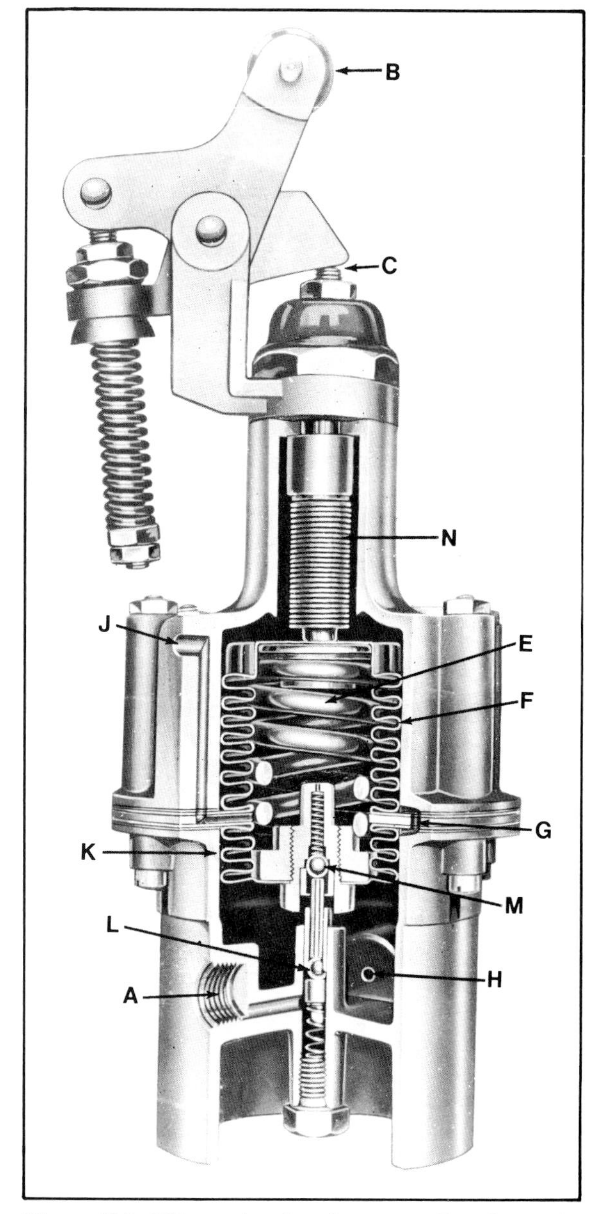

Figure 7.9. Pilot valve for the controller shown in figure 7.8 (*Courtesy Mason-Neilan*)

Friction in the rotary-shaft stuffing box can be a source of error, but with modern Teflon packing and the use of thinner shafts, made possible by the decreased load, the error is likely to be insignificant. Possible change in specific gravity of the liquid under control can be another source of error. A decrease in specific gravity of the liquid will allow the float to sink deeper, thus producing an error. An increase in specific gravity will cause the level indicator to read high. Both density and specific gravity change with temperature; this change for many liquids is significant and must be taken into account. Perhaps more significant than changes due to temperature variations is the fact that the density of the fluid varies with its quality. This is particularly significant in the case of petroleum and its products.

Instruments Using Displacement Elements

Indicating and control devices using displacement elements resemble the buoyancy type but use a float that is heavier than the maximum amount of liquid that it can displace. The advantages of using displacer floats with torque tubes or flexure tubes include the absence of stuffing boxes and the friction they can cause, the wide range of level that can be controlled, and the accuracy of control. Special applications can also be achieved. For example, the level of an interface between two immiscible liquids like gasoline and water can be determined and controlled. Such use requires that a difference in relative density exists between the two liquids and that the displacer float is completely submerged. A similar application, again with totally submerged displacer float, enables the specific gravity of liquids to be measured.

Displacer Float and Torque Tube

One displacement device consists of a displacer float installed in the float cage and connected to a float rod (fig.7.10). Figure 7.11 shows a sectional view and details of the torque tube that is an important part of this type of indicator and control system.

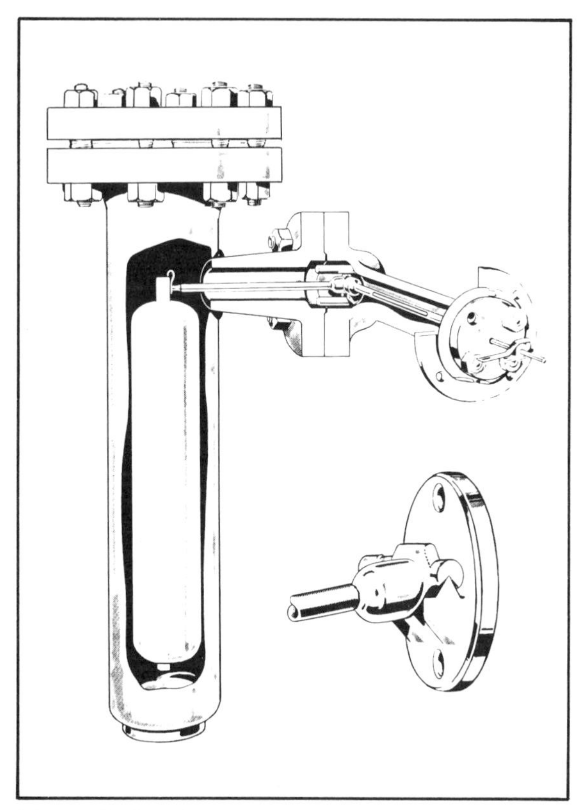

Figure 7.10. Displacer-float assembly of a torque tube displacer-type level indicator and controller (*Courtesy Fisher Controls*)

The typical design of a displacer float produces a force of about 5 pounds in air and a force of about 1 pound when fully submerged in water. The downward force on the float rod thus varies from a maximum of about 5 pounds when there is no water in the cage to about 1 pound when water fills the cage to a point above the displacer float. Obviously the value of the downward force existing when the cage contains liquid will be an inverse function of the density of the liquid; that is, the force will increase as the density of the liquid decreases.

The important action that accomplishes the change in force exerted by the displacer float on its float rod may be visualized with the aid

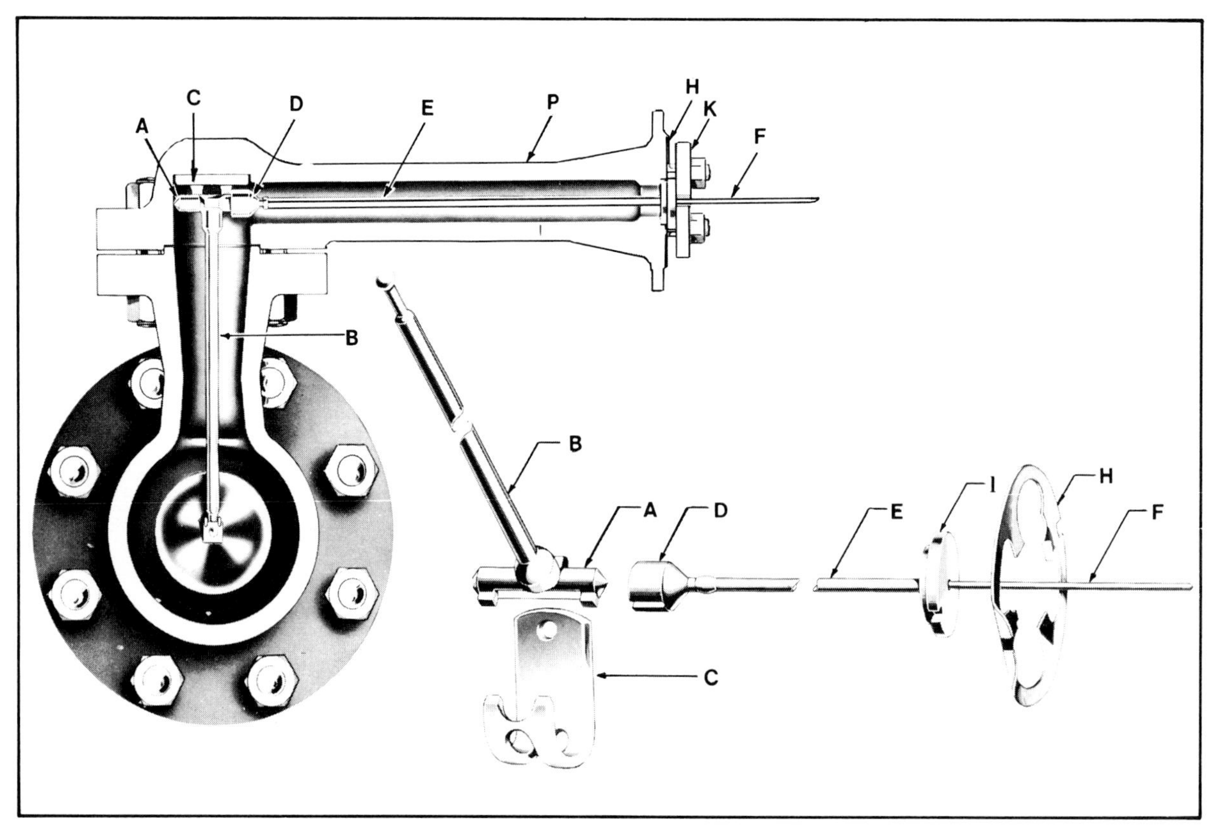

Figure 7.11. Torque-tube unit of a displacer-type level indicator and controller (*Courtesy Fisher Controls*)

of figure 7.11. *A* and *B* are the float-rod driver and the float rod, respectively, while *C* is a bearing plate containing knife-edges on which *A* rests. *D* is a female socket that fits tightly over *A*. *E* is a hollow torque tube welded to *D* and to *I*, the outer torque tube flange. *F* is a thin rotary shaft that fits through *I* and the torque tube and is welded to socket *D*. *H* is a steel plate used to position the torque tube and the float, while *P* and *K* are torque-tube housing and retaining flange, respectively. The important action occurring because of the weight of the displacer float is the twisting of torque tube *E* and the rotation of rotary shaft *F*, which is free to turn at outer tube flange *I*. The rotary shaft *F* extends into the housing of a pneumatic controller, where it positions the flapper of a nozzle-flapper assembly.

A complete control system using the displacer-float and torque-tube idea is shown schematically in figure 7.12. Flapper *B* is attached to the end of the rotary shaft shown as *F* in figure 7.11. The nozzle assembly is unusual, as it receives air from a tube that is fitted inside a Bourdon tube. Air from the supply-line reducer is fed to the air relay, which contains the fixed orifice through which air flows to chamber *L* and to the nozzle. Chamber *L* contains nozzle back-pressure, and this pressure is exerted on diaphragm *M*, causing inlet valve *O* to open, admitting air to chamber *N* and the control valve. To establish equilibrium, the pressure in *N* works against diaphragm *P*, moving it up until inlet valve *O* is closed. If the nozzle-flapper clearance should increase, causing a drop in back-pressure, the equilibrium would be upset, and the pressure in N would push up diaphragm P and open bleed-off valve K, allowing air to escape to the atmosphere until equilibrium was again established.

The same air pressure that is applied to the control-valve actuator is also applied to three-way valve assembly H. A manual adjustment of this valve positions the plug between inlet port I and outlet port G. When G is shut off completely, the full control valve diaphragm pressure is also applied to Bourdon tube C. The Bourdon tube therefore pulls the nozzle away from the flapper. If the three-way valve is positioned to shut off I completely, the Bourdon tube will remain stationary for any pressure applied to the control valve, and a very sensitive control action will be observed. The three-way valve is used to adjust the proportional band from 0 to 100%. The proportional band adjustment will vary considerably with the specific gravity of the liquid whose level is to be controlled. Charts are supplied by manufacturers for obtaining correct settings.

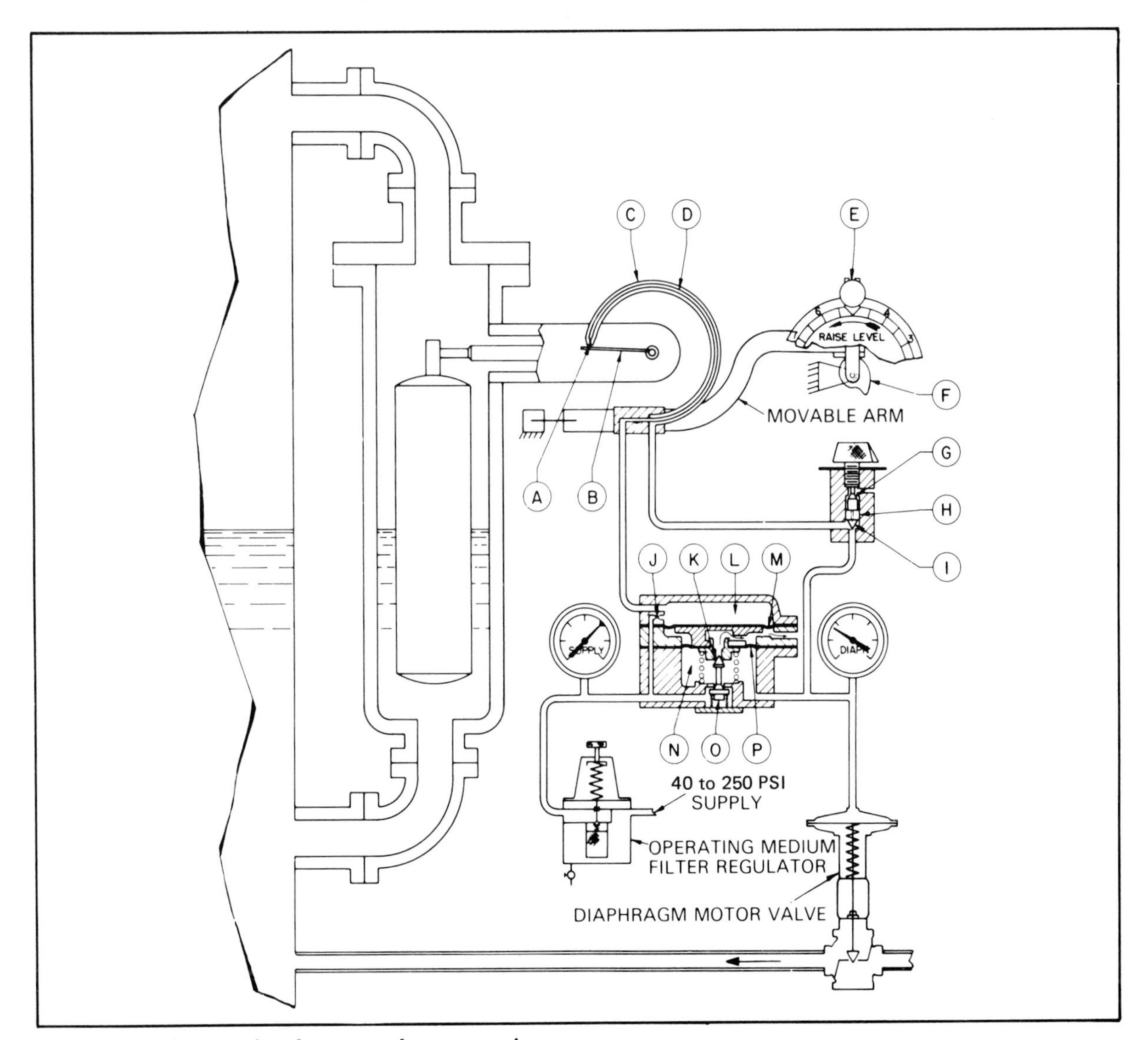

Figure 7.12. Schematic of a control system using displacer-type level indicator and controller (*Courtesy Fisher Controls*)

The level-set dial operates a cam that changes the relative position of the nozzle with respect to the flapper. The liquid level can be controlled to any value that lies between the top and the bottom of the displacer float. Floats are made in various lengths from about 1 to 15 feet.

The system shown in figure 7.12 is controlling the flow *into* the vessel in order to maintain desired liquid level. The block on which the Bourdon tube nozzle assembly is mounted can be easily repositioned to provide reverse action. Liquid level will then be controlled by regulating the *discharge* from the vessel.

Displacer Float and Flexure Tube

In another form of displacer-float device (fig. 7.13), a flexure tube rather than a torque tube is used. The shaft within the flexure tube serves a purpose similar to the rotary shaft in the torque-tube device, but it does not rotate. Its free end is positioned according to the position of the displacer float and can be made to move a flapper or to operate a pilot valve. This simple system gives excellent results when an on-off action is acceptable.

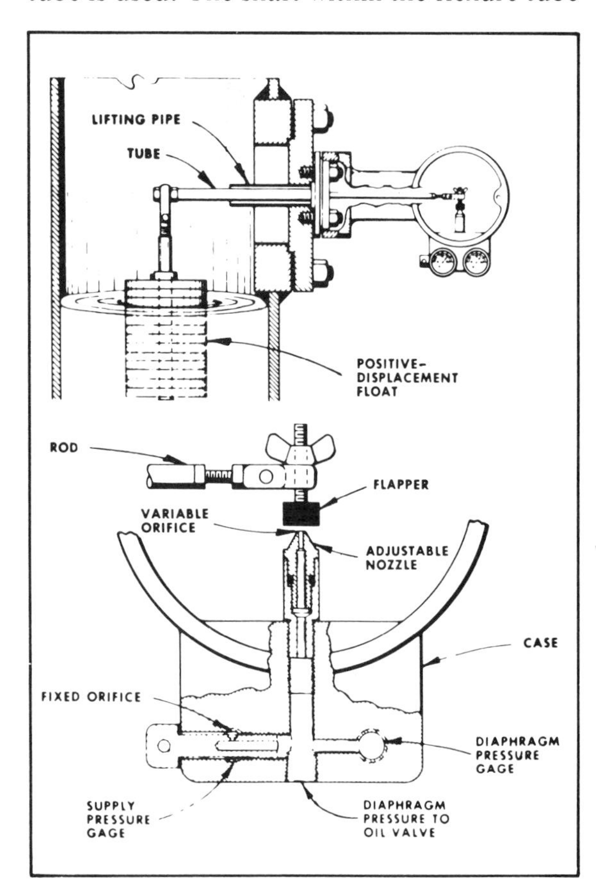

Figure 7.13. Displacer-type indicator and controller with flexure tube (*Courtesy C-E Natco*)

Instruments Using Hydrostatic Pressure

Many instruments using hydrostatic pressure to measure and control liquid level exist. Their operation depends on the basic physical laws of hydrostatic pressure.

Principle

Pressure caused by the height of a liquid column, that is, hydrostatic head, can be used to measure and to control liquid level. For each foot of vertical height, a column of water will produce a pressure of approximately 0.433 psi. Other liquids, of course, will also produce pressure under these conditions, and the value of the pressure will vary with the specific gravity of the particular liquid. A simple formula for predicting the hydrostatic pressure (p) is—

$$p = 0.433 \text{ psi/ft} \times G \times H$$

where G is the specific gravity of the liquid and H is its height in feet above the measuring point. Since the specific gravity is a factor in the pressure value, temperature or any other change that affects specific gravity will also affect the accuracy of the measurement.

Although hydrostatic pressure due to liquid column height can be used with closed and pressurized vessels, it is necessary that a particular arrangement be used to neutralize the effect of positive gauge pressure bearing on the surface of the liquid.

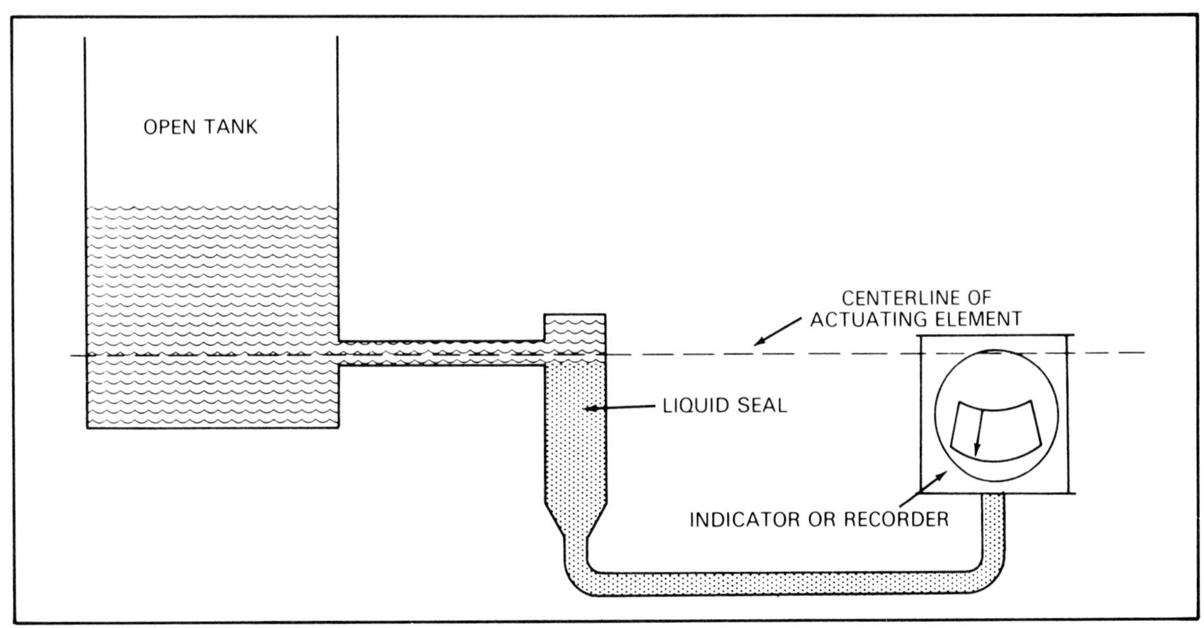

Figure 7.14. Liquid-seal type of hydrostatic pressure measuring system

Measurements in Open or Vented Tanks

Simple method. Figure 7.14 is representative of a simple system of liquid-level measurement in an open tank or in one vented to the atmosphere. A seal is installed in the line from tank to level indicator-recorder to prevent tank liquid from reaching the latter. This seal is necessary if the tank liquid is corrosive, viscous, or in any way likely to clog the lines leading to the instrument. The centerline of the actuating element—probably a bellows or diaphragm pressure unit calibrated in liquid-level units—is located at the point of minimum level to be measured, although the indicator can be mounted lower if the additional head is accounted for. Obviously, if the actuating element of the indicator-recorder is also made to position a flapper, this system can be used with a pneumatic system to control liquid level by regulating the flow into or out of the tank. In a similar way the actuating element can operate a mercury switch and electric control system to achieve the same purpose.

Diaphragm-box method. Another open-tank method of liquid-level indication and control is shown in figure 7.15. The hydrostatic pressure detecting element in this case is a *diaphragm box.* An open type is shown in figure 7.15*A*, a closed type in 7.15*B*. They are similar, but one has either an entirely open lower case or a case containing several vent holes, while the other, intended for mounting external to the vessel, has a closed lower case with a pipe fitting. Hydrostatic pressure in either box is applied against a soft Neoprene

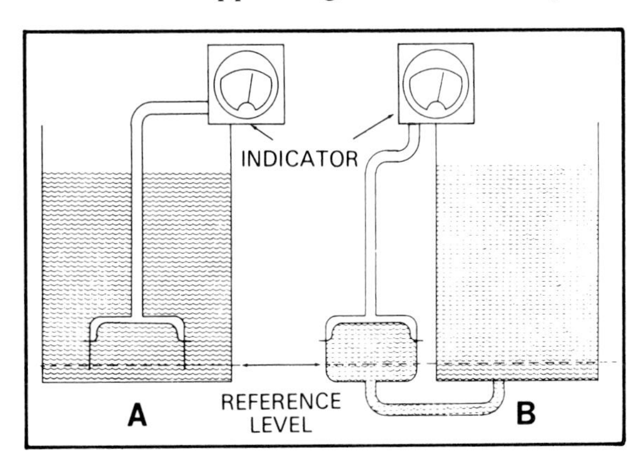

Figure 7.15. Diaphragm-box type of hydrostatic measuring system

diaphragm. Piping or tubing leads from the upper diaphragm to a sensitive pressure-type indicator, recorder, or controller, calibrated for liquid level.

The diaphragm-box system can be placed in operation most simply by starting with zero liquid level, that is, by lowering the liquid level to the same elevation as the diaphragm. The air trapped in the upper chamber of the diaphragm box, the tubing, and the actuating element is then allowed to assume atmospheric pressure by opening a vent valve for a few seconds. After the air in the system has become equalized with the atmosphere by venting, the vent valve is tightly closed, because it is important that a constant quantity of air remain in the system.

As the liquid level in the vessel rises above the zero reference line, the hydrostatic head will exert a pressure on the diaphragm in direct proportion to the rise in level. The pressure will be 0.433 psi/ft $\times G \times H$, where G is the specific gravity of the liquid and H is the height of the liquid level in feet above the zero reference line. Assuming that the Neoprene diaphragm is flexible enough to offer only negligible resistance to the hydrostatic pressure, the diaphragm will be deflected until the air pressure in the upper chamber becomes equal to the hydrostatic pressure bearing on the underside of the diaphragm. The actuating element in the indicator or controller will respond to the air pressure, which in turn indicates the liquid level.

An advantage of the diaphragm-box method of liquid-level measurement and control lies in the fact that the actuating element of the indicator or controller is not critical in regard to its elevation with respect to the liquid level being monitored, because the weight of the air in the system is insignificant and thus presents no appreciable head.

The open-type diaphragm box is not suitable for use in corrosive liquids, but the externally mounted closed type is satisfactory if a liquid seal is used to isolate it from direct contact with the corrosive liquid. This type of box can also be adversely affected by the temperature of the liquid in contact with the diaphragm. Temperature above 150°F is apt to have a deleterious effect on diaphragm material.

Air-trap method. The air-trap method is similar to that using a diaphragm box but overcomes the objections of internal mounting in corrosive liquids and the effects of high temperature (fig. 7.16*A*). It involves an open box with no diaphragm. Air is simply trapped under the box, and as the level rises in the vessel its rise in the box is resisted by the trapped air that must be compressed in proportion to liquid-level rise. Since the box can be built of material that is immune to attack by high temperature and corrosive liquids, this system can operate in very rough environments. However, the liquid may absorb some of the trapped air, and thus accuracy may be affected. This method, as well as the diaphragm-box method, requires a very tight system to prevent the escape of the air trapped in the measuring components. Any loss of air in the diaphragm-box system will destroy the accuracy of the system, and loss of air in the air-trap system may allow contaminating or corrosive liquid to reach the tubing or actuating element of the indicator or controller.

Air-bubble or air-purge system. A final example of an open-tank liquid-level measuring and control method based on hydrostatic pressure is the air-purge system, more commonly called the air-bubble system (fig. 7.16*B*). The same form of indicator, recorder, or controller as used with other hydrostatic methods is used with the air-purge system, and relative elevation of the instrument is not important.

A pipe made of material capable of resisting corrosive action of the liquid is placed in the vessel, its outlet end safely above the sediment level of the vessel. The outlet end represents the zero reference level. Regulated air pressure is applied to the bubble tube

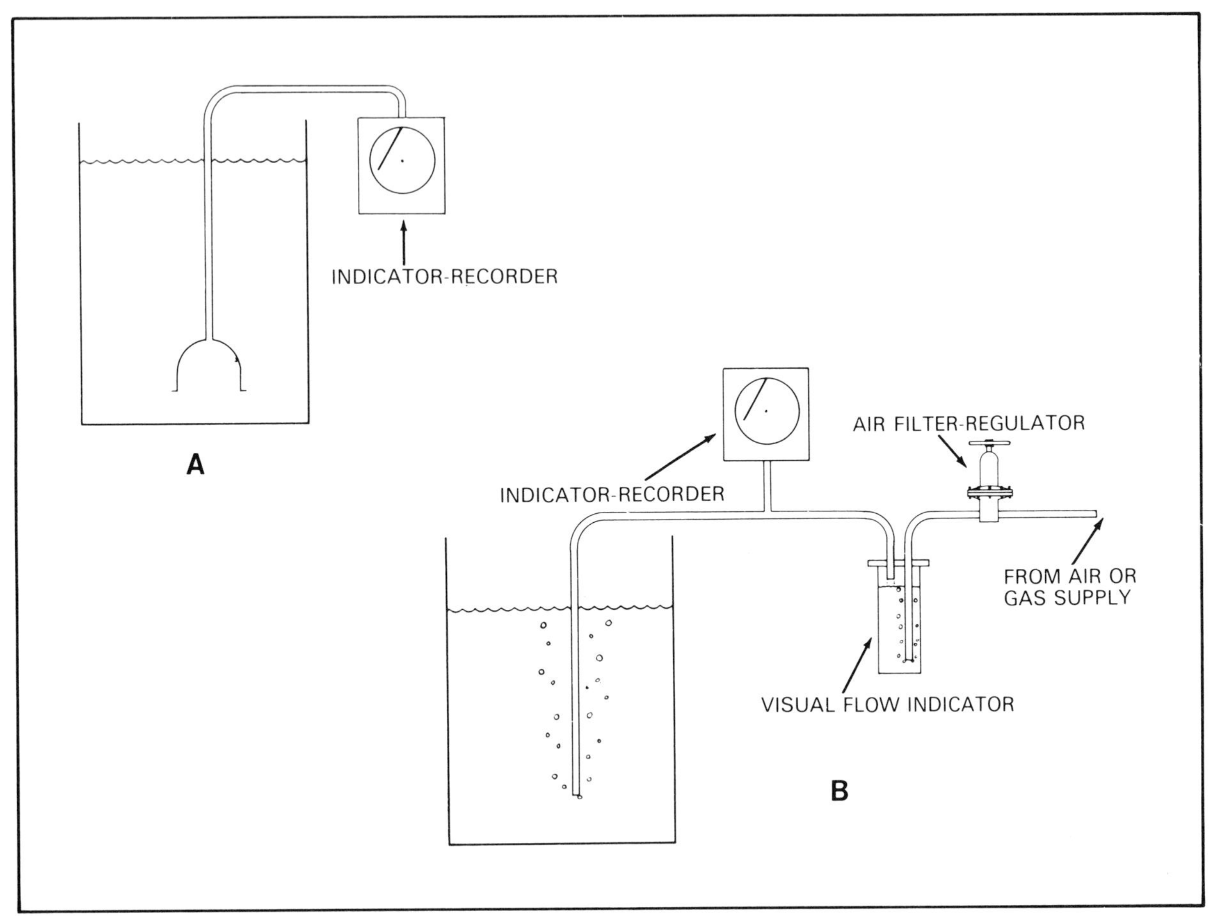

Figure 7.16. Two types of hydrostatic pressure systems: *A*, air trap; *B*, air bubble.

through a positive flow indicator. Some bubble systems use a rotameter to indicate flow. The positive flow indicator shown in the figure is a simple device made of a short length of glass tubing tightly sealed at either end with caps and partially filled with liquid. Air from the regulated supply bubbles to the surface of the liquid from the tube that extends nearly to the bottom cap. The accumulation of air at the surface then flows out to the bubble tube and to the indicator or controller.

The bubble tube is large enough to pass air freely if its open end is not obstructed, but as water rises in the vessel above the outlet end of the tube, an increasing value of pressure is required to overcome the hydrostatic head and allow the air to escape at the open end of the tube. Here again, the pressure needed to equal that due to the hydrostatic head is 0.433 psi/ft $\times$ G $\times$ H. If there is a slow, regular flow of air through the tube, the pressure necessary to maintain this flow will be reflected in the indicator.

The importance of the positive flow indicator lies in the fact that if the hydrostatic head should exceed the regulated pressure applied to the bubble tube, the flow of air through it would cease, and the indicator-recorder would indicate the value of the regulated air supply. Any value of hydrostatic pressure higher than the pressure of air supplied to the system would go undetected by the indicator, and indications would be false until the liquid level returned to the point at which hydrostatic pressure would be equal to or less than the air supply pressure. There will be a regular flow of bubbles when air is flowing through the indicator, and if no bubbles are visible, then probably something has happened to the regulated air supply, and liquid-level values reflected by the indicator-recorder are not necessarily correct.

Measurements in Pressurized Vessels

In using hydrostatic pressure to measure liquid level in a pressurized tank or other vessel, it is desirable to use some form of differential-pressure gauge. In the sense used here, differential pressure means the difference between two pressure values, neither of which is necessarily atmospheric pressure.

Figure 7.17 shows three situations—*A, B,* and *C.* Two conditions are common to all situations: (1) a differential pressure gauge similar to one described in chapter 5 has its P_2 inlet tap connected to the vessel, and (2) a given liquid level, *H,* exists in each situation. In figure 7.17*A,* inlet tap P_1 of the gauge is vented to the atmosphere, and so is the vessel. The gauge in this instance measures as in an open-tank condition, and it will read a pressure value of 0.433 psi/ft $\times$ G $\times$ H.

For situation *B,* (fig. 7.17*B*), fluid pressure p is applied to the top of the liquid in the vessel. Each increment of gauge pressure applied will be reflected on the gauge by adding it to the hydrostatic pressure caused by the height of the liquid. If the pressure p is known, the liquid level can be inferred by subtracting p from the gauge reading and using the relationship between level and hydrostatic pressure.

Using the full capability of the differential pressure gauge will allow an automatic subtraction of the pressure p from the total pressure appearing at P_2. P_1 of the gauge is now connected to the upper portion of the tank (fig. 7.17*C*). With this arrangement, each increment of pressure above the liquid

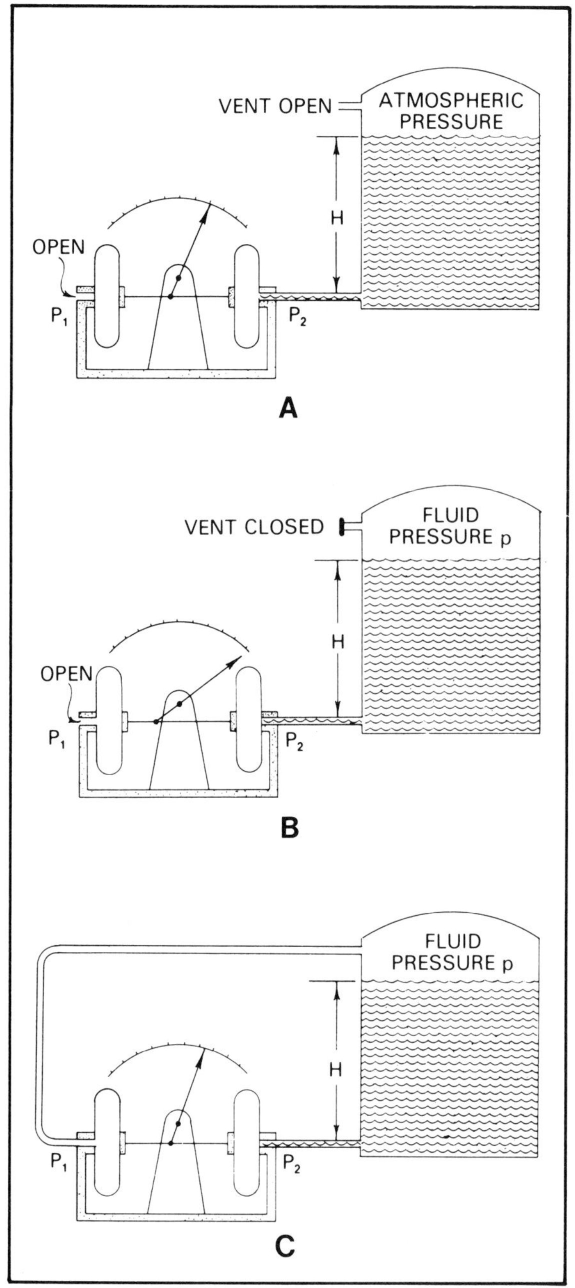

Figure 7.17. Pressurized tanks requiring differential pressure measurement

surface is applied to both capsule assemblies of the gauge, and since they are in opposition, the increment is cancelled as far as the gauge is concerned. Only the hydrostatic pressure, which is applied to tap P_2, is effective in causing any response by the gauge.

Regulators

Several interesting forms of regulator-type liquid-level controllers are effective and are in use for particular applications. In a straightforward arrangement using a diaphragm motor valve (fig. 7.18), a small pipe leads from the discharge line of the vessel to the underside of the diaphragm and exerts an upward force that tends to open the double-ported valve. This force is directly proportional to height H. The upward force is opposed by the loading head of liquid in the loading chamber. The loading liquid need not be the same as that in the vessel and in fact should be a liquid that is not likely to freeze or evaporate. As liquid flows in and out of the upper level of the diaphragm chamber, the level of liquid in the loading chamber will rise and fall. To reduce the amount of change in level caused by this action, the loading chamber should be made relatively broad.

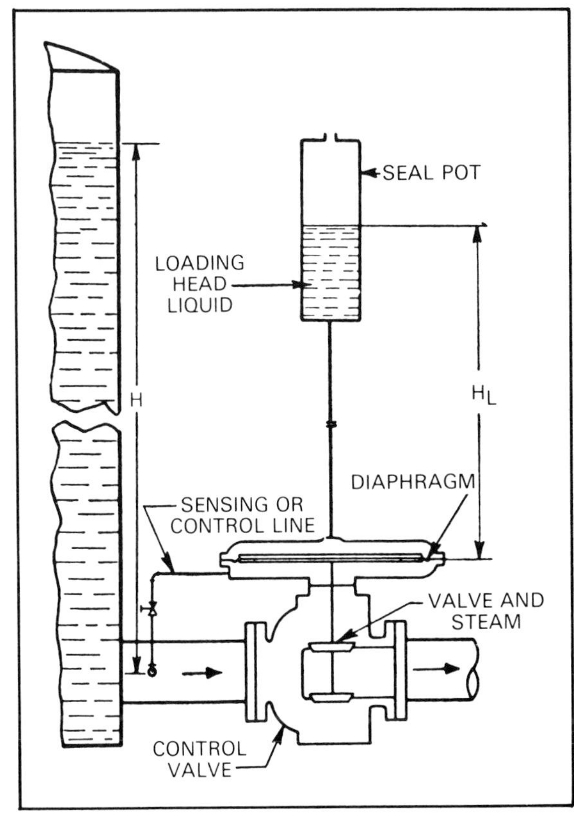

Figure 7.18. Liquid-loaded diaphragm valve used to regulate liquid level

The force exerted on the lower side of the diaphragm is given by the equation

$$\text{Force} = H \times G_v \times 0.433 \text{ psi/ft} \times A + p$$

where

H = height of vessel liquid above the diaphragm, in feet;
G_v = specific gravity of vessel liquid;
A = area of diaphragm;
p = gauge pressure on surface of vessel liquid.

P is zero for an open tank or vented vessel.

The force on the upper side of the diaphragm is given by the equation

$$\text{Force} = H_l \times G_l \times 0.433 \text{ psi/ft} \times A + p$$

where

H_l = height of loading liquid, in feet;
G_l = specific gravity of loading liquid;
A = area of diaphragm;
p = gauge pressure on surface of vessel liquid.

Operation of the control system depends on the balance of forces existing on either side of the diaphragm. If the level in the vessel rises above the set point determined by the level in the loading chamber, force on the underside of the diaphragm will increase and the valve will open, unless frictional forces prevent its prompt action. Once open, the valve will release liquid until the loading-liquid head exceeds that from the vessel, at which time the valve should close.

An arrangement similar to that shown and described uses an adjustable spring in place of the liquid loading and has advantages such as ease of adjustment, more compactness, and probably less cost. However, since the force required to compress a spring is proportional to the amount of compression, valve stem action is comparatively insensitive. The liquid loading provides a nearly constant force for any valve stem position and is therefore more sensitive. Because of the friction in the valve stem stuffing box, the liquid level will probably have to change an appreciable amount before valve action can be initiated in either of

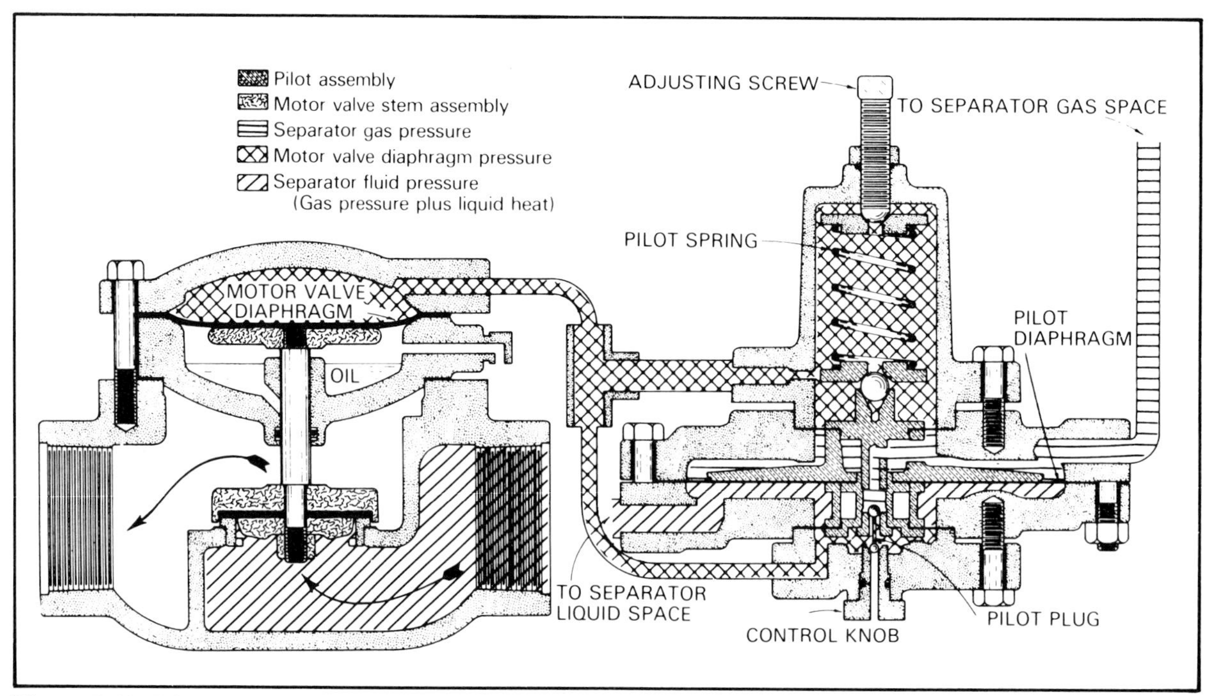

Figure 7.19. Diaphragm valve controlled by a pilot valve used to regulate liquid level (*Courtesy Kimray*)

these forms. Although sensitivity can be improved by using very large diaphragms, neither of these forms is capable of providing fine control.

A level regulator suitable for pressurized vessels (fig. 7.19) finds popularity in controlling liquid level in oil and gas separators. It is somewhat more sensitive than the type previously described. The diaphragm valve makes use of discharge pressure to force its plug away from the seat, while a diaphragm actuator drives the valve closed.

The pilot valve operates on the differential pressure caused by the head of liquid in the vessel. Gas or other fluid pressure existing above the liquid of the vessel is applied to the top side of the pilot diaphragm, while pressure due to the combined effects of the hydrostatic head of liquid in the vessel and the fluid pressure at the surface of the vessel liquid is applied to the lower side of the diaphragm. The pilot spring exerts a downward force on the diaphragm.

The pilot plug is a miniature barbell assembly—two stainless steel balls rigidly connected by a stiff, thin shaft. The upper ball controls flow of gas from the upper diaphragm chamber to the motor valve actuator and to the pilot spring chamber, while the lower ball controls venting of the motor valve actuator. Pressure existing in the chamber between the two balls exerts an upward force on the internal pilot assembly. In order to neutralize this force, the same pressure is also applied to the pilot spring chamber, where it may exert an equal downward force.

Should liquid level rise in the vessel, the pilot diaphragm assembly will be driven upward against the pilot spring. The upper ball of the pilot plug will prevent flow of gas to the motor valve actuator, while the lower ball will be lifted from its seat, allowing the motor valve diaphragm pressure to bleed off. Simultaneously, pressure will bleed off from the pilot spring chamber. The reduced pressure in the motor valve actuator will permit the liquid pressure from the vessel to force the valve plug away from its seat.

As the liquid level in the pressurized vessel decreases, so does the force acting on the lower side of the pilot diaphragm. Eventually the gas pressure and the pilot spring force the diaphragm assembly lower, restoring equilibrium. The pilot valve is capable of providing either semisnap action or a narrow throttling range by adjusting the position of the seat for the lower of the two balls of the pilot plug.

Electric Devices

Many different systems can easily be adapted to producing an electric signal for operating indicators or controllers. Any of the pressure-sensitive devices or the buoyant elements can be made to operate microswitches or mercury switches, just as they are made to position a flapper in a pneumatic controller. Most of the control action produced in this way, that is, with mircroswitches and mercury switches, is of the on-off variety. For those applications in which on-off action is suitable and electricity is available, electric systems are reliable and inexpensive.

Figure 7.20. Electric liquid-level controller with a microswitch actuated by the float and flexure tube (*Courtesy Instruments, Inc.*)

Microswitch Elements

Microswitch is a trade name that has gained general acceptance in designating a particular form of switch that is characterized by its swift action in closing or opening a circuit in response to a very small movement of an actuator. The actuator may be a pinlike shaft or a heavier plungerlike device that triggers the snap-acting and spring-loaded contacts. Switches capable of handling 15-ampere circuits need only 10 to 14 ounces of force to depress the pin or plunger a distance of about $\frac{1}{64}$ inch—the movement necessary to cause operation. Releasing the pin or plunger and allowing reverse travel of a fraction of an inch will allow the switch to reverse its position.

A commercial form of liquid-level control (fig. 7.20) uses a float attached to a flexure tube similar to the arrangement used to position a flapper. In this case, however, a microswitch replaces the flapper-nozzle assembly.

Mercury Switch Elements

Mercury switches, like microswitches, are devices capable of opening and closing circuits and require a minimum amount of force for operation. They require less force

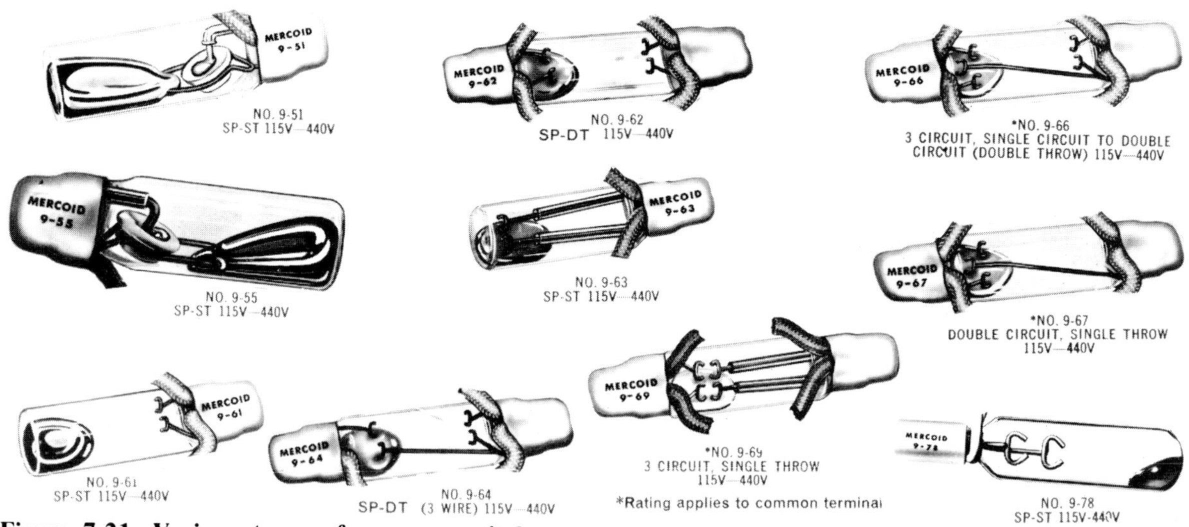

Figure 7.21. Various types of mercury switches (*Courtesy Mercoid*)

than microswitches but need a greater movement to cause operation. Figure 7.21 shows several forms of mercury switches. Their operation depends on tilting the entire switch assembly. The switch is made by placing contacts and a small amount of mercury in a section of glass tubing sealed at either end. Electric leads from the contacts are brought out of the glass in a way that forms a metal-to-glass bond that defies leakage. The mercury and the contacts enjoy an environment that is free of dust, corrosion, and other contaminants.

The sensitivity of mercury switches is frequently described by the amount of tilting needed to cause the mercury "puddle" to move to or away from the contacts. Tilting is usually expressed in degrees of rotation; sensitivity of about 3° is possible, although 5° and greater are more common and give better performance.

Figure 7.22 is a form of buoyancy-type control using a mercury switch. The switch assembly includes a pivoted plate and a permanent magnet. The buoyant float of the mechanism is attached to a soft iron plunger that travels in a vertical tube made of nonmagnetic material. When the plunger reaches

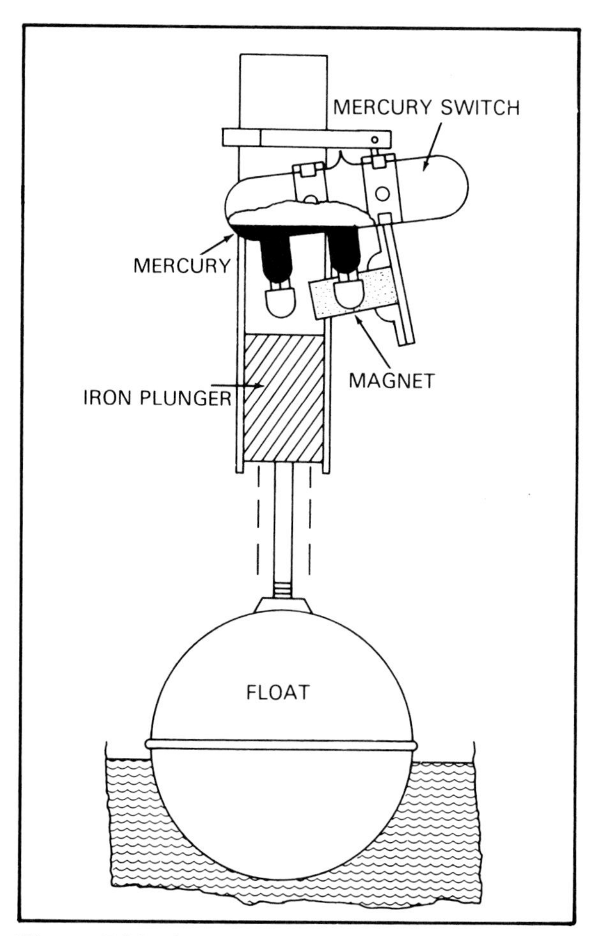

Figure 7.22. Schematic of a liquid-level control system with a mercury switch

a critical point determined by the liquid level, the magnet swings quickly against the nonmagnetic tube and in doing so causes the entire switch assembly to rotate about its pivot. This tilting action is more than sufficient to operate the mercury switch. As the liquid level declines, a point will be reached at which the force of attraction between plunger and magnet will be overcome, and the magnet will fall away rapidly, thus causing a reversal of the previous switch action.

Immersion Electrodes

Immersion electrodes are popular for maintaining water level in open or closed tanks and for protecting pumps against dry runs. The liquid used must be conductive enough to form a part of the electrical circuit, but flammable liquids are unsuitable.

One type of immersion electrode control (fig. 7.23) is wired to provide either of two services: (1) controlling liquid level by

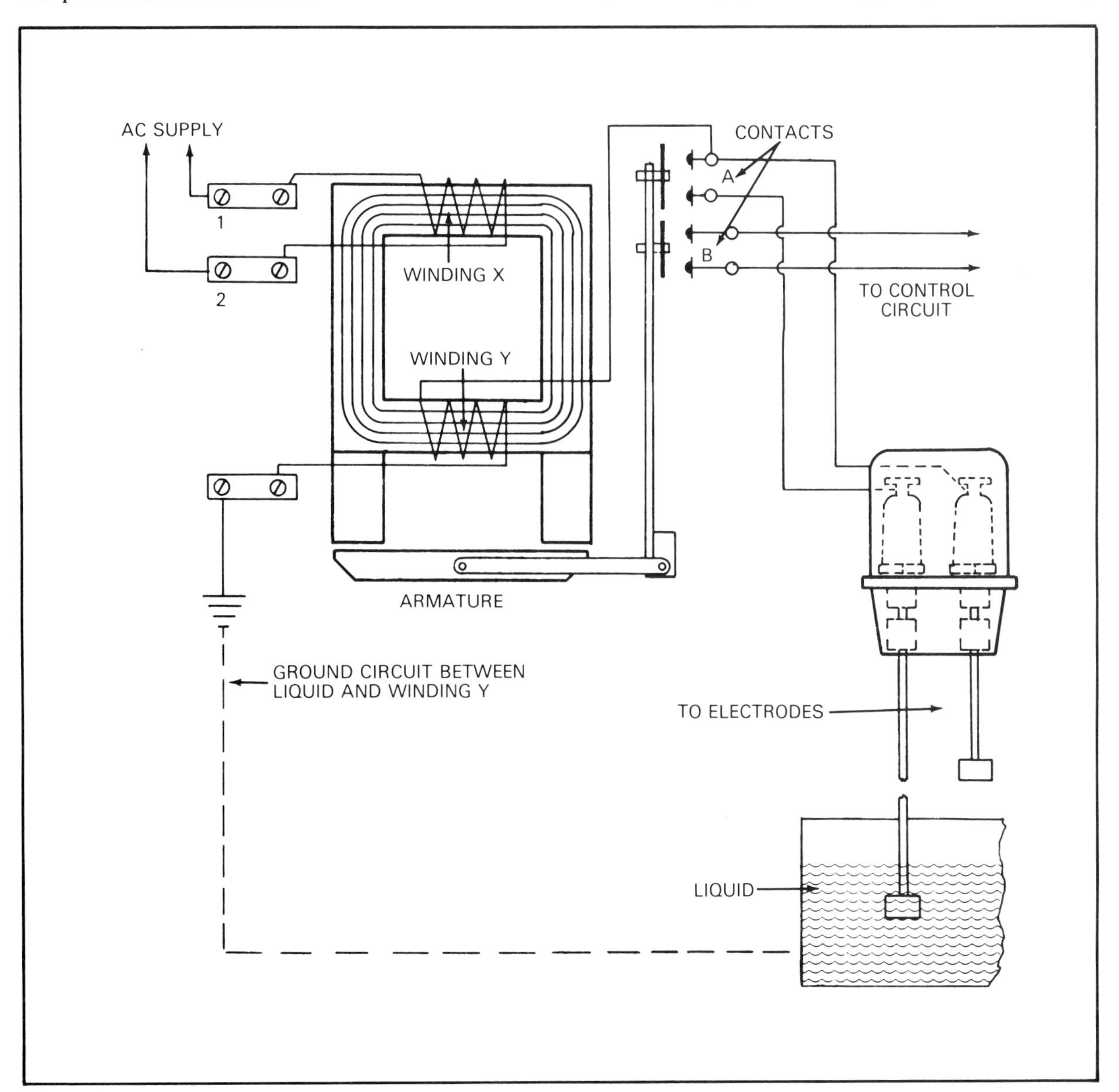

Figure 7.23. Schematic of a liquid-level control system with conductive immersion electrodes

operating a valve or pump in the discharge line of the vessel or (2) protecting against discharge below the level of the lower electrode. The heart of this device is a combination transformer and relay. The transformer action provides a means of reducing the operating voltage as well as isolating the control circuit from the alternating-current supply line. The magnetic path in the iron core is shown for the transformer action. When current flows in secondary winding Y, which is also the relay coil, the magnetic path includes the relay armature, and motion of the armature closes or opens contact sets A and B.

Contact set B of the device may be assumed to be in the circuit of a solenoid-type or other electric valve in the discharge line. As shown in the figure, contact set B is open, so the discharge valve is closed and will remain closed until liquid level reaches the upper electrode in the vessel. Once liquid level rises to the upper electrode, the circuit through the relay is completed, with the liquid and ground circuit figuring prominently in the process. The relay closes, and this action not only opens the discharge valve but also closes contact set A so that the upper and lower electrodes are effectively shorted together. Even after the level falls below the upper electrode, the circuit to the relay remains complete because of the closed condition of contact set A. When the level falls below the lower electrode, the relay circuit is opened and the armature moves to open both sets of contacts.

Capacitor Gauges

Capacitor gauges have particular advantages for some applications. They are readily adaptable to remote transmission of signals, have no moving parts, and can be made to resist the effects of temperature, pressure, and corrosion. They are particularly useful where temperature and composition of the liquid remain rather uniform. Errors due to changes in temperature and composition are apt to be more severe than with other types of measurement.

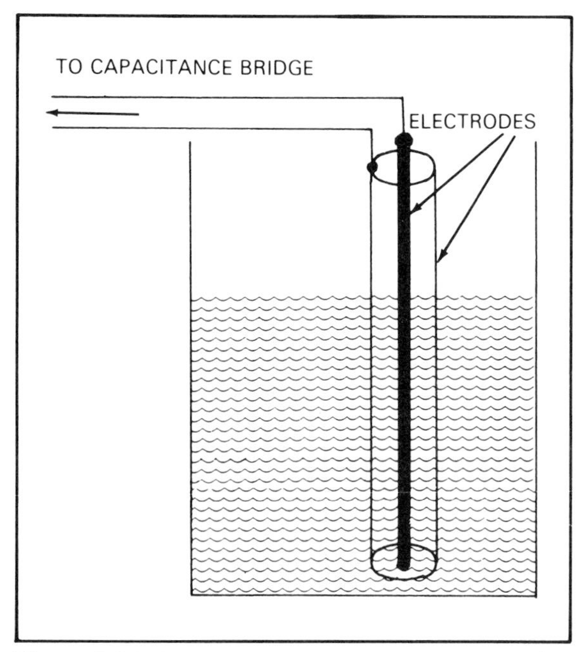

Figure 7.24. Schematic of a capacitor-type device used for liquid-level control

Figure 7.24 is representative of the system. An inner and an outer conductor are provided, and the process liquid is permitted to act as a dielectric material as it rises in the space between the electrodes. A *capacitor* is formed by the two electrodes, and the capacity of the capacitor, that is, the amount of electric charge it is capable of taking on, is directly proportional to the *dielectric constant* of the substance separating the electrodes. Air and most other gases have a dielectric constant of 1, and other substances have values ranging upward. Assuming that the capacity present between the electrodes is C when the vessel is empty, if the vessel is filled to the top of the electrodes with a liquid having a dielectric constant of 3, the capacity existing between the electrodes will be $3C$.

The problem of using this method to measure liquid level becomes a matter of measuring capacity. Changes in the dielectric constant of the liquid can destroy the accuracy of the method. For that reason it is important to maintain uniform quality of the liquid. The effects of temperature on the value of the dielectric constant can be compensated for if temperature variations are significant.

Figure 7.25 is a diagram of a bridge circuit capable of measuring capacity by the *null* method, that is, by adjusting the variable resistors for zero response at the detector. The detector may be headphones, an oscilloscope, or a sensitive alternating-current voltmeter. The alternating-current generator can be an audio oscillator, preferably one producing a frequency of about 1 000 Hz.

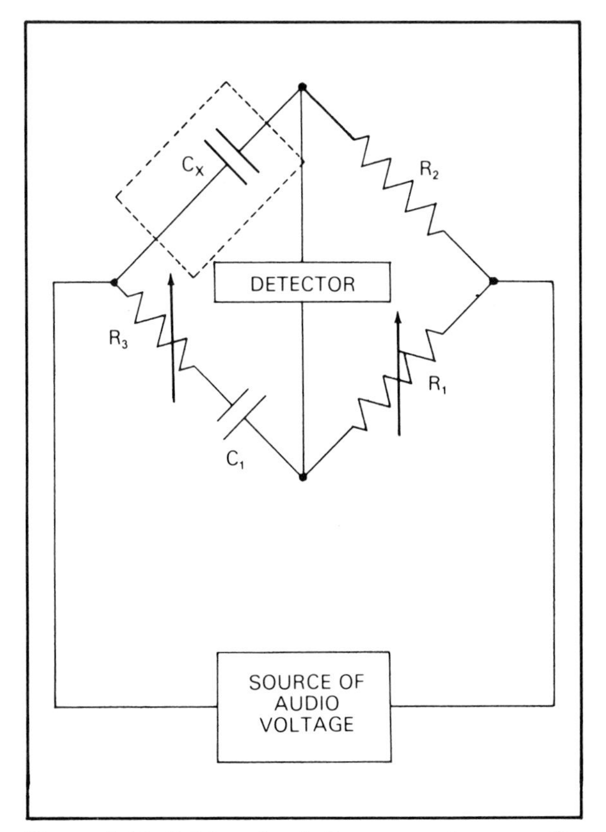

Figure 7.25. Bridge circuit for measuring capacity

In the figure, R_1 and R_3 are calibrated variable resistors, and R_2 is a precision fixed-value resistor. C_1 is a fixed-value capacitor, while C_X is the capacity existing between the electrodes at any given liquid level. When resistors R_1 and R_3 are adjusted to produce a balanced condition—no signal at the detector—C_X may be determined from the equation

$$C_X = \frac{R_1}{R_3} \times C_1.$$

It will be necessary to have a set of values for C_X covering the range of liquid-level heights in the vessel. These values can be obtained through trial runs with each type of liquid to be gauged.

Sonic Devices

Measurement using sonic devices finds only limited use, usually in applications where other methods are totally inadequate. Use as depth finders on seagoing ships and other watercraft is a typical case in point.

The principle of operation involves the transmission of ultrasonic sound waves and the measurement of the time interval required to receive the echo of the waves from the surface of the liquid or the bottom of the vessel, depending on the particular arrangement of the installation. Ultrasonic waves are those above audible frequencies.

When an ultrasonic transmitter is placed at the bottom of a liquid-filled vessel and emits ultrasonic waves toward the surface, some of the energy of these waves will travel back toward the transmitter after being reflected at the surface. Once the speed of sound in the particular liquid is known, measurement becomes a matter of determining the time required for the round trip of the sound to surface and back. Electronic instruments are capable of measuring the short time interval with such precision that liquid level can be determined with an accuracy of 0.01 inch for each foot of depth.

Besides their use as depth finders in the sea, sonic devices find some use in very large storage tanks, in mines, and in oilwells.

Vibrator Device

An interesting instrument that finds use as the sensing element for liquid-level control is a vibrator device (fig. 7.26). The device is small and designed to be mounted to the vessel through a 20 mm pipe fitting. Its operation is simple, and it is durable enough for most liquid-level applications.

A 60 Hz source of alternating current is applied to a solenoid coil in the driver end of the instrument. An iron armature vibrates in response to the varying magnetic field and transmits some of its kinetic energy to the paddle through the stiff, thin rod. The paddle is also connected to a similar rod that drives an armature in the pickup end. The pickup end contains a permanent magnet and pickup coil, and the vibration of the armature causes a voltage to be induced in the pickup coil. This voltage is proportional to the amplitude of the vibrations of the armature.

When the paddle is allowed to vibrate in air, the maximum amount of energy is transmitted to the armature in the pickup end, and voltage output will be a maximum—perhaps 0.5 V. If liquid surrounds the paddle, its amplitude of vibration will be sharply dampened, and voltage output from the pickup coil will decrease significantly, perhaps to almost zero for most liquids.

The alternating-current low voltage output from the pickup coil can be amplified and made to actuate an on-off type of control system.

Radiation Devices

Certain chemical elements and their compounds experience a *decay,* or change in basic composition, by emitting fundamental particles from their nuclei. Radium, for example, can decay to radon by emitting an *alpha particle.* An alpha particle is equivalent to the nucleus of a helium atom; that is, it contains two protons and two neutrons. An important side effect that accompanies some decay action is the emission of *gamma rays.* These rays are electromagnetic radiations, similar to ordinary light but of very much shorter wavelength. Gamma rays have incredible penetrating ability and will pass through the metal walls of vessels and pipes and through the liquids they might contain. Some of the gamma rays will be absorbed by the metal walls and

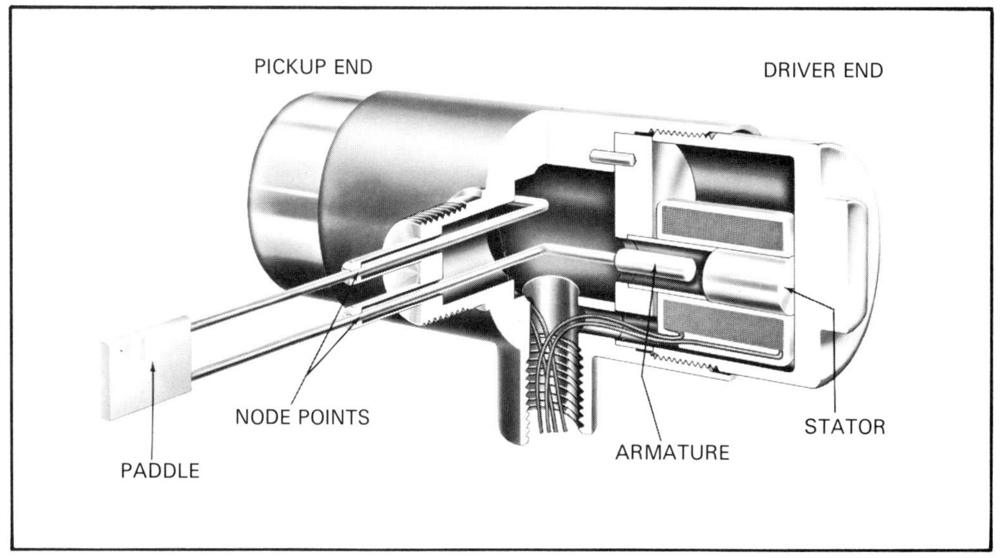

Figure 7.26. Vibrator-type liquid-level control device (*Courtesy Automation Products*)

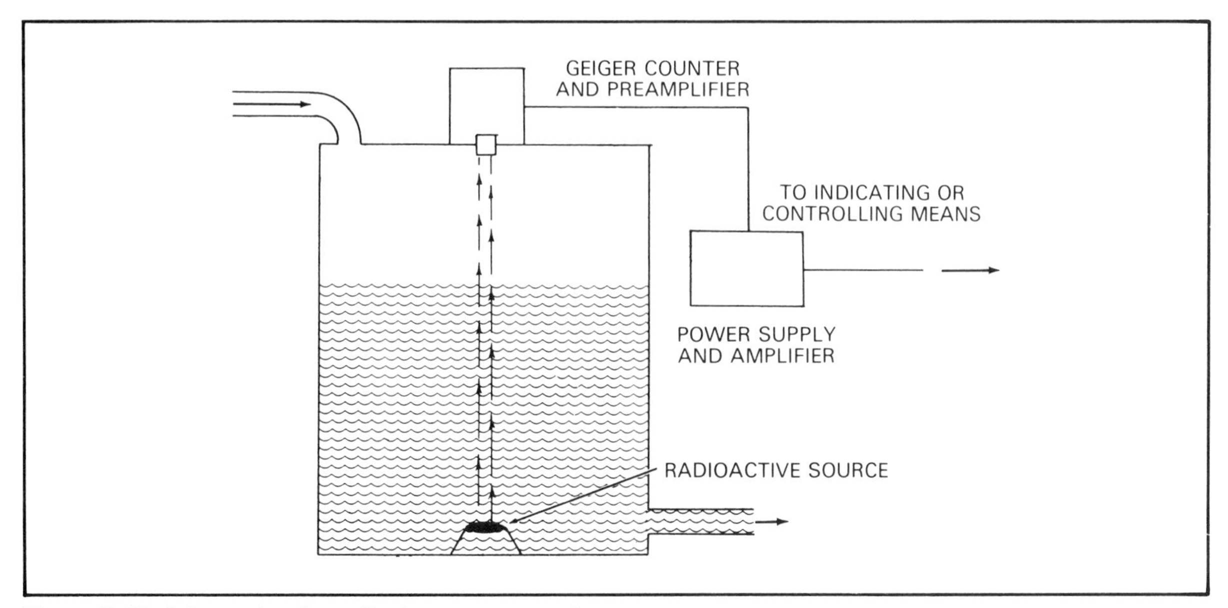

Figure 7.27. Schematic of a radiation-type control system

liquid, and the amount of absorption can be used as the basis for liquid-level measurement.

Figure 7.27 shows one method of using a radioactive source to measure liquid level. The gamma ray source can be a minute quantity of radium salt placed at the bottom of the vessel. The receiver is mounted above the vessel top. Gamma rays penetrate the radium container, the liquid, and the vessel top. Absorption of gamma rays by the radium container and the vessel top is a constant quantity, and only the composition of the liquid and its level height are variable factors in the radiation measurement.

The receiver is a Geiger counter, which is an instrument capable of detecting gamma rays by the ionization they cause in a Geiger tube. In addition to the Geiger tube, the receiver contains a means of amplifying, integrating (accumulating or adding together), and rectifying the individual pulses formed by the ionization process. The resulting direct current produced in the receiver is directly related to the gamma ray count. The count is inversely proportional to the liquid level, so the direct current can be used to operate an indicator or a controller.

Radiation-type measuring and controlling devices are capable of great accuracy and have the advantage of not having to be exposed to the environment of the vessel interior. Most radioactive sources suitable for this application are expensive, and the auxiliary equipment needed is also likely to be more costly than that for most of the other systems of measurement and control considered in our studies. For that reason radiation-type systems are to be found only in specialized situations.

CHAPTER

8

Flow Measurement

When fluid flow is controlled in order to regulate temperature, pressure, or liquid level, it is the *manipulated variable.* The fluid is the *control agent,* and temperature, pressure, or liquid level is the *controlled variable.* In flow measurement, the fluid flow is treated as the *controlled variable;* it is measured and controlled for the purpose of determining the *quantity of fluid* that is used or produced in a system or process.

Units and Dimensions

Units

Flow measurement is the process of measuring the quantity of fluid that passes a particular point in a given interval of time. Thus, gallons of water per minute, cubic feet of gas per hour, and barrels of oil per day are all valid measurements of flow.

A quantity of fluid can be expressed as a volume or as a mass. Expression as volume is often flawed. For example, a gallon of gasoline at 40°F becomes more than a gallon at 100°F. Many automobile owners in past years had the experience of filling fuel tanks with cool gasoline and noting that after the car had remained in the sun a short time, gasoline ran out the vent hole of the filler cap. Modern environmental practices prohibit the venting of gasoline or its vapors into the atmosphere.

The wide variations in volume that accompany temperature change in a fuel present a problem so troublesome in many instances that volume measurement has been abandoned. The mass of a quantity of gasoline (or other fluid) does not, on the other hand, change with temperature. For that reason, the quantity of gasoline or other fuel an aircraft can carry or will consume under certain conditions of flight is expressed in terms of mass by those in military and commercial aviation. Mass measurement represents a much more accurate indication of the energy available from the fuel than volume measurement.

In many areas, however, volume measurement of fluids is still the most prevalent method, despite its deficiencies. Gasoline is still sold by the gallon and natural gas by the cubic foot. But when large quantities of fluid are transported and sold, the conditions of temperature and pressure are stated, amounting to a provision for determining the mass of the fluid.

Dimensions

Liquid-level measurement has one simple dimension: length. Flow measurement is more complex. It has two dimensions—volume and time, or mass and time. Mass can be determined if density and volume are known, for mass of a fluid equals its density times its volume.

Sometimes only the total quantity of fluid transported, produced, or used is important. In this case, time is ignored as a factor or dimension, for the quantity of the fluid is more important than the speed with which it is transported or used. Many meters, such as those used for measuring quantity of natural gas, register the amount of fluid that passes, but not the time-rate of its passage. Such devices are logically called *quantity meters,* but their operation depends on a flow of fluid through them. Meters that measure the flow in terms of volume or quantity per unit of time are called *rate meters.*

Differential Pressure Flowmeters

Perhaps the most popular method of measuring flow when huge quantities of gas are involved is that using differential pressure. Other methods are more popular for measuring liquids. The metering devices employed incorporate a constriction placed in the flow line to cause a change in the form of energy possessed by the flowing fluid. Certain physical laws allow this change in energy form to be expressed in terms of pressure change, or the differential pressure that exists across the constriction. This differential pressure is called *head,* and the meters are therefore called *head meters.*

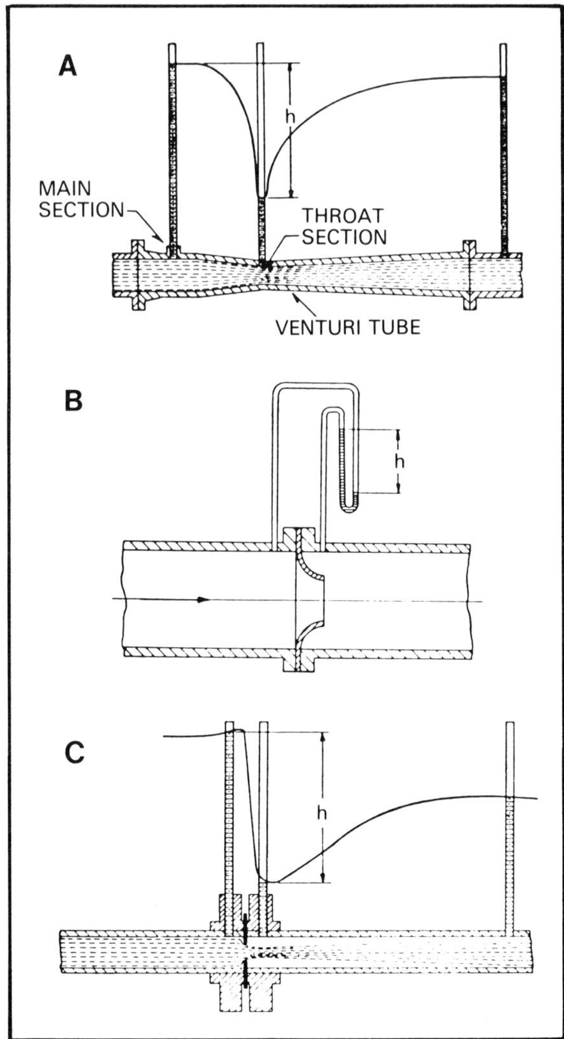

Figure 8.1. Restrictive elements for head flowmeters. *A*, venturi section; *B*, flow nozzle; *C*, flange-type flat plate orifice.

Forms of Restrictive Elements

Three forms of head meter elements that place a restriction in the flow line are the venturi section, the flow nozzle, and the orifice plate (fig. 8.1). In each instance, simple indicating tubes show the differential pressure, or head. Each of the forms enjoys popular use, but the orifice plate, probably because of its simple design, low cost, and ease of installation and replacement, is the most widely used.

Factors that enter into the choice of head meter include the following:

1. accuracy and stability over long periods of use;
2. initial cost and cost of maintenance or replacement;
3. ease of replacement (considered as a factor when replacement is required to meet changes in the fluids or the flow conditions);
4. performance over wide variations of flow rates and other variables such as density and viscosity;
5. freedom from fouling by the settling of solids; and
6. the efficiency with which the element transmits fluid—the orifice place, for example, having the greatest energy loss and the venturi section the least.

In head meters the differential pressure is measured at points called *pressure taps.* The point upstream is the high-pressure tap, and the point downstream, the low-pressure tap. Mercury manometers are the most popular means for measuring the differential pressure, but other pressure-measuring elements may be used.

Venturi section. The venturi section—either a venturi tube or a Dall tube—is one of the most efficient of the primary constrictive elements. It is also very difficult and expensive to manufacture to uniform quality standards and has the added disadvantage of not being easy and quick to replace.

In the line drawing of a venturi tube (fig. 8.2), pressure taps and throat section are

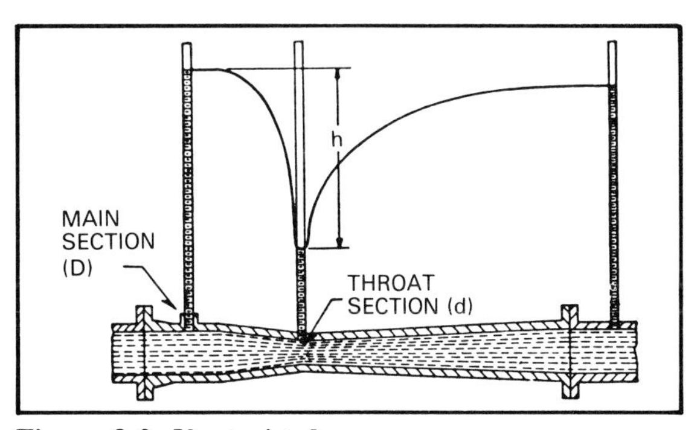

Figure 8.2. Venturi tube

shown. The throat is a section having diameter *d*. The input side of the tube begins with a uniform section of diameter *D*, then a sharply tapering cone that leads to the throat, while the discharge side of the tube is a cone section of greater length and less slope. Pressure taps for venturi tubes are typically installed in the straight section upstream and at the throat. The radial location of the taps is of some importance, depending on the type of fluid being carried and the position of the venturi, whether vertical or horizontal.

The efficiency of the venturi tube is due to the gradual velocity changes that the flowing fluid experiences. Such a transition permits smooth flow, so little energy is wasted. Efficiency is reflected by the fact that a very high percentage of the fluid pressure is recovered downstream of the element.

A *Dall tube* is similar to the venturi tube, but its efficiency is even greater (fig. 8.3). Its design differs significantly from the venturi in two respects: (1) a short uniform section is followed by an abrupt shoulder that begins the inlet cone, and (2) the low-pressure tap at the throat leads to an annular groove that encircles the throat. Although the efficiency of the Dall tube is difficult to account for, empirical data indicate that, for a given set of conditions, it produces about twice the differencial pressure of good venturi design, and its pressure recovery is superior to any other form of head meter primary element.

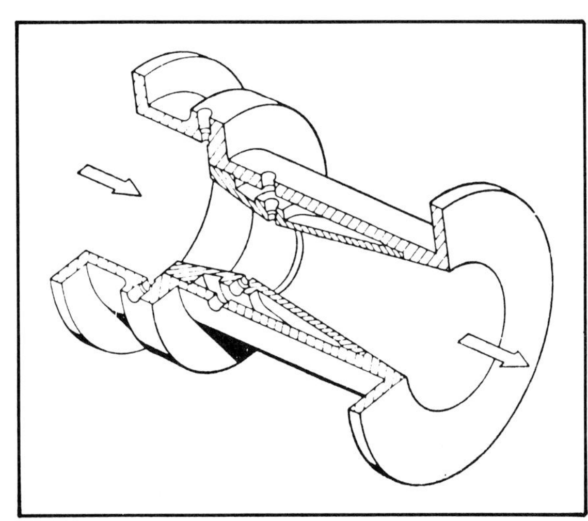

Figure 8.3. Dall tube

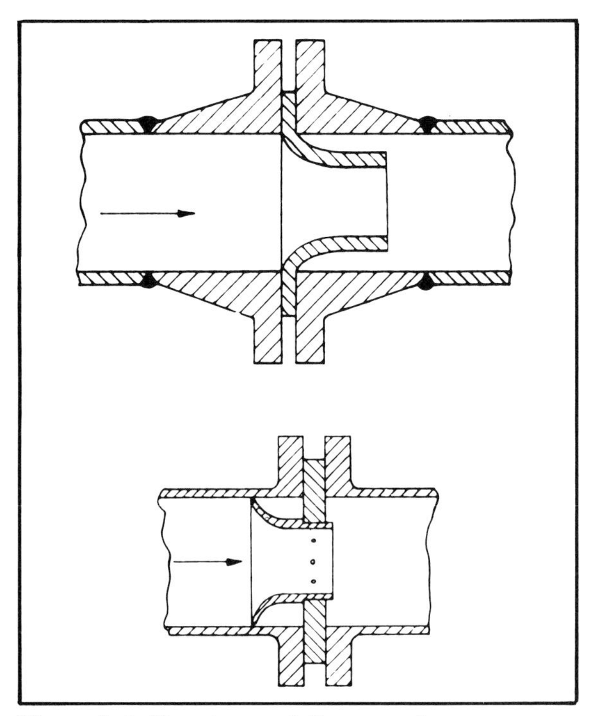

Figure 8.4. Two types of flow nozzles

The annular groove in the Dall tube precludes its use for fluids that contain any form of matter that might clog the groove, since clogging would ruin calibration. The shoulder offers a restriction that also may cause some fouling. The great sensitivity of the Dall tube requires that the piping leading to the pressure taps is absolutely leakproof, because the differential pressure value is extremely sensitive to any flow occurring at the pressure taps.

Flow nozzle. Flow nozzles (fig. 8.4), because of their smooth inlet, approach the overall efficiency of venturi tubes, but they must be made to very accurate specifications. Since the design includes the careful shaping of curved surfaces, manufacture must be carried out in shops equipped with expensive machine tools.

Orifice plate. Because of its simplicity and ease of manufacture, installation, and replacement, the orifice plate has become the most popular primary element for head meters in many flow measurement applications. However, it has disadvantages such as the following:

1. It compares unfavorably with the flow nozzle and the venturi tube in ability to pass fluid. For equal values of throat and orifice diameters, the venturi tube will pass about 65% more fluid than the orifice plate.
2. It is more likely to cause a pileup of solid matter upstream of the element.
3. Its sharp edges are eroded rapidly by fluids containing abrasive materials, causing inaccurate meter readings.

Despite its adverse characteristics, the orifice plate possesses advantages for many applications that make it immensely popular. For large-scale metering of natural gas, for example, its permanent pressure loss can be tolerated without serious consequences. Any erosion or buildup of solid matter is fully compensated for by the easy and relatively inexpensive replacement. Figure 8.5 shows how an orifice plate is installed between two flanges in a line. In the figure, the plate is shown positioned within the circle formed by the flange bolts (*A*). A cross section through flange and orifice plate is shown, as well as the pressure tap connections and drain connections (*B*). The orifice flange and one of the two jack studs that are used to force the flange apart far enough to slip the orifice plate in and out of position are given in side view (*C*).

Orifice plates can be manufactured or duplicated in machine shops equipped with the most common machine tools. The plates are made of metal chosen for its ability to resist erosion, corrosion, and distortion caused by differential pressure. Three principal dimensions are of importance in the design of the plates: diameter of the orifice, thickness of the plate, and nature of the orifice edge—rounded, beveled, or square.

The diameter of the orifice is closely related to the diameter of the pipe in which the orifice is to be used. If the diameter of the orifice is denoted by d and the inside diameter of the pipe by D, then the ratio of orifice to pipe diameter, d/D, is called the *beta factor*, and is logically designated by the Greek letter β (beta). Beta factors are numerically less than 1.0, and values ranging from 0.125 to 0.75 are common.

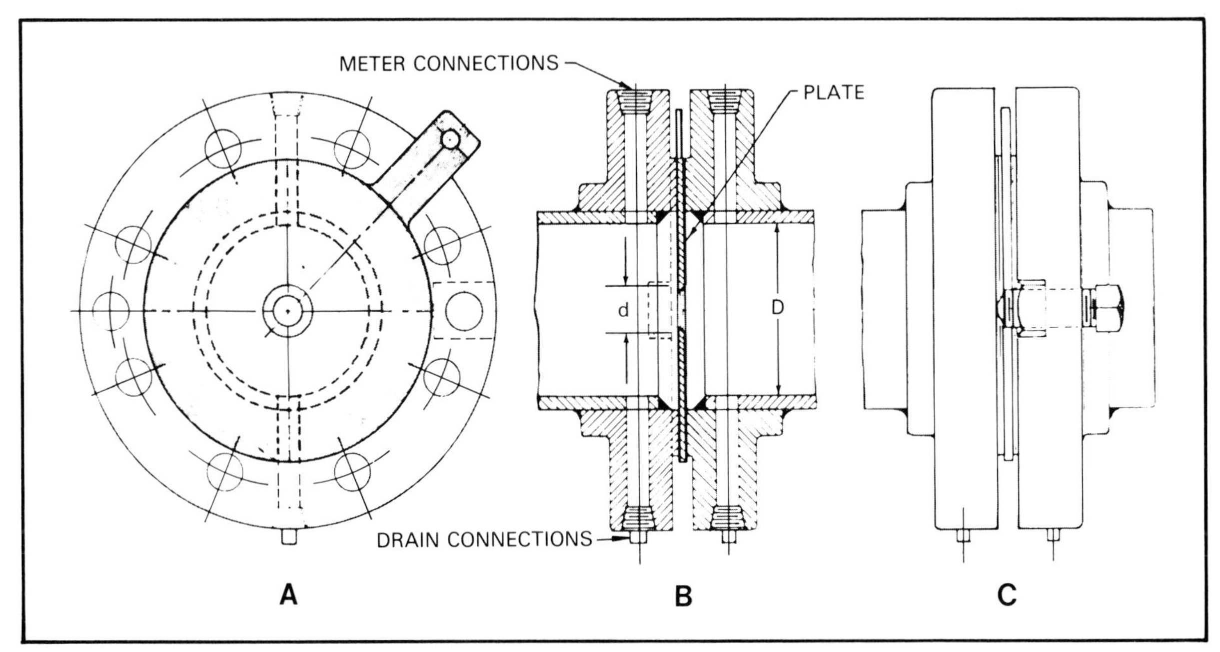

Figure 8.5. Three views of an installed orifice plate

The thickness of the orifice plate is kept as small as practicable without sacrificing the stiffness needed to resist forces that might cause buckling or other distortion. A very thin plate will produce a predictable flow pattern. The thickness should conform to certain standards. It must not exceed $\frac{1}{30} \times D$, $\frac{1}{8} \times d$, or $\frac{1}{8} \times (D - d)$, where D is the inside pipe diameter, and d is the orifice diameter.

Most orifice plates are made with square edges, because these are easiest to build and to reproduce. If a plate is thin enough, the square edge has no adverse effect on the formation of the fluid jet, and the jet shape will be a little different from that issuing from a sharp-edged orifice. A thicker plate may have the downstream side of the orifice beveled so that fluid from upstream encounters a sharp edge. Rounded edges are undesirable but are sometimes formed by erosion or corrosive action. Rounding of the upstream edge has the same effect as though the orifice diameter were increased.

Besides the popular *concentric* orifice, which is distinguished by its simple symmetry and ease of manufacture, *annular, eccentric,* and *segmental* orifice plates are also used to a limited extent. Figure 8.6 shows concentric, eccentric, and segmental plates. An annular orifice plate allows flow of fluid between the plate and the pipe wall. Each of these orifice plates possesses some advantage for a particular application.

Installation Arrangements for Primary Elements

Head meters are affected adversely by flow disturbances that occur near the upstream inlet. As used here, *near* means within 100 pipe diameters. Disturbances can be caused by obstructions in the pipe, and obstructions can take the form of throttling valves or rough and uneven joints in the pipe. Disturbances can also be caused by elbows or other deviations from straight line flow. Unless the disturbance is of extremely severe nature—such as the rapid opening and closing of a throttling valve—a few yards of straight, smooth pipe will remove the effects before they reach the head meter primary element.

The length of straight-run pipe necessary to take care of flow-line disturbances is related to the diameter of the pipe, D, and to the beta factor, d/D. The particular cause of the

Figure 8.6. Three types of orifice plates

disturbance is important in determining the length of straight run. Curves available in many handbooks indicate the necessary lengths needed to offset the effects of disturbances caused by elbows, valves, and changes in pipe diameter.

Sometimes the requirements for straight runs of pipe cannot be met because of installation problems. For instance, a length of straight pipe equal to 60 to 80 diameters may be necessary for corrective action. Most of the difficulties that could be removed by straight runs can at least be alleviated by *straightening vanes* (fig. 8.7). They consist of a bundle of comparatively small tubes welded together. Straightening vanes are placed in the piping ahead of the primary element.

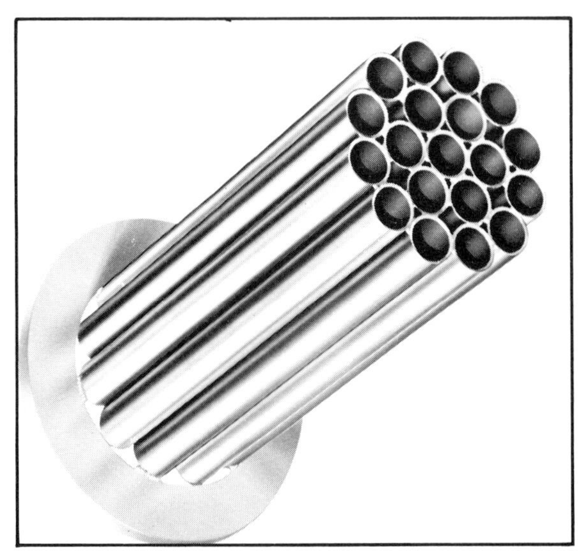

Figure 8.7. Straightening vanes *(Courtesy Foxboro)*

Straightening vanes break up swirls and vortices by forcing the fluid to pass through numerous small-area passages. For maximum benefit, the individual pieces of tubing or other paths for the fluid should have a flow area no greater than 10% of the pipe in which they are fitted, but flow areas of 25% of the pipe are commonly used. A criterion that seems satisfactory to use is that the cross section of the largest of any of the passages should be no greater than 25% of the inside diameter of the pipe, and the length of the vanes should be at least ten times the dimension of the greatest transverse opening in the vane assembly.

Although straightening vanes assist in solving problems by making long straight runs of pipe unnecessary, they may not be installed nearer than 6 pipe diameters to the upstream pressure tap, so head meters will in almost all cases require a considerable length of straight pipe preceding their inlet.

The obstructions and disturbances to flow that occur downstream from head meters have much less effect on meter performance than those occurring upstream. In fact, elbows, valves, and pipe diameter changes occurring more than 5 pipe diameters downstream from the nearest pressure tap have only insignificant effect on the meter.

A graph of several functions relating to the installation of an orifice plate in a line containing two elbows upstream and one downstream is shown in figure 8.8. The two

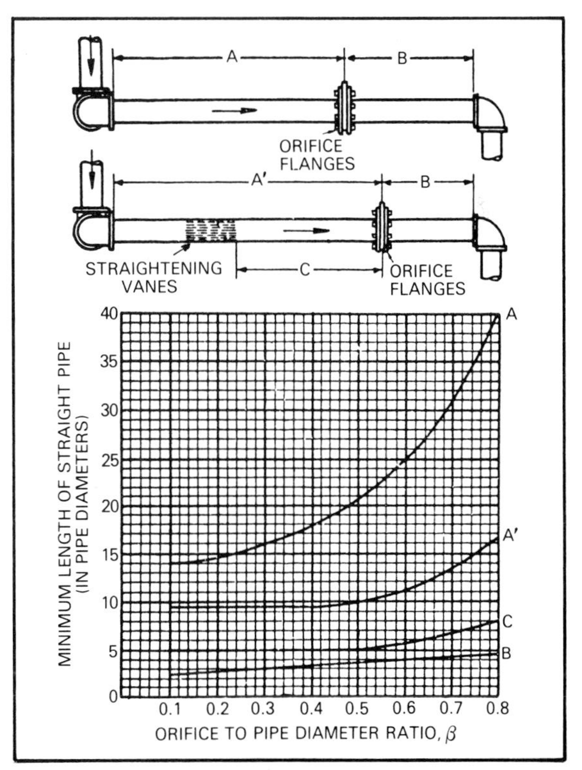

Figure 8.8. Comparison of straight-run and straightening-vane installations

upstream elbows are in different planes, producing an effect that requires longer runs than would be the case for two elbows in the same plane. The straight run of pipe required for correction increases steeply as the beta factor increases. For a beta factor of 0.6, straightening vanes will reduce the straight run from 25 to 11 pipe diameters.

Orifice fitting. A special fitting provides for rapid removal and replacement of orifice plates in a pipeline, in some cases without interrupting service. Flange-mounted orifice plates have a real advantage in the comparative ease with which they can be removed and replaced; nevertheless, the procedure still requires removing the flange bolts and using jack screws to force the flanges far enough apart to free the orifice plate. In addition, the procedure requires removing the line from service.

One form of orifice fitting allows the changing of orifice plates without shutting down the line (fig. 8.9). The orifice plate rides in a plate carrier that can be cranked up and down by a pinion-and-shaft arrangement. For example, if a change or inspection of the orifice plate is desired without interruption of service along the line, the following steps are taken:

1. Valve *7* is opened by applying a crank wrench to pinion-and-shaft assembly *8*, allowing access between the lower and the upper parts of the fitting.
2. The plate carrier is cranked from its lower to its upper position by applying the crank wrench to shaft *9*, later to shaft *5*.
3. Once the carrier has been lifted clear of valve *7*, it is closed by operating shaft *8*.
4. Bleeder valve *4* is opened to allow pressure in the upper part of the fitting to leak off.
5. Clamping-bar setscrews *1* are loosened, and the bar is slid out of its groove.

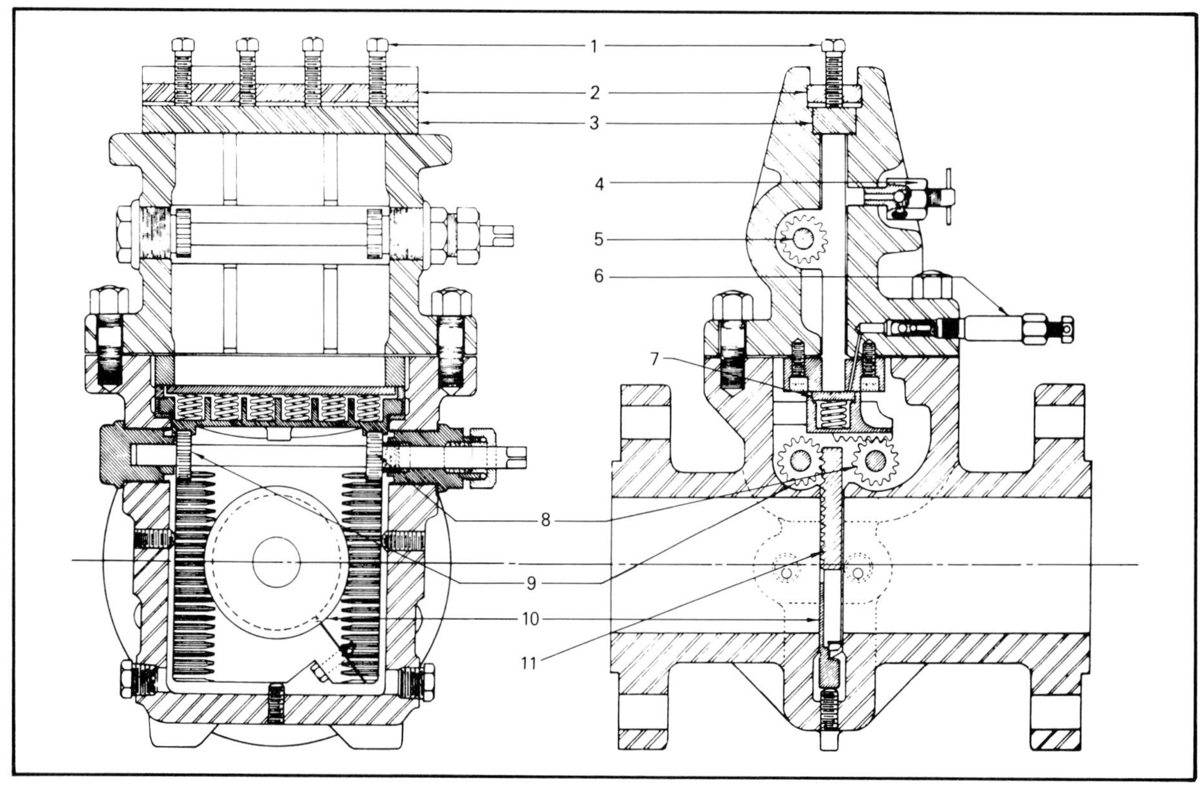

Figure 8.9. Orifice plates that can be changed without interrupting service

6. Sealing bar *3* is removed, and with shaft *5* the plate carrier is cranked out the top of the fitting.

To replace the orifice plate and carrier, the procedure is carried out in reverse.

Location of pressure taps. In measuring differential pressure, one point of measurement will be upstream of the orifice or other restriction, and the other downstream. The location of the two points is a matter of great importance in the design of flow-metering arrangements. The points of measurement are called pressure taps. The point upstream will be the high-pressure tap, and the point downstream will be the low-pressure tap. Pressure taps for head meters are mounted with the following considerations common to all forms of primary elements. For horizontal piping arrangements the taps are located as follows:

1. for gas lines, on the top of pipe, flange, or element;
2. for liquid lines, on the side of pipe, flange, or element; and
3. for steam lines, in the top of pipe, flange, or element when the differential pressure measuring instrument is located *above* the line, and to the side of pipe, flange, or element when the instrument is *below* the line.

For vertical piping arrangements, peripheral orientation of the pressure taps is not important.

Pressure taps for venturi sections are usually located in the uniform section upstream ahead of the cone for the high-pressure measurement, and at the throat for the low-pressure measurement.

Flow nozzles typically have the upstream (high-pressure) tap located 1 pipe diameter from the face of the nozzle and the downstream (low-pressure) tap ½ pipe diameter from the nozzle face. The downstream tap should never be located beyond the end of the flow nozzle. One arrangement (fig. 8.10) is popular for small-diameter pipes. The upstream tap is 1 pipe diameter from the face of the nozzle, while a flange tap, somewhat less than ½ pipe diameter from the face, is used downstream.

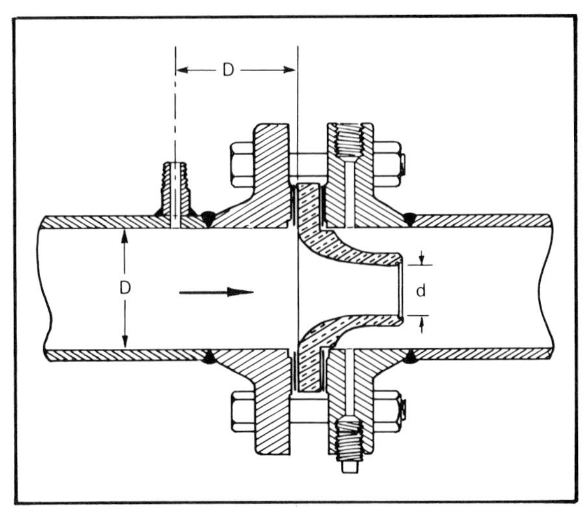

Figure 8.10. Location of pressure taps for flow nozzles ***(Courtesy Foxboro)***

Orifice plate primary elements may use any of several pressure-tap arrangements. Four common locations are designated as (1) vena contracta, (2) flange, (3) corner, and (4) pipe.

Vena contracta taps are located so that maximum differential pressure is obtained. The downstream tap is positioned at the point of minimum area of flow, the vena contracta, and therefore at the point of minimum pressure. The upstream tap is located about 1 pipe diameter from the orifice plate. The apparent advantage of the implied sensitivity gained by using points that give maximum pressure differential is overshadowed by other considerations. For example, the location of the vena contracta will vary for a given installation with rate of flow and characteristics of the fluid.

Flange taps are popular because they not only afford good pressure differential but also offer the advantage of precise location, since they are included as a part of the manufacture of the flanges. Since the two taps are located at equal distances from the orifice plate, measurement of flow in either direction can be accomplished with a minimum amount of

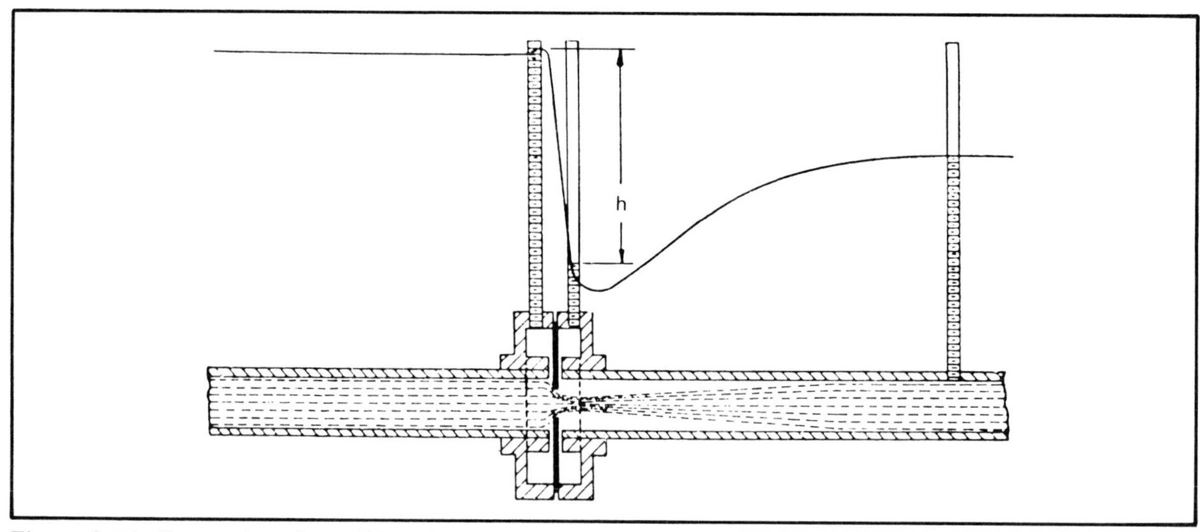

Figure 8.11. Corner taps for orifice plates

difficulty. Not so obvious is the advantage that an enormous amount of experimental work has been done to establish a large collection of data relating to all possible applications of this form of pressure tap.

Corner taps (fig. 8.11) resemble flange taps when viewed externally, but annular cavities have been formed in the flanges, and these cavities are free to assume the pressure existing on the immediate upstream side of the orifice plate on the one hand, and the immediate downstream side on the other hand. Pressure taps connect to the flange cavities. Corner taps are popular in Europe but are not favored in Canada and the United States because of their alleged susceptibility to freezing and other forms of clogging.

Pipe taps (fig. 8.12) are so called because they are located in the piping adjacent to the orifice plate. The upstream (high-pressure)

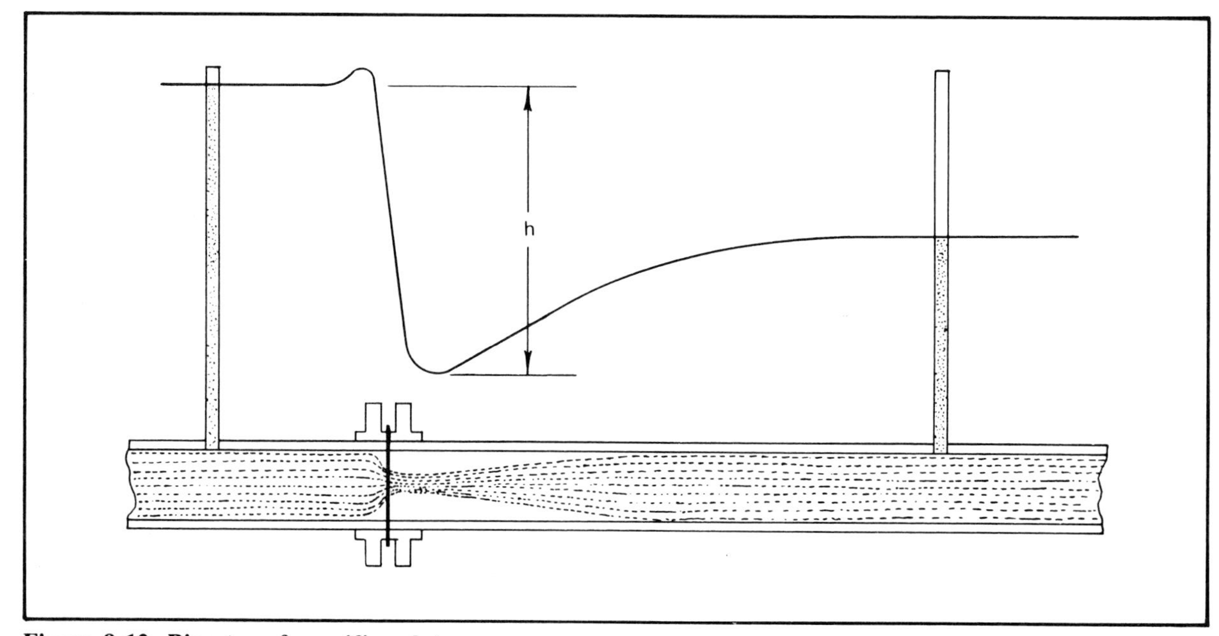

Figure 8.12. Pipe taps for orifice plates

tap is located 2½ pipe diameters ahead of the orifice plate, while the low-pressure tap is 8 pipe diameters downstream. These taps are sometimes called full-flow taps because they are at locations in which (1) the upstream tap is at a point ahead of any pressure change caused by the orifice plate, and (2) the downstream tap is at a point where maximum pressure recovery has been achieved. Pipe taps suffer from several disadvantages: (1) they are usually made at the time of piping installation, meaning that factory precision and smoothness of the pipe area around their location is not assured; (2) their smaller pressure differential usually means less accuracy of measurement; and (3) the considerable lengths of piping between the taps mean longer straight runs on either side of the orifice plate and allow a greater chance for error due to pipe roughness. Because pipe taps show smaller differential pressure for a given flow rate, they can provide a wider range of flow measurement with a particular meter range than other types. Pressure between the pipe taps (fig. 8.12) is very much less than the maximum, *h*, available. Although still used extensively, pipe taps are losing popularity at the same rate that flange taps are gaining it.

Calculation of Flow Velocity

Velocity of flow equation. The basis for flow calculations for head meters is an equation derived from two equations from the laws of physics and known as the *velocity of flow equation*. Derivation of the equation is as follows:

(a) $$v = gt$$

where

v = velocity, in feet per second;
g = acceleration due to gravity, in feet per second squared;
t = time, in seconds.

(b) $$h = \tfrac{1}{2}gt^2$$

where

h = distance, in feet;
g = acceleration due to gravity, in feet per second squared;
t = time, in seconds.

From equation (a), t may be stated in terms of v and g, as

(c) $$t = v/g.$$

Substituting this value of t in equation (b),

(d) $$h = \tfrac{1}{2}g\,(v/g)^2 = v^2/2g.$$

From this relation, v^2 and v may be found.

(e) $$v^2 = 2gh \text{ and}$$
$$v = \sqrt{2gh}. \qquad (8.1)$$

The tank shown in figure 8.13 is nearly filled with water and contains a sharp-edged round hole in its side. The hole is d feet in

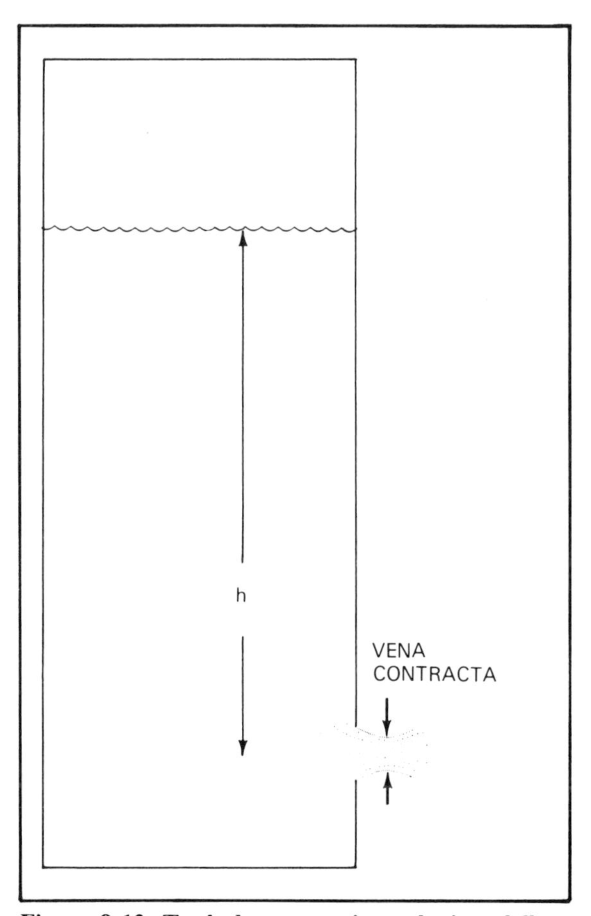

Figure 8.13. Tank demonstrating velocity of flow

diameter and is located h feet below the surface of the water in the tank. By using equation 8.1, the velocity of water spurting from the hole can be determined. It will be very nearly the value of $\sqrt{2gh}$, or $\sqrt{64h}$. Actually, the velocity will be about 0.98 or 0.99 of this value, and this factor of 0.98 or 0.99 is called the *velocity coefficient*. It is one of many corrective factors used in such calculations.

Equation for volume rate of flow. In determining the velocity of flow from the tank (fig. 8.13), hole diameter d had no significance. Only the constant number $2g$ and the height h of the water level above the center of the hole were factors. However, knowing the diameter of the hole (and thus being able to calculate its area) enables a good measurement of the volume of water flowing out of the hole to be made. Since the area of the hole is $\pi d^2/4$, multiplying both sides of equation 8.1 by this area factor will result in a volume measurement. Since

$$v \times \pi d^2/4 = \text{ft/sec} \times \text{ft}^2 = \text{ft}^3/\text{sec},$$

then

$$V/\text{sec} = \pi d^2/4 \times \sqrt{2gh},$$

where V = volume.

Despite the fact that the velocity error noted before is reasonably small (less than 2%), the volume measurement indicated by this last equation is apt to have a serious error, particularly for the sharp-edged orifice (hole) used in the example. Most of the volume error is caused by the fact that the jet of water escaping from the sharp-edged orifice actually contracts in diameter, so at a distance of about one-half the orifice diameter away from the tank, the jet attains a minimum value of cross-sectional area. The point at which the jet has its smallest diameter is called the *vena contracta,* and the cross-sectional area of the jet at this location is about 0.62 that of the orifice itself. The factor 0.62 is called the *contraction coefficient.*

The product of velocity and contraction coefficients is called the *discharge coefficient.* Its use in the equation allows derivation of a much more accurate value for volume rate of flow. Thus,

$$V/\text{sec} = 0.98 \times 0.62 \times \pi d^2/4 \times \sqrt{2gh}$$
$$= 0.61 \times A \times \sqrt{2gh}$$

where

A = area of the orifice.

The discharge coefficient given applies only to sharp-edged orifices; it is not suitable for application to venturi tubes and flow nozzles.

Equation 8.1 was simply derived from two other equations that relate to free-falling bodies. Bernoulli's theorem, which states that the total energy possessed by an element of fluid flowing without resistance in a streamline remains constant, is the real foundation of equation 8.1 for purposes of flow measurement. Although Bernoulli's theorem is straightforward and not difficult to follow mathematically, the result derived is similar to equation 8.1. For those interested in the theorem, almost any general textbook on physics contains a satisfactory treatment of the topic.

In equation 8.1, the factor h under the square root sign represents merely the vertical height of a liquid surface above an orifice, or other restriction, in the line. In deriving an equation by means of Bernoulli's theorem, the factor h takes on much more significance. It becomes the *differential pressure head,* measured in feet.

Corrective Factors for Velocity of Flow Equation

It has been noted that because of the contraction of the jet after fluid has passed through a sharp-edged orifice, a correction factor is needed to make the theoretical rate of flow agree with reality. In fact, a number of conditions that are not true have been assumed, among which the most important are the following:

1. The velocities (v) existing at the upstream and the downstream points of

measurement are related inversely as the areas (A) of the upstream pipe and the constriction; that is,

$$v_1A_1 = v_2A_2 \text{ and}$$
$$v_1/v_2 = A_2/A_1$$

where the subscripts *1* and *2* denote upstream and constriction values, respectively.

2. No energy is lost in the primary element or between the measuring points.
3. The velocity of flow at all points *across the diameter* of the pipe is the same; that is, the velocity of flow near the wall of the pipe is the same as that at the pipe center.
4. The density of the fluid does not change between measuring points.

It must be remembered that these assumptions are *false,* prompted by ignoring disturbances and some natural conditions that cannot be entirely overcome by design. Each condition that contributes to error can be corrected by including a factor whose numerical value is based upon hundreds of tests conducted under controlled conditions. The results of such tests are called *empirical data.*

The false assumptions will cause serious errors in calculating flow through head meter primary elements, particularly orifice plates. As noted earlier, the stream of fluid contracts after passing through an orifice, and its cross-sectional area at the vena contracta is equal to about 0.62 that of the orifice area. This value becomes a factor in measurement and is called the *contraction coefficient.* Also noted earlier was that, because of frictional losses, the true velocity of the jet, or stream, is only 0.98 to 0.99 of the theoretical value, so there is another factor, equal to about 0.98 or 0.99, called the *velocity coefficient.* The product of the *velocity* and *contraction* coefficients is called the *discharge coefficient.*

A factor that arises quite naturally in developing the flow equation from Bernoulli's theorem is the *velocity of approach factor.* It is expressed as

$$\text{Velocity of approach factor} = \frac{1}{\sqrt{1 - \beta^4}}$$

where β is the beta factor.

The velocity of approach factor can take on very significant values under certain conditions. If, for example, there is a 4-inch line with a 2-inch orifice plate, the beta factor d/D is $2/4$, or $1/2$. The fourth power of $1/2$ is $1/16$. The quantity $\sqrt{1 - 1/16}$ is 0.968, and this divided into 1 gives a velocity of approach factor of 1.033. Although that value is not very great, if the orifice plate had been 3 inches instead of 2, the factor would have been 1.21!

The velocity of approach factor can be combined with the discharge coefficient to produce the *efficiency factor, E;* that is, E is the product of the velocity of approach factor and the discharge coefficient. Letting Q be the volume rate of flow and A the area of the orifice plate or other constriction,

$$Q = EA\sqrt{2gh}. \qquad (8.2)$$

It is necessary to keep in mind that the value of the contraction coefficient mentioned earlier (0.62) applies to sharp-edged orifices and is not acceptable for venturi tubes or flow nozzles. The contraction coefficients for well-designed venturi tubes and nozzles approach much nearer to unity: 0.98+ for venturi tubes, 0.95+ for flow nozzles.

This equation is satisfactory for measuring water flow over a fairly wide range of variables—flow rates, temperature, and pipe diameters—but for measuring gases, steam, and many liquids, additional corrective factors must be taken into account. The viscosity and specific gravity of the liquids and the compressibility of gases are examples of factors to be considered. In addition to these considerations is the problem of *units of measurement.* Should the flow be measured in weight per minute, gallons per hour, barrels per day? Each of these forms of measurement will require the use of one or more factors to convert the *cubic* measurement inherent in the present equation. How is the differential pressure to be measured? The choice is important.

Development of a Working Equation

The basic equation can be simplified and put into a more usable form by taking care of certain constants and using the following criteria:

1. Pipe diameters and orifice diameters are expressed in inches.
2. The factor $2g = 2 \times 32.17 = 64.35$ feet per second squared.
3. The differential pressure, h, is measured in inches of water column.
4. Standard density of water is 62.34 pounds per cubic foot.
5. The rate of flow, Q, is expressed in cubic feet per hour.

The rate of flow in the basic equation is cubic feet *per second* because the factor g is expressed in feet and seconds (32.17 ft/sec^2) and area A of the orifice in square feet. In order to express Q in cubic feet *per hour,* the right side of the equation must be multiplied by 3,600, the number of seconds in an hour.

The area, A, of the orifice requires simplification and expression in square inches. Thus,

$$A = \pi d^2/4.$$

In the basic equation, d^2 is in square feet, so in order to be expressed in square inches, it is divided by 144, the number of square inches in a square foot. With Q representing flow rate in cubic feet per hour, the equation up to this point is

$$Q = \frac{3{,}600\pi d^2}{4 \times 144} \times E \times \sqrt{64.35h}.$$

E is simply a pure number without dimensions of length or time, so only h is left to convert. In the basic equation, h is the differential pressure, or head, expressed in feet of the flowing fluid, but it must be expressed in inches of water column. The conversion is made by using the factor 12 (the number of inches in a foot) and the relation between the density of water and that of the flowing fluid. Now,

$$h = \frac{G_W \times h_W}{G_f \times 12}$$

where

h = head, in feet of the flowing fluid;
h_W = head, in inches of water column;
G_W = weight density of water, in pounds per cubic foot;
G_f = weight density of the flowing fluid, in pounds per cubic foot.

The equation is now written as

$$\begin{aligned} Q &= \frac{3{,}600 \times \pi d^2}{4 \times 144} \times E \\ &\quad \times \sqrt{64.35 \times 62.34 \times h_w/12G_f} \\ &= 359Ed^2 \times \sqrt{h_w/G_f} \qquad (8.3) \\ &= \text{flow in } \textit{cubic feet per hour.} \end{aligned}$$

Calculating Liquid Flow

The specific gravity of a liquid is the weight density of the liquid divided by the weight density of water, with some standard temperature such as 60°F used as a reference. In working with liquids, it might be desirable to convert the flow equation so that the specific gravity of the flowing liquid is used instead of the weight density ratios. Note that in the equation for h, the quantity G_W/G_f is the reciprocal of specific gravity of the flowing fluid. Therefore, the equation for h can also be written as

$$h = h_w/12G$$

where G is the specific gravity of the flowing liquid. With this plan, the flow equation for liquids can be written:

$$Q = \frac{3{,}600\pi d^2}{4 \times 144} \times E \times \sqrt{64.35h_w/12G.}$$

Simplified, this equation becomes

$$Q = 45.47Ed^2\sqrt{h_w/G.} \qquad (8.4)$$

This equation is satisfactory for some applications in liquid flow, but for maximum accuracy a factor related to viscosity and rate of flow is needed. Some insight into this factor will be gained in the discussion of Reynolds' number. Sometimes it may also be desirable to include factors that correct the volume of liquid to certain standards of temperature, pressure, and specific gravity.

Calculating Gas Flow

For measuring the flow of gases, it is desirable to develop an equation that includes provision for important corrective factors and is easy to work with. It is best to begin with the basic equation, 8.3:

$$Q = 359Ed^2\sqrt{h_w/G_f}.$$

The quantity G_f must be converted to specific gravity for gases. The specific gravity of a gas is the ratio of weights of equal volumes of gas and air at standard temperature (32°F) and pressure (14.7 psia). Under these conditions the weight density of air is 0.08073 pounds per cubic foot, so $0.08073G$ would be the weight density of any gas where G is the specific gravity of the gas *with respect to air*. However, since the density of gas varies directly with absolute pressure and inversely with absolute temperature, corrective factors for temperature and pressure are as follows:

$$1/G_f = 1/0.08073G \times 14.7/p_f \times T_f/492$$

where

14.7 = standard absolute pressure, in psia;
492 = standard absolute temperature, in °R;
p_f = actual absolute pressure;
T_f = actual absolute temperature, in °R;
G = specific gravity of flowing gas.

When these values are substituted for $1/G_f$, the new equation becomes

$$Q = 218.44Ed^2\sqrt{h_wT_f/p_fG}. \quad (8.5)$$

In gas flow measurement the volume of gas is usually expressed at certain base values of temperature and pressure. For example, the volume of gas delivered according to equation 8.5 is reckoned at standard pressure and temperature, but thought must be given to the idea that the gas might be delivered or sold on the basis of other values of pressure and temperature. Designating the *base* values of absolute pressure and temperature as p_b and T_b and the *actual* values as p_f and T_f, respectively, the corrective factor becomes

$$p_f/p_b \times T_b/T_f.$$

This corrective factor takes into account Boyle's law and Charles's law; that is, the volume of a quantity of gas varies inversely as the absolute pressure and directly as the absolute temperature. A new equation containing this correction becomes

$$Q = 218.44Ed^2 \times p_f/p_b \times T_b/T_f \times \sqrt{(h_wT_f)/(p_fG)}.$$

This is a handy equation as it stands, particularly when there is a need to deal with various base values of temperature and pressure. It can be reduced to a simpler form by assuming the flowing gas to be air, p_b to be 14.4 psia, and both T_b and T_f to be 520°F absolute (60°F). These numerical values, when substituted in the above equation, bring about the following:

$$Q = 345.92Ed^2\sqrt{h_wp_f}. \quad (8.6)$$

In this equation, p_f is the flowing pressure of the gas. In some instances the pressure will be measured at the upstream tap; in other instances, at the downstream tap. p_f is sometimes referred to as the *static pressure*. The values of 14.4 psia pressure and 520°R absolute temperature are those chosen as bases by the gas industry. Obviously, when bases other than these are used, corrective factors must be applied. The same is true of the specific gravity factor G. Countless tables, graphs, and other data are available to assist in correcting for various base values.

The quantity $345.92Ed^2$ is the basic orifice factor and is designated as F_b. Tables are available that will provide an overall value of the quantity for beta factors ranging from

0.125 to 0.750, for pipe diameters from 2 to 15 inches, and for pipe and flange pressure taps. Tables are accurate to five significant figures. Table 8.1 is a set of F_b values for flange taps for various combinations of pipe and orifice diameters.

Expansion factor. Noted earlier was a list of false assumptions that were made in arriving at the initial basic flow equation. Item 4 of that list concerns the change in density of the fluid as it moves from one point to another in the primary element. The density of incompressible liquids is virtually unaffected by the differential pressure experienced in the primary element of a head meter, but the density of a gas is proportional to the pressure to which it is confined. In passing from the high-pressure upstream tap to the area of lower pressure in the primary element, gas will expand and its density will decrease in proportion to the loss of absolute pressure.

The error in a head meter caused by gas expanding in the primary element is a function of the following conditions:

1. the beta factor: the error tends to increase as this factor increases;

TABLE 8.1
BASIC ORIFICE FACTORS (F_b) FOR FLANGE TAPS

Orifice Diameter (inches)	*Standard Pipe Size with Actual Diameter (inches)*							
	2" Std. 2.067	*3" Std. 3.068*	*4" Std. 4.026*	*6" Std. 6.065*	*8" Std. 8.071*	*10" Std. 10.136*	*12" Std. 12.090*	*15¼" Std. 15.25*
0.250	13.002	12.996	12.974	-	-	-	-	-
0.375	29.079	29.026	28.989	-	-	-	-	-
0.500	51.680	51.444	51.367	51.328	-	-	-	-
0.625	81.129	80.426	80.218	80.080	-	-	-	-
0.750	117.77	116.16	115.67	115.30	-	-	-	-
0.875	162.11	158.70	157.81	157.08	156.82	-	-	-
1.000	215.04	208.21	206.62	205.45	204.97	-	-	-
1.125	277.93	264.97	262.21	260.39	259.69	259.27	-	-
1.250	353.04	329.41	324.72	321.93	320.94	320.35	320.00	-
1.375	443.44	402.10	394.35	390.11	388.73	387.95	387.47	-
1.500	554.71	483.80	471.35	464.98	463.10	462.07	461.44	-
1.625	-	575.47	556.06	546.62	544.07	542.73	541.93	541.06
1.750	-	678.61	648.93	635.11	631.64	629.94	628.93	627.85
1.875	-	794.99	750.50	730.55	725.90	723.73	722.48	721.15
2.000	-	926.76	861.42	833.06	826.86	824.11	822.56	820.95
2.125	-	1076.6	982.50	942.83	934.59	931.10	929.21	927.26
2.250	-	1251.2	1114.9	1060.0	1049.1	1044.7	1042.4	1040.1
2.375	-	-	1259.9	1184.9	1170.5	1165.1	1162.3	1159.5
2.500	-	-	1419.0	1317.7	1299.0	1292.1	1288.7	1285.4
2.625	-	-	1593.9	1458.7	1434.4	1425.8	1421.7	1417.9
2.750	-	-	1786.7	1608.3	1577.1	1566.4	1561.5	1557.0
2.875	-	-	2000.3	1767.0	1727.1	1713.8	1707.9	1702.6
3.000	-	-	2245.3	1935.2	1884.6	1868.2	1861.0	1854.9

NOTE: Conditions are 60°F base temperature, 60°F flowing temperature, 14.4 psia base pressure, 1.0 specific gravity.

2. the ratio of differential pressure to static pressure: the error tends to increase as this ratio increases;
3. the point at which the static pressure measurement is made: if the measurement is made downstream, *indicated* flow will be greater than *actual* flow, and vice versa;
4. the ratio of the specific heat of the flowing gas at a *constant pressure* to its specific heat at a *constant volume:* the error tends to increase as the ratio increases; and
5. the choice of pressure taps: the error is apt to be more severe when pipe taps are used.

A careful choice of beta factor and efforts to keep the ratio of differential pressure to line pressure as low as practicable will reduce the error caused by expansion of gas in the primary element. Empirical equations have been devised to determine the *expansion factor* needed to correct the error that might arise from this condition, but tables are available in many handbooks and should be used when available.

Supercompressibility factor. The laws that deal with variations of gas volume with temperature and pressure are not exactly correct in practice because ideal, or perfect, gases are not being worked with. According to Boyle's law, the volume of a gas varies inversely as its absolute pressure at a given temperature. In reality, when a typical gas is compressed, its volume contracts and its density increases by an amount somewhat greater than Boyle's law would indicate. This lack of agreement between Boyle's law and reality is referred to as the *supercompressibility* of the gas. A related event occurs when gas is allowed to expand to a lower pressure, but in this case the new volume is greater and the density is less than straightforward computation would indicate. This deviation from theory is called *superexpansibility*.

The supercompressibility varies with the type of gas and is not easy to determine accurately, although accurate curves and tables are available for many common gases. The error caused by the deviation from Boyle's law is determined by (1) specific gravity of the gas: the error increases with specific gravity; (2) temperature of the gas: error decreases as temperature increases; and (3) static pressure of the flowing gas: error increases with pressure. The corrective factor for supercompressibility is designated by the symbol F_{pv}. An empirical equation for its determination, one suitable for values of specific gravity between 0.50 and 0.75 is

$$F_{pv}=\sqrt{1+\frac{p_{fg}\times 3.444\times 10^{5}\times 10^{1.785G}}{T_f^{3.825}}} \qquad (8.7)$$

where

p_{fg} = indicated gauge pressure, in psi;
G = specific gravity of the gas;
T_f = absolute temperature, in °R.

A table of logarithms or a calculator is needed to deal with the exponents in this equation.

The error due to neglecting the supercompressibility factor may range from as little as $1/10$ of 1% for low static pressures and specific gravity (p_{fg} = 10 psig, G = 0.56, T_f = 520°R) to as high as 10% for high values of pressure and specific gravity (p_{fg} = 500 psig, G = 0.83, T_f = 520°R).

Reynolds' Number

Reynolds' number is a numerical ratio between the dynamic forces of a flowing fluid and its viscosity. The dynamic forces are related to the density and velocity of the flowing fluid and the diameter of the pipe in which it flows. Reynolds' number provides a clue as to whether flow is *viscous* or *turbulent*.

Broadly speaking, viscous flow, also called laminar flow, occurs at low velocities, and turbulent flow occurs at high velocities of flow. If a liquid tends to "wet" or adhere to pipe walls or other material, then flow inside

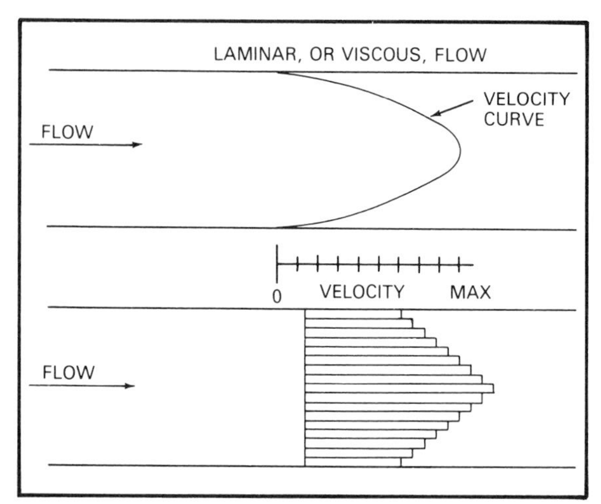

Figure 8.14. Probable velocity distribution for viscous flow

the pipe at low velocities will be characterized by a nonuniform velocity distribution that has a maximum value at the centerline of the pipe and decreasing values at points away from the centerline. Figure 8.14 shows the probable velocity distribution for viscous flow; it can be described as resembling a parabola. It is as though the liquid were made up of many telescoping sleeves, with the center sleeve moving at maximum and the outermost sleeve at lowest velocity. It may be recalled that in the list of false assumptions made in arriving at the initial flow equation, item 3 noted the fact that a *uniform* velocity distribution is assumed. If the velocity pattern in a pipe or other flow line is a true parabolic curve, then the *average* velocity of flow will be between 57% and 58% of the maximum value. The actual shape of the velocity curve will vary widely according to pipe diameter, viscosity of the fluid, and velocity.

As the velocity of flow increases, a critical value is eventually attained at which the sleevelike pattern of flow is distorted, and a bodily sliding of a turbulent mass of fluid through the pipe takes place. The effects of viscosity are still in evidence, and a certain velocity pattern exists beyond this critical value, but the ratio of *average* velocity to *maximum* (centerline) velocity is greater than in the case for viscous flow. In fact, if the velocity is increased somewhat above the critical value, the ratio of average to maximum velocity will quickly approach a stable figure of perhaps 0.8 or higher and additional increase in velocity will have little effect on the ratio. A stable value for this ratio is highly desirable for flow measurement purposes, because accuracy of flow measurement is a direct function of the accuracy with which the average velocity of the flowing fluid is measured or inferred. Stating that the shape of the velocity distribution curve or viscous flow will vary in accord with several factors at least implies an uncertainty in the measurement of average velocity for this form of flow.

The critical value of velocity at which flow can be said to change from viscous to turbulent is dependent upon four factors: (1) rate of flow, (2) pipe diameter, (3) weight density of the fluid, and (4) viscosity of the fluid. The determination of Reynolds' number involves these factors and can be defined as

$$R_D = \frac{v \times D \times S}{\mu}$$

where

R_D = Reynolds' number with respect to pipe diameter D;
v = flow velocity, in feet per second;
D = pipe diameter, in inches;
S = weight density;
μ = absolute viscosity of the fluid, in centipoises.

In order to convert this equation to a more usable form, let pipe diameter D be expressed as $2r$ where r is pipe radius, and multiply numerator and denominator by πr. Thus,

$$R_D = \frac{r \times v \times 2r \times S}{r \times \mu}$$
$$= 2\pi r^2 \times vS/\pi r\mu.$$

Since πr^2 is equivalent to pipe area, and the product of velocity, area, and weight density is equal to weight per unit time, the equation becomes

$$R_D = \frac{2 \text{ lb/sec}}{\pi r \mu} = 0.636 \times \frac{\text{lb/sec}}{r\mu}.$$

To measure flow by the hour rather than by the second and to measure pipe diameter in inches rather than pipe radius in feet require further conversion steps. Multiplying the flow rate by 3,600 will convert it from seconds to hours; multiplying the numerator by 2 will give $2r$ in the denominator ($2r = D$), and the product $12 \times D$ will express diameter in inches. The new equation is

$$R_D = \frac{2 \times 3{,}600 \times 0.636 \times \text{lb/sec}}{2 \times r \times 12 \times \mu}$$

$$= \frac{2 \times 0.636 \times \text{lb/hr}}{D \times 12 \times \mu}$$

$$= \frac{0.106 \text{ lb/hr}}{D\mu}$$

The absolute viscosity, μ, has dimensions of force, time, and area. In common units, it would be

$$(\text{lb} \times \text{sec})/\text{ft}^2,$$

but viscosity is usually measured in a unit called a poise. This unit is frequently divided into smaller quantities called centipoises (1 poise = 100 centipoises). Clearly, in order to measure absolute viscosity in poises or centipoises and the other quantities in such units as feet and pounds, conversion factors must be incorporated in the equation to convert pounds to dynes and feet to centimeters. Basing the equation on the use of centipoises for measuring viscosity, and incorporating the necessary conversion factors, the usable form of the equation becomes

$$R_D = 6.32v/D\mu \qquad (8.8)$$

where

R_D = Reynolds' number with respect to pipe diameter;
v = flow velocity, or rate of flow, in pounds per hour;
D = pipe diameter, in inches;
μ = viscosity, in centipoises.

Now that a practical way of determining Reynolds' number has been devised, how can it be used? First, it should be realized that Reynolds' number can be derived on the basis of constriction diameter of venturi tubes, nozzles, and orifice plates just as well as pipes. In such instances, the number is sometimes symbolized by R_d, where d signifies the constriction diameter. Reynolds' number can provide evidence as to whether viscous or turbulent flow exists in a line (fig. 8.15). Note that for Reynolds' numbers greater than 20,000 (2×10^4), the ratio of average to maximum velocity becomes and stays quite stable. Low values of Reynolds' number, representing viscous flow, indicate great uncertainty concerning the average velocity of flow.

In working with the flow equation to develop accurate data for the efficiency factor E, it has been found that greatest consistency of results is obtained when data are plotted against the Reynolds' number for the particular form of head meter installation. For maximum accuracy, a Reynolds' number correction factor must be applied to the basic flow equation. These correction factors are available in table form in many handbooks relating to orifice meter measurements.

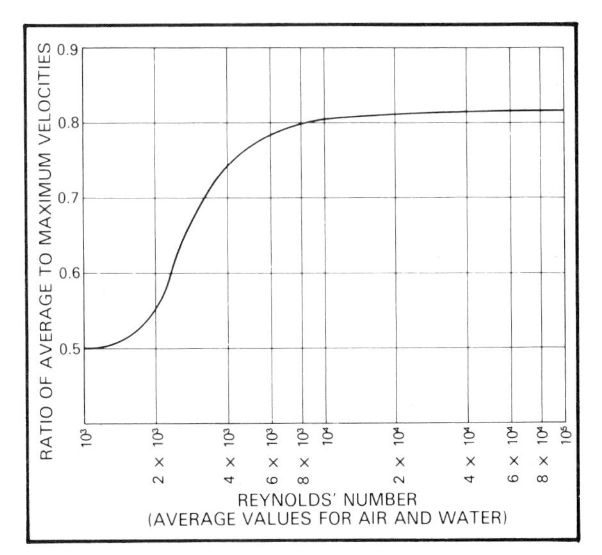

Figure 8.15. Relation of Reynolds' number to velocity of flow

TABLE 8.2
REYNOLDS' NUMBER FACTORS, F_r
FLANGE TAPS FOR 2-INCH LINE

Orifice Diameter (inches)	$\sqrt{h_w p_f}$									
	4.0	*4.5*	*5.0*	*5.5*	*6.0*	*7.0*	*8.0*	*10*	*12*	*15*
0.250	1.0232	1.0206	1.0185	1.0168	1.0154	1.0132	1.0116	1.0093	1.0077	1.0062
0.375	1.0181	1.0161	1.0145	1.0132	1.0121	1.0104	1.0091	1.0073	1.0060	1.0048
0.500	1.0147	1.0131	1.0118	1.0107	1.0098	1.0084	1.0073	1.0059	1.0049	1.0039
0.625	1.0126	1.0112	1.0101	1.0092	1.0084	1.0072	1.0063	1.0051	1.0042	1.0034
0.750	1.0118	1.0105	1.0094	1.0086	1.0079	1.0067	1.0059	1.0047	1.0039	1.0031
0.875	1.0119	1.0106	1.0095	1.0087	1.0080	1.0068	1.0060	1.0048	1.0040	1.0032
1.000	1.0129	1.0114	1.0103	1.0094	1.0086	1.0074	1.0064	1.0052	1.0043	1.0034
1.125	1.0144	1.0128	1.0115	1.0104	1.0096	1.0082	1.0072	1.0057	1.0048	1.0038
1.250	1.0161	1.0143	1.0129	1.0117	1.0108	1.0092	1.0081	1.0065	1.0054	1.0043
1.375	1.0179	1.0159	1.0143	1.0130	1.0119	1.0102	1.0089	1.0072	1.0060	1.0048
1.500	1.0193	1.0172	1.0155	1.0141	1.0129	1.0110	1.0097	1.0077	1.0064	1.0052

Table 8.2 is a sample Reynolds' number factor table. Factor tables are computed on the basis of a constant value of absolute viscosity (the absolute viscosity of most gas is approximately 0.0103 centipoise and varies little from this value). Other factors involved in the computation of Reynolds' number, such as density and velocity, are functions of the beta factor, the differential pressure, and static pressure. The tables are arranged so that for given combinations of pipe diameter and orifice dimensions, a series of columns headed by progressive values of $\sqrt{h_w p_f}$ contain the corrective factor.

A study of a full range of tables reveals that the maximum error likely to be incurred by neglecting the Reynolds' number factor, F_r, is about 3%. The error increases inversely as the beta factor and inversely as the quantity $\sqrt{h_w p_f}$. For an orifice meter installation using pipe taps in a 2-inch line, a beta factor of 0.125, and a value of 4.0 for $\sqrt{h_w p_f}$, the error caused by neglecting the factor F_r will be approximately 2.7%. On the other hand, changing the orifice plate to give a beta factor of 0.5 and increasing the static pressure so as to give a value of 100 for $\sqrt{h_w p_f}$ will result in a set of conditions that will need no application of Reynolds' number factor, F_r.

Resume of Corrective Factors for Flow Equations

The equations that have been developed in connection with gas flow measurement, beginning with the form designed to measure flow in terms of cubic feet per hour, equation 8.2, and through the form that assumes certain constant values for specific gravity, base temperatures, and base pressure, equation 8.6, vary in their usefulness according to the particular problem of flow measurement. Equation 8.2 is applicable to any fluid flow problem. It is the equation one would use in measuring flow of fluids under conditions that are not well documented with experimental data. This equation is rather bare in respect to corrective factors. Sometimes corrective factors, if there are enough of them, have a way of offsetting one another, and the results obtained from uncompensated calculations using equation 8.2 might be remarkably accurate.

Equation 8.6 is a specialized form for gas flow measurement. As it stands, it too is bare of corrective factors, and it would be risky to measure flow without including at least several of the more important factors. It may be recalled that the equation is based on the flow of air (specific gravity = 1) at a temperature of 520°R.

The flow measurement of a gas having specific gravity of 0.64 needs a corrective factor of 1.2500. Obviously, specific gravity, which is not a factor in equation 8.2, takes on enormous importance in equation 8.5. Temperature variations from the base value do not cause nearly so severe an error—about 1% error for each 10°F deviation from the base value of 60°F.

It is enlightening to study the tables of factors to be used with equation 8.6. It is particularly interesting to note the effect each factor has on the indicated flow value. Does the use of the factor produce a larger number or a smaller one—that is, is the factor greater than or less than unity?

Some of the factors and their derivations may be looked at. Equation 8.6 assumes a base pressure of 14.4 psia. If a different base pressure is desired, a corrective factor, F_{pb}, is used to convert the equation to the new base. The factor is derived as follows:

$$F_{pb} = p_b/p_{b1} \tag{8.9}$$

where p_b = 14.4 psia, and p_{b1} is the desired base pressure.

Clearly, if the desired base pressure is greater than p_b (14.4 psia), the factor F_{pb} will be *less than unity,* and vice versa.

The factor needed to change from one temperature base to another is F_{tb} and is derived from the relation

$$F_{tb} = T_{b1}/T_b \tag{8.10}$$

where T_{b1} is the new base, and T_b is the original base temperature.

In this case a desired base *greater* than the one assumed in the equation (520°R) produces a factor *greater than unity,* and vice versa.

The factors that correct for deviations in specific gravity and for the difference between actual and base temperature are square root functions. The derivation of the flowing temperature factor is

$$F_{tf} = \sqrt{T_f/T_{f1}} \tag{8.11}$$

where T_f is 520°R and T_{f1} is the absolute temperature of the flowing gas.

The factor for specific gravity correction is

$$F_g = \sqrt{G/G_1} \tag{8.12}$$

where G = 1 and G_1 is the relative density of the gas being measured.

As a final note, equation 8.6 might be rewritten as

$$Q = C'\sqrt{hp_f}. \tag{8.13}$$

In this form C' is a compound factor representing the product of all the individual corrective factors and the basic orifice factor, $F_b = 345.92Ed^2$. This is a total of eight factors, as follows:

$$C' = F_b \times F_r \times Y \times F_{pb} \times F_{tb} \times F_{tf} \times F_g \times F_{pv}$$

where

F_b = basic orifice factor = $1.1325Ed^2$;
F_r = Reynolds' number factor;
Y = expansion factor;
F_{pb} = pressure base factor;
F_{tb} = temperature base factor;
F_{tf} = flowing temperature factor;
F_g = specific gravity factor;
F_{pv} = supercompressibility factor.

Each of the above factors has been the object of extensive investigation, with the result that tables and curves of their values are readily available for an almost unlimited number of practical combinations. There has been, and there continues to be, a mighty effort to achieve absolute accuracy of measurement in the head-meter field. The quantities involved in flow measurement are often so enormous that accuracies on the order of 0.1% are highly desirable. Consider, for example, the dollars and cents aspect of an error of 0.5% for a natural gas system that transmits 250 million cubic feet of gas per week. Assuming the value of the gas to be $3 per thousand cubic feet (Mcf), this apparently small error amounts to $3,750 per week! In the matter of measuring natural gas, even the numerous corrective factors do not assure accuracies approaching 0.5%. In fact 2% is probably more nearly the correct figure, but for purposes of accounting, moneywise, it is well to apply as much effort as practicable to achieve absolute accuracy and then assume accuracy of near perfection. This will please the accountants who balk at rounding off money amounts in the thousand-dollar range.

Measurement of Differential Pressure

A mercury manometer, or U-tube (fig. 8.16), is a simple means of measuring the differential pressure developed across an orifice plate or other head-meter primary element. Pressure is measured in inches of mercury, but the head in inches of water may be obtained by multiplying by the specific gravity of mercury (about 13.56). This particular form of manometer is not of great use in actual practice because it provides only visual indications, similar to a thermometer or voltmeter. Usually some form of recording or integrating device is needed to maintain a record of total volume of flow.

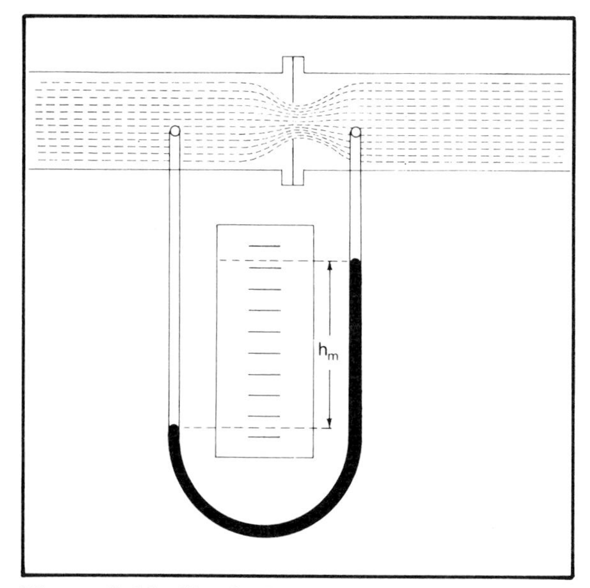

Figure 8.16. Simple U-tube manometer used to measure differential pressure

Despite its lack of utility, the simple U-tube manometer illustrates the need for another corrective factor in using manometer-type instruments for differential pressure measurement. Assuming that water is flowing in the pipe (fig. 8.16), the differential pressure head represented by the symbol h_m will be zero when the flow rate is zero. For this condition, mercury in the left and right columns of the U-tube are at equal levels, and pressure from two sources is exerted on the surface of each column—(1) the fluid static pressure of the line and (2) the pressure due to the head of water above the columns. The total pressure on one column is equal to that on the other column.

Assuming now that flow in the line is of such rate that a differential pressure exists across the column and that level difference in the columns is h_m, most of this level difference is due to difference in fluid pressure existing at the pressure taps, but at least a small part of it is caused by the fact that the left column now has a greater head of water on its surface. The head is greater by the vertical height of h_m, and the error it causes is equal to the numerical value of h_m in *inches of water*. For example, if $h_m = 5$ inches of mercury, then the error caused by the greater head of water on the left column of mercury is 5 inches of water.

Correction for weight of fluid above mercury. The example cited above indicates that error caused by weight of fluid immediately above the mercury in a manometer can be serious—about 4% in the example above. The error factor will be a constant quantity that depends on the specific gravity of the fluid immediately above the mercury. If h_w is the *indicated* differential pressure (in inches of water) and h_{aw} the *true* differential pressure across the pressure taps (in inches of water), the following equations can be written:

$$h_w = \frac{h_{aw} \times 13.56}{13.56 - G_m},$$

and

$$h_{aw} = \frac{h_w \times (13.56 - G_m)}{13.56}$$

where G_m is the specific gravity of the fluid above the mercury and 13.56 is the specific gravity of mercury. Specific gravity is measured with respect to water.

Remembering that h is a square root function in the flow equation, the corrective factor can be written as

$$F_m = \sqrt{\frac{13.56 - G_m}{13.56}}$$

where F_m is the symbol for this factor.

This equation makes it clear that the factor is very important for liquids, particularly those having appreciable values of specific gravity, and is of less importance when the fluid above the mercury is a light gas. If air is the fluid above the mercury, the corrective factor can be calculated on the basis of the specific gravity of air, which is roughly 0.081. Thus,

$$F_m = \sqrt{\frac{13.56 - 0.081}{13.56}} = \sqrt{0.994}$$
$$= 0.997$$

The error in this case is insignificant. For water, as in the earlier example, the corrective factor will be

$$F_m = \sqrt{\frac{13.56 - 1.00}{13.56}} = \sqrt{0.926} = 0.962.$$

The error caused by neglecting the factor would be about 4%.

Orifice meter: mercury manometer type. A practical variation of the simple U-tube manometer is a popular means of measuring differential pressure for head meters (fig. 8.17). Its widespread use with orifice plate installations has led to its being called an orifice meter. Three connections are provided for receiving upstream, downstream, and static pressures. These are the variables needed to account for the rate of flow.

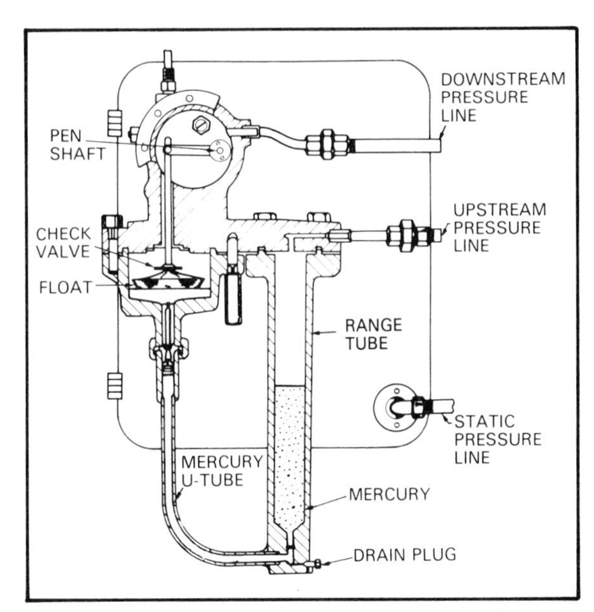

Figure 8.17. Orifice meter with a modified U-tube manometer

The upstream pressure line is applied to the range tube, or high-pressure chamber. The downstream pressure is applied to the low-pressure or float chamber, which is very much larger in cross-sectional area than the range tube. A float that rises and falls with the mercury level in the chamber actuates the recording pen shaft through a special differential linkage. The static pressure line is attached to a Bourdon tube or other pressure-actuated device that drives a separate recording pen.

Many instruments of this type are capable of being fitted with several different range tubes, each differing from the others in internal diameter. This choice of range tubes is an effective means of changing the differential pressure range of the instrument. In order to explain this, it may be assumed that total float travel is 1 inch and internal diameter of the float chamber if 4 inches. The cross-sectional area of the chamber is $\pi(2)^2 = 12.56$ square inches. Since float travel is 1 inch, the volumetric change within the chamber is 12.56 cubic inches. It may also be assumed that a range tube having a diameter of 2 inches is being used. Then for each inch of range tube length there are 3.1416 cubic inches of mercury. Thus mercury level in the range tube must be driven down 4 inches in order to operate the float through vertical travel of 1 inch. Now, since the mercury in the float chamber has risen 1 inch and that in the range tube has fallen 4 inches, the total differential is 5 inches of mercury, or 67.8 inches of water (5×13.56). Evidently the meter range can be increased by selecting a range tube with a smaller diameter, or the range can be reduced by increasing the diameter of the range tube. Range tubes are typically designed to produce ranges of 20, 50, or 100 inches of water.

Manometer-type orifice meters are usually quite rugged instruments and are capable of withstanding large static pressures. Protective features that prevent the likelihood of mercury being driven entirely out of the chambers are almost invariably a part of good meter design. Should the float in figure 8.17 be driven beyond a certain height, a check valve disk mounted atop the float will close off the upper access to the chamber, thus preventing mercury from being forced into the downstream pressure line. Another check valve of the ball-and-seat variety closes the lower access to the float chamber should the mercury level fall below a certain low level in the chamber.

Orifice meter: bellows type. The mercury manometer and instruments using bellows or diaphragm elements are the only popular means for measuring differential pressure. The bellows-type meter has been developed into a remarkably durable and accurate instrument in recent years.

Figure 8.18 is a sectional view of a bellows assembly designed to drive a recorder pen. The meter body consists of two opposed bellows that are screwed to a center plate and sealed against leakage by O-ring gaskets. After the bellows are assembled in this way, their interior spaces and the interior space of the center plate are evacuated with a vacuum pump. The interior void is then completely filled with a noncompressible liquid having a very low freezing point temperature.

During operation, upstream and downstream pressures are applied to the high-pressure and low-pressure housings, respectively. These are tight, rugged housings that cover the exterior of the bellows. A differential pressure across the meter assembly will tend to compress the high-pressure bellows, driving more of the noncompressible liquid into the low-pressure bellows. Passages for this liquid transfer include overrange valves and a pulsation dampener, shown lightly in the drawing as a needle valve in the lower part of the center plate. Motion of the free end of the low-pressure bellows is transmitted to the drive arm through an assemblage of overrange valve plugs and linkages.

Excessive travel of the drive arm is prevented by conelike plugs that close off the passage of liquid fill between one bellows unit and the other when certain limits of movement are attained. Temperature compensation is required in this particular form of meter because of the volume variation of the liquid fill with temperature change. A bimetal compensator fixed in the high-pressure bellows controls the expansion and contraction of the assembly to the required amounts to compensate for ambient temperature changes over a wide range.

The bellows-type orifice meter can be adapted to a number of ranges by changing the range-spring assembly. This assembly may include special springs that are insensitive to temperature variations. Action of the springs is to oppose the force of the upstream pressure on the high-pressure bellows. From the figure it is clear that action of the range springs tends to keep the low-pressure overrange valve closed.

Bellows-type meters of this form can be used for measuring the flow of liquids or gases, although care must be taken to prevent the damaging effects of having corrosive or extremely viscous fluids enter the pressure housings. The error of such meters can be as low as 0.5% of the meter range.

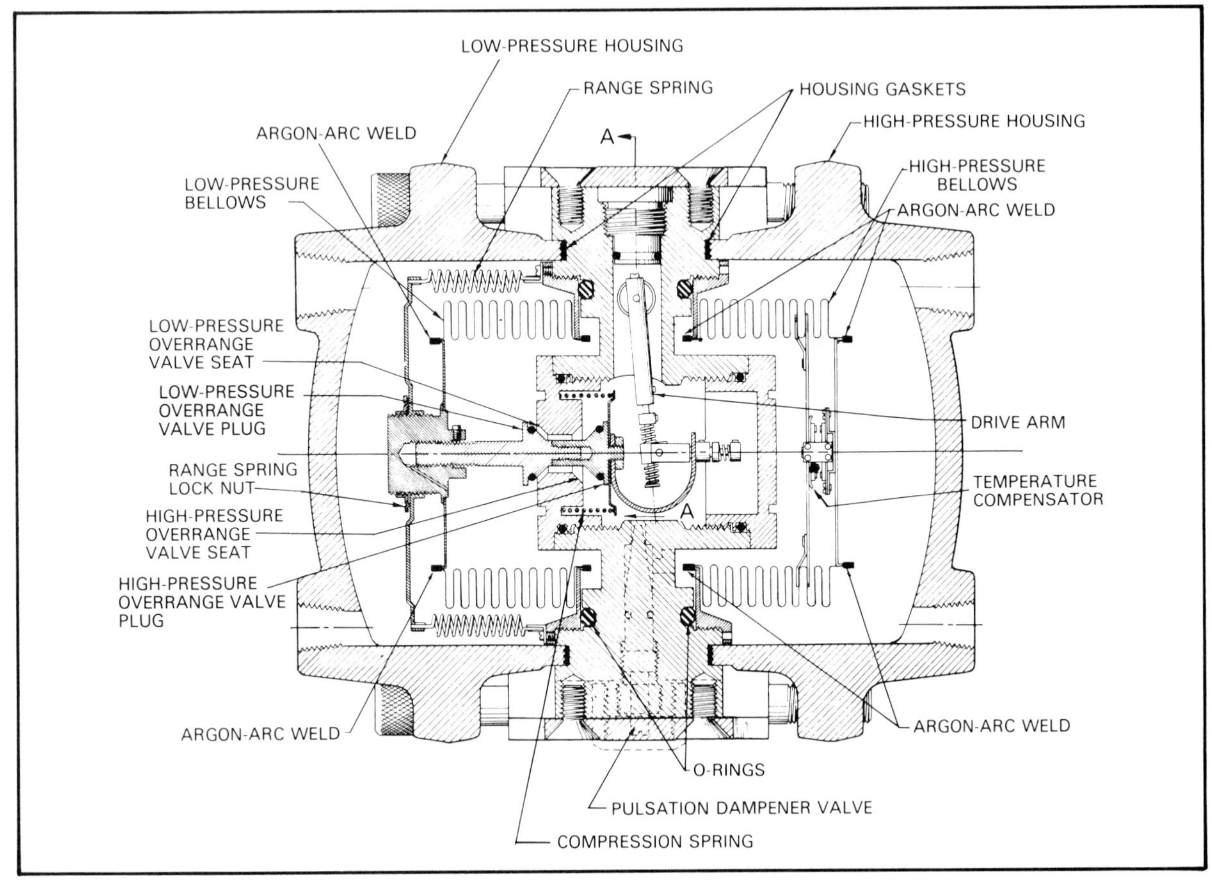

Figure 8.18. Bellows assembly for a bellows type of orifice meter ***(Courtesy Industrial Instrument Corporation)***

Variable-Area Flowmeters

Variable-area flowmeters contain a device that moves in response to the rate of flow, and this movement acts to vary, in some way, the cross-sectional area occupied by the flowing fluid at one or more points in the flow line. One simple design uses a piston to vary the area of flow (fig. 8.19).

If there is no flow of fluid, pressures p_1 and p_2 have equal values, and the piston will fall of its own weight to the lowest position. As flow begins, the pressure on the bottom of the piston exceeds that on top, and the piston is driven to higher elevations, thus exposing more flow area until equilibrium is established. The height of the piston is an

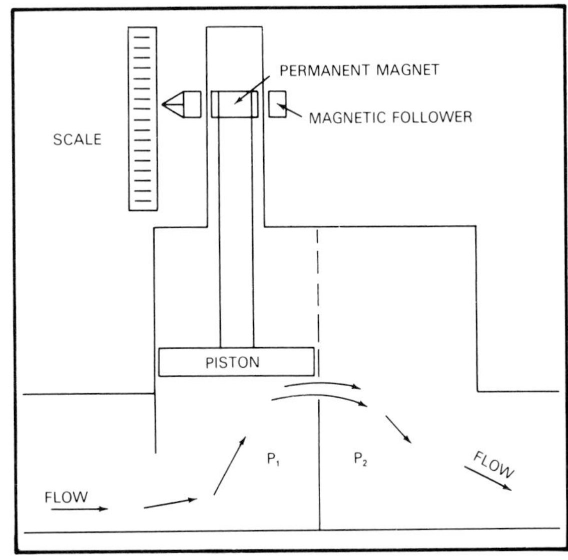

Figure 8.19. Schematic of a variable-area flowmeter

Figure 8.20. Variable-area flowmeter capable of remote transmission ***(Courtesy Honeywell)***

indication of the rate of flow. Figure 8.20 is a commercial form of variable-area flowmeter that is capable of remote transmission of signals to indicating, recording, or control equipment.

The most common variable-area flowmeter, the *rotameter,* is popular for measuring very low flow rate and the rates of viscous liquids. The rotameter has been mentioned previously in connection with its use to indicate the feeble flow of air or gas through a bubbler system for measuring liquid level.

A simple and effective form of rotameter (fig. 8.21) consists of a tapered metering tube, that is, a special glass tube that has a uniformly increasing cross section from bottom to top. The tube is tightly fitted between metal ends that form the inlet and outlet pipe fittings. Inside the tapered metering tube is the metering float, a plummet-like device that must be somewhat heavier than the amount of liquid it can displace. The float is limited in its vertical travel by stops that prevent it from reaching the ends of the tube.

The rotameter must be installed vertically, and flow must be from bottom to top. When there is no flow through the metering tube, the float rests on the lower stop, but as flow increases from zero, a net force is exerted *upward* on the float, tending to lift it higher in the tube. As the float rises in response to the force caused by the flowing fluid, the area formed in the annular space between the float and the tube becomes greater because of the taper of the tube.

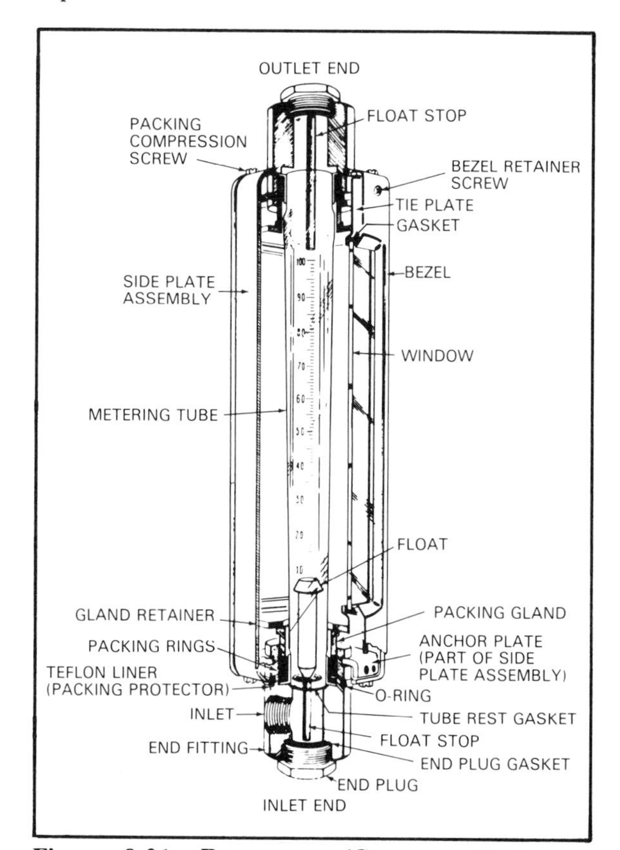

Figure 8.21. Rotameter ***(Courtesy Fischer & Porter)***

Several forces act on the float when fluid is flowing in the tube: (1) buoyant force of the fluid, (2) the weight of the float, and (3) force caused by the velocity of the fluid flowing in the tube. The net effect of (1) and (2) is a constant value for a given fluid density, and it is this constant downward force (or weight) that is balanced against the upward force caused by the flowing fluid. As the float rises in the tapered tube, the area of the annular space increases, and the velocity of the fluid decreases. Thus, if the flow rate is within the range of the rotameter, the float will settle at some point of equilibrium, at which the net force acting on it is zero.

An equation quite similar to that used for head meters applies to rotameters.

$$\text{Volume-rate} = A\sqrt{2gh}$$

where

A = area of the annular space;
h = differential pressure;
g = acceleration due to gravity.

An important difference is that A, not h, is the variable measured to determine flow. A significant advantage over the head meter now becomes apparent, for the rotameter responds *linearly,* whereas the head meter is a *square root* instrument. The linear response of the rotameter makes it suitable for measuring wide ranges of flow—ranges for which the maximum rate might be as much as 10 to 20 times the minimum rate. The ratio of maximum to minimum flow measurement capability is several times that of a practical orifice meter.

In addition to the advantages noted above, rotameters have other desirable characteristics.

1. Proper design of the metering float provides an instrument that remains accurate over wide variations of viscosity.
2. Most rotameter designs are unaffected by piping characteristics, and lengths of straight pipe ahead of the meter are not necessary.
3. The rotameter can handle difficult liquids, ranging from molten sulfur and viscous tars to strong acids.
4. The instrument can measure extremely small flow rates—flows lower than 0.1 cubic centimeter per minute, for example.

The rotameter (fig. 8.21) is an indicating device that would be rendered almost useless if opaque liquids were passed through the metering tube. However, several methods are used, not only to provide indications of flow in such applications but also to transmit the flow-rate values to recorders or controllers. In some units the float is attached to a rod containing one or more permanent magnets. The rod and magnets move inside a glass or other nonmagnetic tube, while a linkage system fitted with a yoke-type permanent magnet operates outside the tube. As the float, rod, and magnets move up and down inside the tube, the external yoke-magnet follows the motion because of the strong attraction between the magnets.

Variable-area meters are usually designed and calibrated to measure flow of a particular fluid. If the design has been of such nature that viscosity of the fluid is of no significance, then two corrections are necessary if the instrument is to be used with fluids other than that for which calibration is applicable: (1) specific gravity and (2) buoyancy of the metering float.

The corrective factor for change in specific gravity G is $\sqrt{G_2/G_1}$, where subscripts *1* and *2* refer to the fluid for which initial calibration is valid and the new liquid, respectively.

The corrective factor for change in buoyancy is $\sqrt{W_2/W_1}$, where subscripts *1* and *2* refer to the apparent weight of the float in the liquid for which calibration is valid and the new fluid, respectively. The apparent weight of the float is that measured while it is totally submerged in the liquid whose flow is to be measured (also defined as the true weight minus the force of buoyancy).

The capacity of area meters can be altered by changing the density of the float. Thus, by increasing the density of the float, a greater velocity of flow is required to lift the float a given distance in the metering tube, and vice versa.

Magnetic Flowmeter

A magnetic flowmeter operates on the principle that an electric conductor moving across a magnetic field has a voltage induced in it. The conductor is the fluid (liquid) flowing in the line, and the magnetic field is established by field coils mounted on diametrically opposite sides of a special section of line. Special features must be provided in order to use this principle for measuring fluid flow. An exploded view of a commercial magnetic flowmeter (fig. 8.22) shows that it consists of a short section of flange-fitted stainless steel tubing lined with a durable insulating compound such as Neoprene. Two electrodes are insulated from the tubing but fit through it and its lining so that electrical contact can be made with the flowing liquid. Two coils, located in such a position as to produce a strong magnetic field perpendicular to the diameter represented by the line between the electrodes, and a weatherproof cover complete the assembly.

Operation of the meter is in accord with Faraday's law, which can be expressed by the equation

$$E = khLv$$

where

E = induced voltage;
k = a constant determined by the system of units;
h = the magnetic field strength;
L = length of the conductor (the distance between the electrodes in this case);
v = velocity of the conductor in the magnetic field.

The fluid to be measured must be a conductor, although it need not be a very good one, since the voltage induced is independent of

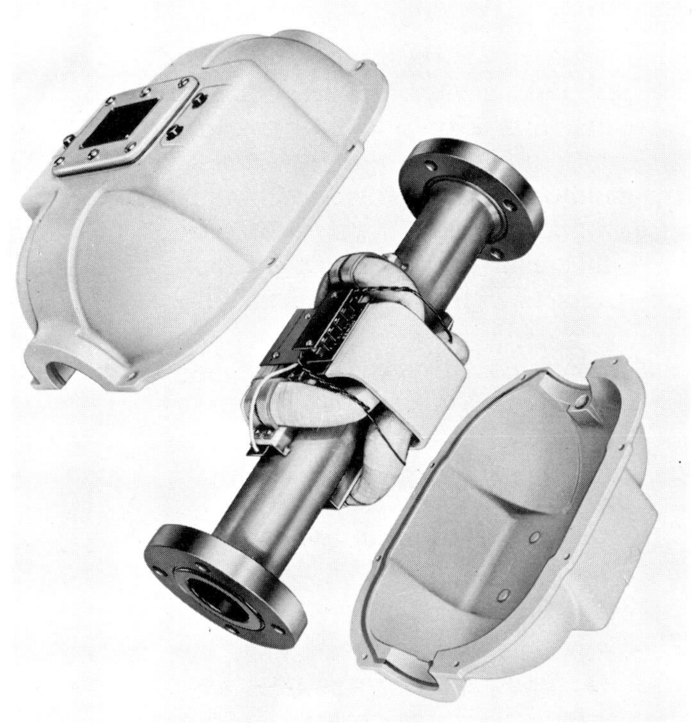

Figure 8.22. Exploded view of a magnetic flowmeter (*Courtesy Foxboro*)

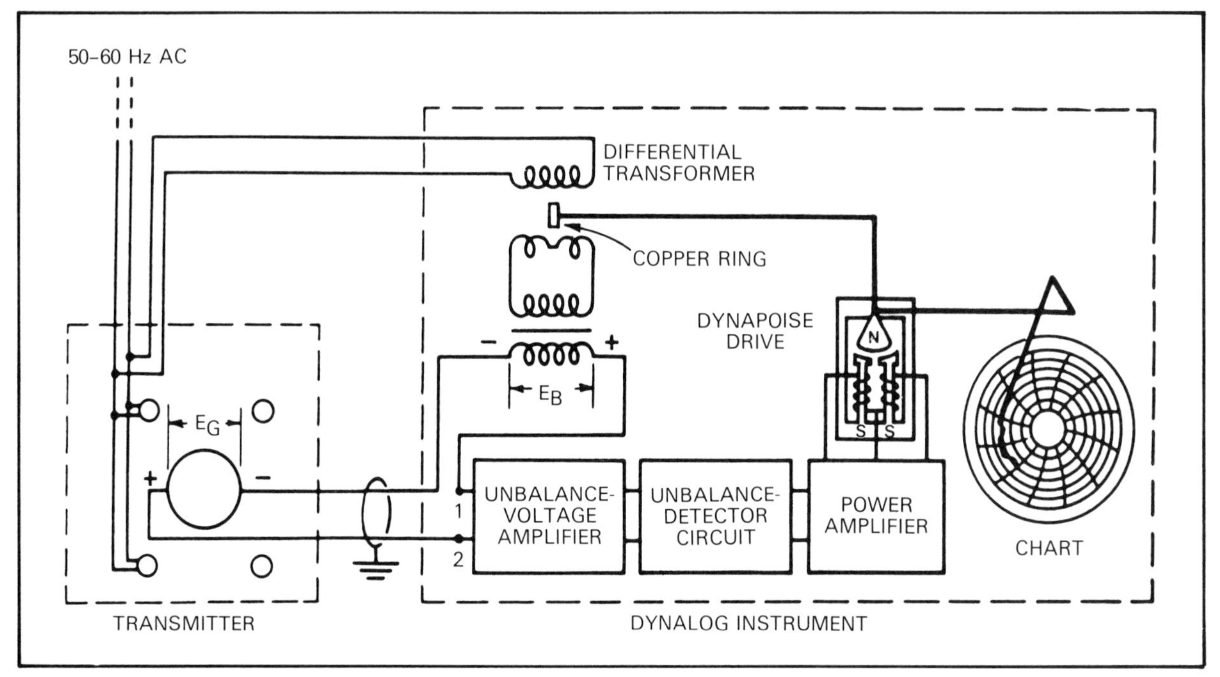

Figure 8.23. Schematic of measuring instrument for a magnetic flowmeter ***(Courtesy Foxboro)***

the resistance of the conductor. Resistance does pose a problem, however, because any load, that is, any current drain on the circuit, will cause a large voltage drop to occur between the two electrodes, leaving little or nothing to be measured. The measurement of induced voltage can be achieved without loading the feeble circuit (fig. 8.23).

A source of alternating current is used to excite (energize) the magnetic field, and this same source also is applied to the input winding of a *differential transformer*. The fact that alternating current is used to generate the magnetic field assures that the voltage induced in the moving conductor (liquid) will also be alternating. This voltage is applied to the output side (E_b) of the differential transformer in series with the high-impedance input of the *unbalance-voltage amplifier*. The output side of the differential transformer also has a voltage induced by the primary, or input, winding, and this voltage is in opposition to that from the electrodes. If the two voltages in the output winding are equal and opposite, their sum is zero, and consequently no signal will appear at the input to the unbalance-voltage amplifier. Should the flow rate decrease or increase, this change will be amplified by the unbalance-voltage amplifier and applied to the unbalance-detector circuit. The phase of the unbalance voltage will be compared with a reference signal in the unbalance detector to distinguish between increased and decreased flow rates.

The power amplifier is a push-pull circuit that receives two direct-current voltage signals from the unbalance-detector circuit. Each of these voltages controls one side of the push-pull circuit, its value and phase determining whether current in that side of the circuit is to be comparatively large or small. The output circuit of the power amplifier connects to the solenoid motor. This motor consists of two solenoids energized by the push-pull power amplifier, and a pivoted soft iron element whose position is determined by the amount of current flowing in each solenoid coil. The motor has more than sufficient torque to drive an indicator or recorder pen and to position the copper ring in the differential transformer.

The position of the copper ring in the transformer has a direct effect on the value of voltage induced in the output winding, so, when a change in flow rate causes a change in voltage E_g, the chain of events ultimately causes the solenoid motor to reposition the copper ring in order to bring about a condition of equilibrium, or zero voltage, in the output winding of the transformer.

This arrangement—using alternating current to excite the magnetic unit and to operate the amplifying, detecting, and other circuit components—results in a system whose accuracy is unaffected by ordinary line voltage variations. This is true because a variation in the induced voltage caused by change in the magnetic field strength will be almost perfectly compensated by the action of the differential transformer circuit.

According to the equation given earlier, $E = khLv$, the meter normally responds to the velocity of the flowing fluid, v, since the other factors, khL, properly represent a constant. Quantity or volume rate of flow is given by Av, where A is the cross-sectional area of the pipe, and v is the average velocity of flow. Volume rate can therefore be expressed as a function of voltage as follows:

$$\text{Volume rate} = \frac{AE}{khL}$$

where k is apt to be a constant that is tied in with the calibration of the meter for a particular application.

Many of the advantages of the magnetic flowmeter are obvious. For example, it offers no restriction to flow, because only the small area represented by the contact surface of the electrodes is exposed to the flowing fluid. Less obvious advantages include its ability to handle very rough flows such as slurries and sewage; its ruggedness and immunity to clogging (as with pressure tap installations); its linear response to flow rate; and its accuracy and ease of remote transmission of information.

Disadvantages are few, but fluids that have little or no conductivity cannot be measured. If 1 part of salt is added to 40,000 parts of pure water, the resulting solution will be sufficiently conductive to operate most magnetic flowmeters. Ordinary tap or drinking water has slightly greater conductivity than such a solution, and seawater typically contains about 1 part of salt for 3 to 5 parts of water. Despite the favorable conductivity of most liquids, many chemical and petroleum process liquids are definitely too low in conductivity to operate magnetic flowmeters.

Mass Flowmeters

Many forms of meters respond to the *mass* of the flowing fluid (rather than volume, area, or velocity) in order to indicate or record flow. One type imparts angular momentum to the flowing fluid and then measures the *torque* required to remove this momentum. Torque is related to force and mass, and its measurement enables flow to be expressed in terms of mass.

In a simplified drawing of a mass flowmeter, (fig. 8.24), a synchronous motor at the left drives an impeller at some constant speed. The impeller resembles a cylinder containing many individual fluid passages just under its outer surface. Since the impeller fits closely in the housing, fluid approaching from the left is forced to flow through the longitudinal passages and in doing so is given rotary motion by the revolving impeller. The rotary motion imparted to the fluid in each passage can be expressed as angular momentum $I\omega$, where I is the *moment of inertia* and ω is *angular velocity* equal to $2\pi f$, f being the number of revolutions per second of the impeller.

The moment of inertia is a function of the *mass* of the flowing fluid in the passages and the *radius* of this mass from the center of rotation of the impeller. Accurate calculation

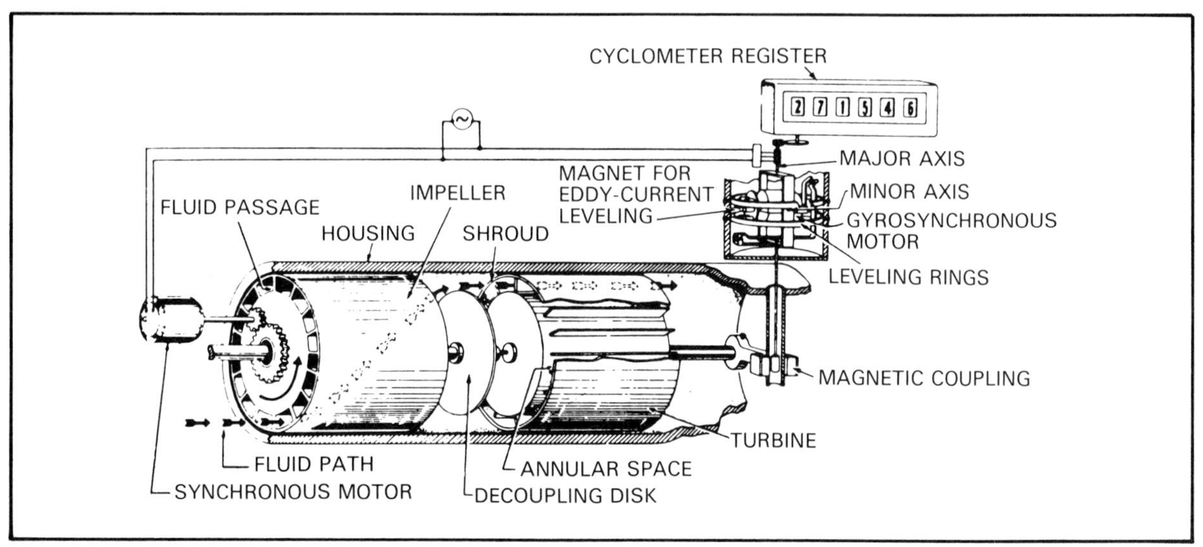

Figure 8.24. Mass flowmeter that measures torque *(Courtesy General Electric)*

of this value requires integrating the mass and its radius from minimum to maximum values. The net result is the same as though the total mass were concentrated in a thin ring having a given constant radius. The angular momentum can be expressed as $m\omega r^2$, where m is the total mass of the fluid in the impeller at a given instant, r is the radius at which it would be correct to assume all the mass concentrated, and ω is $2\pi f$, as noted earlier.

Immediately to the right of the motor-driven impeller is a stationary *decoupling disk*. The purpose of this disk is to reduce the drag that might exist between the rotating impeller and the turbine at the right. Such drag would be particularly troublesome with liquids and very low or zero flow rates.

After leaving passages in the impeller, the fluid retains the momentum it has gained and flows into similar passages of the turbine. The kinetic energy represented by the moment of inertia and angular velocity of the fluid will tend to set the turbine in motion, but the turbine is not entirely free to rotate, being restrained in its travel by coupling to an external device. In flowing through the passages of the turbine, the fluid gives up all the angular momentum it gained in the impeller and once again flows in a straight line. Its angular momentum has changed from maximum to zero. The *rate of change* in angular momentum is called torque. Torque is also known more familiarly as the product of force and radius, force being applied at right angles to the radius. The radius in this case is the distance from the turbine axis to the magnetic coupling shown in the figure.

Relation between Mass Flow Rate and Torque

Torque can be shown mathematically to be directly proportional to the rate of flow of mass through the turbine.

(a) $$T = \frac{dH}{dt}$$

where T = torque, and dH/dt is the time rate of change of angular momentum.

The angular momentum of the small element of mass, dm, that leaves the impeller in the interval of time, dt, can be expressed as

(b) $$dH = \omega dI$$

where ω is the angular velocity, and dI is the moment of inertia of the small element with respect to the axis of rotation.

Since the total mass is concentrated in a ring having a constant radius (sometimes called the radius of gyration), dI can be written as

(c) $$dI = r^2 dm.$$

Combining equations (b) and (c) and converting them to derivatives by dividing by dt,

(d) $$dH/dt = \omega r^2 (dm/dt) = T.$$

The term dm/dt is the time rate of flow of mass, so it is clear that torque is directly proportional to the mass rate of flow. Assuming ωr^2 to be an acceptable constant determined by the angular velocity and dimensions of the impeller, equation (d) can be integrated to give

$$M = T \times t/\omega r^2.$$

where

M = mass;
T = torque;
t = time;
ωr^2 = a constant.

Measurement of Torque

A number of methods can be used to detect, measure, and record the torque produced by the loss of angular momentum experienced by the fluid flowing through the turbine. A restraining spring (fig. 8.25), is a practical system. The spiral spring would have a force constant k, and the total force required to wind it through an angle θ would be θk. It

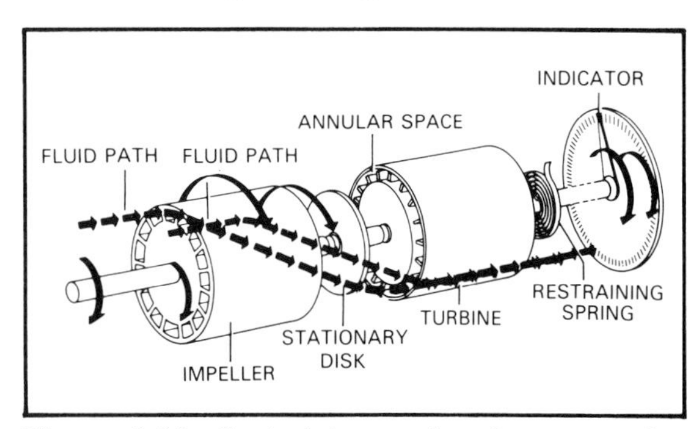

Figure 8.25. Restraining spring for measuring torque *(Courtesy General Electric)*

would be a simple matter to include an indicator dial calibrated in pounds of flow per minute, hour, and so forth. Or, instead of the dial, an integrator or recorder could be driven.

The flowmeter system (fig. 8.24) incorporates a gyroscope to respond to torque application. The response of the gyroscope is a rotation about its major axis, and the speed of this rotation is proportional to the applied torque. Most people are familiar with gyroscopic action, having observed it in the case of spinning tops and gyroscope toys. The behavior of any rapidly spinning heavy wheel subjected to forces applied at certain points is usually surprising. For example, assuming such a wheel is rotating about an axle that is horizontal, one may visualize the reaction of a wheel and axle if an effort is made to tilt the axle to a nonhorizontal position. The effort to move the axle in this way will be resisted with amazing force, and the spinning wheel and its axle will attempt to turn about an axis that is perpendicular to the applied torque. This turning action is called *precession,* and its axis is called the *major axis* of the gyroscope. An axis that is perpendicular to both the major axis and the axis of rotation of the wheel is the *minor axis.*

The use of a gyroscope in the mass flowmeter depends upon the fact that a spinning gyroscope having equal forces applied to its axle ends will not precess (rotate about its major axis), but once a tilting force is applied at axle ends, rotation (precession) about the major axis will take place, and the speed of this rotation will be proportional to the applied tilting force. As shown in the figure, a counter is driven by the rotating gyroscope assembly.

Because the synchronous motor and the gyroscope wheel are both driven by the same source of alternating-current power, this system is self-compensating as far as variations in line frequency are concerned, and it is insensitive to reasonable voltage changes. This and similar mass flowmeters have noteworthy advantages over other systems of flow measurement. Mass measurement alone is a vast improvement over volumetric, area, and other forms, because variables like density, temperature, and pressure are of no consequence to accurate measurement. The accuracy of mass flowmeters of this design is commendable.

Turbine Meters

A turbine meter is another device that uses the factors of area and velocity to determine flow rate. If fluid passes through a conduit having a cross-sectional area of A square feet at a velocity of v feet per second, then the rate of flow is simply

$$A \text{ ft}^2 \times v \text{ ft/sec.} = Av \text{ ft}^3\text{/sec.}$$

In a cutaway drawing of a turbine meter (fig. 8.26), its simplicity of construction is apparent. The effective cross-sectional area is a constant value, and the turbine wheel is the principal component for determining the velocity of the fluid passing through the meter. If the effects of friction are ignored, the turbine wheel will revolve at a speed directly proportional to the velocity of the flowing fluid. For a given size of meter there will be a range of flow values for which it will be remarkably accurate. For example, a 2-inch meter will function well with flow of 40 to 400 gallons per minute, and the factor of accuracy at any point in this range will probably be better than 99%.

As shown in the figure, the turbine meter is provided with an electrical pickup to determine the speed of the turbine wheel. It would be possible, of course, to arrange for a mechanical drive system to count the number of revolutions, and this is done in some turbine meters. Mechanical systems are apt to introduce an appreciable amount of friction, however, and affect the accuracy of the meter. In the electrical pickup system a permanent magnet and coil of wire are located in the pickup element. As each blade end passes the pickup, it alters the magnetic circuit of the element, thereby causing a pulse of voltage to be induced in the pickup coil. Evidently it is

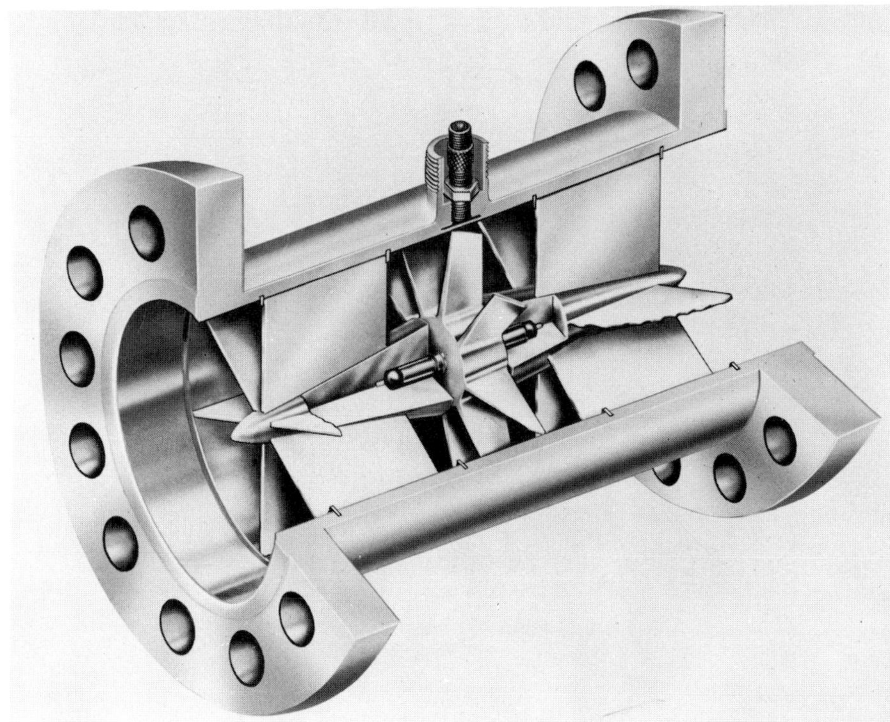

Figure 8.26. Wheel in a turbine meter made to revolve by the flow of fluid (*Courtesy Halliburton*)

necessary only to count the number of pulses per second in order to determine the velocity of flow through the meter.

Some auxiliary equipment can and must be used with the meter in order to determine rate of flow and the accumulated total flow volume (fig. 8.27). Such units contain solid state circuitry to count the pulses and to provide actuating force to drive the cumulative counter. The meter dial indicates the instantaneous flow rate at any moment, and the counter indicates the total volume of fluid that has passed through the metering element.

The turbine meter can be adapted to measuring flow of almost any fluid. It is particularly successful in measuring sand-laden liquids and other abrasive fluids that tend to ruin elements of many other types of instruments. Like most meters that are not positive-displacement devices, they work best when a particular range of flow value is maintained through them. A 1-inch meter element has a useful range of approximately 5 to 50 gallons per minute, an 8-inch element, 350 to 5,000 gallons per minute. The ratio is at least 10 to 1 between maximum and minimum rates, and the meters retain extremely high accuracy over these ranges. Should the flow rates fall appreciably below the minimum useful values, however, the accuracy is likely to suffer greatly.

Displacement Meters

The types of flowmeters discussed so far are ideally suited for particular applications, but they possess disadvantages that rule out their use in some instances. The orifice meter, for example, must be carefully chosen for the range of differential pressure values expected in the measurement. Thus, it would hardly be suitable for measuring flow rates ranging from 1 cubic foot per hour to 500 cubic feet per hour. Magnetic and mass flowmeters have disadvantages that become serious when large numbers of meters are needed, as is the case in water or gas distribution systems where a separate meter is needed for each customer. These meters are expensive and require a source of electric power, and while these factors are not serious for high-volume installations, they become disqualifying considerations for distribution systems.

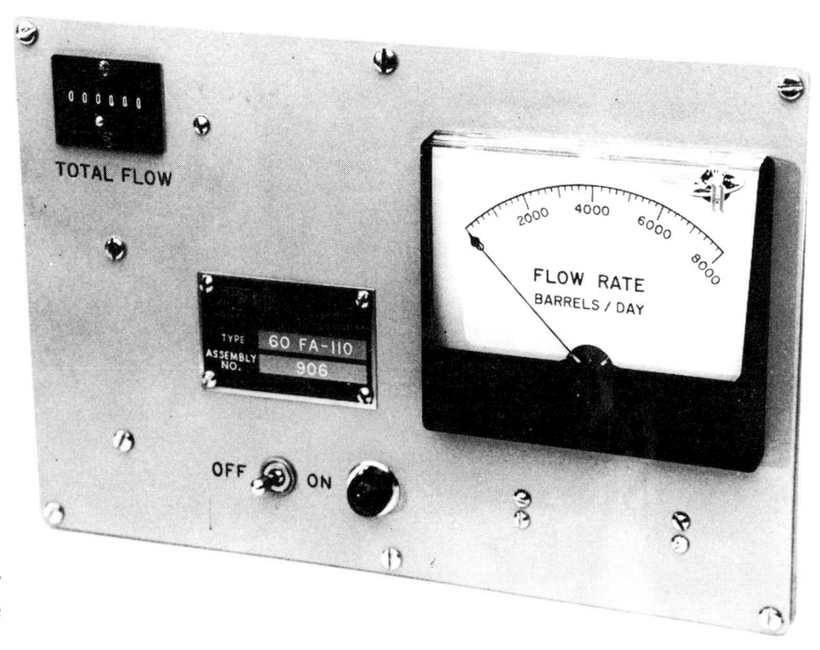

Figure 8.27. Flow rate indicator and recorder for the turbine meter (*Courtesy Halliburton*)

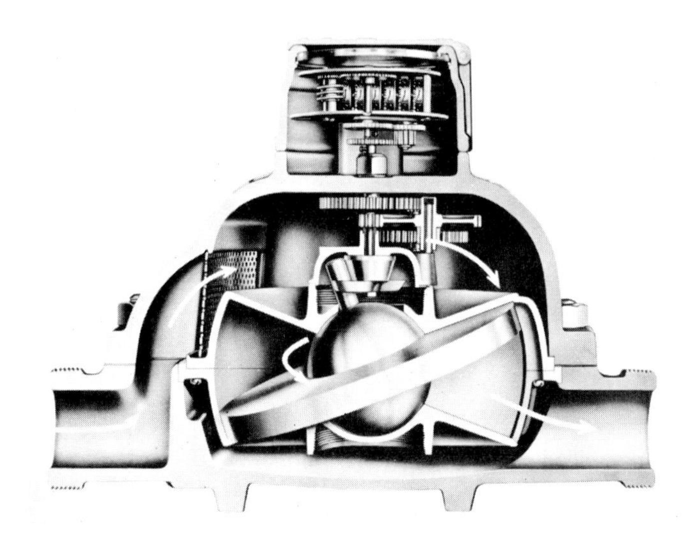

Figure 8.28. Nutating piston positive-displacement meter (*Courtesy American Meter Controls*)

Displacement meters are characterized by their ability to use energy from the flowing mass of fluid to operate indicating or integrating mechanisms and are used extensively in water and gas distribution systems where maximum flow rates are usually small, but actual flow is subject to frequent changes from zero to maximum. In addition to the small-capacity meters used for utility services, high-capacity meters are also available and are being used increasingly to measure the transfer of petroleum liquids. A high-capacity meter is one capable of passing liquid at rates of at least 2,000 gallons per minute.

Nutating Piston Meters

One positive-displacement meter is employed extensively for measuring liquid flow (fig. 8.28). It is called a *nutating piston* meter, and its widespread use in municipal water distribution systems accounts for its being the most popular flowmeter in use.

The piston resembles a disk more than it resembles our usual concept of a piston; the meter is sometimes called a disk-type quantity meter. The piston is limited in its motion by three features: (1) a ball-like center section rides in a socket and prevents vertical motion of the piston; (2) the disk contains a single deep and narrow slot into which a rigid separator plate fits, an arrangement that prevents free motion of the disk and limits it to rocking motion; and (3) a pin protruding upward from the ball rides against a fixed conical surface, and this action forces the disk to assume a cocked position at all times.

At the position of the disk, shown in the bottom left-hand drawing of figure 8.29, liquid from a previous cycle is trapped under the right side of the disk and can go only out the discharge port, which is located on the right side of the rigid separator. Liquid from upstream entering the inlet port (left side of the rigid separator) flows under the disk and tends to lift it. The disk responds to this lifting

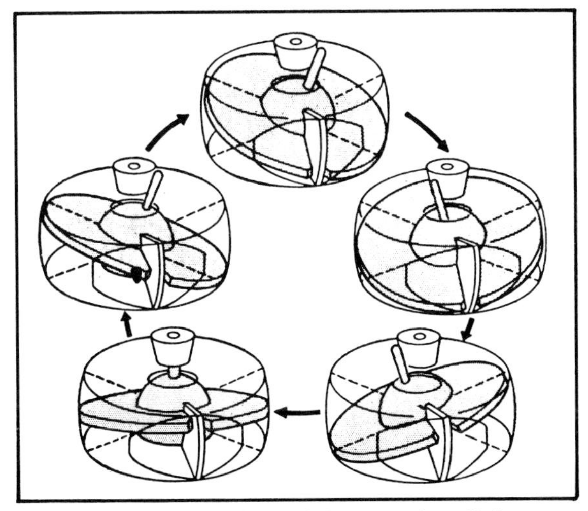

Figure 8.29. Operation of the nutating disk

force by *squeezing* liquid out through the discharge.

In the upper left-hand drawing of the figure, liquid is also flowing in above the disk, and in the upper center drawing, most of the fluid trapped from the previous cycle has now been forced out into the downstream line. Another quantity of liquid is about to be trapped below the disk (upper right-hand drawing). Fluid is now free to travel into the cavity *above* the disk and begin a new cycle.

If the clearances in the ball and socket and between the disk and its cavity and separator plate are close enough that flowing liquid forms a good seal against leaks, this form of meter will be reasonably accurate—approaching a factor of 1% or 2%. In order to measure quantity, it is a fairly simple matter to attach a counter to the nutating pin by means of a slotted crank lever.

Figure 8.30. Gas meter ***(Courtesy American Meter)***

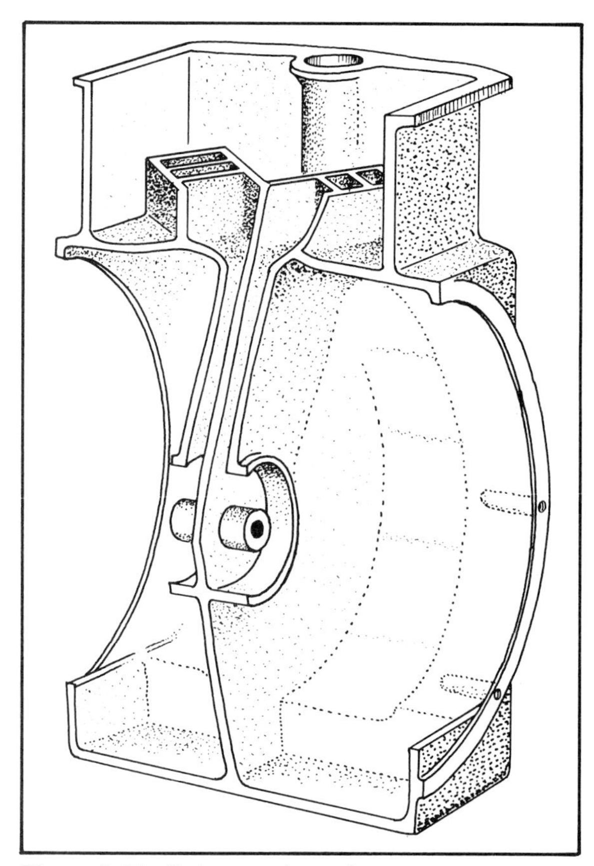

Figure 8.31. Cutaway view of meter case

Gas Meters

Just as the nutating piston meter is used in water distribution systems, so the gas meter is used to measure gas consumed by individual, commercial, and residential customers. It is necessary to have a meter that is capable of accurately measuring volumetric quantities over a wide range of flow values, and it is desirable for it to have no dependence on external sources of power. A particular problem arises in low-pressure gas meters because only feeble pressure is available to operate the counter functions.

Figure 8.30 shows a complete gas meter; figure 8.31 shows a cutaway section of its main case. Figure 8.32 is a set of drawings that help explain operation of the meter. This form of meter is called a two-diaphragm, four-compartment type because there are, in

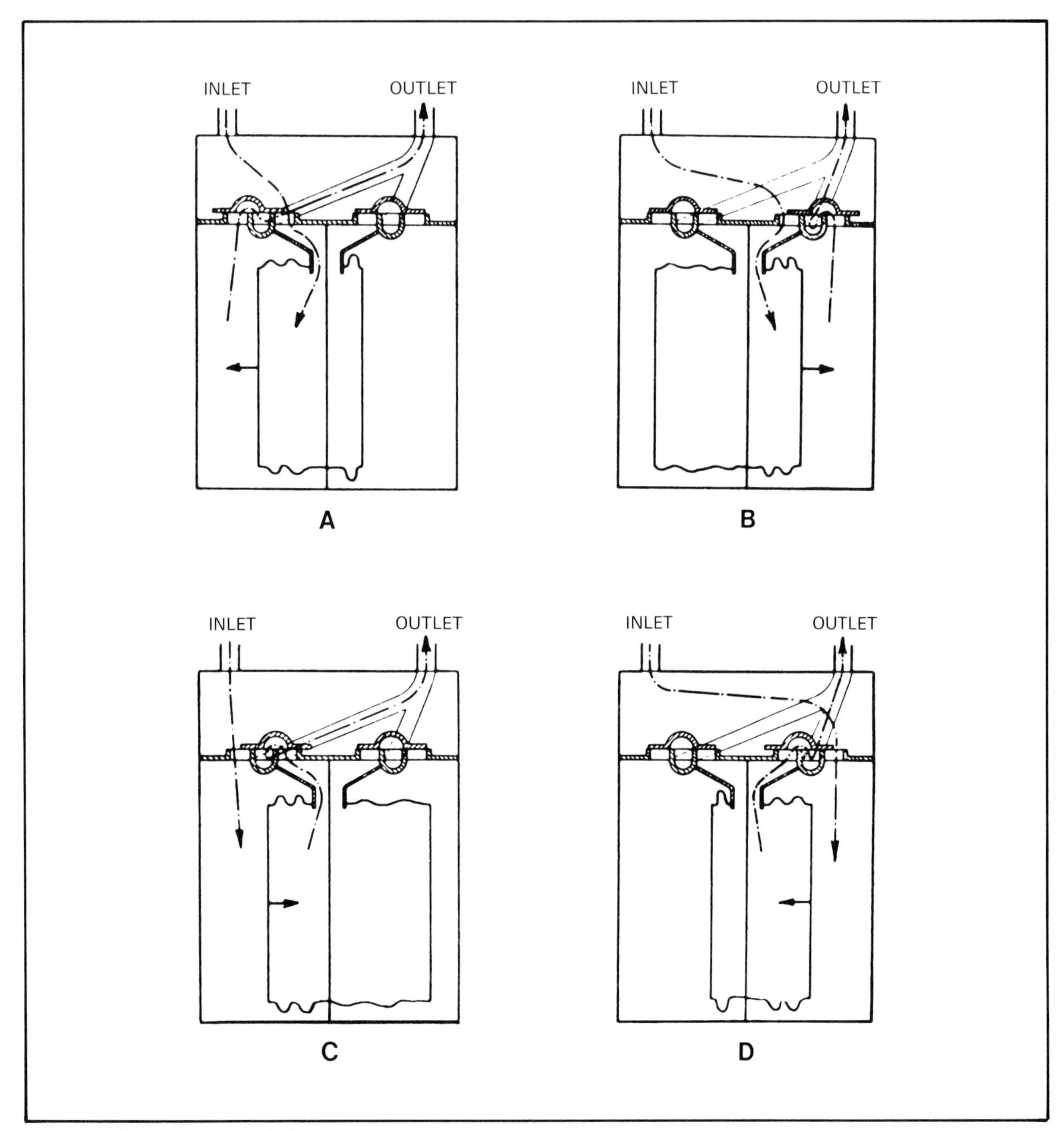

Figure 8.32. Operation of the gas meter

fact, two diaphragms, each acting in a separate compartment. The spaces enclosed by the diaphragms form the other two compartments. Each compartment can be connected to the inlet piping or the outlet piping, or can be closed off by sliding valves. Two sliding valves work together in a definite relation through a system of levers that are ultimately attached to free ends of the diaphragms.

In drawing *A* (fig. 8.32), gas is flowing into the interior compartment of the left diaphragm assembly, and expansion of this assembly is forcing gas out of the exterior compartment into the outlet piping. When the free end of the left diaphragm approaches its maximum travel on the expansion phase, a lever attached to it will operate the two interconnected sliding valves, driving them to the position shown in drawing *B*. This action shuts off the entire left side of the meter and simultaneously opens the appropriate right-hand compartments to the inlet and outlet piping. Gas entering the right-hand interior compartment expands the assembly and forces gas from the exterior compartment into the outlet piping.

Full expansion of the right-hand diaphragm assembly operates the slide valves, the action driving both valves toward the center. The entire right side of the meter is closed off, while the left side receives gas into its exterior compartment, causing a compression of the diaphragm assembly and forcing its gas into the outlet piping. The position of the valves is shown in drawing *C*.

Full compression of the left diaphragm again actuates the valves, driving them to the positions shown in drawing *D*. Gas then flows into the right-hand exterior compartment, compressing the diaphragm and operating the valves to the positions shown in drawing *A*. The cycle of operation is then complete.

The use of two diaphragms and four compartments provides a satisfactory delivery of gas. The output will not contain serious pulsations of pressure, because valve operation is so timed that the rather rapid closing off of a compartment near the end of its delivery phase is accompanied by equally rapid opening of the compartment next in line for delivery.

A phantom view of a gas meter (fig. 8.33) shows the location of the Bakelite sliding valves (appropriately called D-valves) and the arrangement of levers used to operate them. It is difficult to explain or to show in drawings the clever system of valve operation. Only motion pictures or observing the actual operation of a partially disassembled meter can furnish adequate detail.

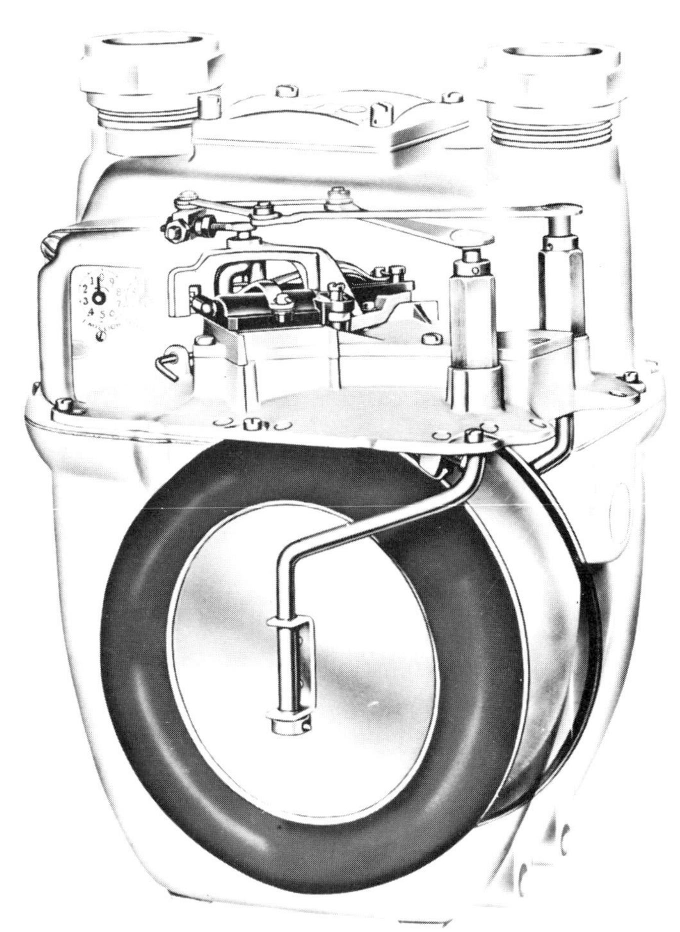

Figure 8.33. Phantom view of a gas meter (*Courtesy Rockwell Manufacturing*)

The coordinated in-and-out motion of the diaphragms causes a *tangent crank* to rotate a shaft that operates the valves and drives the counter. A counter is commonly called an *index* when used on flowmeters.

CHAPTER

9

Flow Control

Controlling the flow of fluid is by far the most important method for controlling other variables such as pressure, temperature, and liquid level. When used in this manner, flow is referred to as the *manipulated variable.* When it is used to produce a change in the rate of flow from a set point in order to bring about corrective action in a control system, it is called the *controlled variable.* As such it has a limited number of applications.

Flow Control Devices

Sometimes very simple devices are used to effect flow control. The adjustment of a water tap to control the rate of flow through a garden hose is an example of manual control. In this case the quantity of water applied to the lawn or garden is important. The rate at which it is applied is important, too, because it should not be applied so rapidly that much of it is wasted by runoff.

Flow Beans

Fixed flow bean. Fixed flow control can be provided by a *flow bean,* or choke, used in the piping from a well producing natural gas. A fixed flow bean (fig. 9.1) is merely a constriction placed in a special nipple that forms part of the piping. It is a metal plug having a hole drilled through it and fitted with a section of external threads and an allen wrench socket. The threads and the wrench socket provide for rapid and easy changing of the plug.

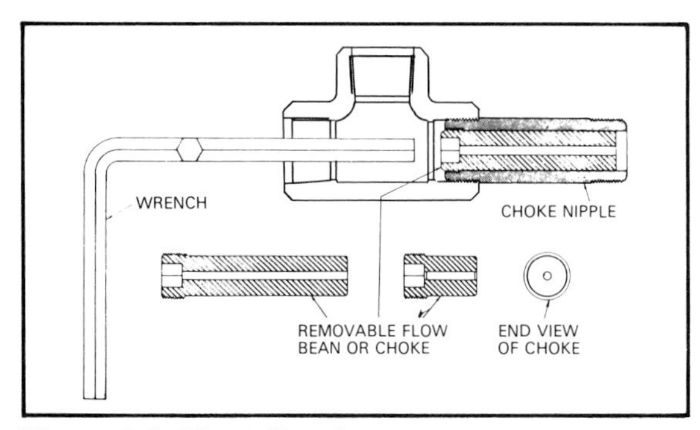

Figure 9.1. Fixed flow bean

Adjustable flow bean. Another form of flow bean—an adjustable type (fig. 9.2)—is essentially a needle valve in a right-angle body. Flow beans are fitted into piping between producing wells and low-pressure vessels that represent the initial stages of processing equipment. The size of hole in the fixed bean or the adjustment of the manual type are determined by well pressure and handling capacity of the processing equipment. Well pressures can be in excess of

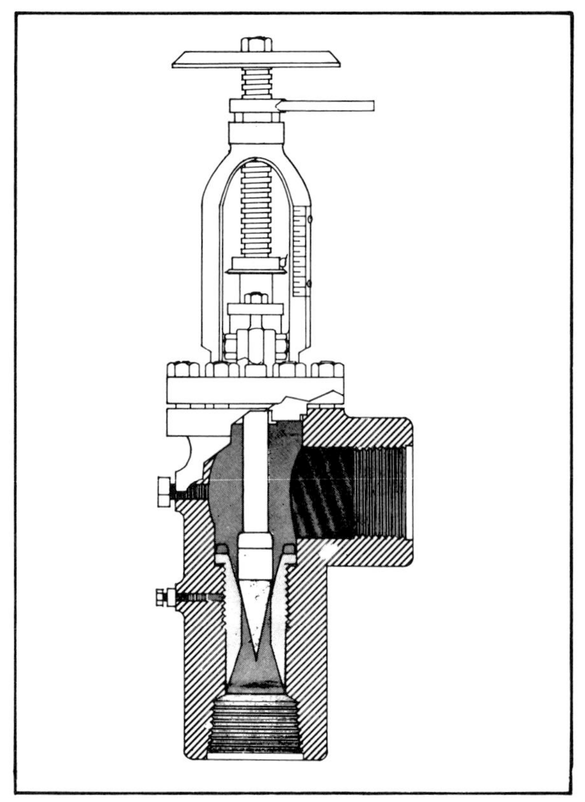

Figure 9.2. Adjustable flow bean

10,000 psi, and pressures of 4,000 to 6,000 psi are common. Reducing the flow rate at the wellhead allows the use of a less expensive type of piping leading to the process vessels.

Differential Pressure Devices

Processes and systems that have a need for accurate control of the flow rate of fluids almost invariably use a head-type primary element in the measuring means. Other instrument types, such as the magnetic flowmeter, are quite capable of providing the necessary signals to be used in flow control, but orifice plates operating with suitable differential pressure devices represent the most important combination for flow control. Many differential pressure measuring instruments are readily adaptable to control functions, and it is not unusual to find an instrument that combines the functions of recording and controlling flow rate.

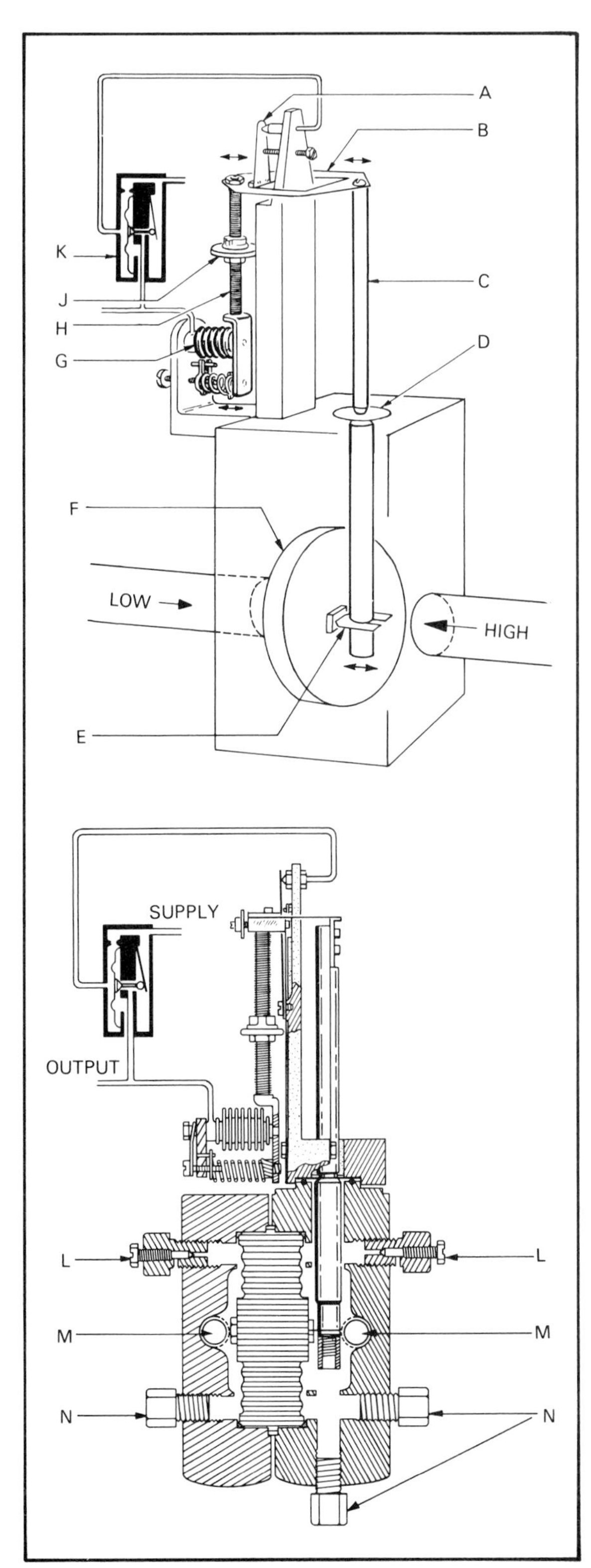

Figure 9.3. Differential pressure cell with flapper-nozzle assembly, air relay, and feedback bellows *(Courtesy Foxboro)*

Differential pressure cell. Differential pressure cells enjoy popular use in several measurement and control applications, particularly liquid-level and flow control. Its construction (fig. 9.3) incorporates several familiar features—flapper-nozzle assembly, air relay, and feedback bellows. The unit shown is of small size but very strong construction, capable of use in systems employing static pressures of 6,000 psi. The differential pressure range is adjustable, the lowest range being 0 to 20 inches of water, the highest being 0 to 850 inches of water. A differential pressure cell is shown installed on a gas line in figure 9.4.

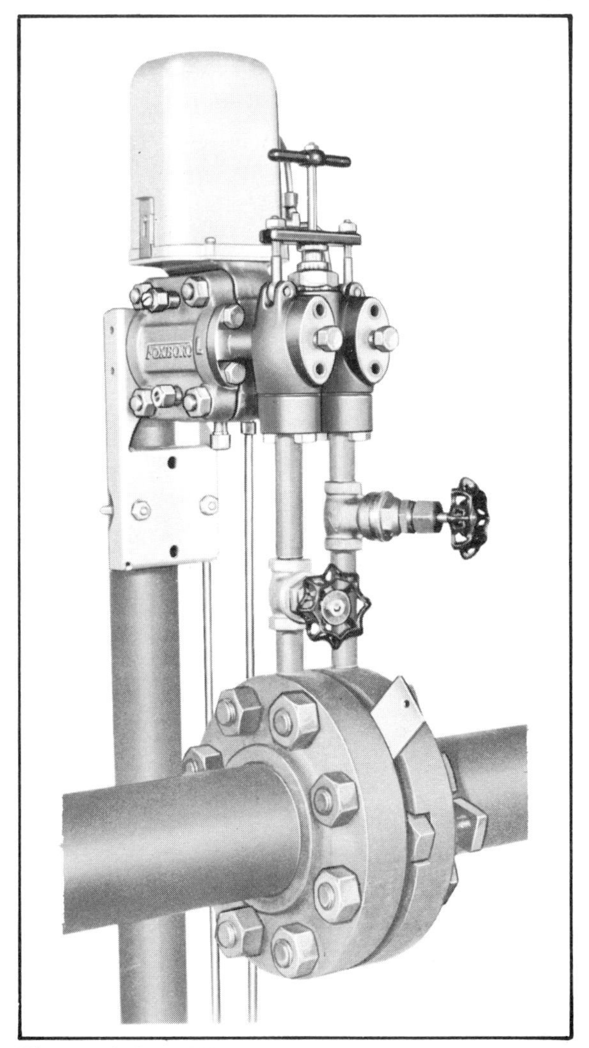

Figure 9.4. Differential pressure cell installed *(Courtesy Foxboro)*

Upstream and downstream pressures are applied to fittings that lead into cavities on either side of the silicone-filled diaphragm capsule, *F,* which is rigidly connected to the force bar, *C,* by a C-type flexure fitting, *E* (fig. 9.3). A special-alloy diaphragm, *D,* acts as a seal and as a fulcrum for the force bar. When a differential pressure exists across the diaphragm capsule, the force bar transmits a force that is exactly proportional to this pressure through the flexure plate, *B,* to the range bar, *H.* The range bar pivots about a range wheel, *J.* The range wheel can be set to any point along the threaded portion of the range bar; this setting determines the measurement range.

Motion of the range bar pivoting on the range wheel causes a displacement of the flapper-nozzle relation at *A,* and the nozzle back-pressure disturbance caused by this action is amplified by the air relay, *K.* Output of the air relay (3–15 psi) is the analog representation of the differential pressure in inches of water column. The force relations are as follows:

1. The force exerted by the feedback bellows is exactly proportional to the force applied to the range bar, *H,* by the force bar, *C.*
2. The force exerted by the force bar is exactly proportional to the differential pressure. The output pressure may be used to drive a valve actuator, a recorder, an indicator, etc.
3. Because of the foregoing relations, the air pressure in the bellows and in the output line is exactly proportional to the differential pressure.

If the differential pressure across the capsule element increases—that is, the force bar, *C,* is driven to the left at point *E*—the action will cause the flapper-nozzle clearance to decrease, and the nozzle back-pressure will rise. The air relay will respond by transmitting more air to the output and the feedback bellows. Expansion of the bellows will increase the flapper-nozzle clearance and thus reduce the nozzle back-pressure. Very quickly, a balanced condition will be established; that is, the torque caused by the force of the bellows acting on the lower end of the range bar will exactly equal that caused by the force acting on its upper end.

Plugs, *N,* are drain plugs. Fine-wire screens, *M,* prevent large solid particles from entering the cell. Vent screws, *L,* are used for relieving pressure in the cell cavities before draining or disassembly.

The differential pressure cell is considered a form of *force-balance* instrument. The operation of force-balance instruments depends on balancing one force against another to achieve a condition of equilibrium. In the unit described, the forces exerted on opposite ends of the range bar are related, but actually, a balance of torques rather than forces is used. It may be recalled that the equilibrium established in a pneumatic controller is attained by balancing the linear displacement of certain components (See chapter 4).

True force-balance sensing device. Figure 9.5 is a drawing of a differential pressure sensing instrument that operates on a true force-balance basis. Upstream, *Hi,* and downstream, *Lo,* pressures are admitted to two cavities separated from one another by a sensing diaphragm. The diaphragm is clamped between a pair of baffle plates equipped with drive rods that extend out of the cavities through sealing bellows. The rod on the right side contains a flanged fitting on which rests the zero-adjusting spring. The flange of the fitting serves as an overrange and underrange stop.

The rod from the left cavity is attached to a baffle and feedback diaphragm, and the outer surface of the baffle forms part of a nozzle-flapper assembly. Changes in differential pressure applied to the instrument cause motion of the two diaphragms and variation of

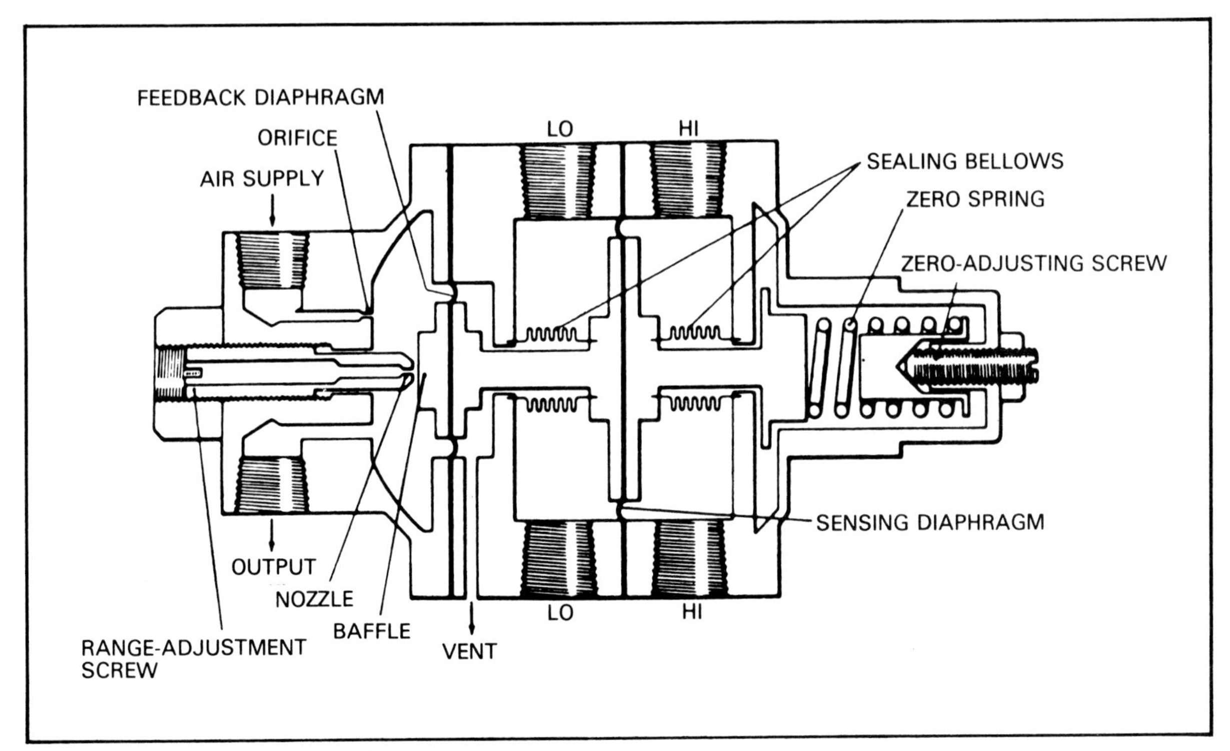

Figure 9.5. Force-balance differential pressure sensing instrument ***(Courtesy Taylor Instrument)***

the nozzle-flapper clearance. Air pressure from the supply line is applied through a fine orifice to the left side of the feedback diaphragm. This air leaks off through the nozzle at a rate determined by the flapper-nozzle clearance and the position of the range-adjustment screw.

During a condition of equilibrium, forces acting on the sensing diaphragm and on the feedback diaphragm have a net value of zero. Should the differential pressure rise or fall, the change in force acting on the sensing diaphragm will be opposed by a change in force acting on the feedback diaphragm.

The zero-adjusting screw is used to establish the set point of the system, while the range-adjustment screw varies the sensitivity of the instrument in much the same way as the range wheel described previously. The air output from the instrument can be used to actuate a remote recorder or an air relay for control purposes.

General Considerations of Flow Control

Controlling flow can be done by several methods, and the method chosen depends largely on what is to be accomplished. Certain general considerations must be taken into account in order to establish relations between the flow rate and the variable conditions that may occur upstream or downstream of the control or measuring point.

It should be recalled that the flow equation can be expressed as

$$Q = k\sqrt{hp}$$

where

k = a single constant representing all the corrective factors discussed in chapter 8;
h = differential pressure across the primary element;
p = static pressure in the line.

Two variables—h and p—will thus affect the problem of control. For the discussions to follow, p is assumed to be taken at the downstream tap.

Control of Gas or Steam Flow

A particular installation for gas or steam flow may require that a constant flow rate be maintained into a downstream line having a constant pressure; that is, as long as the flow into the line is constant, the downstream pressure will remain stable. This installation will function best if the throttling valve is placed upstream of the orifice plate, because the valve will then keep the pressure just ahead of the orifice plate at a constant value, regardless of pressure fluctuations upstream of the valve (fig. 9.6). The same arrangement can be used if there is a need to deliver gas into a fluctuating downstream pressure and it is desirable that the delivery rate is definitely related to the downstream pressure.

Another installation may have a constant upstream pressure and a fluctuating downstream pressure. In order to maintain a constant flow rate, the throttling valve should be downstream of the orifice plate. In this way a constant pressure is maintained on either side of the orifice plate, so p remains constant (fig. 9.7).

A third installation may have variable pressures both upstream and downstream. In this case, the throttling valve should be installed *upstream* and a pressure-compensated flow controller used to control

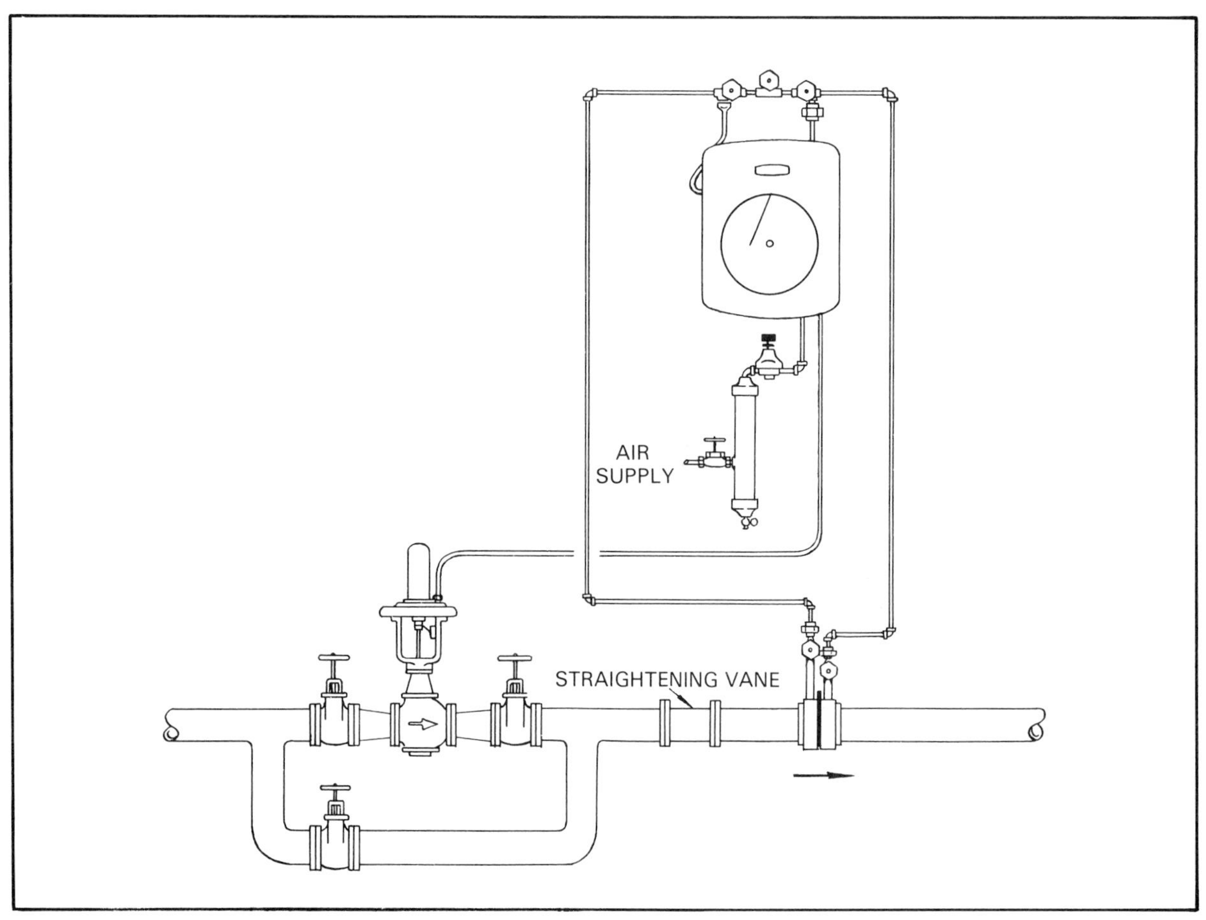

Figure 9.6. Installation for maintaining constant downstream pressure *(Courtesy Foxboro)*

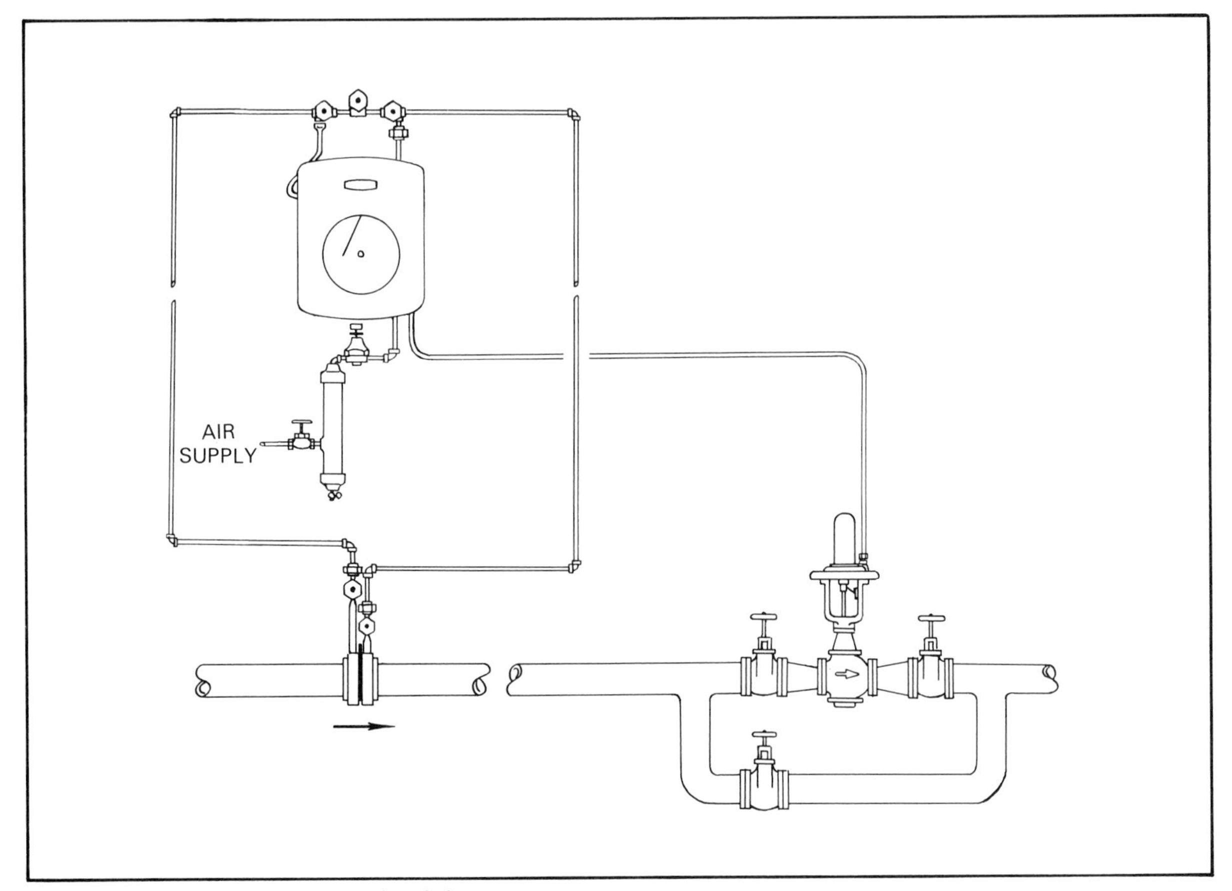

Figure 9.7. Installation for maintaining constant upstream pressure ***(Courtesy Foxboro)***

it. Pressure compensation in a controller is accomplished by a pressure-actuated device that automatically corrects for variations in downstream pressure, *p*.

Control of Liquid Flow

In the control of gas flow, the control valve and the primary element are both located in the line in which flow is to be controlled. Some installations involving the control of liquids cannot use this system. Consider the case of a liquid flow line in which the liquid is driven by a reciprocating or other positive-displacement pump. It would be risky to install a throttling valve in the discharge line of a positive-displacement pump because the line pressure that would build up against a closed valve would be limited only by the power source driving the pump and the rupture strength of the line or valve.

Controlling the flow rate of liquid from the discharge of positive-displacement pumps can be achieved in several ways.

1. For steam-driven reciprocating pumps, the orifice plate or other primary element is installed in the discharge line, and the throttling valve is placed in the supply line of the steam end of the pump. In this way there can be no danger of valve or controller failure blocking the discharge line and causing damage to the system. Regulating the energy input to the steam end of the pump effectively controls its speed and discharge rate. The pressure-sensing

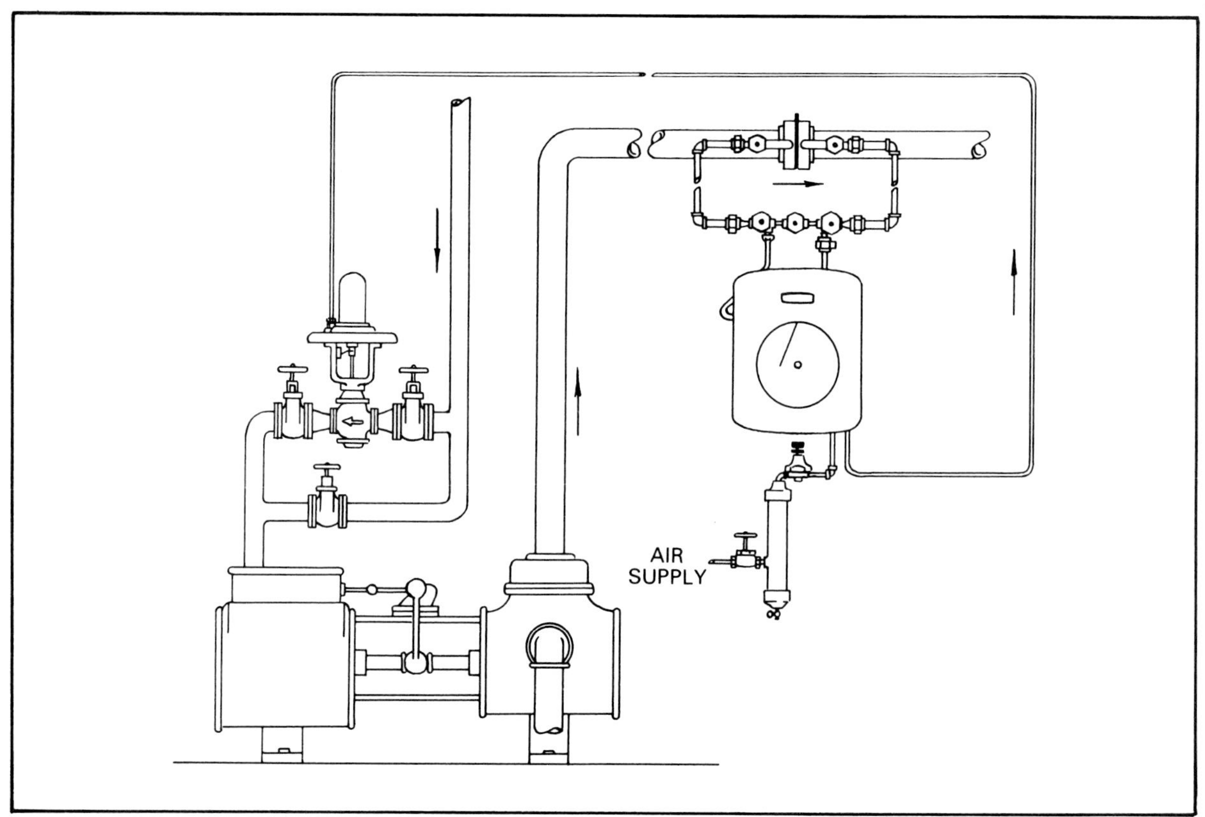

Figure 9.8. Downstream pressure sensing device for a steam-driven pump ***(Courtesy Foxboro)***

device downstream provides feedback to a throttling valve in the steam line of the pump engine (fig. 9.8).

2. For steam, gas, or diesel engine driven pumps, the orifice plate is installed in the discharge line, and the controller drives a diaphragm motor that either resets an engine governor or controls the steam or fuel feed to the engine.
3. For any positive-displacement pump, regardless of drive, and particularly for constant-speed electric motor drive, the orifice plate is located in the discharge line, and a throttling valve located in a bypass line from pump discharge to pump intake. In this case, a constant flow through the pump occurs for constant speed, and the discharge into the line is determined by the degree of closure of the throttling valve (fig. 9.9).

If a centrifugal pump is used to drive liquid through a line, a control valve can be installed in the discharge line (fig. 9.10). Centrifugal pumps are characterized by the fact that, for a given pump and speed, a certain maximum discharge pressure will not be exceeded, even against a tightly closed-off discharge line. As long as the line, control valve, and other components of the system involved are capable of withstanding the maximum discharge pressure of the pump, the direct throttling of the discharge is quite suitable.

Discharge from centrifugal pumps that are steam-turbine driven is sometimes controlled by throttling the steam supply to the turbine, or a diaphragm-motor device is used to reset a turbine-speed governor. Such arrangements are usually found where the saving of input energy (steam) achieved in this way is significant.

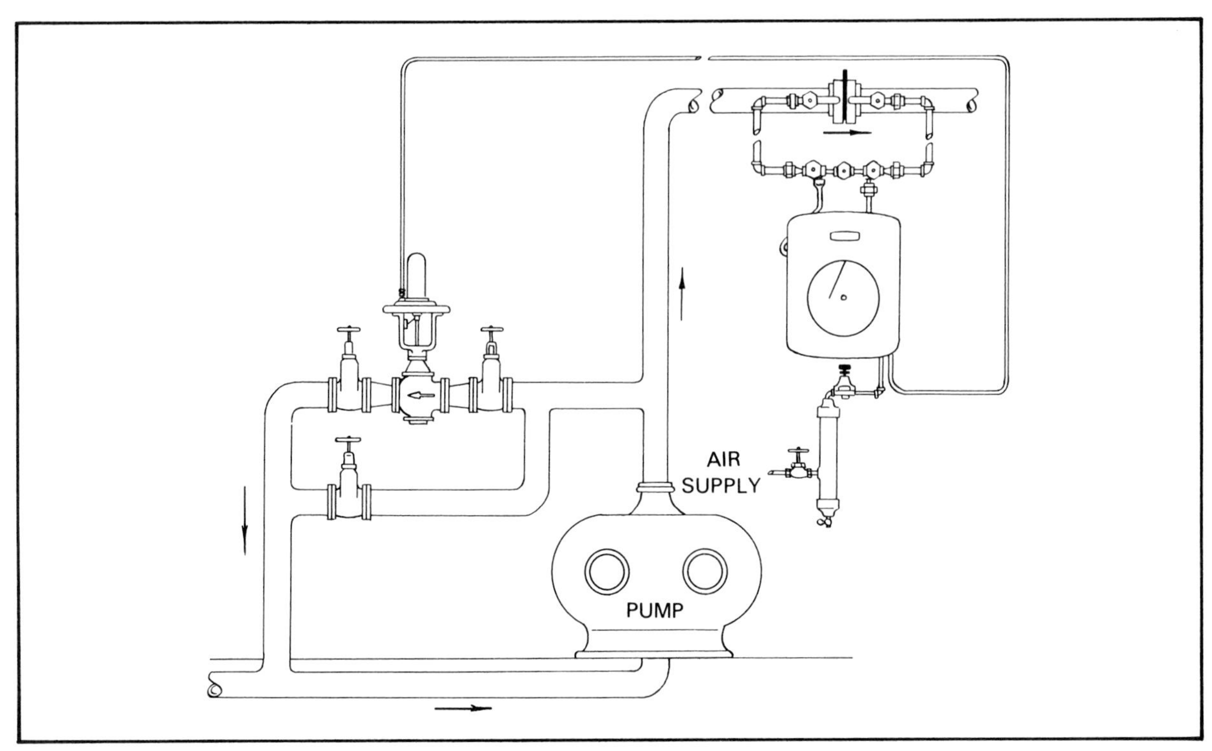

Figure 9.9. Throttling valve in bypass line for a positive-displacement pump ***(Courtesy Foxboro)***

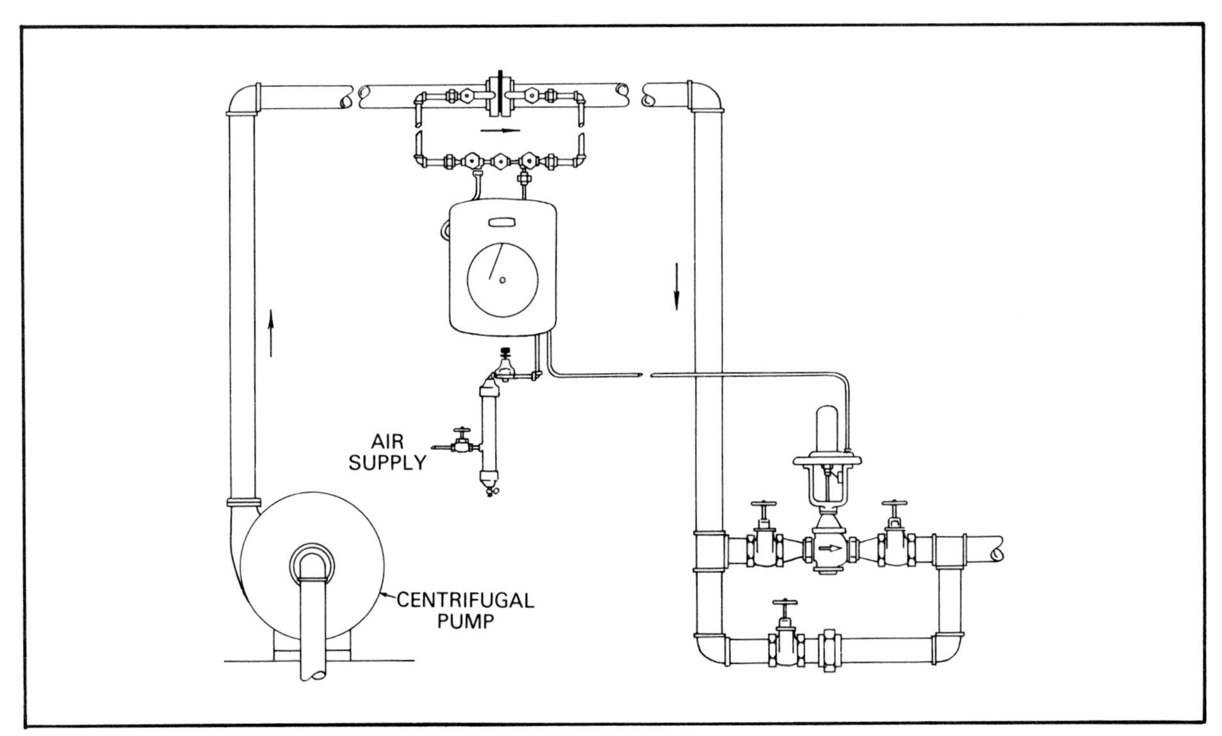

Figure 9.10. Control valve installed downstream from a centrifugal pump ***(Courtesy Foxboro)***

Flow Control in Fractionating Columns

A fractionating column is one of several basic refinery devices used to process petroleum. The simple forms of columns are capable of receiving a steady input of petroleum consisting of two or more miscible compounds and of separating this input into two products, or fractions. During proper operation of a column, the fractions are produced at a fairly steady rate. Separation of the petroleum into two fractions is based on the differences in boiling temperatures that exist among the hydrocarbon compounds that make up the petroleum.

Control of a fractionating column sometimes involves an intricate and integrated system of instruments for regulating temperature, pressure, and flow. Flow control is applied to the feed supply line to the column and to one or more discharge lines carrying the fractions from the fractionating system. Frequently it is possible to use a flow control system that consists of individual control units not integrated with other units; that is, they control entirely on the basis of the flow rate determined by their measuring means.

Feed Rate Control

The feed rate controller (fig. 9.11*A*) is made possible by a feed supply source of such magnitude that feed rate demand is not likely to tax it to any extent; thus the fractionator could run indefinitely without exhausting the feed supply. This form of control is likely to be found on the initial column of a fractionating train, or series of columns. The actual control of flow consists of an orifice plate as primary element, a differential pressure sensing device that positions a flapper in a proportional or proportional plus reset controller, and a diaphragm-motor valve. A fractionating column using the feed rate control described requires a certain period of time to settle down, even when very uniform sources of feed and heat supply are available. Once a condition of equilibrium is achieved, however, the column has little trouble maintaining the balance of temperature, pressure, and rate of output.

When the feed supply to a column is relatively limited or sporadic in nature, as may be the case for an intermediate column in a fractionating train, the form of control described above is usually not satisfactory. It is not satisfactory because the feed supply may be temporarily exhausted or because the

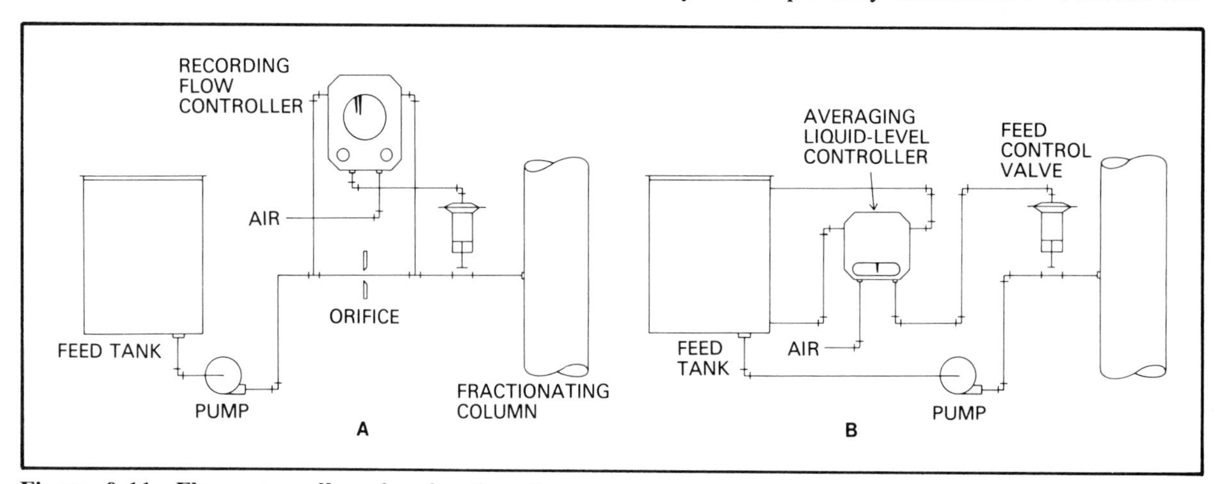

Figure 9.11. Flow controllers for fractionating columns. *A*, feed rate controller for initial column; *B*, liquid-level controller for succeeding columns.

available supply may not be used at an optimum rate. As an example, the second column in a series may be supplied by an output from a previous column. It is economically desirable to feed this second column at about the same rate as the rate at which the first column makes the feed available.

The system shown in figure 9.11*B* controls flow according to the liquid level in a feed tank that is supplied from the initial fractionating column. It would be a mistake to assume that the liquid level is being controlled by regulating discharge from the feed tank. Flow control is still of primary interest, and its rate should be uniform and consistent with the feed supply available.

The liquid level in the feed tank determines the set point of the flow controller. Pneumatic controllers (discussed in chapter 4) are equipped with set-point dials or screws that enable the desired value of the controlled variable for a particular application to be chosen. Most of the set-point arrangements involve a means for varying the nozzle-flapper clearance in the controller. An attachment consisting of a bellows and lever system can be used to actuate the set-point mechanism. A controller equipped with this facility is called either a pneumatic-set or a control-set controller, and a system using such controllers is called a *cascade* system or an *interlocked* system.

The form of liquid-level instrument used in a cascade system is usually a displacer-type primary element actuating a proportional-type liquid-level controller, or indicator-transmitter. The 3 to 15 psi pneumatic output of the instrument is transmitted to the pneumatic-set mechanism of the flow controller. A cutaway photograph of a pneumatic-set controller is shown in figure 9.12; figure 9.13 is a schematic drawing of the pneumatic-set mechanism. Proper adjustment of the proportional band and the ratio travel

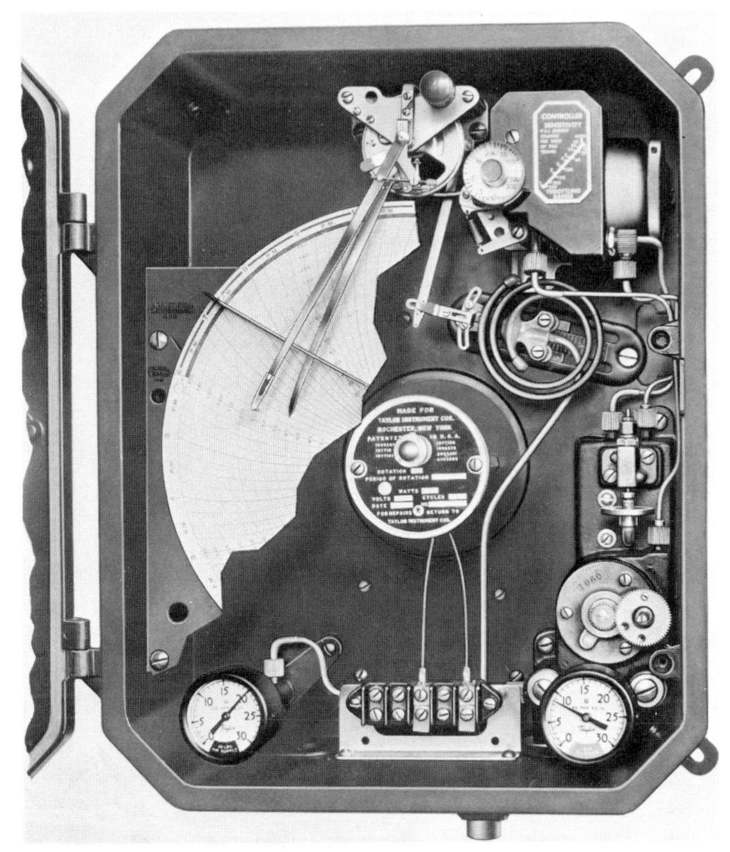

Figure 9.12. A type of pneumatic-set controller (*Courtesy Taylor Instrument*)

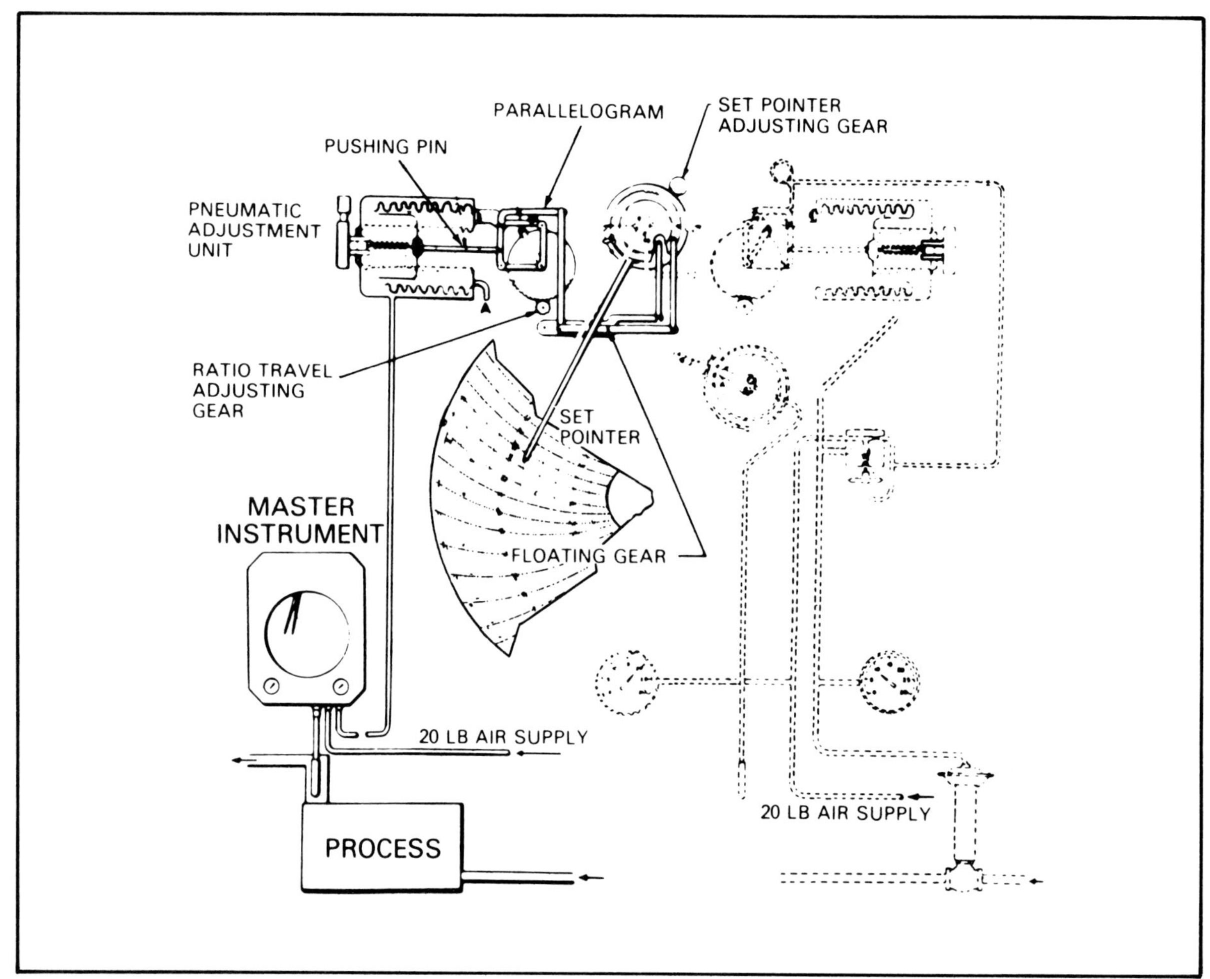

Figure 9.13. Schematic of a pneumatic-set mechanism actuated by the controller ***(Courtesy Taylor Instrument)***

adjustment results in a smooth control of the feed rate to the column. Changes in the input to the feed tank affects its liquid level, but action of the cascaded level and flow instruments smoothly adjusts to the new conditions.

It is logical to suppose that a flow controller could be dispensed with in this system and a liquid-level controller used to control the flow rate to the column directly. There are several valid reasons for using a separate flow controller; an important one will become apparent when the rate of withdrawal of the upper fraction from the column is discussed.

Control of Fraction Withdrawal Rate

In a simple fractionating column, the feed is split into two fractions, one of which is withdrawn from the bottom and the other from the top of the column. Withdrawal rate from the top of the column can be controlled on the basis of temperature and pressure or by using flow rate as the controlled variable. Withdrawal rate from the bottom of a column is usually controlled as a function of liquid level in the column bottom, but a cascade system between feed rate and bottom product withdrawal rate is feasible.

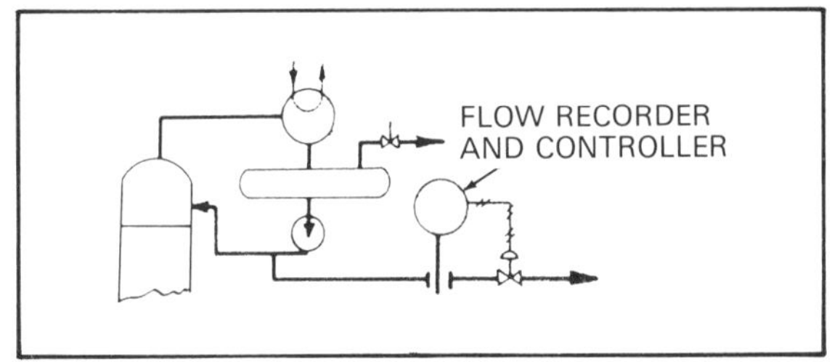

Figure 9.14. A system with simple flow rate control

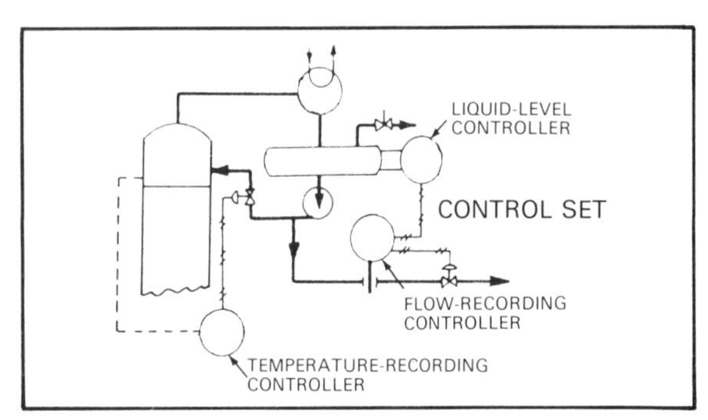

Figure 9.16. Liquid-level control system with accumulator tank

Control of Top Product Discharge Rate

If the column is operating with a constant feed rate, then the system shown in figure 9.14 can be used to control the discharge of the product from the top of the column. Here, a proportional controller or a proportional plus reset controller is used to maintain a steady output of the product. The primary element, as before, is an orifice plate. Some of the product is allowed to flow back into the top of the tower. This is called *reflux,* and a certain amount of the condensed product is returned to the column for the purpose of enriching the output.

For controlling the output of columns that do not have a constant rate of feed, it is usually necessary to provide a cascade control system for the output. In one practical method of control (fig. 9.15), the pneumatic-set signal from the liquid-level transmitter of the feed tank is applied to the flow controller. In another form of flow control of output (fig. 9.16), the set point of the flow controller

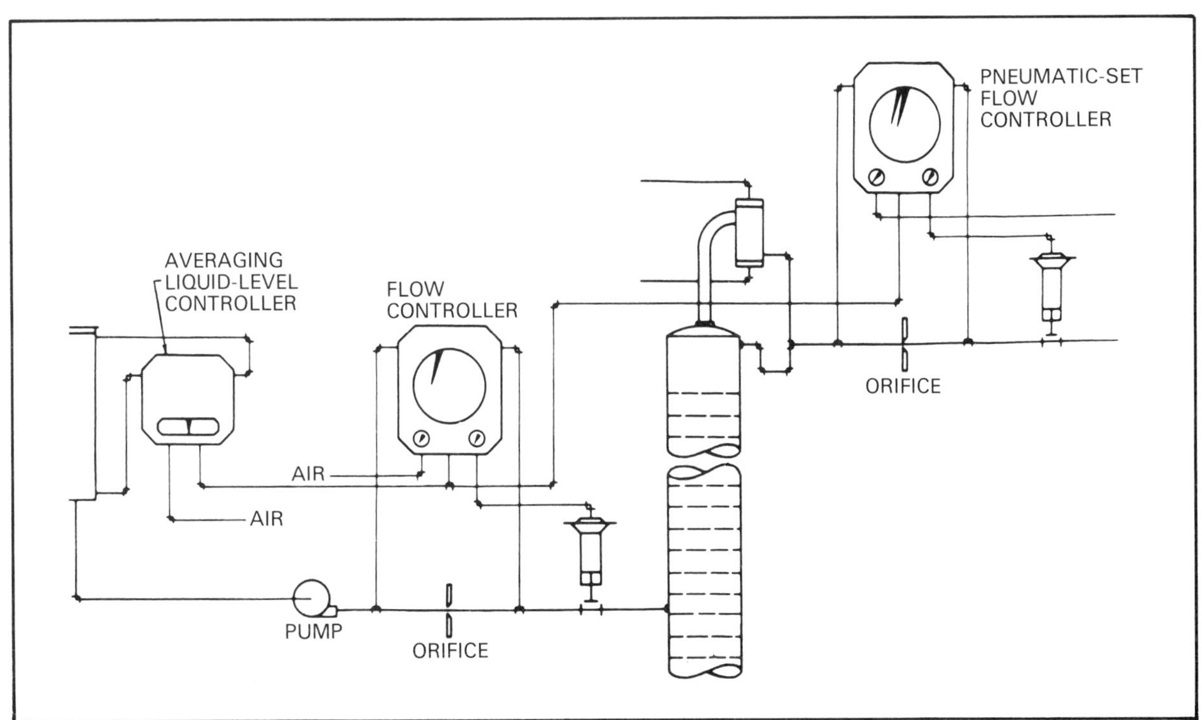

Figure 9.15. Liquid-level control for a system without a constant rate of feed

is established by the liquid level in an accumulator tank. The amount of product returned to the tower as reflux is controlled by the temperature existing near the upper end of the column interior or at some other chosen point in the column. Thus, the reflux rate has an indirect controlling effect on the flow rate of product output.

Control of Bottom Product Discharge Rate

A possible method of using an interlocked feed rate and bottom discharge rate is shown in figure 9.17. The bottom product of the column is gathered in an accumulator tank and is discharged at a rate determined by the feed rate into the column. This system of bottom product discharge control is perhaps less desirable than one based on liquid level existing in the reboiler.

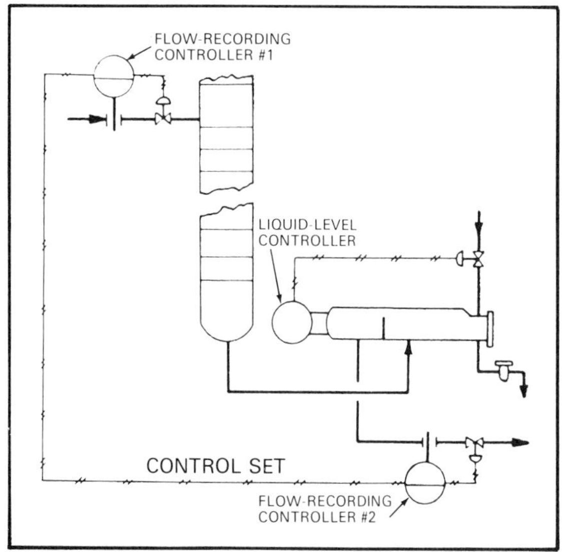

Figure 9.17. System using bottom product discharge control

Ratio Flow Control

The interlocked, or cascaded, control system (fig. 9.15), in which a definite relation exists between the flow rates of feed and upper column product discharge, is a form of *ratio control.* Assuming, for example, that the column divides the feed into two equal fractions and 50% of the upper fraction is fed back into the column as reflux, 25% of the original feed material is left available for discharge as the upper column product. Obviously, for every quantity of liquid passed by the feed control valve, one-fourth of that quantity must pass through the discharge as the upper product. Thus, this system is a true example of ratio flow control.

Ratio flow control is used in many processes, and its accomplishment is usually a matter of allowing a primary flow rate to establish the set point of a secondary flow controller. The primary rate varies according to the needs of the process, and the secondary, or ratio-controlled, rate is automatically maintained in proportion to the primary.

Regulation of Controller Set Point by Vapor Pressure Differential

As noted previously, liquid level can be used to establish the set point of one or more flow controllers, or one flow controller can be used to establish the set point of another controller. Pressure and temperature are also used to adjust the set point of flow controllers. An interesting case of flow control arises in the operation of certain fractionating columns, in which vapor pressure differential is used as a standard for the set point. A column that produces isobutane might use this method.

The separation of isobutane from other fractions depends, of course, on temperature and pressure, but the accuracy with which the temperature must be controlled for high purity is so fine that it is best to regulate it on the basis of vapor pressure differential. Figure 9.18 is a schematic diagram of the control system. A vapor pressure bulb is located on one of the upper trays of the column, and this bulb is filled with liquid that would exist at this tray for normal operation. Now, if the temperature at this tray varies appreciably from the value necessary for acceptable

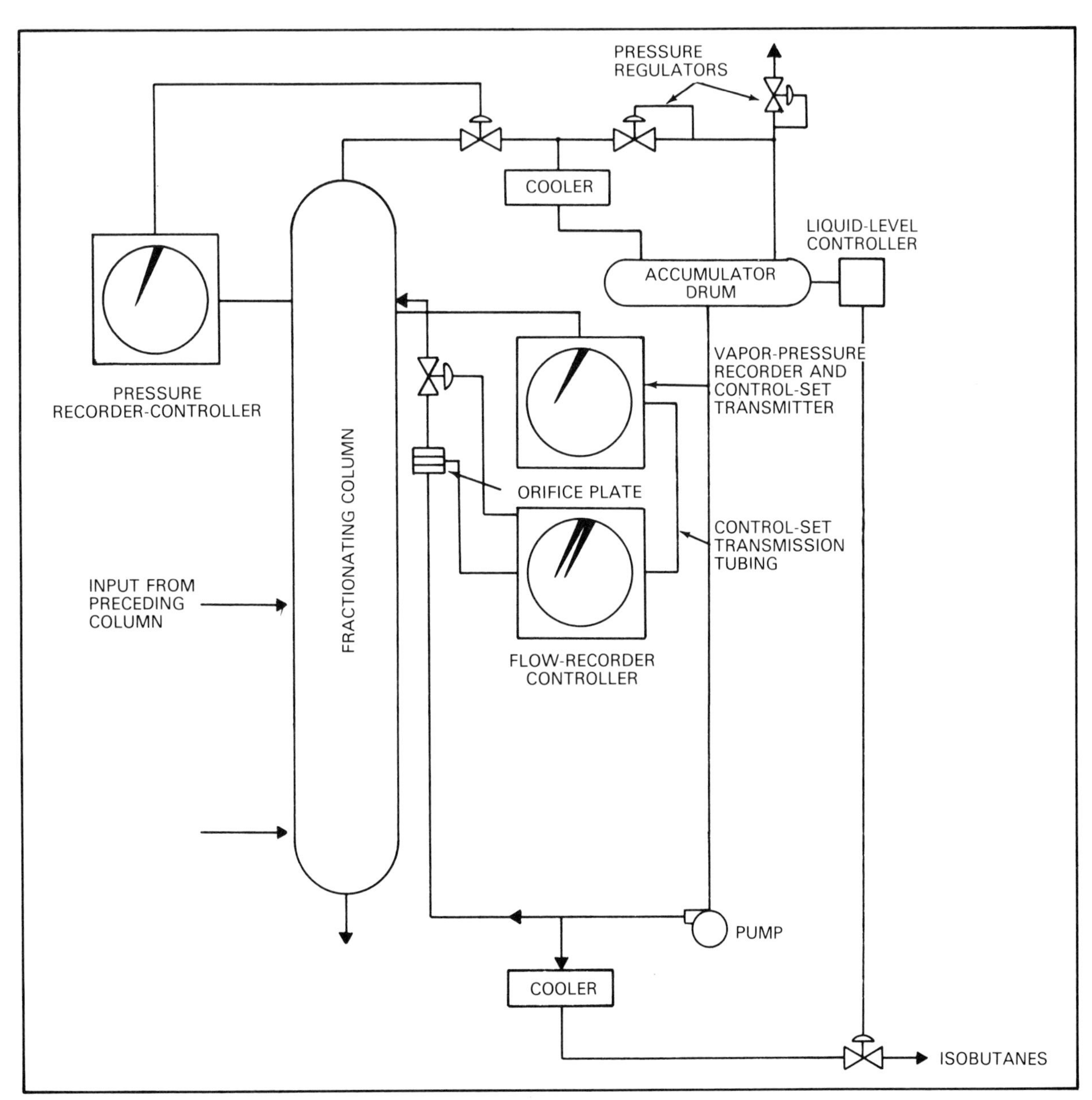

Figure 9.18. System using vapor pressure differential to establish set point of controller

operation, the composition at the tray will change and so will the vapor pressure in the bulb. Consequently, a differential vapor pressure will exist, and it will be detected readily by the differential pressure measuring means. A signal will be transmitted to the control set of the flow controller in the reflux line so that more or less reflux is returned to the column to bring about the needed correction.

The interesting point here is why vapor pressure differential is used instead of temperature variation to make the set-point change. The reason lies in the fact that in this case an appreciable temperature variation amounts to a fraction of 1°F. Obviously, such close control would be difficult to maintain with a temperature controller. On the other hand, a small variation in conditions can be expected to produce a pressure change of about 1½ psi per degree Fahrenheit. This change is much easier to measure and control than temperature variations. Of course, the end result is control of the temperature at the particular tray containing the vapor pressure bulb.

Installation Arrangements for Flow Measurement and Control

It may be enlightening in the study of flow control to examine a few arrangements of actual installations for particular applications. It may be helpful to note what auxiliary apparatus is needed in particular situations and what arrangements are made for mounting this apparatus and the main controller components.

Gas Flow Installation

Figure 9.19 shows a measuring means installation that is satisfactory for noncorrosive dry-gas systems. The meter may be located below the horizontal line, in which case the pressure taps can be at the side of the flange. For vertical flow lines, the meter may be above or below the taps.

The installation in figure 9.19 includes apparatus that is common to almost all flow measuring and controlling systems. The purpose of many of the fittings is obvious. The pipe unions, for example, facilitate the installation and removal of instruments and piping. The purpose of the valves located in the piping may need some discussion. *P* and *Q* are simply shutoff valves to be used when instrument and piping are removed, but they are also used when the instrument is undergoing zero adjustment.

Valves *A, B, D, P,* and *Q* are manipulated in a particular sequence when the instrument is being adjusted for a zero setting or when

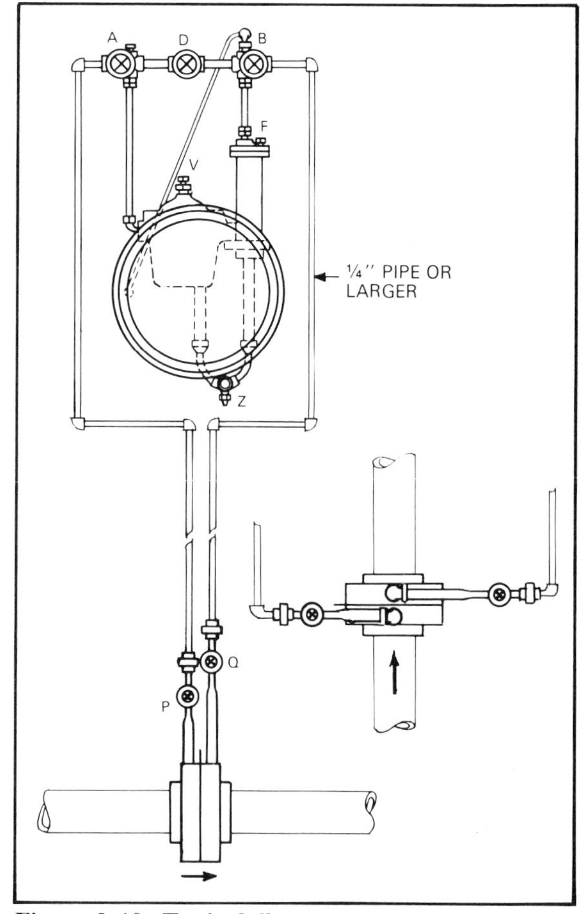

Figure 9.19. Typical flow measurement and control installation for noncorrosive dry-gas systems *(Courtesy Foxboro)*

the instrument is being placed in or out of service. Valves *A, B,* and *D* form a part of the manifold assembly usually supplied by the meter manufacturer. *A* and *B* are line stop valves, while *D* is a bypass valve used for equalizing pressure between upstream and downstream sections of the meter. Some bypass valves are vented. Line stop valves are sometimes accompanied by fittings that allow for attaching static pressure lines. The diagonal pipeline leading from the fitting adjacent to valve *V* in the figure is the static pressure line for the meter.

Instructions for zero adjustment of meters and for placing them in or out of service are provided by the manufacturer of the meters and should be followed.

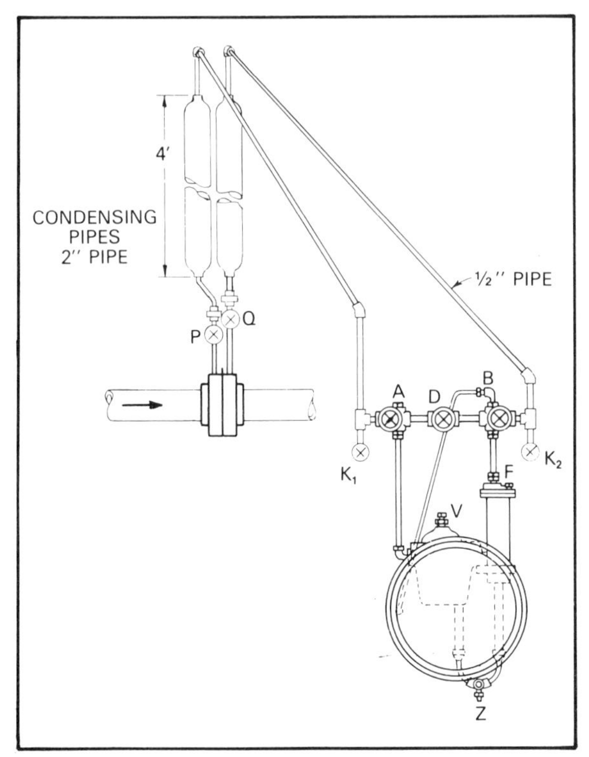

Figure 9.20. Condensing pipes for wet-gas systems *(Courtesy Foxboro)*

Figure 9.20 shows a meter installation suitable for some wet-gas measurement applications. Wet gases are those containing excess water vapor or condensable hydrocarbons. The meter can be installed above or below the line, but the condensing pipes and pressure taps must be installed at the top of the lines, as shown. The two condensing pipes are made of 2-inch pipe, each section being about 4 feet long. Condensable vapors tend to condense in the pipe sections and fall back into the main line. Valves K_1 and K_2 are for the purpose of draining out liquid that has gotten past the condensing pipes. They are of importance when the meter is placed in service, but regular drainage through these valves is unnecessary. Condensing pipes can be used for meters of the manometer type or diaphragm type.

Sometimes space problems arise, making the use of condensing pipes impractical. For installations having this problem, drain valves are located at low points in the lines and are opened at intervals frequent enough to prevent moisture from being trapped in the lines leading to the meter.

Liquid seals for corrosive gases. At times gases contain sulfur or other substances that may react with certain metals of the measuring system or combine with the mercury of a manometer-type meter. In such installations a liquid seal is often employed. The liquid used in a seal that is suitable for a manometer-type instrument (fig. 9.21) can be properly prepared kerosine or any of certain special liquids like dibutyl phthalate (trade name: Aquaseal). The proper choice of seal liquid will be specified by the manufacturer of the meter, and such specification should be observed.

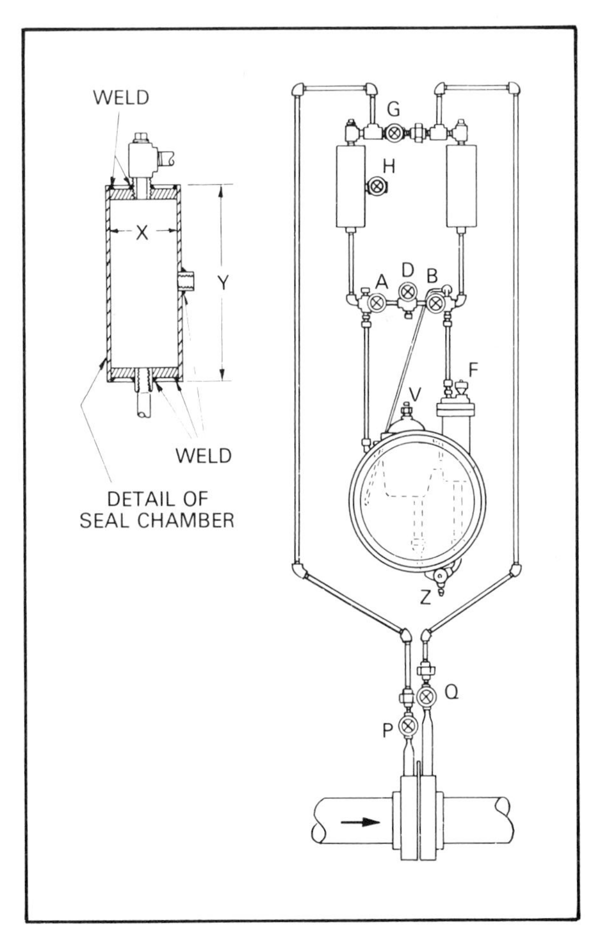

Figure 9.21. Liquid seal required for most corrosive systems *Courtesy Foxboro)*

The seal chamber, shown in detail in the figure, can be made from seamless steel tubing. The dimensions X and Y can be calculated. The length in inches is easily determined, and its value is not critical as long as a certain minimum length is used. Dividing the differential pressure range of the instrument (in inches of water) by the specific gravity of mercury, then adding 2 or 3 inches to the result will provide a satisfactory value for Y.

The value of X is critical if the proper calibration of the instrument is to be maintained. At full-scale differential pressure, the drop in liquid level of the upstream seal chamber plus the rise in level of the downstream chamber must equal the difference in mercury level in the chambers of the meter. This condition will ensure that no differential is introduced into the system by the seal liquid.

To find the value of X,

$$D_m/A + D_m/A = R/13.57$$

where

A = cross-sectional area of each seal chamber, in square inches;

D_m = volume of displacement of the meter mercury, in cubic inches;

R = range of the meter without seals, in inches of water;

13.57 = specific gravity of mercury.

From this,

$$A = 27.14\, D_m/R.$$

Since A is also equal to $\pi X^2/4$,

$$X = 5.88\sqrt{D_m/R}.$$

It should be clear from this equation that a set of seals suitable for one meter and range is not apt to satisfy the needs of another meter or range.

A considerable shifting of mercury between the float and range chambers of a mercury manometer instrument occurs as the differential pressure varies. This fact makes it necessary that the chambers for the liquid seals be carefully designed to avoid error caused by the change in head of sealing liquid and to ensure that the chambers be large enough to accommodate the rising and falling of liquid level without overflowing one chamber and driving the other dry. When diaphragm-type differential elements are used, the liquid seals can be much smaller and simpler because of the smaller displacement of sealing liquid that occurs in such instruments.

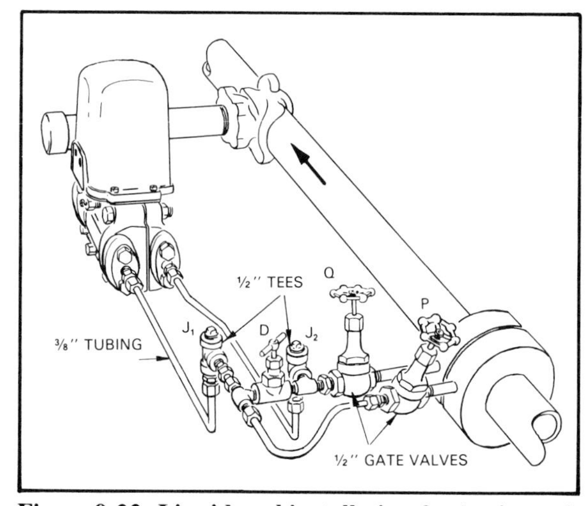

Figure 9.22. Liquid-seal installation for horizontal flow lines *(Courtesy Foxboro)*

Figures 9.22 and 9.23 contain flow measuring means installed for vertical and horizontal flow lines and fitted for liquid seals. In each case the seals are formed as part of the piping, and the only additional components needed are two ½-inch pipe tees that provide for easy filling of the seals. In both installations the differential pressure cell is mounted somewhat lower than the primary element. To prepare the seals, plugs J_1 and J_2 are removed from the tees, and appropriate sealing liquid is poured into the piping until it becomes level in the tees. To avoid error, the level in the two tees should be equal, but since the forward and back motion of the liquid with change in differential pressure is quite small, there is no significant error due to the changing head of sealing liquid in this particular differential pressure measuring element. This situation is

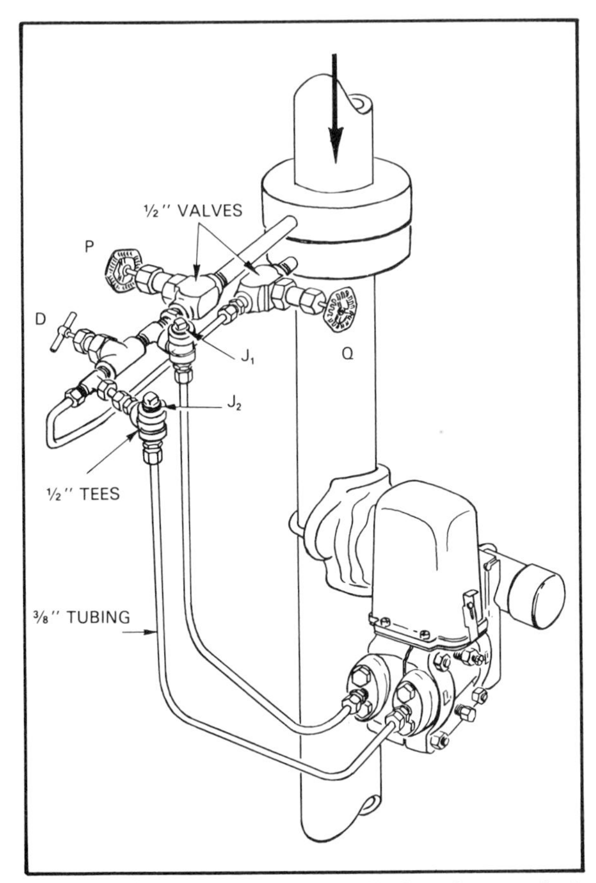

Figure 9.23. Liquid-seal installation for vertical flow lines ***(Courtesy Foxboro)***

not necessarily true for other diaphragm-type elements, for in some cases the change in head due to the in-and-out flow of sealing liquid will be significant, in which case a pair of chambers having shallow depth and broad surface area for the liquid should be used.

Pipe size for gas flow measuring means. Many of the drawings accompanying this section indicate minimum pipe sizes for the particular application represented by the drawing. Satisfactory performance of a measuring means requires certain minimum sizes. Table 9.1 gives satisfactory standards.

Copper tubing of ⅜-inch size can be substituted for the pipe if the length can be kept below 15 feet. Greater lengths of tubing are not recommended because of the difficulty in keeping tubing straight and free of troublesome sags.

TABLE 9.1
PIPE SIZE FOR GAS FLOW MEASURING MEANS

Pipe Length (feet)	*Minimum Size (inches) Wet Gas*	*Minimum Size (inches) Dry Gas*
0–10	¼	¼
10–50	½	¼
Over 50	¾	½

Steam Flow Installation

Installations for measuring and controlling steam flow are quite similar to those for gas measurement, but there is an important addition that provides a liquid seal between flow line and meter. The purpose of the liquid is not necessarily to protect against corrosive effects of the steam, but rather to provide equal liquid heads in the high- and low-pressure lines between primary element and meter.

Some of the details associated with primary elements in a horizontal and a vertical steam line are shown in figure 9.24. The important additional equipment is a pair of bottlelike condensing chambers, which are mounted horizontally at equal elevations. In the case of the vertical steam line, the two condensers are at the same level as the upper pressure connection.

Provision for a solid liquid fill. Considerable care must be taken to ensure that a solid liquid fill exists in the measuring system from the top of the mercury in the chambers of the meter to the neck of the bottlelike condensers. Air or vapor entrapped in the piping system between condensing chambers and meter causes errors. In order to preclude the likelihood of air or vapor pockets in the system, special steps are taken in filling the instrument and instrument piping with liquid. In the straightforward system shown in the figure, where the pipe runs are short and contain no troublesome sags, or low points, the

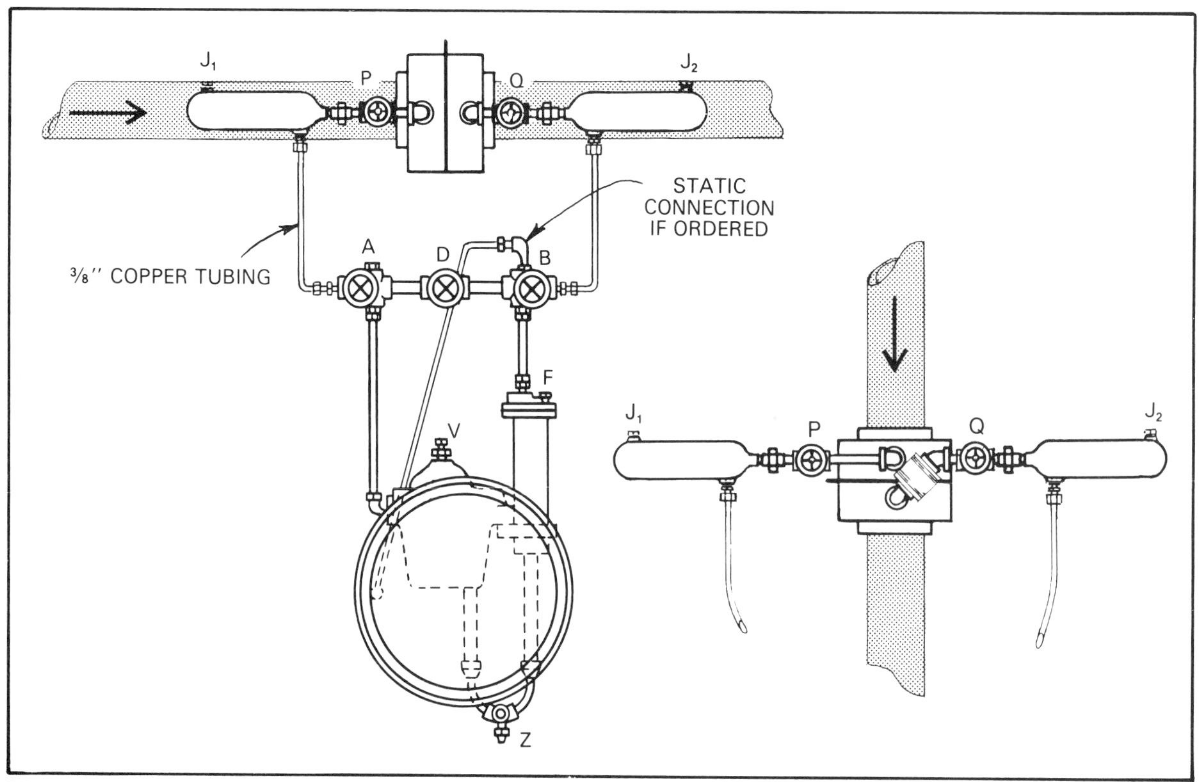

Figure 9.24. Condensing chambers used for steam flow measurement ***(Courtesy Foxboro)***

procedure for filling with water is as follows:

1. Remove plugs (or needle screws) J_1, J_2, *F,* and *V.*
2. Open valves *A, B,* and *D,* and make sure that valves *P* and *Q* are closed tightly.
3. Pour water into openings at *V* and *F* until the chambers are full; then replace the needle screws.
4. Pour water into J_1 until it runs out both J_1 and J_2.

To check for entrapped air or vapor, start with all valves closed; then open *D* to equalize the meter. Relieve pressure from the meter by opening *V* or *F;* then close *D* and *V* or *F.* Open *P* and *A*. If the recording pen or other indicating means shows any appreciable movement, air or vapor is trapped in the system somewhere below valve *B*. If the system contains a solid liquid fill below *B,* the liquid, being virtually incompressible, will stand firm against the line pressure admitted through *P* and *A*. On the other hand, pen movement is an indication that air or vapor in the system is being compressed.

A similar check is necessary to ensure that the upstream part of the system between valve *A* and the meter is free of air or vapor. The same procedure is carried out, except that valves *Q* and *B* are operated instead of *P* and *A*.

Some installations, particularly those having long runs of piping between steam line and meter, will have low points that tend to accumulate liquids and make the simple procedures outlined above ineffective for removing air or vapor pockets. In such instances, it is usually necessary to force the water fill into the piping at certain low points of the system. Instruction manuals provided by the instrument manufacturer should be consulted.

Pipe size for steam flow measuring means. Copper tubing is recommended for short runs between primary element and meter. If the pipe run exceeds 15 feet, but is less that 50 feet, then ½-inch brass pipe is recommended. For greater lengths, ¾-inch brass pipe should be used. As noted in the discussion for gas installations, copper tubing is difficult to maintain in a straight path for long runs, as it tends to sag unless extensive use of supporting fixtures is employed.

Liquid Flow Installation

The installation requirements for liquid flow systems differ slightly from those for gas and steam flow systems. In practically all cases there is a solid liquid fill between primary element and meter. The same is true for steam systems, but not for gas. Condensing chambers are not ordinarily used, but liquid seals find useful application in systems containing corrosive or viscous liquids.

The simplest form of installation is one with very short runs of pipe between a primary element and a meter mounted below this element. Figure 9.25 shows this kind of installation, along with a cut showing the arrangement for a vertical flow line. In all cases, it is essential for accuracy that equal heads of liquid exist in each line above the meter. The instruction manual accompanying the particular type of meter is the best source of information for placing the instrument in operation. It is worth pointing out that vapor traps are used in installations having the meter above the flow line.

Vapor pockets tend to rise to the highest point in the system, and in an installation having the meter below the line, the pockets eventually rise into and are drawn off by the flowing liquid in the line. This situation will not happen if the meter is above the primary element, so vapor traps are placed at the highest points in the measuring system. They are small chambers into which bubbles of air or vapor accumulate. Valves are provided at the tops of the traps for venting the trapped vapors occasionally.

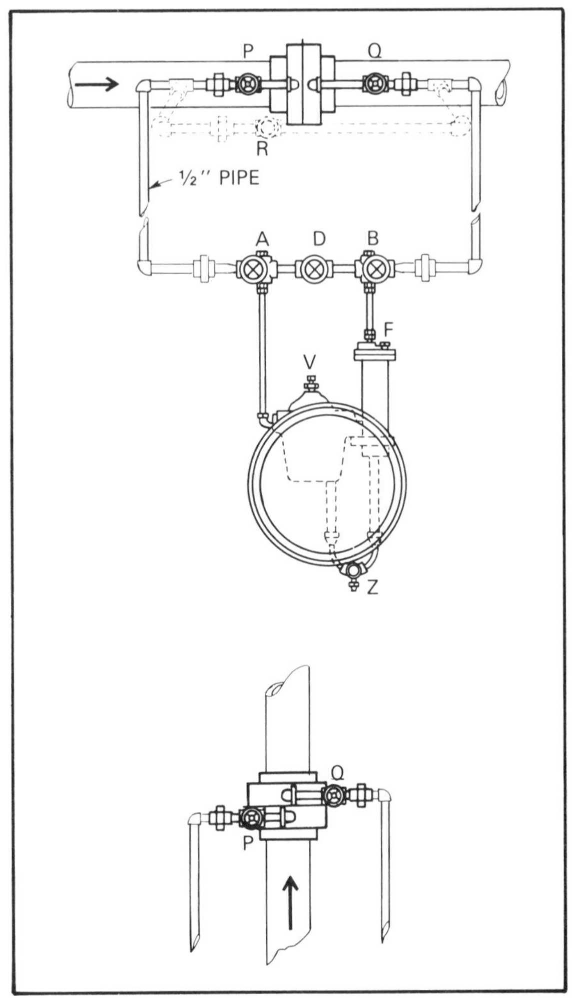

Figure 9.25. Simple installation for liquid flow measurement and control ***(Courtesy Foxboro)***

Liquid seals. Sometimes it becomes necessary to measure or control the flow of liquids possessing characteristics that rule out the simple expedient of allowing these liquids to enter the piping and instruments of the measuring means. Viscous oils, corrosive liquids, highly volatile liquids, and liquids easily frozen are likely to require the use of seals to isolate them from the meter and some of the piping. Viscous liquids are usually pumped through lines at temperatures high enough to make them flow easily, but if allowed to enter

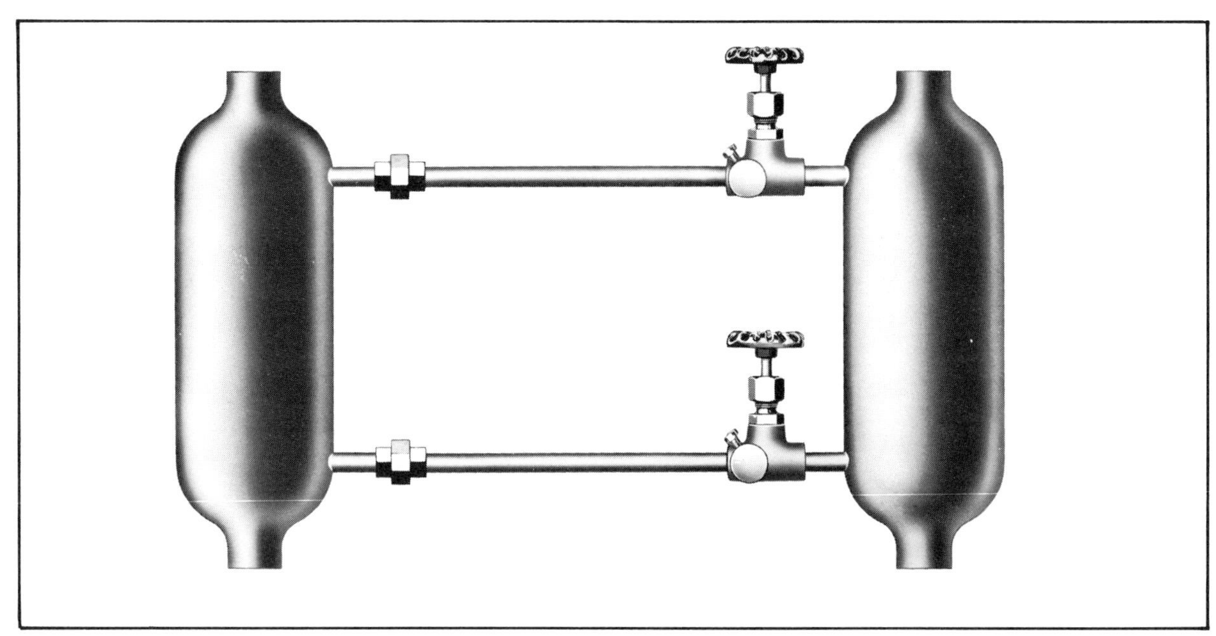

Figure 9.26. Double-seal assembly for differential pressure applications ***(Courtesy Foxboro)***

measuring means piping and instruments, they will cool sufficiently to become troublesome. Very volatile liquids easily form vapor pockets in piping or instrument. Liquids with a relatively high freezing temperature may freeze in meter and piping during winter months.

A double-seal assembly is useful for such differential pressure measuring applications

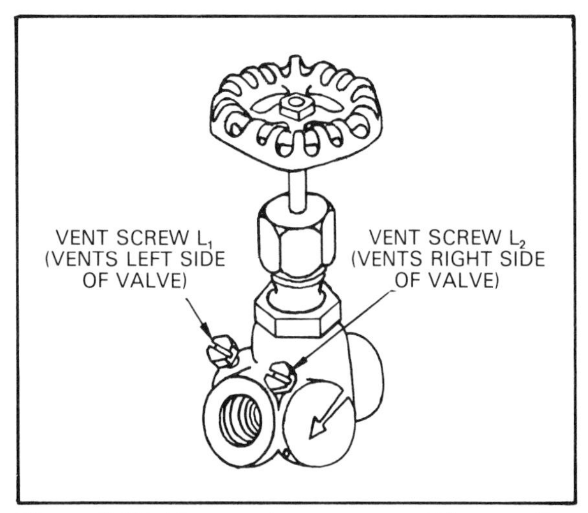

Figure 9.27. Two vent screws permitting venting of line on either side ***(Courtesy Foxboro)***

(fig. 9.26). The bulblike chambers are typically about 8 inches long and 4 inches in diameter. The ends are threaded to take standard pipe. Two parallel sections of pipe, each containing a special hand valve, connect the two chambers. The special feature of one such valve (fig. 9.27) is the inclusion of two vent screws. With the main flow line of the valve closed, the vent screws enable the line to be vented on either side of the valve. The dual-chamber assembly is always mounted in an upright position, as shown in the figures.

Several points should be considered in using liquid seals between meter and flowing fluid.

1. Why is a seal needed—to guard against corrosive action, to avoid freeze-up in the meter, or to counteract high viscosity?
2. Is the sealing liquid heavier or lighter than the flowing liquid?
3. Will the meter be above or below the seals and the flow line?

The choice of sealing liquid will depend upon what is to be guarded against. If the seal

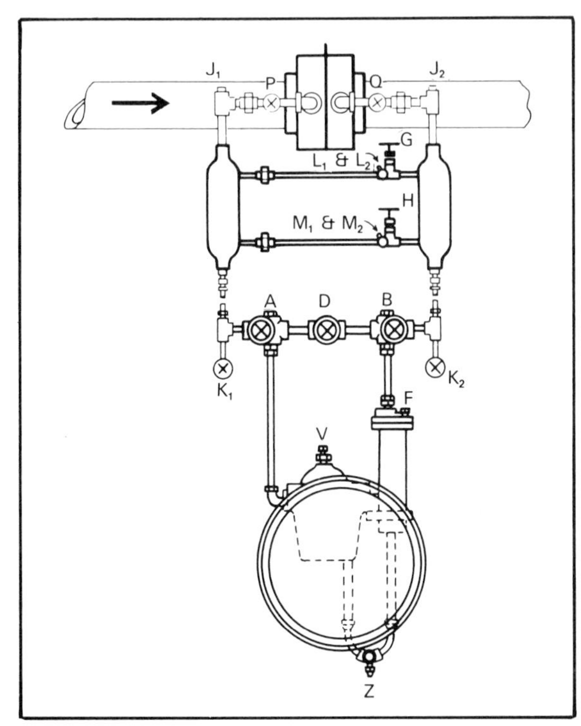

Figure 9.28. Meter mounted below the line; seal liquid heavier than line liquid

is to be against viscous petroleum oil, for example, and no problem of freezing exists, then water will be used as the seal liquid. If there is likelihood of freezing, then a mixture of permanent-type antifreeze and water can be used. Sealing against corrosive liquids can be provided by using a proper quality of kerosine or any specially prepared commercial product.

In cases in which the sealing liquid is heavier than the line liquid, the piping between meter and chambers will be connected to the bottom of the chambers. If the sealing liquid is lighter than the line liquid, the piping from the meter must connect to the top of the chambers.

Figures 9.28–9.31 are four drawings of installations in which the meter is either above or below the flow line and the sealing liquid is either lighter or heavier than the line liquid.

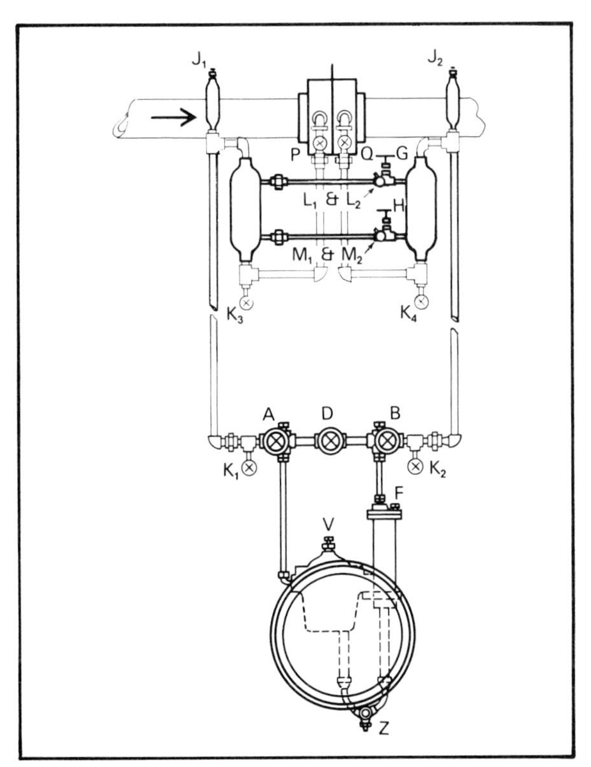

Figure 9.29. Meter mounted below the line; line liquid heavier than seal liquid

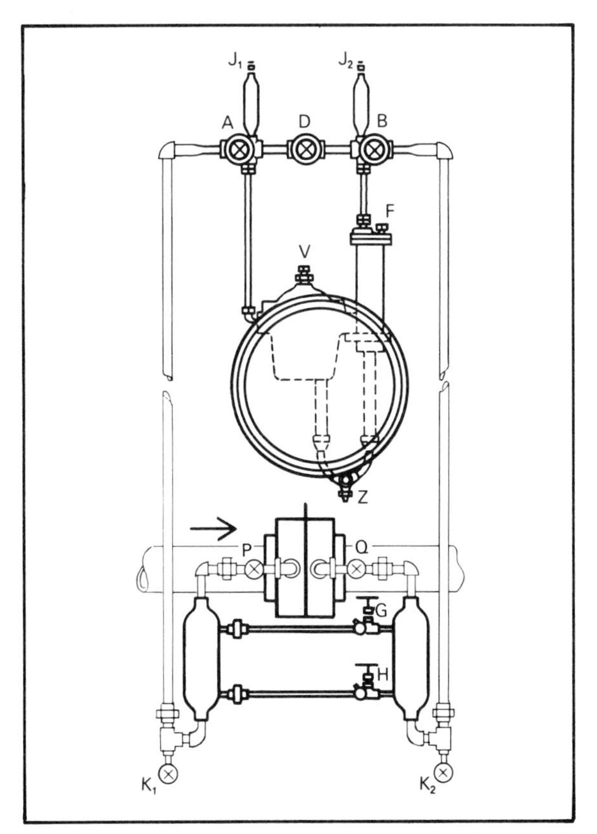

Figure 9.30. Meter mounted above the line; seal liquid heavier than line liquid

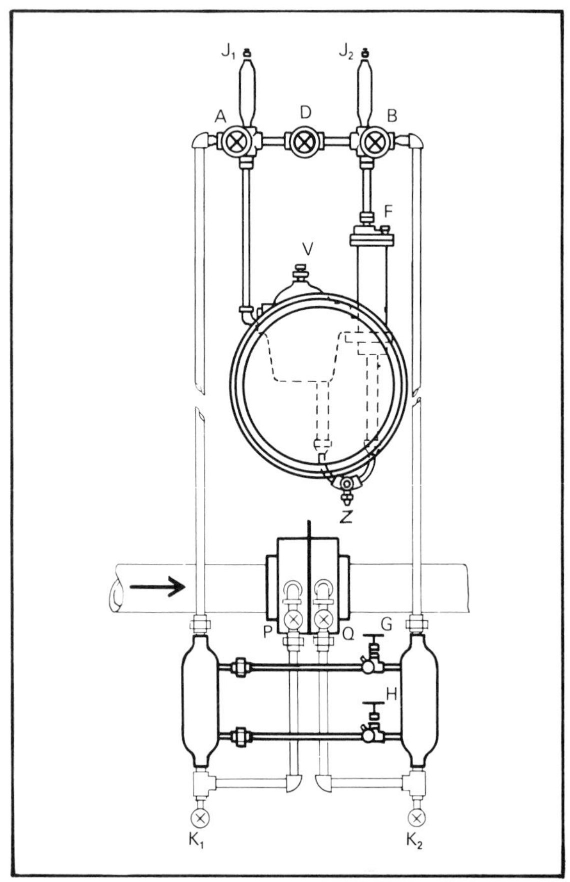

Figure 9.31. Meter mounted above the line; line liquid heavier than seal liquid

The use and location of vapor chambers in the measuring system should be noted.

The procedure for putting into service any of the installations is too lengthy to be included here. The manufacturer's directions should be followed, along with a very important warning. When the system is in operation, valves *A, B, P,* and *Q* are open, and all other valves and vents are closed. Under these conditions, it would be a serious mistake to open any one of the three valves *D, G,* or *H,* since this would cause the seal liquid to be forced into the flow line and would permit line liquid to get into the instrument piping and possibly into the instrument itself.

When the force-balance differential pressure cell is used for liquid flow measurement, the same simple liquid seals as those for gas flow measurement can be used (fig. 9.22).

Pipe size for liquid flow measuring means. The use of ½-inch pipe for runs up to 50 feet is satisfactory for liquid flow measuring systems. For longer runs of pipe, the size should be ¾-inch. Tubing is not recommended for runs greater than about 15 feet because of the extensive support needed to prevent sagging. If tubing is used for short runs, it should be at least ⅜-inch material.

Summary

The flow control applications discussed, while not exhaustive, are representative of all methods commonly used. Piping arrangements and installation practices covered in the latter part of the chapter can be applied to many forms of measuring and controlling functions that are not necessarily concerned with the flow of a fluid. It should be remembered that manufacturers' literature is the best source of information about installation and maintenance of instruments.

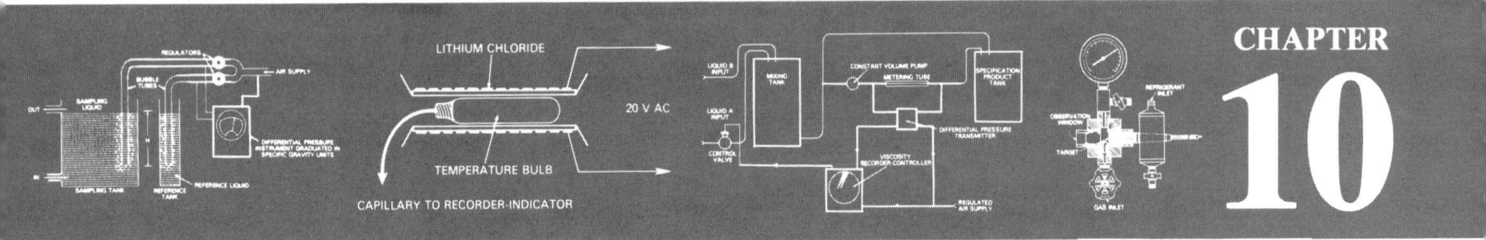

CHAPTER 10

Measurement and Control of Other Variables

Many processes and systems require only the automatic control of pressure, temperature, liquid level, and flow rate; in the calculations variable factors are often involved. Specific gravity and viscosity, for example, are taken into consideration when calculating flow measurements. Sometimes specific gravity, viscosity, and humidity, as well as other lesser known manifestations, become important variables in a process. In such instances, their accurate measurement and often their control become necessary.

Humidity and Dew Point

Humidity is well known, if not well understood, because it is so often associated with human comfort. High humidity and either high or low temperatures are charged with causing physical discomfort. Humidity is also important in many industrial processes. The preparation or manufacture of some industrial gases requires measuring and controlling moisture content, or humidity. The textile and paper manufacturing industries also have need for humidity control.

Absolute and Relative Humidity

Absolute humidity is an expression for the weight of water dispersed as vapor in a unit *weight* of dry air or other gas. A common system of units for measuring it is *grains of water per pound of dry air.*

Relative humidity is the most familiar term, because it is used in connection with weather reports. The amount of water that can exist as a vapor in the air varies with temperature—the higher the temperature, the more water the air will support. Relative humidity is expressed as a percentage of the total moisture that the air can support at a given temperature. Thus, as the temperature rises, the relative humidity decreases, and vice versa, assuming that the absolute moisture content remains constant.

In order to understand relative humidity, a quick review of vapor pressure of water is necessary. In a laboratory setup that might be used to determine vapor pressure (fig. 10.1), a container is partially filled with water, and a vacuum pump reduces the absolute pressure in the space above the water to very near zero. If the chamber is sealed off from the pump, it will be noted that the absolute pressure gauge will soon reach a stable indication. If the temperature is 10°C, the indication will be about 9.2 millimetres of mercury. At this temperature and pressure, a state of equilibrium exists between liquid and vapor—for every molecule of water that escapes the liquid surface and becomes a part of the vapor, a molecule of the vapor returns to the liquid state. The space above the water under the

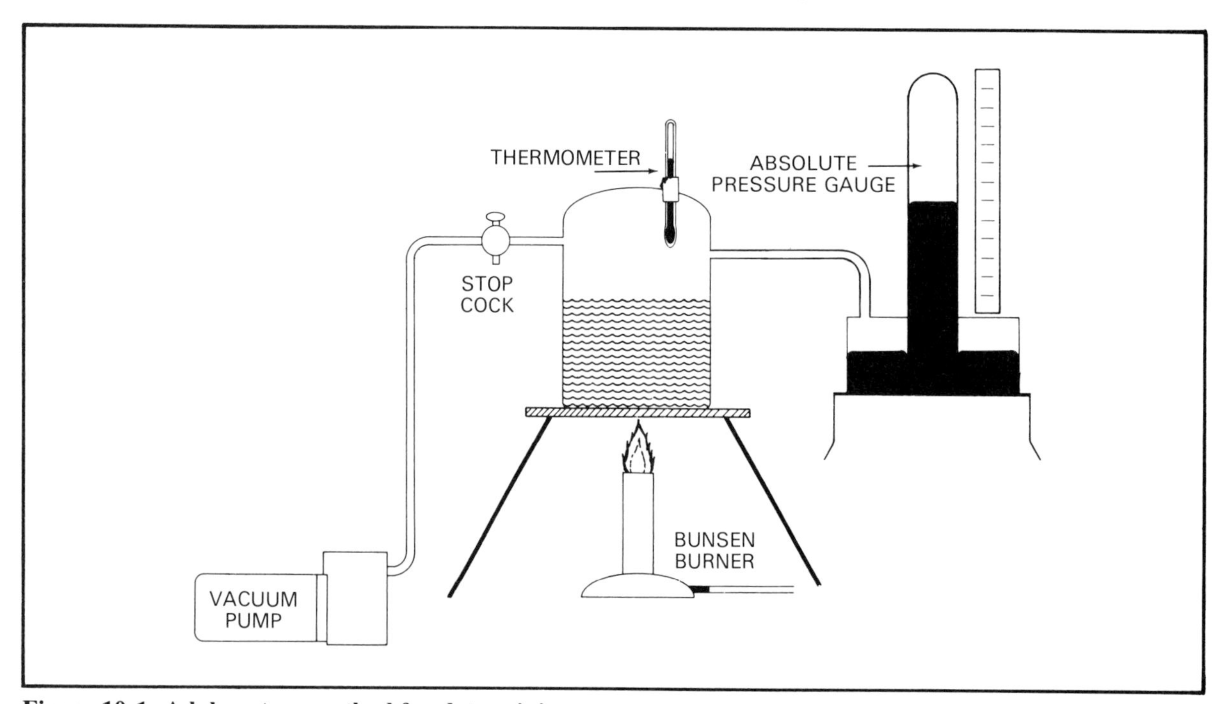

Figure 10.1. A laboratory method for determining vapor pressure

conditions of equilibrium is said to be *saturated* with water vapor.

If the temperature is varied from the value assumed, the vapor pressure will change, increasing with temperature, and vice versa, but it is by no means a linear function. At 20°C, for example, the vapor pressure will be 17.5 millimetres of mercury; at 100°C, it will be 760 millimetres of mercury. Tables of vapor pressures for a wide range of temperatures are available in many handbooks, the *Handbook of Chemistry and Physics,* for example.

Ordinarily, air is not saturated with water vapor, although saturation occurs frequently and accounts for the dew sometimes found on cool mornings. In such cases the temperature has fallen below the dew point, so the air is no longer capable of maintaining its moisture content. Later in the day, with rising temperatures, the dew evaporates and again becomes vapor in the atmosphere.

When the relative humidity of air reaches 100%, the air is said to be *saturated,* and water vapor will begin to condense and fall out as liquid. Fog and dew are examples of this. The *dew point,* a term frequently used in weather reports, *is the air temperature at which water vapor begins to condense.*

Measurement of Relative Humidity

The effects of water vapor in the air or other gas are manifested in several ways, and a measure of relative humidity is *inferred* from these effects. Many materials are *hygroscopic,* that is, they have an unusual tendency to absorb moisture from the air. Some of these materials undergo significant changes in linear dimensions as their moisture content varies. Another effect that relates to moisture content in the air is evaporative cooling.

Among the substances that change dimensions as a function of temperature and moisture content, *human hair* is particularly noteworthy. Natural hair increases in length with increasing relative humidity and has long been in popular use as the sensitive element in humidity indicators (fig. 10.2). To attain reasonable accuracy with such an arrangement, each instrument should be calibrated individually.

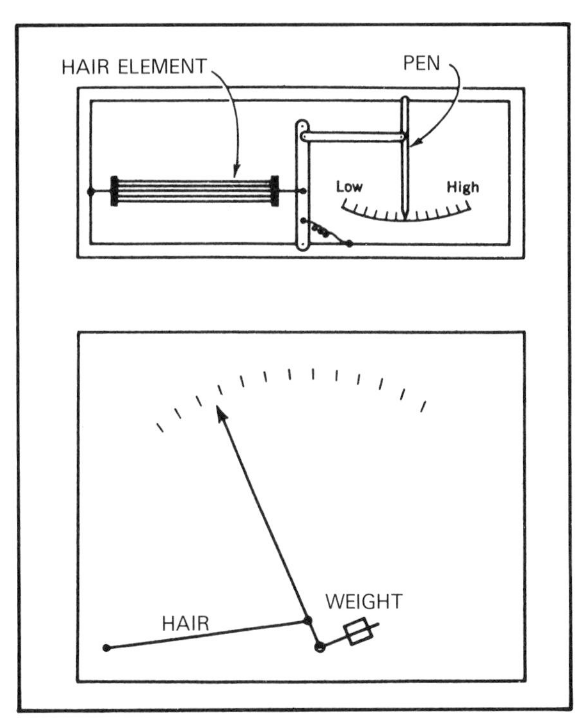

Figure 10.2. Sketch showing principles involved in hair-type indicator used to measure relative humidity (*Courtesy Honeywell*)

The human body uses evaporative cooling. It emits perspiration, and the evaporation of the perspiration cools the body surface. In humid weather the perspiration is apt to accumulate profusely, because the air is already heavily moisture laden and does not eagerly accept additional concentrations of moisture. If the bulb of an ordinary mercury thermometer is surrounded with an absorbent material that has been soaked in water, evaporation of this water will cause the thermometer to indicate a temperature somewhat lower than that of a similar but dry thermometer placed beside it. Thermometers that have a "wetting" device attached to their bulbs are called *wet-bulb* thermometers.

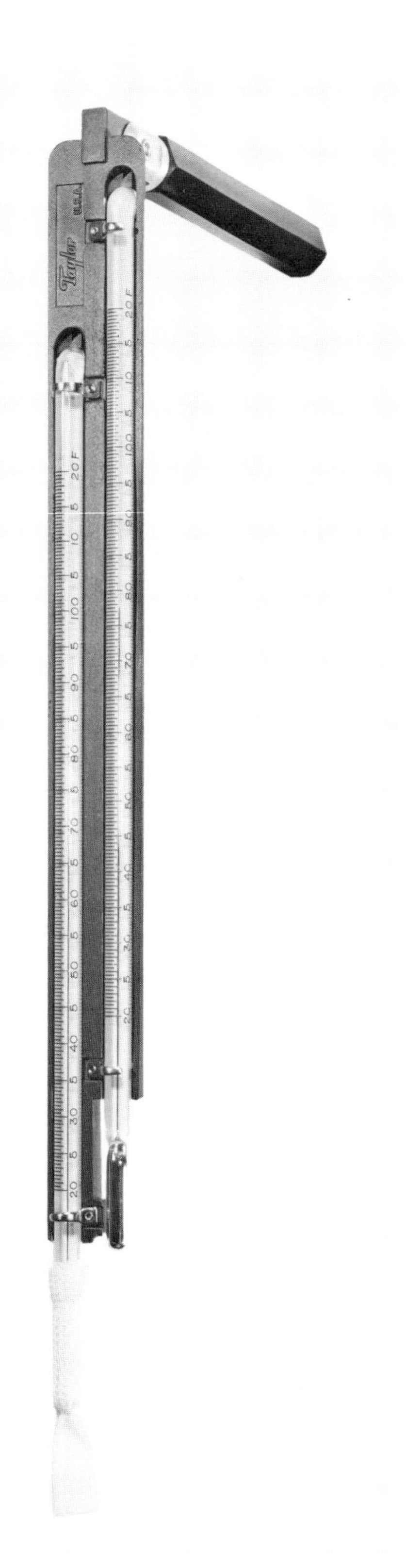

Figure 10.3. A sling psychrometer (*Courtesy Taylor Instrument*)

Evaporative cooling is based on the fact that heat is required to evaporate (or vaporize) water and other substances. Thus, the vaporizing of moisture in the wet material around the thermometer bulb causes a heat loss at that point, so the temperature decreases.

The evaporation principle is used in a *psychrometer,* which is, in broad terms, an instrument that uses two temperature measurements to determine relative humidity. A sling psychrometer is an arrangement of two thermometers fitted into a frame (fig. 10.3). One of the thermometers has a sleevelike piece of cloth over its bulb. This thermometer is the *wet-bulb* thermometer, while the other member of the pair is the *dry-bulb* thermometer.

Operation of the sling psychrometer involves wetting the cloth sleeve thoroughly with clean water, then whirling the assembly rapidly for 20 to 30 seconds so that a steady flow of air acts to vaporize some of the moisture. The wet-bulb thermometer is then read quickly, followed by a reading of the dry-bulb unit. The use of tables or a special slide rule enables a quick determination of the relative humidity.

An installation arrangement for one form of recording psychrometer (fig. 10.4) has wet and dry bulbs that actuate pens in the recorder, so that a record of the two temperatures is maintained. Of particular interest is the porous sleeve that fits over the wet bulb and serves the same purpose as the cloth sleeve in the sling psychrometer. A source of relatively pure water is obtained from the filter and is applied to the porous sleeve at a constant low pressure. The low pressure is maintained by the 12 to 18 inches of head provided by the standpipe. With this small pressure, the sleeve is kept moist without being flooded with excess water.

The arrangement shown in the figure would be quite satisfactory for recording the wet- and dry-bulb temperatures in a ventilation duct or

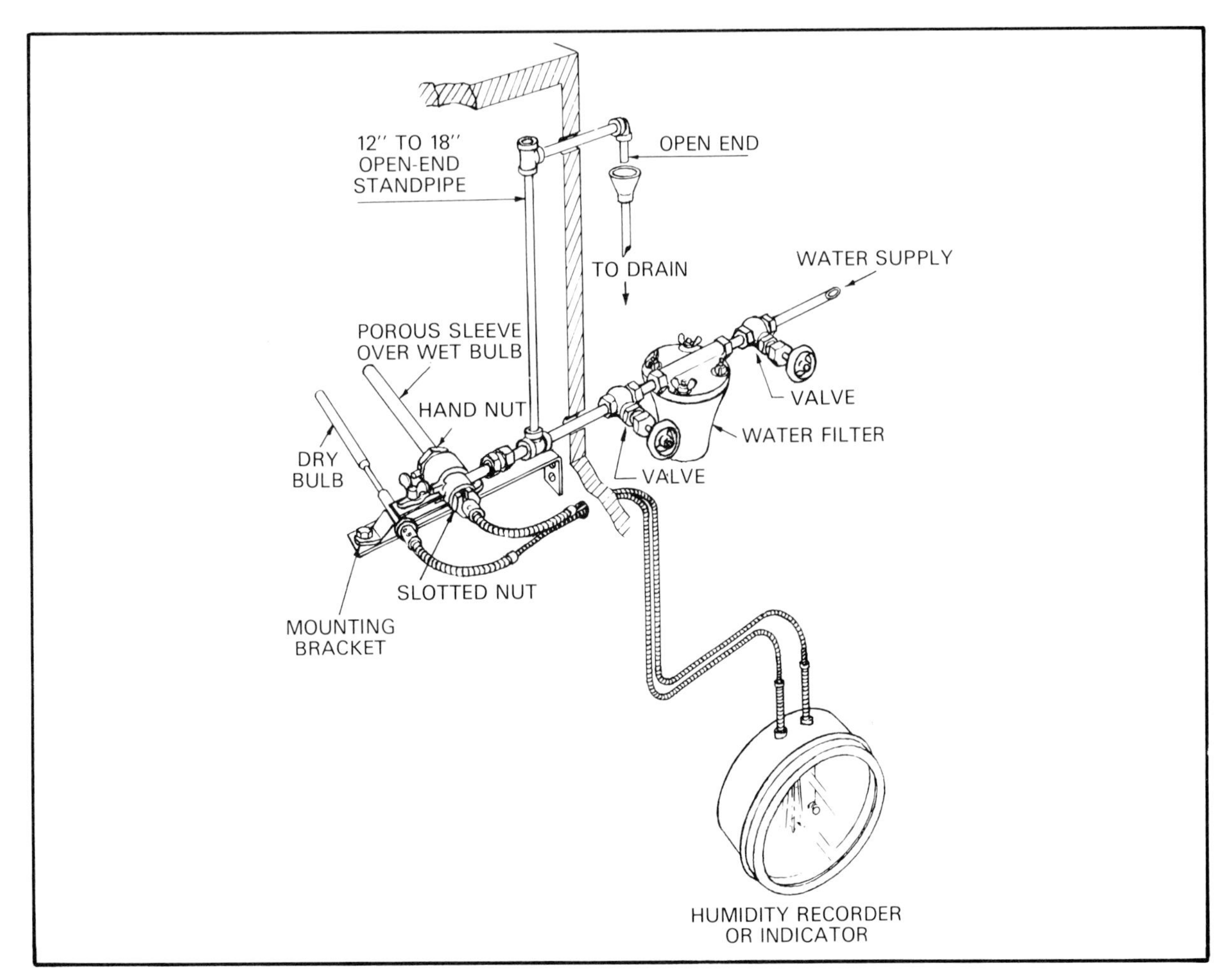

Figure 10.4. A recording psychrometer for measuring and recording humidity (*Courtesy Foxboro*)

other space where there is sufficient movement of air to ensure accurate wet-bulb response. Where there is insufficient air movement, fans or blowers are used to stir the air near the wet-bulb sensitive element.

A psychrometer chart, or graph, is used for determining relative humidity from two temperature values (fig. 10.5). The graph shows the nonlinear relation between relative humidity and temperature and serves as an aid in dew-point measurement.

Electric elements are utilized for inferential measurement of relative humidity. One such device uses a pair of gold-leaf grids electrically separated from one another by a coating of

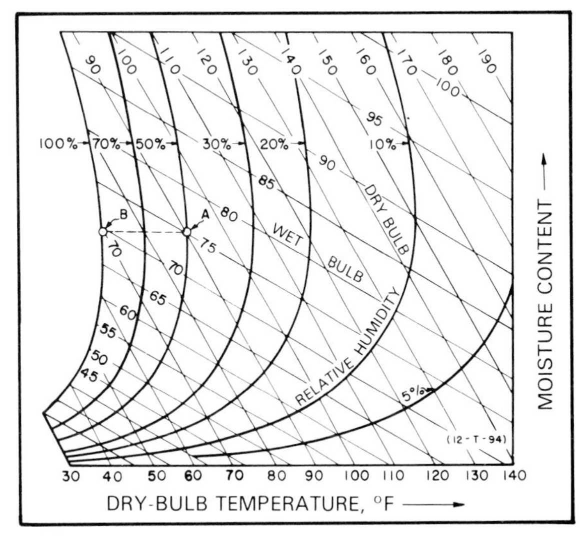

Figure 10.5. A psychrometer chart

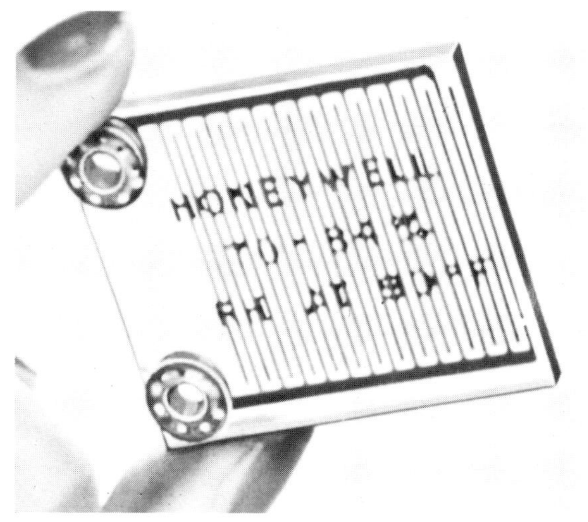

Figure 10.6. A gold-leaf grid, lithium chloride element used for measuring humidity (*Courtesy Honeywell*)

lithium chloride (fig. 10.6). When perfectly dry (humidity zero), the lithium chloride coating is a nonconductor of electricity but conducts current quite effectively when moist. The hygroscopic nature of lithium chloride causes it to absorb moisture from the surrounding atmosphere, and the amount it absorbs bears a definite relation to the humidity of the atmosphere. The electrical resistance between the two gold-leaf grids varies with the moisture content of the lithium chloride and therefore also with the humidity. Electrical elements of this type must have proper compensation for temperature variations in order to indicate relative humidity.

Dew-Point Measurement

As noted earlier, dew-point temperature is that at which moisture saturation of the air, or other gas, exists. For many applications dew-point measurement is more important than humidity measurement, and a number of methods for carrying out the measurement are in widespread use. Some of the methods involve elaborate systems of heating and cooling a polished surface and photoelectric devices to detect the act of condensation as it occurs on the surface.

Dew point is of great importance in the natural gas industry. A dew-point tester called the Bureau of Mines type (fig. 10.7) is capable of being used at gas pressures as high as 3,000 psi and at any attainable temperature below 140°F. Gas is allowed to bleed into the instrument where it impinges on a polished mirror surface. Temperature of the mirror is lowered until moisture in the gas condenses on it. The instrument operator observes the mirror surface through a system of lenses that forms a gas-tight window in the central cavity containing the mirror.

The temperature of the mirror is controlled by carefully regulating the flow of a refrigerant that contacts its back side. A thermometer protrudes from the instrument opposite the window, and the bulb of this thermometer is in contact with the mirror. The refrigerant may be ice water if temperature values above 35°F are satisfactory. Lower temperatures may be obtained by allowing compressed propane gas to expand in the cooling chamber or by using a solution of dry ice and acetone.

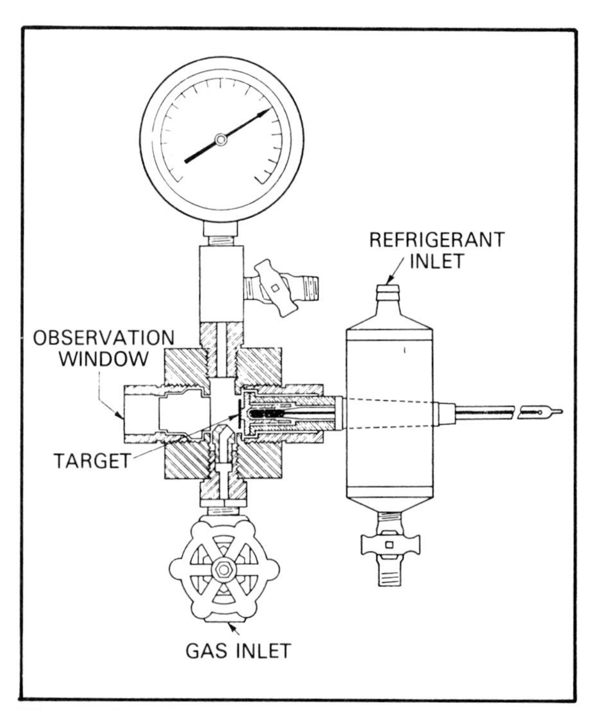

Figure 10.7. A Bureau of Mines type of dew-point tester

Great care on the part of the operator is required for accurate measurement. He must be vigilant to detect the first sign of condensation, and he must be capable of distinguishing between condensate of hydrocarbons—butane, for example—and that of water.

Many of the principles used in the Bureau of Mines type of dew-point tester have been incorporated in automatic systems capable of measuring and recording dew-point temperature. Some systems use automatic refrigeration and heating units that adjust the temperature of the mirror to the dew-point value. A photoelectric device controls the heating and cooling effects by observing the formation of dew on the mirror, while a thermocouple circuit monitors the mirror temperature and transmits its value to a recorder-indicator.

A recent method of achieving unattended indication and recording of dew-point temperature uses a bobbinlike tube covered on the outside with a durable cloth material impregnated with lithium chloride (fig. 10.8). Two open-ended coils are wound over the impregnated material and energized from a low-voltage (about 20 volts AC) source. The coils are of silver or other corrosion-resistant bare wire wound on the bobbin so that a turn of one coil lies between two turns of the other coil. The only electrically conductive path between the coils is that offered by the lithium chloride. A temperature-sensing element in the form of a thermocouple or one of the filled-system bulbs is located inside the bobbin and connected to the indicating-recording unit.

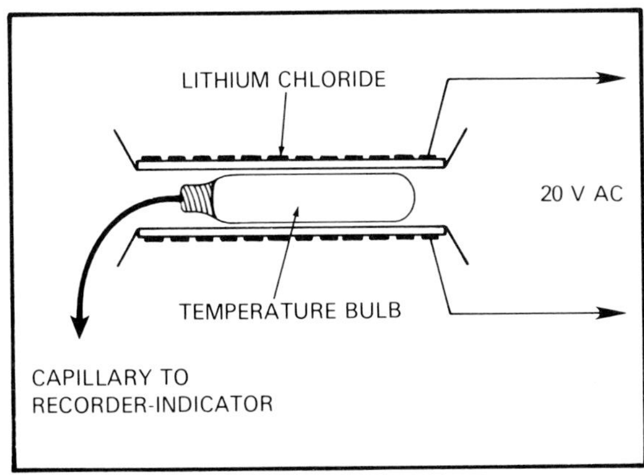

Figure 10.8. Sensitive element for a dew-point recorder-indicator system

Dry lithium chloride is not a good conductor of electricity. Lithium chloride begins conducting electricity progressively well when the relative humidity of the surrounding air or gas reaches and exceeds about 12%. With increased flow of current in the coils of silver wire, temperature of the bobbin assembly increases. A state of equilibrium will be reached at which time the temperature of the bobbin and current flow in the lithium chloride will remain stable at a given value of temperature. The equilibrium temperature at the bobbin will vary in step with the dew point, so that the indicator-recorder can be calibrated in dew-point temperature.

The dew-point temperature may be obtained readily from a psychrometer chart like that of figure 10.5 if the wet- or dry-bulb temperatures are known. Note that point A is at the intersection of lines for wet- and dry-bulb temperatures of 75°F and 90°F, respectively. If now a line parallel to the baseline is drawn from point A, its intersection with the 100% relative humidity line will indicate the dew-point temperature, which is 70°F. Note that wet- and dry-bulb temperature lines of equal value intersect at the 100% relative humidity line.

Specific Gravity

Specific gravity is usually thought of as being a number that expresses a comparison between the densities of a particular substance and water. Water, in this case, is the reference substance and has a specific gravity of 1. In the study of gas-flow measurement, air was used as a reference for expressing the specific gravity of gases. Water and air are used almost exclusively for specific gravity measurements, although oxygen is sometimes used for the critical scientific measurement of gases.

Density is affected by temperature and pressure, and this fact must be carefully considered when making specific gravity measurements, although ordinary pressure is of no importance when dealing with noncompressible liquids. Laboratory and scientific work usually specifies double-distilled water at 4°C (39.2°F) for accurate measurements, or air at standard temperature and pressure of 0°C and 760 millimetres of mercury, respectively. Engineering applications usually specify 60°F and 14.73 psia for temperature and pressure, respectively, although deviations from these values are common.

Measuring Scales

The measurement of specific gravity in everyday life is not nearly so commonplace as the measurement of temperature, humidity, or even atmospheric pressure, all of which are important variables in weather forecasting. The average person is probably aware of specific gravity measurement as a means of determining the charge in a lead-acid storage battery or as a measure of the strength of antifreeze solutions in cooling systems. In scientific and medical analysis, the measurement of specific gravity takes on great importance.

In the preparation of solutions in industrial processes, specific gravity is often the simplest and most accurate measurement for determining the correct composition. The petroleum industry, the Bureau of Mines, and others use a system of measurement based on specific gravity, known as the API scale, to judge the quality of petroleum, although it must be understood that quality is contingent upon many other factors. The strength of acid solutions is most readily determined by specific gravity, and it is this method by which the charge of a lead-acid storage battery is inferred.

API scale. During the early 1920s the American Petroleum Institute devised and adopted a scale of measurement units called *degrees API.* Although the scale is very different from the ordinary specific gravity scale, it bears a definite relation to it as follows:

$$^{\circ}\text{API} = \frac{141.5}{G} - 131.5$$

where G is specific gravity of the petroleum with reference to water, both at 60°F. On this scale, water has a value of 10°API, and as the actual specific gravity of the petroleum *decreases,* the degrees API *increase.*

The API scale has advantages for its particular applications. For one thing it provides finer gradations between whole number units. Consider a change of 1°API (from 20° to 21°), for example. The equivalent change in specific gravity will be from 0.9340 to 0.9279. Clearly, the API scale is much easier to deal with. In addition to this advantage, the scale also lends itself to schemes for correcting to a temperature standard of 60°F.

Baumé scale. Another scale that has been in use for almost two centuries is the Baumé scale. This scale is widely used to measure acids and certain other liquids such as syrup. One form of the Baumé scale used for measuring light liquids is expressed as

$$^{\circ}\text{Baumé} = \frac{140}{G} - 130.$$

The other form of Baumé scale is used for measuring heavy liquids, that is, those heavier than water. It is expressed as

$$^{\circ}\text{Baumé} = 145 - \frac{145}{G}.$$

In both Baumé scales, specific gravity (G) is usually assumed to have been measured at 60°F.

Other scales. Numerous other scales are used for converting specific gravity to special needs. These may be found in handbooks or texts devoted to the particular subject. Almost invariably a simple equation is used, as in the preceding examples, to provide a system of units that is uniquely suited to the needs of the application.

Measuring Devices

The most common device for measuring the specific gravity of liquids is the hydrometer. Also, some of the devices used in connection with liquid-level measurement can be readily adapted to specific gravity measurement. The air-bubbler system, displacer-float devices, and methods using hydrostatic head are examples of possible adaptations. Each of these can be made to operate an indicator-recorder, so that a record of specific gravity can be maintained if desired.

Hydrometer. A hydrometer is an instrument that contains a large bulblike section on one end and a thin, straight stem on the other (fig. 10.9). The stem is graduated according to the particular scale adopted for its use. A hydrometer works on the principle of balancing the total weight of the hydrometer against the weight of the liquid it displaces when upright and floating in the liquid. The bulb of the hydrometer displaces a large portion of the liquid, while the thinness of the stem provides a considerable spread of the scale for accurate determination of specific gravity. The weight and other design considerations of a hydrometer depend upon the range of values to be measured. An instrument to be used for measuring storage batteries is apt to have a scale range of 1.050 to 1.310, since this somewhat more than encompasses the significant range between a very dead battery and one that is fully charged. A hydrometer scale for degrees API might contain any portion of the range from 0°API to 100°API. The two hydrometers shown in figure 10.9 range from 29°API to 41°API.

Some hydrometers contain a thermometer so that temperature of the liquid can be obtained along with specific gravity. The range of the thermometer will be appropriate for the purpose of the hydrometer; for example, the hydrometer shown in figure 10.9*B* has

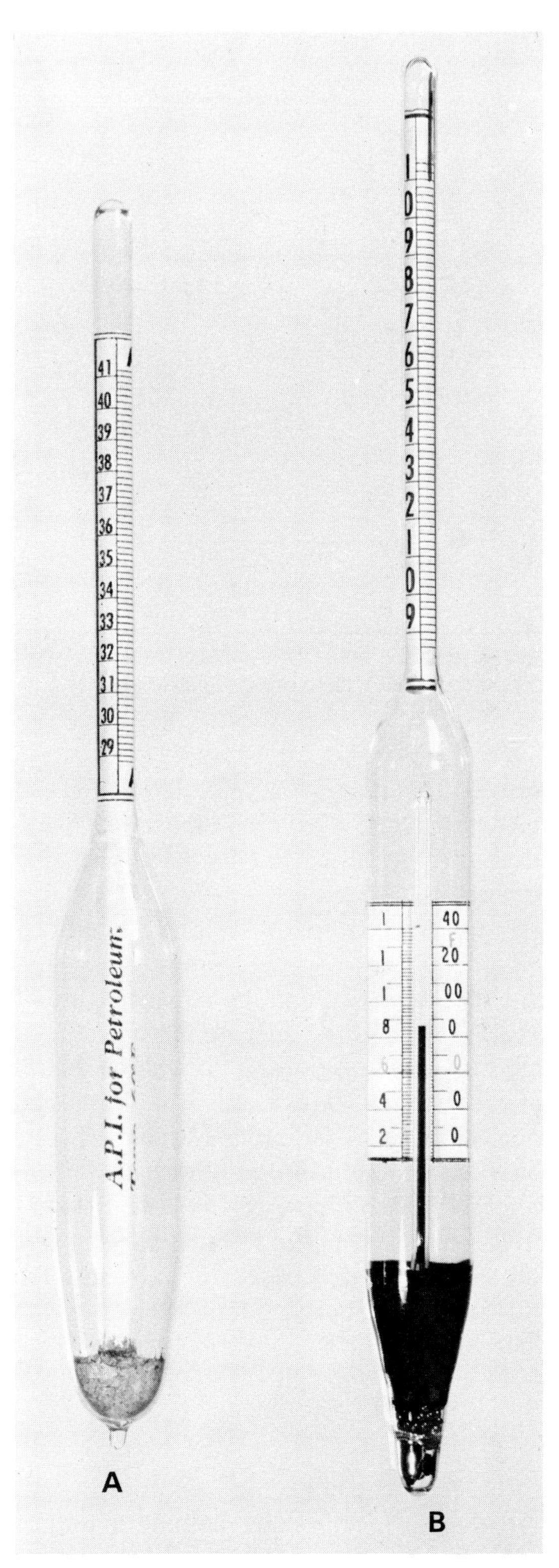

Figure 10.9. Two forms of hydrometers used to measure specific gravity (*Courtesy Taylor Instrument*)

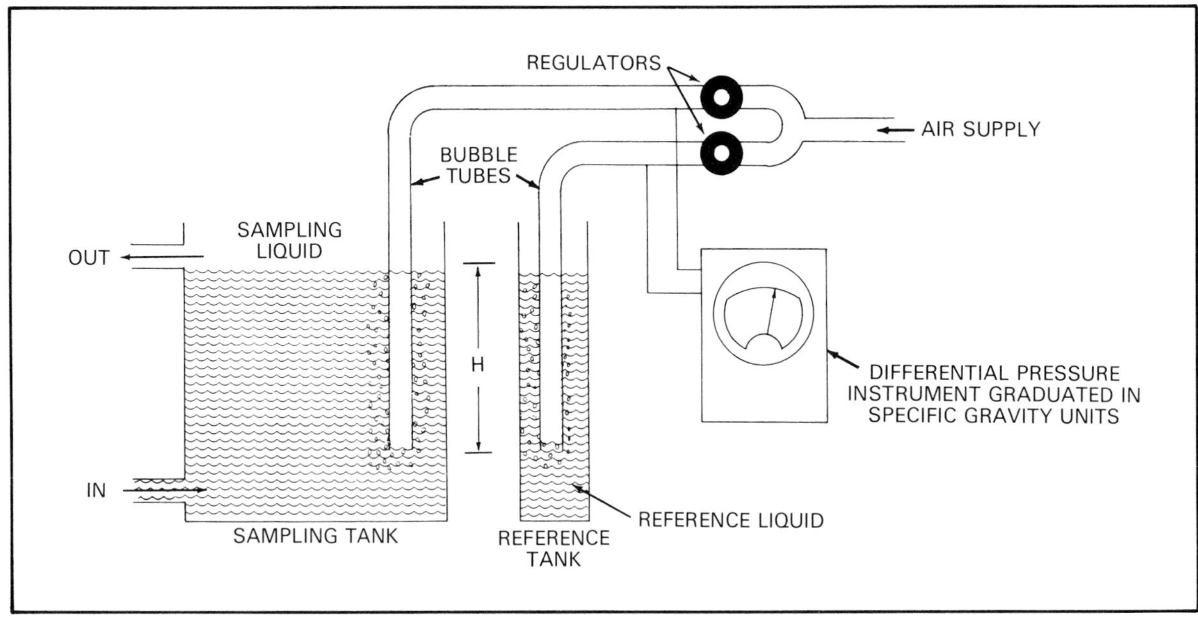

Figure 10.10. An air-bubbler system with reference tank for measuring specific gravity

a temperature range of 20°F to 140°F. Special instructions are usually associated with measuring specific gravity and making temperature corrections.

Air-bubbler system. In an air-bubbler system applied to specific gravity measurement, two tanks containing equal heights of liquid above a common baseline are needed (fig. 10.10). The reference liquid might be water or another liquid suited to the needs. Air is supplied to the bubble tubes at a rate just great enough to produce a steady stream of bubbles. Two inputs connect the differential pressure instrument to the bubble tube lines. The pressure needed to overcome the hydrostatic head in either tank is given by the equation

$$p = 0.433 \times G \times H$$

where

p = pressure, in psi;
G = specific gravity of the tank liquid;
H = height, in feet of liquid surface above lower end of bubble tube.

To understand how this equation can help determine specific gravity, assume the following:

1. A liquid of unknown specific gravity is in the sampling tank.
2. Water is in the reference tank and has a specific gravity of 1.
3. A differential pressure of 0.5 psi exists, and the direction of the indication is such that the liquid appears to have a specific gravity greater than unity.

From this information:

$$p = 0.433 \times 1 \times H$$

where

p = pressure in the reference tank, in psi;
H = height, in feet.

The differential pressure indicator shows that pressure in the sampling tank is 0.5 psi greater than that shown by the preceding equation for the reference tank. With this knowledge, the following equation can be set up:

$$\begin{aligned} 0.433 \times G \times H &= 0.433 \times 1 \times H + 0.5 \\ &= 0.433 \times H + 0.5. \end{aligned}$$

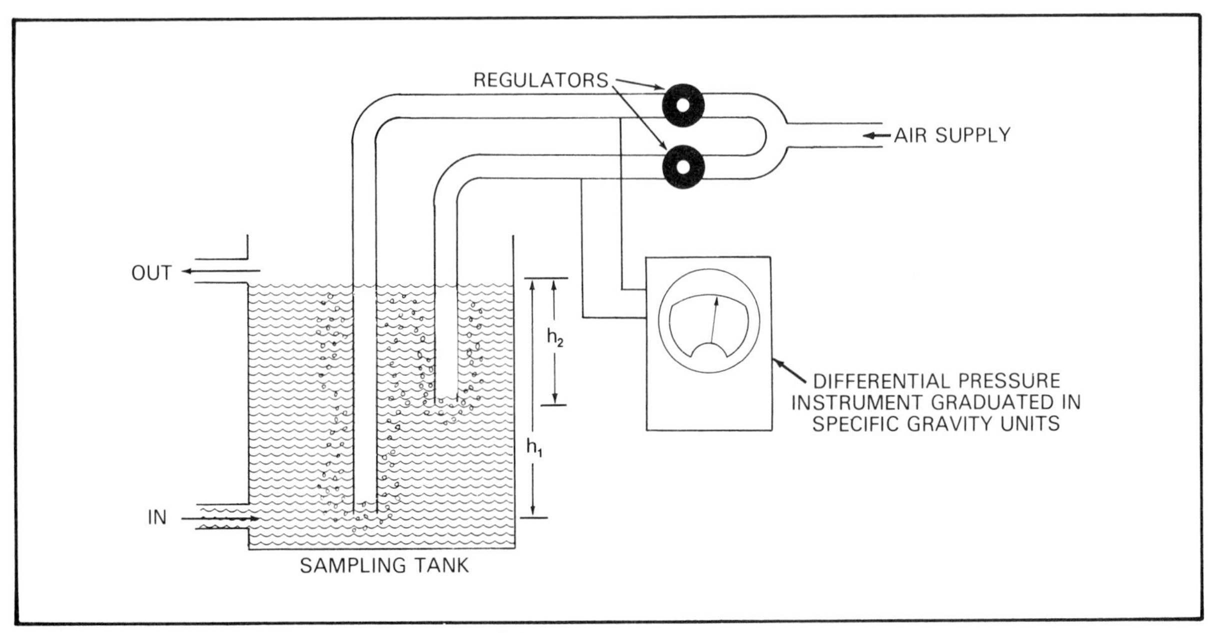

Figure 10.11. Air-bubbler system without reference tank

Dividing through both sides of the equation by $0.433 \times H$,

$$\frac{0.433 \times H}{0.433 \times H} + \frac{0.5}{0.433 \times H} = 1 + \frac{0.5}{0.433 \times H}.$$

Assuming the numerical value for H is 10 feet, the specific gravity of the liquid in the sampling tank would be

$$1 + 0.5/0.433 \times 10 = 1.115.$$

Another bubbler system needs no reference tank (fig. 10.11). Here, again, it is important that a fixed level be maintained in the sampling tank. The principle of operation is based on the fact that differential pressure existing between the two bubble tubes will depend not only on the vertical separation of their open ends, but also on the specific gravity of the liquid. In order to see this clearly, it will help to explore an actual example.

Suppose the bubbler tube ends are separated by 2 feet of water, one being 2 feet from the surface, the other 4 feet. The pressure values required to overcome the heads will be $0.433\ (1) \times 2$ and $0.433\ (1) \times 4$, or 0.866 and 1.732 psig, respectively. Evidently, the differential pressure instrument will indicate $1.732 - 0.866 = 0.866$ psig for this condition. Now, suppose salt or some other compound is dissolved in the water until the specific gravity becomes 1.25. The pressures now needed to overcome the hydrostatic head are $0.433 \times 1.25 \times 2$ and $0.433 \times 1.25 \times 4$, or 1.082 and 2.165 psig respectively, and the differential pressure indication will now be $2.165 - 1.082 = 1.083$ psig. Thus, a change in specific gravity from 1 to 1.25 has produced a change in differential pressure of 0.217 psig. It is interesting to note that the greater the separation of the bubble tube ends in the liquid, the greater will be the differential pressure indication for a given change in specific gravity of the liquid.

Displacer-float devices. In the study of liquid-level measurement and control, the fact was mentioned that displacer-float elements could be used to measure and control specific gravity. Displacer floats, it will be recalled, are buoyant elements that are heavier than the maximum amount of liquid they can displace.

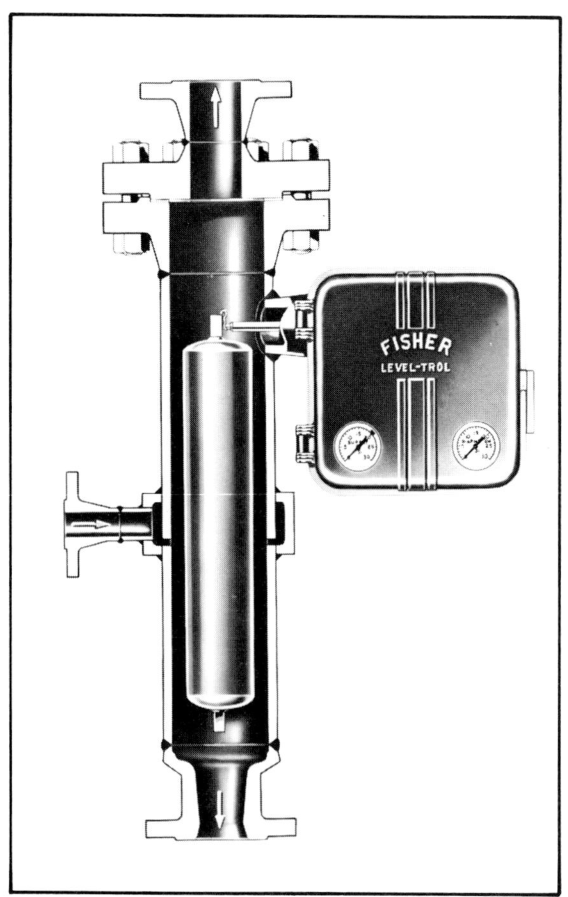

Figure 10.12. A displacer-float specific gravity meter (*Courtesy Fisher Controls*)

In the displacer-float element and cage shown in figure 10.12, the liquid is flowing into the cage at its middle and out the ends. The design arrangement for permitting liquid entry at the middle is called a *piezometer ring,* and its purpose in this case is to reduce or to eliminate entirely the adverse effect rapidly flowing liquid would have on the response of the ordinary instrument of this type.

The specific gravity measuring and control system in figure 10.13 uses the displacer float in a piezometer ring cage. A quick review of the action taking place in this type of controller shows that the displacer float exerts a downward force on the float rod. The rod twists the torque tube in proportion to the *apparent* weight of the float, and this action causes the free end of the rotary shaft to position a flapper in the controller or transmitter. The *apparent* weight of the float is its true weight minus the buoyant force exerted on it by the liquid in the cage.

When used for measuring or controlling specific gravity, displacer floats are entirely submerged in the liquid. The downward force exerted by the element then becomes a function of the specific gravity of the liquid. As an example, consider a displacer float of 100 cubic inches and 4.75 pounds true weight. When submerged in water, the float, of course, displaces 100 cubic inches of water, and since water weighs 0.036 pounds per cubic inch, the total amount displaced weighs 3.6 pounds. The weight of the displaced water is also the buoyant, or upward, force exerted on the float. The *apparent* weight of the float when submerged is then 1.15 pounds $(4.75 - 3.6)$. Now this is the effective force exerted on the float rod, and some definite relation between nozzle and flapper will be established for this particular value of apparent weight.

What happens if the water is replaced by a liquid having a specific gravity of, say, 0.600? If water weighs 0.036 pounds per cubic inch, this new liquid weighs $0.6 \times 0.036 = 0.0216$ pounds per cubic inch, so the new value of downward force, or apparent weight, of the float element will be $4.75 - (100 \times 0.0216)$, or 2.59 pounds, a very considerable change indeed! As the specific gravity exceeds 1.00, an absolute upper limit for this choice of displacer float is reached when the buoyant force of the liquid equals the true weight of the float, that is, when

$$0.036 \times G \times 100 = 4.75,$$

where G is the specific gravity of the liquid and is equal to about 1.32. The practical limit for this arrangement should not exceed about 1.2, but the use of floats having a greater effective density will provide a practical way of measuring any reasonable value or range of specific gravity.

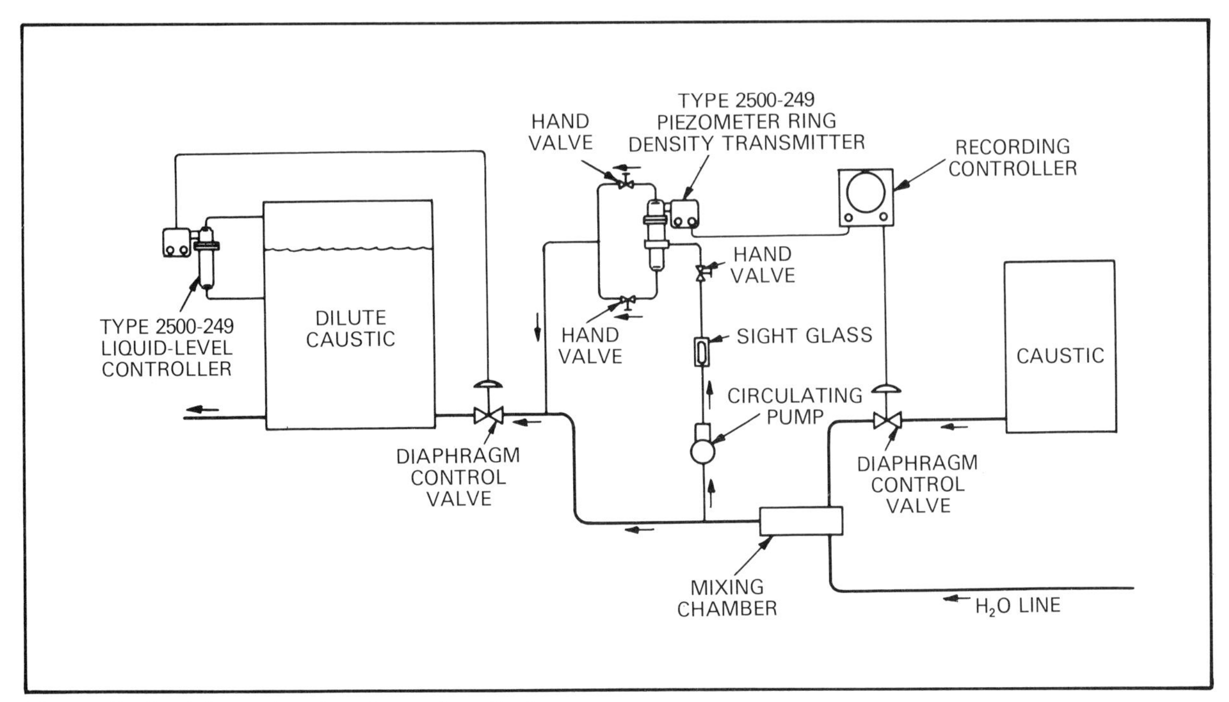

Figure 10.13. Schematic of system using displacer float (*Courtesy Fisher Controls*)

In practice the choice of displacer float will depend on several considerations, among which are (1) the desired proportional band setting; (2) the range of specific gravity values to be measured; (3) the nature of the liquid, that is, whether it is corrosive; and (4) the effects of temperature change and whether it must be compensated.

The control system shown in figure 10.13 is a straightforward example of regulating the strength of a caustic solution by means of specific gravity measurement and control, but two or three items need some comment. Note the use of manual-control flow valves in lines leading from the displacer-float cage. They are adjusted so that a balanced flow comes from top and bottom of the cage, thus assuring a completely filled cage at all times. The sight glass, or rotameter, and the hand valve above it are used to regulate the rate of flow through the float cage. The recording controller supplements the density transmitter, providing continuous indication and a permanent record of specific gravity.

Viscosity

Viscosity is a property that everyone recognizes as belonging to syrup, heavy oils, mucilage, catsup, and all other liquids that seem reluctant to flow or to pour as easily as water. Ordinarily water, gasoline, and air are not thought of as viscous fluids, although these fluids certainly do possess viscosity as a native property.

Dimensions

In the study of Reynolds' number the dimensions of viscosity were noted, and it might seem odd that dimensions of force, time, and area bear any relation to the usual concept of viscosity. The usual concept is one that recognizes viscosity as being a property of fluids that causes them to resist a tendency to flow. It is time to investigate the matters of force, time, and area as they relate to viscosity. What kind of force, for example, and how does area become a reckoning factor?

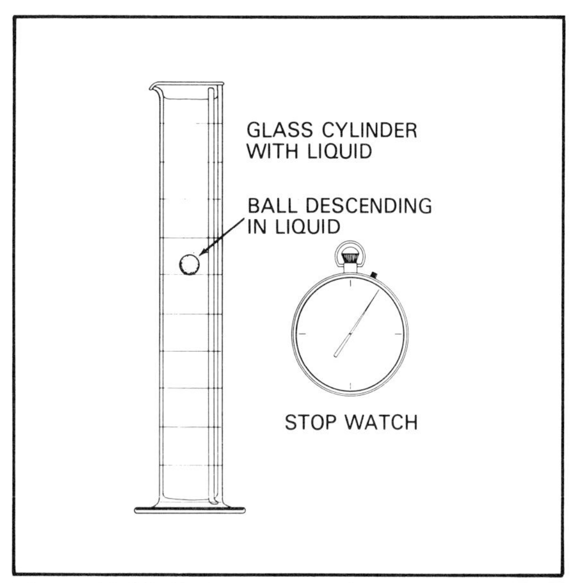

Figure 10.14. Laboratory equipment used over a century ago for making experiments to determine viscosity of liquids

Well over a century ago a man named Stokes carried out experiments to determine the viscosity of various liquids, using simple laboratory equipment (fig. 10.14). This equipment represents a fine way to explain viscosity and the means of deriving the factors of force, time, and area. The long, glass cylinder, its length graduated in centimetres, is filled with the liquid whose viscosity is to be determined. A polished ball, made of material having a density somewhat greater than the liquid, is allowed to fall through the column of liquid. It is general knowledge that if the ball should be permitted to fall freely in a vacuum, its velocity would increase indefinitely according to the equation:

$$v = g \times t$$

where

v = velocity;
g = acceleration due to gravity (about 32 feet per second per second);
t = elapsed time after the free fall began from a position of rest, or zero velocity.

On the other hand, an object falling through air, other gas, or liquid will eventually attain a *terminal velocity;* that is, the velocity will become a constant. This condition occurs when the *weight* of the object is exactly balanced by the frictional forces of the air or other fluid acting against the descent of the object. It might be helpful to remember that weight is the product of mass and acceleration due to gravity and is therefore equivalent to force. Thus, at terminal velocity the *net* force acting on the object is zero.

Stokes carefully chose the length of the liquid column and the density of the ball so that he could accurately observe the time it took the ball to fall a given distance after it had attained its terminal velocity. He found the velocity to be inversely proportional to the absolute viscosity of the liquid and arrived at the following formula, or equation, for determining absolute viscosity:

$$\mu = 2r^2(\rho - \rho')g/9v$$

where

μ = viscosity, in poises;
r = radius of ball, in centimetres;
ρ = density of ball, in grams per cubic centimetre;
ρ' = density of liquid, in grams per cubic centimetre;
g = acceleration due to gravity, in centimetres per second squared;
v = velocity, in centimetres per second.

Now putting these factors in dimensional form:

$$\text{poises} = \frac{2 \times \text{cm}^2 \times \text{g/cm}^3 \times \text{cm/sec}^2}{9 \times \text{cm/sec}}.$$

Care must be taken here to avoid indiscriminate canceling. By rearranging the factors,

$$\text{poises} = \frac{2 \times (\text{cm}^2/\text{cm}^3) \times (\text{g} \times \text{cm/sec}^2)}{9 \times \text{cm/sec}}.$$

Note that $(\text{g} \times \text{cm/sec}^2)$ = force. Saving this quantity and canceling where possible elsewhere,

$$\text{poises} = (2 \times F \times \text{sec})/(9 \times \text{cm}^2)$$

where F is force.

It should be clear that factors of force, time, and area as developed in Stokes' equation came from the gravitational and frictional forces acting on the ball, the time-related factors of acceleration and velocity, and the linear distances represented by the radius of the ball and the terminal velocity. It should be understood, however, that forces and velocities due to causes other than gravity are equally applicable, for example, the force or pressure (force per unit area) needed to push fluids through a horizontal pipe.

Units and Scales

Up to this time only *absolute viscosity,* or *dynamic viscosity* as it is sometimes called, has been discussed. Temperature and pressure have special scales in addition to those based on absolute values, and the same is true for viscosity measurements. In fact, probably no other variable has as many variations in scales or means of expression as viscosity. To understand viscosity better, consider a few definitions.

Kinematic viscosity. The word *kinematics* is a physics term denoting the study of motion without regard to forces or mass, and kinematic viscosity represents a significant departure from the dimensions stated for absolute viscosity. It is expressed as the ratio of the absolute viscosity to the density of the fluid, and therefore has dimensions of length squared divided by time (L^2/t). In the centimetre-gram-second system, a unit of kinematic viscosity is the *stoke,* named for Stokes. A centistoke (0.01 stoke) is also commonly used.

Specific viscosity. Specific viscosity is the ratio of absolute viscosity of a substance to that of a standard fluid, like water, with the viscosity of both fluids being measured at the same temperature.

Relative viscosity. Relative viscosity is similar to specific viscosity, but water at 68°F is used as the standard fluid. Since the absolute viscosity of water at this temperature is approximately 1 centipoise, the relative viscosity of a fluid is practically equal to its absolute viscosity.

Viscosity index. The term *viscosity index* is frequently applied to petroleum-base lubricants and represents a number that indicates the effect a change in temperature will have on the viscosity of the oil. Oils that undergo minimum change in viscosity for a given change in temperature are said to have a high viscosity index.

Fluidity. The reciprocal of absolute viscosity, fluidity is a measure of the ease with which fluids flow. The unit of fluidity in the centimetre-gram-second system is called the *rhe,* and it equals 1/poise.

Saybolt Seconds Universal. The Saybolt Universal is an arbitrary system for measuring viscosity that uses Saybolt seconds for units of measurement. Liquid at a constant temperature and pressure is permitted to flow through a small orifice that forms part of an instrument called a *viscosimeter,* and the number of seconds required for a given volume of the liquid to pass through the orifice is used as the measure of viscosity. Another Saybolt system, called *Saybolt Furol,* is similar to the Saybolt Universal system, but a larger orifice is used.

Redwood, Redwood Admirality, and Engler seconds. Identical with or very similar to the Saybolt system are the Redwood, Redwood Admiralty, and Engler seconds. These are merely the British or European counterparts of Saybolt seconds.

Engler degrees. The Engler degree system expresses viscosity for a given fluid under specific conditions as a ratio of the time taken for a given volume of the fluid to pass through an Engler viscosimeter to the time taken for the same volume of water at 68°F to pass through the same viscosimeter.

Measuring Devices

Viscosimeter is a name that can logically be applied to any instrument used primarily to measure the viscosity of liquids. Stokes' apparatus forms one kind of viscosimeter, and many others quite similar to it exist. One type substitutes a piston for the ball, while another type requires the timing of the rise of a bubble of air through the liquid.

The methods used to measure viscosity on the basis of noting the time required for a specified quantity of liquid to pass through a thin channel, or an orifice, are very popular due to relative simplicity, low cost, and accuracy.

Saybolt viscosimeter. A commercial form of Saybolt viscosimeter (fig. 10.15*A*) is simple in principle but includes features that provide versatility and accuracy. This instrument has a temperature-controlled bath that maintains close limits on temperature. As many as four sample tubes can be placed in the machine and four 60-millilitre flasks positioned to receive flow from them (fig. 10.15*B*). Appropriate orifice fittings, either Universal or Furol size, are screwed into the lower end of the sample tubes. Operation of the instrument involves using a stopwatch to determine the time for 60 millilitres of liquid to flow into the receiving flask.

As noted earlier, Saybolt viscosity is expressed in seconds of time, either as Saybolt Seconds Universal (SSU) or as Saybolt Seconds Furol (SSF). Such expressions can be converted with reasonable accuracy to kinematic viscosity in centistokes (μ_k) by the following relation:

$$\mu_k = \text{SSU}/4.635 = \text{SSF}/0.470.$$

The absolute viscosity in centipoises (μ_a) can be derived from the relation:

$$\mu_a = \mu_k \times \rho$$

where ρ = density, in grams per cubic centimetre.

Direct-reading instruments. Many instruments for field and laboratory use are capable of a direct presentation of viscosity values; that is, no stopwatch or other timer is required. Some of these devices are electrical

Figure 10.15. A commercial Saybolt viscosimeter (*Courtesy Precision Scientific*)

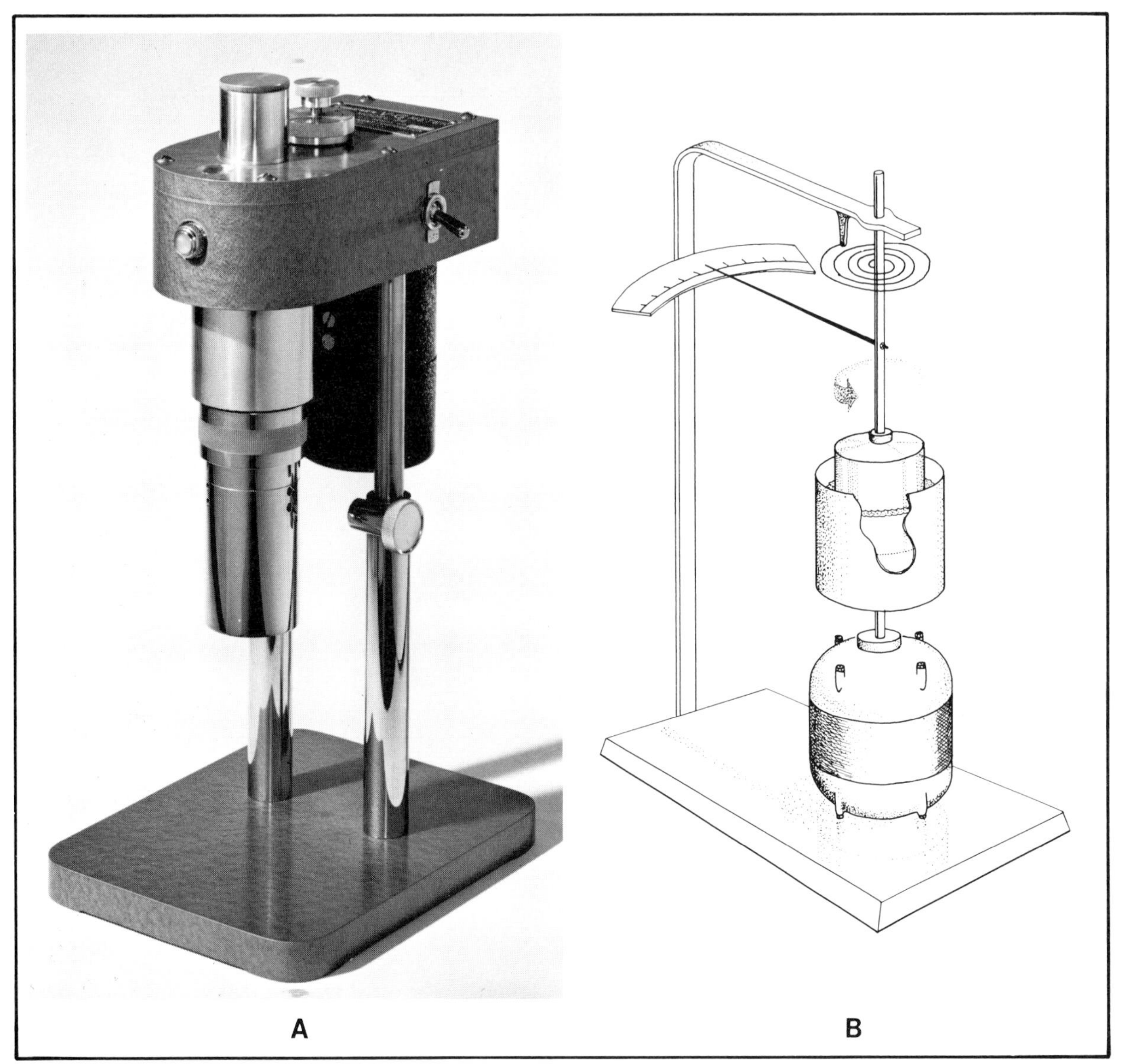

Figure 10.16. Direct-reading instruments for measuring viscosity. ***A,*** **one type of direct-reading instrument;** ***B,*** **a simplified schematic of the principle involved.**

motor driven while others may have hand cranks, but almost all invariably use the idea of relative motion between concentric cylinders separated by annular volumes of the liquid being measured. In an instrument of this type (fig. 10.16) liquid fills a large part of the volume between two cylinders. As the lower cylinder rotates, the liquid it contains is also set in motion and will exert a torque on the upper cylinder, which is restrained by a calibrated spring. The torque will be proportional to the speed of rotation and the viscosity of the liquid.

Controlling and Recording Systems

In some instances a continuous check of viscosity is required, or a permanent record

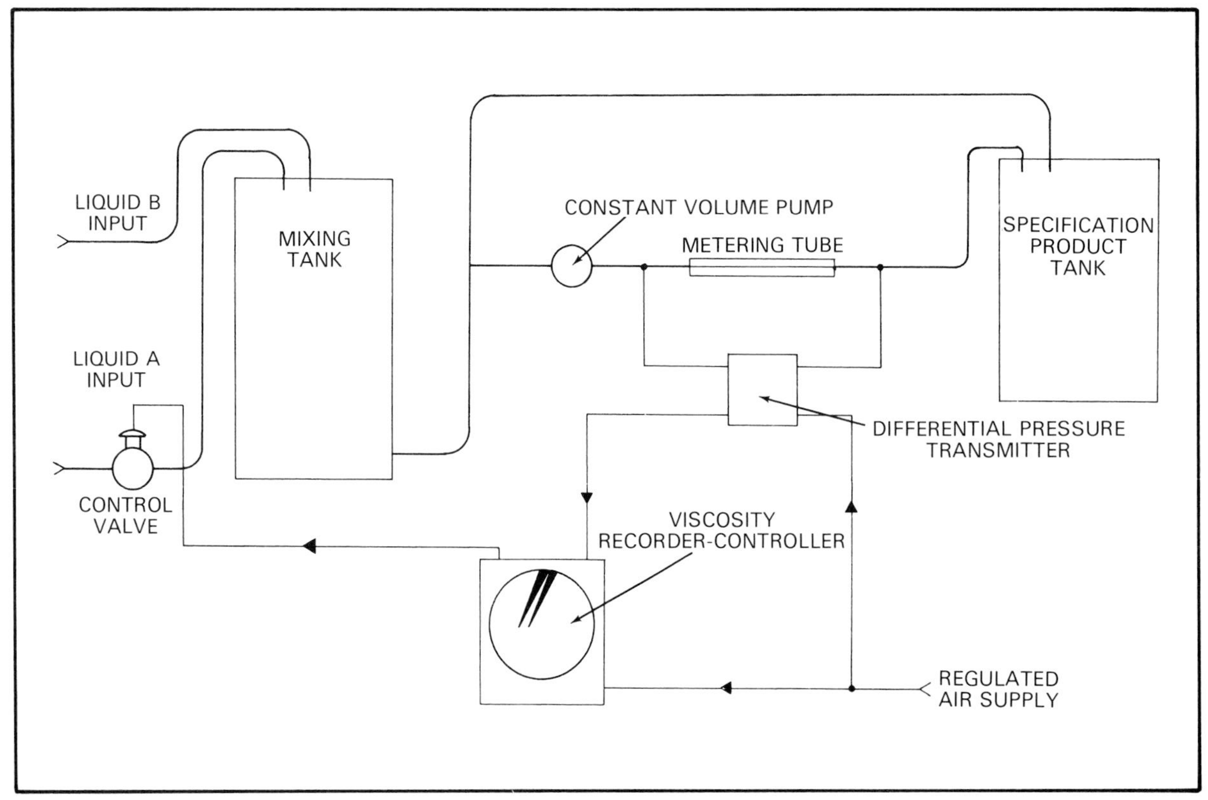

Figure 10.17. Schematic of a system for recording and controlling viscosity

and outright control are needed. One system, shown schematically in figure 10.17, is based on the principle that the differential pressure developed across a thin-channel metering tube is proportional to the viscosity of the fluid passing through it. A controlled-volume pump maintains a uniform flow rate through the metering tube, and a differential pressure cell transmitter detects changes in pressure and relays this information to the recorder-controller in the form of pneumatic signals. The recorder-controller regulates the viscosity by controlling the amount of one of the ingredients entering the mixing tank.

The possibility of using differential pressure across a metering tube to measure viscosity must have been recognized more than a century ago. During the first half of the nineteenth century, Poiseuille, a French scientist, derived an equation that relates flow rate through capillary-like tubes with viscosity. One of the factors in the equation is differential pressure. In its early and most basic form, Poiseuille's equation has the form

$$\mu = \pi \Delta p r^4 / 8Ql$$

where

μ = absolute viscosity, in poises;
Δp = differential pressure, in grams per square centimetre;
r = the inside radius, in centimetres;
l = length of the round tube, in centimetres;
Q = rate of flow through the tube, in cubic centimetres per second.

It can be seen at once that when using a constant volume pump, this equation has only two variables, Q and Δp, and they are directly proportional to one another. It is of interest to note that Saybolt and other similar forms of measurement are based on Poiseuille's equation.

pH Factor

Up to this time, the chemical composition of the fluids or other material that was heated, pressurized, flowing, or in some way connected with a variable in a controlled system was not considered when measuring or controlling. Of course, the corrosive characteristics of certain agents were taken into account, but usually only because of a need to protect equipment against the damage these characeristics, which are really chemical in nature, might cause. For the first time, the chemical nature of substances as a variable will be reckoned with, even if only to the limited extent of determining whether they are acid or base (alkaline) and to what degree. This degree of acidity or alkalinity of a substance is its pH factor.

It is general knowledge that an *acid* will neutralize a *base,* and vice versa. This neutralizing effect explains why a strong solution of baking soda and water is used as a first-aid measure in treating a person who has spilled a harmful acid substance on his skin. The baking soda reacts as a *base* substance. The neutralization caused by mixing acids and bases results in the formation of a salt. A solution of hydrochloric acid (HCl) added to one of sodium hydroxide (NaOH) results in the formation of common table salt (NaCl).

When acid is poured into water, it will break down, or dissociate, and produce *hydrogen ions* (H^+). A base-in-water solution produces *hydroxyl ions* (OH^-). Any solution will have a certain concentration of hydrogen ions and hydroxyl ions, regardless of the strength of the acid or base solution. For very strong base solutions, the hydrogen ion concentration will be quite small, just as the concentration of hydroxyl ions in a very strong acid solution. In pure water the concentrations of hydrogen and hydroxyl ions are equal, so water can be considered as being neither acid nor base. The degree to which hydrogen ions are concentrated in a solution is a measure of the strength of that solution as an acid, and as one might suppose, the greater the concentration of hydrogen ions, the smaller the concentration of hydroxyl ions. Since water is neither base nor acid, it is logical to label anything with a hydrogen ion concentration greater than water as *acid,* and anything with a hydrogen ion concentration smaller than water as *base.* It is an interesting fact that the *product of the hydrogen ion and hydroxyl ion concentrations is a constant for any solution.* Thus, calculating the concentration of one ion is a simple matter if that of the other is known.

The pH Scale

Pure water will ionize only slightly; that is, only a very small percentage of its molecules will dissociate into H^+ and OH^- ions at any one time. In fact, for usual conditions only about 10^{-7} *gram-ions* of H^+ ions exist in a litre of water. *Mole* is one of several characteristic chemical terms related to Avogadro's number (6.023×10^{23}). A mole is simply 6.023×10^{23} ions. Since the dissociation of one molecule of water into ions produces one hydrogen and one hydroxyl ion, it is clear that the concentration of hydroxyl ions is also 10^{-7} mole per litre of water.

The pH factor is derived from a simple expression that enables one to deal with acid-base relations without the need for negative powers of ten:

$$pH = -\log [H^+]$$

where $[H^+]$ designates the mole concentration of $[H^+]$ ions per litre of water or other solution, and log is logarithm to base ten. Since the hydrogen ion concentration of water is 10^{-7}, the pH factor can be calculated as follows:

$$\begin{aligned} pH = -\log [H^+] &= -\log 10^{-7} \\ &= -(-7) = 7. \end{aligned}$$

Sometimes the factor pOH is shown. It is calculated in a manner similar to that shown above, except that hydroxyl ion concentration is used in place of hydrogen ion concentration.

TABLE 10.1
RELATION OF pH FACTOR TO ION CONCENTRATION

Actual H^+ Ion Concentration (Gram-Ions/Litre)	10^0	10^{-1}	10^{-2}	10^{-3}	10^{-4}	10^{-5}	10^{-6}	10^{-7}	10^{-8}	10^{-9}	10^{-10}	10^{-11}	10^{-12}	10^{-13}	10^{-14}
pH Factor	0	1	2	3	4	5	6	7	8	9	10	11	12	13	14

← Increasing Acidity — Neutral — Increasing Basicity →

The relation between pH factor and the actual concentration of hydrogen ions is shown in table 10.1. It is important to recognize that a small value of pH factor actually reflects a relatively large hydrogen ion concentration. This fact arises from the use of the logarithm of the reciprocal of the concentration. As mentioned earlier, the product of hydrogen ion and hydroxyl ion concentration is a constant. A similar relation exists when working with pH and pOH factors; the *sum* of pH and pOH factors is always 14.

It might be interesting to look at the pH factors of several common substances. Lemon juice, as one would naturally suspect, is quite acid and has a pH factor of 2.0 to 2.2; other citrus fruits range from 3.0 to 4.5. Vegetables and melons have pH factors ranging from 5.0 to 7.0, while fresh milk is slightly acid at 6.50 to 6.65.

pH Factor Indicators

Those who have had a laboratory course in elementary chemistry will recall that *litmus* paper was used extensively for determining whether a solution was acid or base in character. Red litmus will change to blue when dipped in a base solution, and the color will change back to red if it is then dipped in an acid solution.

Litmus paper is one of several acid-base indicators used extensively in laboratory work and for rough estimates of pH factor in practical field applications. Other forms of indicating pH utilize special liquid compounds, which are added to a solution and the resulting color change is noted.

Litmus paper and the special liquid compounds operate on the principle that they contain weak acids and bases and their salts. When the salt of a weak acid is different in color from the nonionized acid, the color resulting from mixing them will depend on the degree of concentration of the acid and salt forms. A score or more of special liquids are available to cover the pH range from 0 to 14. These liquids are used with a set of standard colors graduated in steps of 0.2 pH unit. Careful use by an experienced observer will produce results accurate to within 0.1 pH unit. Paper-type indicators (fig. 10.18) are available in sets, with each dispenser covering a particular range of pH values and a color scale for comparison. These indicators are convenient, inexpensive, and accurate enough for many noncritical applications. However, these indicators are useless or capable of only extremely rough approximations when used in strongly colored solutions or those containing powerful oxidizing or reducing agents.

Electrical Method of Measurement

Measuring pH factor by electrical means is possible because a solution can be made to act as the electrolyte in a voltage cell, and the cell voltage will depend on the hydrogen ion concentration. One of the most important

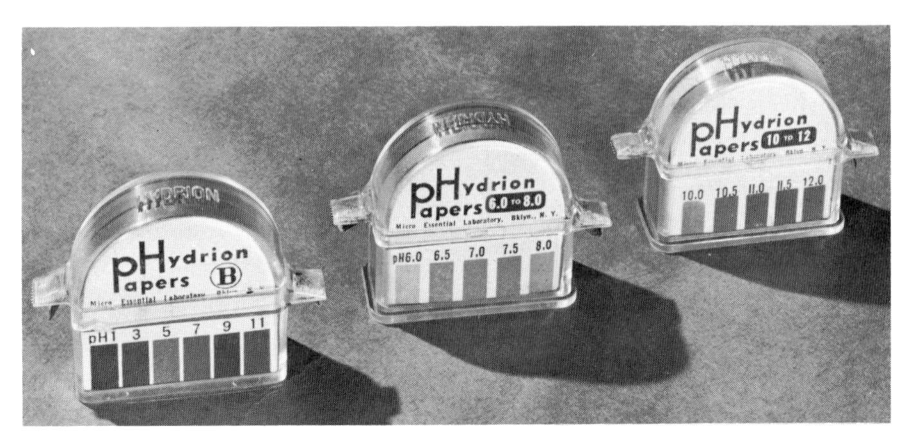

Figure 10.18. Paper pH factor indicators (*Courtesy Micro Essential Laboratory*)

methods for measuring pH factor by electric means employs a *glass electrode* as the sensing element (fig. 10.19). The chemical action responsible for the operation of this means of pH measurement is beyond the scope of this text.

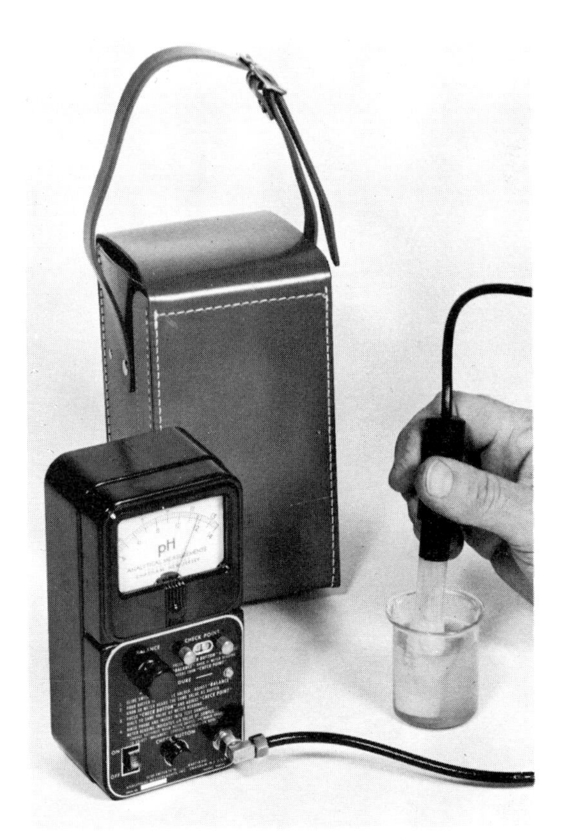

Figure 10.19. A portable electric instrument that uses a glass electrode to measure pH factor (*Courtesy Analytical Measurements*)

Conclusion

Among the quantities that must be dealt with as possible variables in instrumentation are humidity, specific gravity, viscosity, and pH factor. Many other variables have to be considered in measuring and controlling, but with the knowledge gained about these more common variable quantities, it should be easier to take the short step toward the measurement of such things as linear and angular velocity, light intensity, and numerous measurements related to electrical properties. It should also be a simple matter to bridge the gap to the field of analysis, where the chemical and physical properties of matter become the subject of primary interest.

CHAPTER 11

Transducers, Transmitters, and Converters

According to a very broad definition, a transducer is a device that aids in the process of responding to signals from one medium and supplying related signals to another medium. Microphones and loudspeakers are transducers. A microphone is capable of responding to sound waves in the air and converting them to electrical waves or impulses. A loudspeaker (sometimes called a reproducer) accepts the amplified signals from a microphone or other source and converts them to sound waves.

For purposes of instrumentation, a transmitter is an element in a control system that accepts information at some point in the system and conditions it for transmission to a control center, remote terminal unit, or other appropriate location. A transmitter usually conditions a signal to a standard form—4 to 20 milliamperes, 1 to 5 volts, and 3 to 15 psi are examples.

Converters typically change the form of input signals to suit special needs. A millivolt converter unit may accept an input from a thermocouple and produce a proportional output in terms of 4 to 20 milliamperes or 1 to 5 volts.

It is sometimes difficult to distinguish the differences that exist among transducers, transmitters, and converters. Thus, this chapter will cover the principles that account for the operation of a limited variety of transducers, transmitters, and converters without regard for the similarities that tend to obscure one unit from another.

Transducers

Two types of transducers are the current-to-pressure transducer (designated as I/P) and the pressure-to-current transducer (P/I). A current-to-pressure transducer accommodates a current input and produces a pneumatic output; hence, it is often called an electro-pneumatic or current-to-pneumatic transducer. Conversely, the P/I unit is often referred to as a pneumatic-to-current transducer.

Current-to-Pressure Transducer

A common form of current-to-pressure (I/P) transducer accepts current values ranging from 4 to 20 milliamperes (the most prevalent electrical transmission format) and provides for a proportional output in the 3 to 15 psi range.

The desirability of converting electric signals to proportional values of pneumatic pressure stems from a number of valid considerations:

1. Modern techniques for measuring variables frequently employ primary elements that produce an electrical signal; thermocouples and resistance elements for measurement of temperature are two of many examples.
2. Electrical systems are superior to pneumatic systems for the distant transmission of measurement and control signals.
3. Pneumatic systems are reliable and accurate in the matter of positioning final control elements.

Although measuring and transmitting values of temperature, pressure, and other variables are usually more feasible by electrical means, using pneumatic equipment to position throttling valves is often advantageous.

The *force balance principle* is generally used to go from an electric current or voltage format to a proportional pneumatic range of 3 to 15 psi. This principle involves balancing the force in a magnetic circuit with the force produced by a bellows unit under pneumatic pressure.

One electro-pneumatic transducer (fig. 11.1) can be adapted to voltage or current input. Only the windings in the input circuit coils differ in the two forms of the unit. This discussion will center on the 4- to 20-milliampere input to 3 to 15 psi output unit. Important components of the transducer include the coil assembly, permanent magnet with appropriate pole pieces, armature, feedback bellows, a flapper-nozzle arrangement, zero adjustment spring, span adjustment bar, and air relay. The air relay, discussed in chapter four, adjusts the air output signal according to the value of a pressure signal generated by the flapper-nozzle arrangement.

Figure 11.1 shows an armature supported by a torsion bar. The armature is itself a flat bar of magnetic material with rectangular cross section and is positioned between pole pieces of a permanent magnet system. When

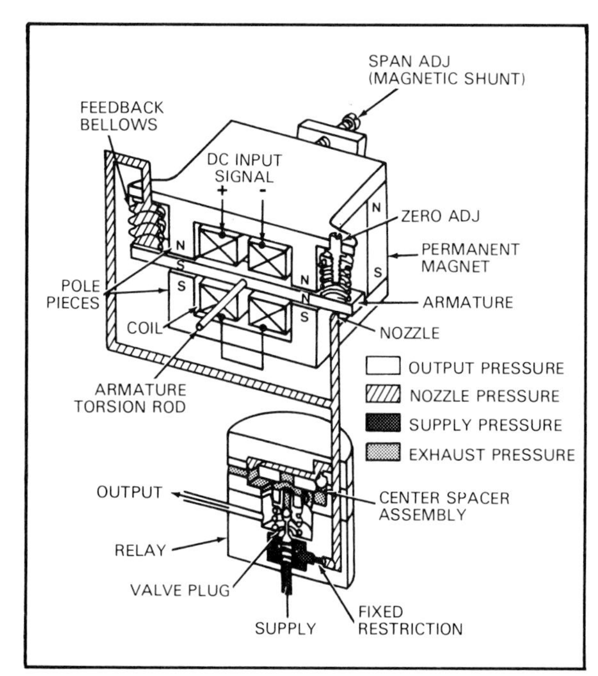

Figure 11.1. Simplified schematic diagram of a current-to-pneumatic transducer (*Courtesy Fisher Controls*)

current in the coils is 12 milliamps, that is, midway in the 4- to 20-milliamp range, the armature should be midway between the magnet pole pieces. For this position the nozzle pressure should be 9 psi, and so should the output pressure of the air relay. This value marks the midway point in the 3 to 15 psi range of pressure. When the current flow decreases, the armature rotates counterclockwise. With an increase in current, the armature moves clockwise.

The armature serves as the baffle in a nozzle-baffle arrangement. With clockwise movement of the armature, the nozzle pressure increases, and this pressure is applied to the feedback bellows. This action introduces a force opposite to the magnetic attraction caused by the increased current flow. A zero adjust spring also exerts a force on the armature. This force is opposite to that produced by the bellows. With proper adjustment the transducer will produce an output pressure that accurately tracks the 4- to 20-milliamp input signals.

As shown in the diagram (fig. 11.1), the system produces a direct-acting output—increased current flow means increased pneumatic pressure output. Reversing the flow of current in the coils produces a reverse-acting output. Transducers of this sort are intended for use in areas requiring intrinsically safe wiring, so equipment may not be reversed in the field. Such units must be purchased as either direct- or reverse-acting units.

Another effective arrangement for accommodating a current input and producing a proportional pneumatic output is one that performs somewhat like an overgrown galvanometer meter movement (fig. 11.2). This unique force motor consists of a rectangular coil wound with fine copper wire and surrounding a cylindrical permanent magnet. There is no contact between coil and magnet, but there is strong interaction between the magnetic fields when the coil is energized.

When energized with direct current, the coil will attempt to rotate but is restricted somewhat by flexures that permit only slight rotation, about 7° maximum, against their spring action. Some damping of coil movement is provided by a single shorted turn, which produces a form of electrodynamic braking. Other factors such as careful balancing, stiff flexures in the feedback mechanism, and overall low mass of moving parts contribute to the stability of the instrument.

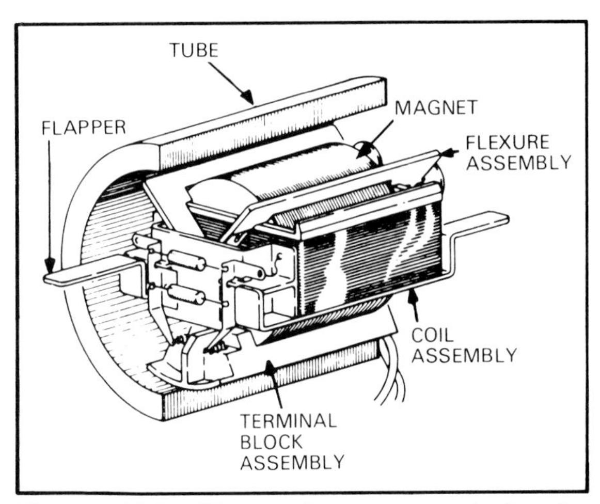

Figure 11.2. The main component in the galvanometric motor device for going from a current input to a proportional pressure output (*Courtesy Foxboro*)

The flapper component of the transducer is an integral part of the coil structure (fig. 11.3). With increasing current input, the

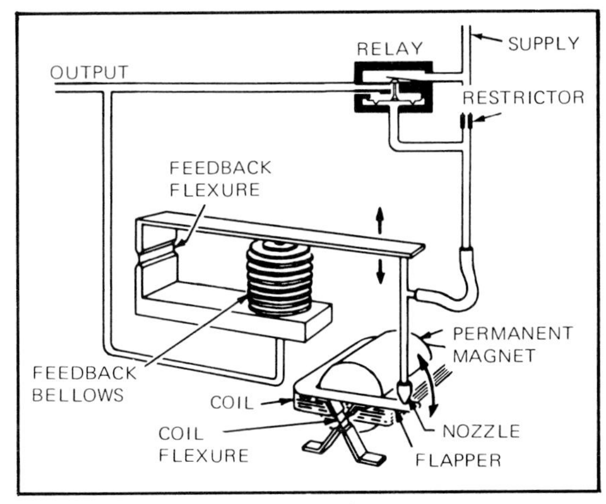

Figure 11.3. The pneumatic components contained in the galvanometric I/P transducer (*Courtesy Foxboro*)

flapper moves to cover the nozzle, thus increasing nozzle back-pressure. In typical fashion this back-pressure is applied to an air relay, which will supply air signals of 3 to 15 psi to remote instruments, as well as to the feedback bellows.

This type of unit is used extensively in transmitters that accept voltage or other input and ultimately produce a pneumatic output.

Pressure-to-Current Transducer

The principles and the electric circuitry involved in a pressure-to-current (P/I) transducer are more complex than those encountered in a current-to-pressure (I/P) unit. The I/P transducer provides an easier method of obtaining the opposing forces that need to be balanced. Another important aspect of the I/P is the ease of providing a negative feedback signal, represented by the bellows and its supply of nozzle pressure. In the study of pneumatic controls (chapter four), it was found that some form of negative feedback is useful in providing a degree of stability to a control system.

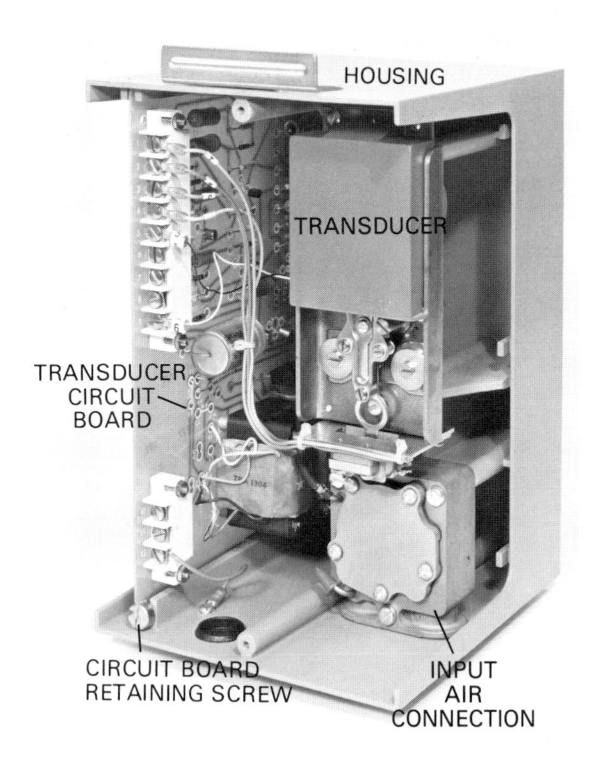

Figure 11.4. A complete pressure-to-current (P/I) transducer, including regulated power supply with protective cover removed to expose components. The section identified as *transducer* includes the parts shown in figure 11.6 (*Courtesy Taylor Instrument*)

One complete pneumatic-to-current (P/I) transducer (fig. 11.4) includes a regulated power supply adaptable to 120- or 230-volt AC, 50 or 60 hertz. A simplified diagram of the instrument (fig. 11.5) will serve to explain the principles of operation. The 3 to 15 psi input signal is applied to a bellows that acts against one end of a force lever. This input bellows can be moved toward or away from the force lever pivot by a dual cam arrangement. This movement serves as the span adjustment; span increases with movement of the bellows toward the pivot. A zero spring applies force to the force lever to zero the instrument.

A detector plate is mounted on the input bellows end of the force lever and rides in the air gap between the primary and secondary coils of the detector assembly (figs. 11.5 and 11.6). The primary coil of the detector is excited by an AC signal from the oscillator, and the secondary coil output is amplified by the AC amplifier. The output voltage of the secondary coil is determined to some extent by the efficiency of its coupling with the primary coil. The efficiency of the coupling is affected by the reluctance of the air gap between the coils, and that reluctance varies with the position of the detector plate.

The oscillator simultaneously feeds an AC signal to the detector primary and the demodulator stage, making the latter phase sensitive. Output of the secondary winding of the detector passes through the two-stage AC amplifier, then is applied in proper phase relation to the demodulator. Output of the demodulator is a DC voltage that provides bias for the current regulator and limiter.

The output of this transducer is in the standard range of 4 to 20 milliamperes. Note

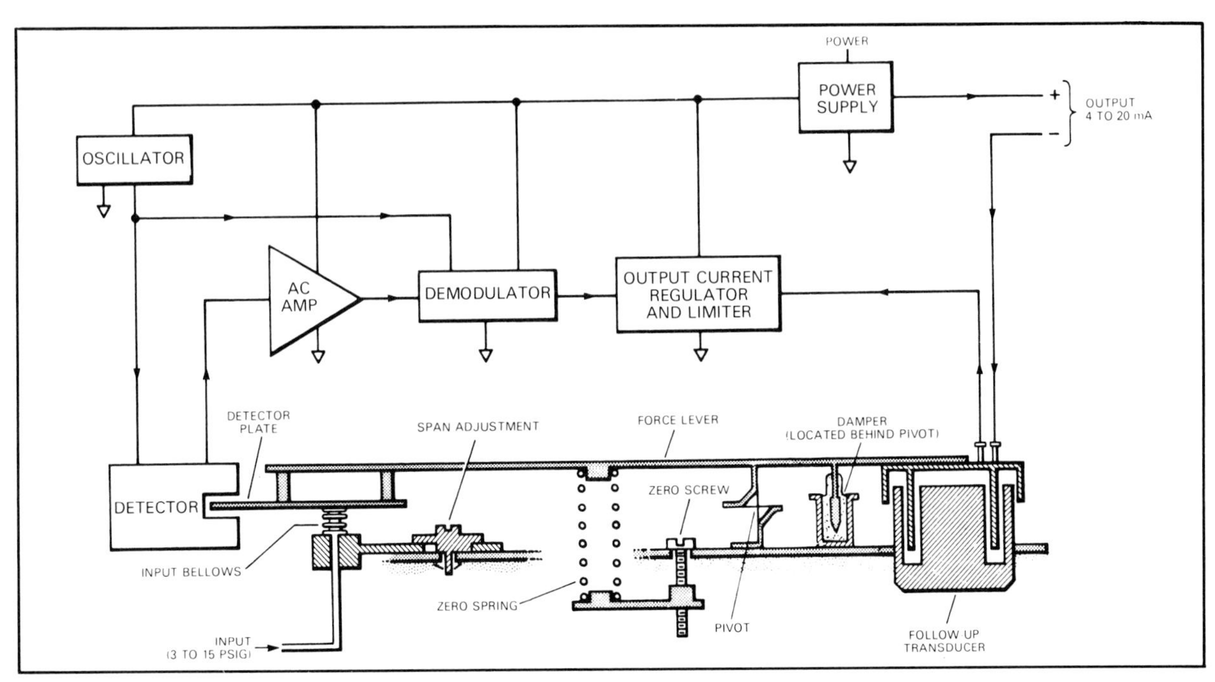

Figure 11.5. Simplified schematic of the P/I transducer shown in figure 11.4 (*Courtesy Taylor Instrument*)

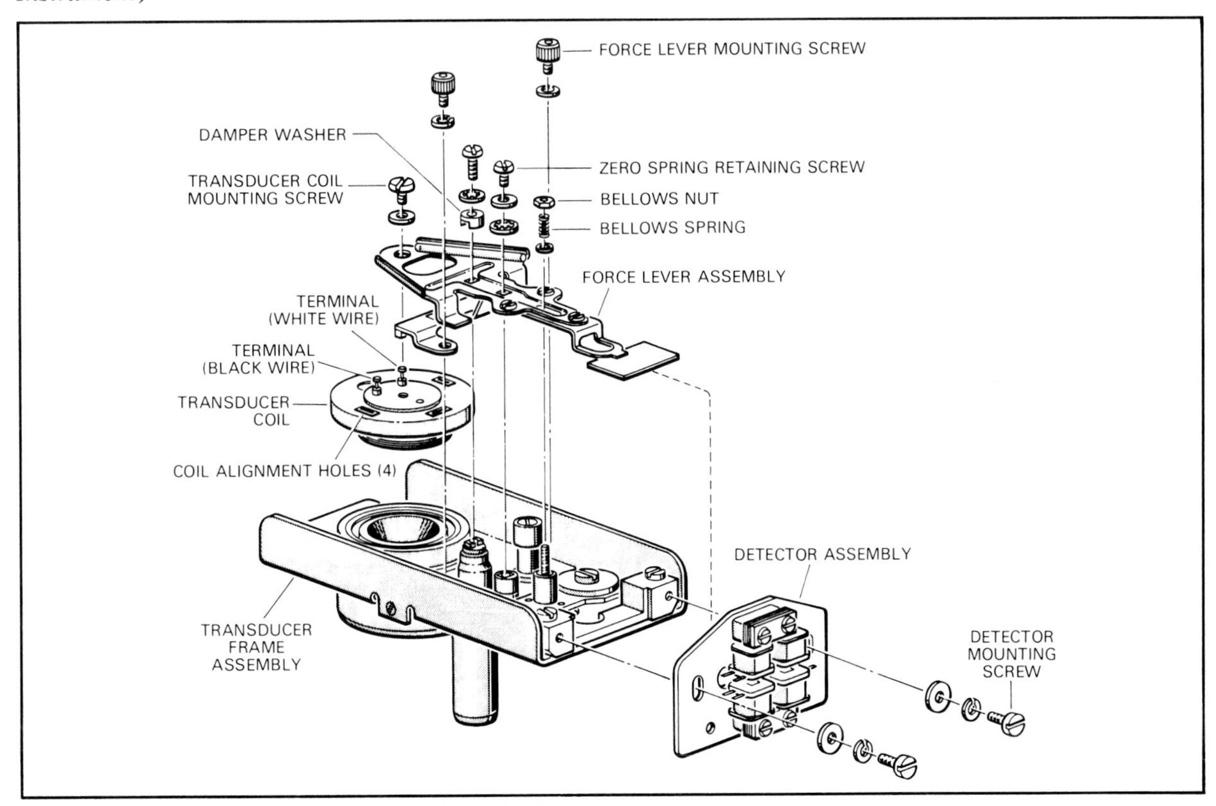

Figure 11.6. An exploded view of the principal parts that make up the transducer of figure 11.4 (*Courtesy Taylor Instrument*)

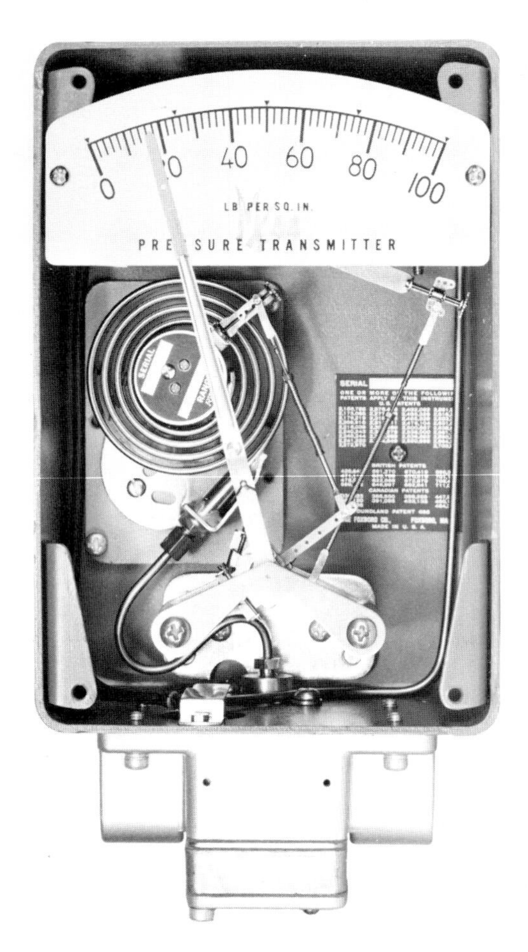

Figure 11.7. A pneumatic transmitter with front cover removed (*Courtesy Foxboro*)

that this current flows through the moving coil of the follow-up transducer. Principal components of the follow-up transducer are the moving coil and its associated permanent magnet. Flow of current through the moving coil brings about a repulsive force between coil and magnet that opposes the force of the input bellows, thus providing the negative feedback so essential to stability.

The damper serves to assure smooth motion of the force lever, thus reducing any tendency to overshoot the mark when responding to varying input pressures.

Calibration and adjustment of this transducer are straightforward tasks. However, instruments capable of measuring pressure and voltage with great accuracy, plus a precision resistor (0.1%) of 62.5 ohms, are needed. Calibration and adjustments should be carried out by experienced personnel using the manufacturer's technical instructions.

Transmitters

Straightforward Pressure Transmitter

A straightforward pressure transmitter (fig. 11.7) can be used to indicate values of pressure or temperature and to transmit the analog of these values to a remote indicating or controlling point. The analog values will be in the standard 3 to 15 psi pneumatic control range.

A principal member of the instrument is a spiral bourdon tube that drives an indicating pointer and simultaneously acts through a linkage system (fig. 11.8) to position the flapper in a flapper-nozzle arrangement. The bourdon tube can be tailored to match the

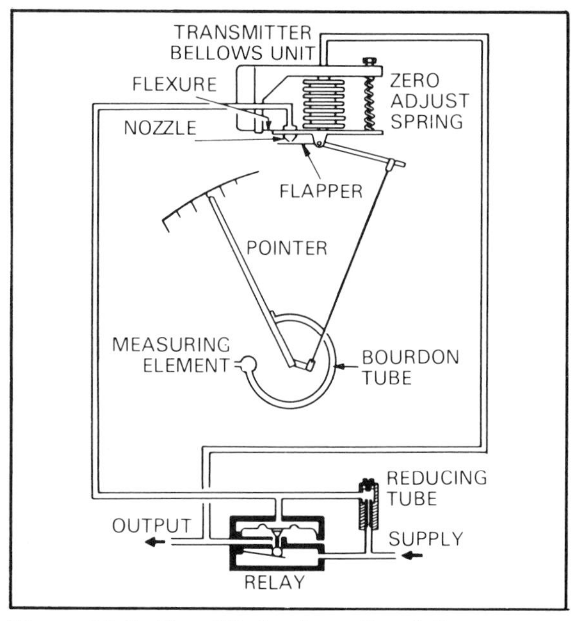

Figure 11.8. Simplified schematic of the pressure transmitter shown in figure 11.7. The reducing tube is the fixed orifice in the flapper-nozzle system. The zero adjust spring adjusts the transmission system. Another adjustment (not shown) zeros the pointer. (*Courtesy Foxboro*)

range of pressure to be monitored or controlled.

A useful variation of the instrument contains a vapor pressure system for measuring and/or controlling temperature. The bulb, capillary tube, and spiral bourdon spring are a complete closed system and must be installed as a unit. Care must be observed to avoid kinking, cutting, or twisting the capillary line in its run from instrument to point of measurement.

Other parts of the transmitter include familiar items such as a flapper-nozzle, feedback bellows, and the relay that responds to the nozzle pressure and sends high-volume pneumatic signals (3 to 15 psi) to controllers, recorders, or indicators (covered extensively in chapter four).

As with other instruments, transmitters require care in their installation, and special instructions that accompany new purchases should be followed. Name plate data also include important installation requirements, for example, the relative height of the instrument and the temperature bulb.

Adjustments and calibration of these instruments, when used either as pressure or temperature indicator-controllers, should be carried out by technicians familiar with such procedures and in accord with the manufacturer's instructions.

EMF-to-Pneumatic Transmitter

In the section on electronic controls (chapter four) a method of converting voltage signals to current signals was described. It was a simple method and provided at least a theoretical way to go from input signals of 1 to 5 volts to proportional output signals of 4 to 20 milliamperes. An EMF-to-pneumatic transmitter is a device capable of accepting a wide range of EMF signals (from millivolts to hundreds of volts) and producing a proportional pneumatic output. The circuitry and components for such a conversion are more complex than the simple method used in chapter four (fig. 4.33).

At the outset, one part of figure 11.9 can be dealt with quickly. (The output transducer is

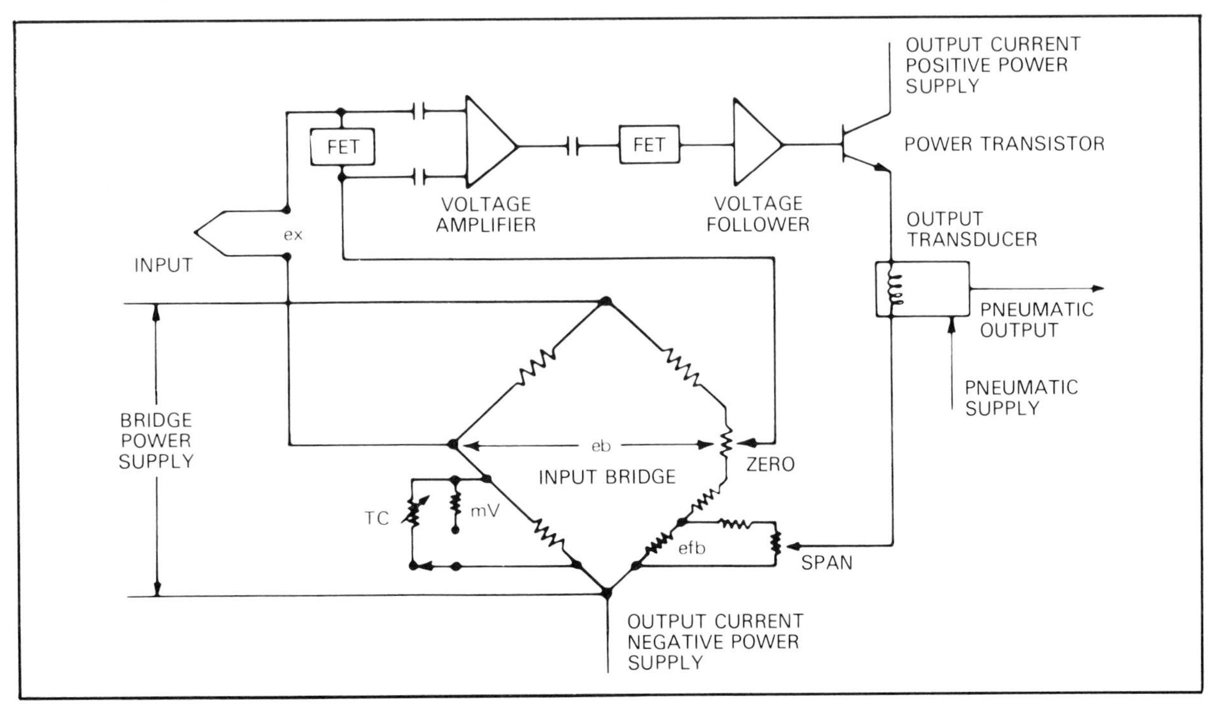

Figure 11.9. Elementary circuitry diagram of an EMF/P transmitter (*Courtesy Foxboro*)

the same unit discussed in connection with figures 11.2 and 11.3.) That part of the transmitter is merely a current-to-pneumatic (I/P) transducer. The rest of the transmitter will require more attention.

The simplified circuitry of the transmitter (fig. 11.9) does not show details of the major sections that include (1) the input bridge, (2) a chopping stage (FET), which takes the net voltage output from the bridge and conditions it, in effect changes it to AC for application to (3) a differential voltage amplifier, (4) a field effect transistor (FET) stage to demodulate the AC signal from the differential amplifier, (5) a voltage follower stage that provides bias to the base of (6) a power transistor.

The power transistor responds to the voltage values in the range span of measurement, producing a proportional range of current values. It should be recalled from chapter four that a voltage follower is an op amp connected to produce unity gain, with high input and low output impedance.

The bridge power supply, not shown in the diagram, is a closely regulated source of direct current. If the voltages developed across each leg of the bridge are equal, the bridge output, e_b, will be zero. This condition could prevail if, for example, a thermocouple was the input at e_x and the range span was set for 10 to 100 millivolts. The zero potentiometer of the bridge would be adjusted to produce equal inputs to the differential voltage amplifier when the thermocouple *emf* equaled 10 millivolts. For such a condition the bias for the power transistor would be the correct value to cause a current flow of 4 milliamperes (or 10 milliamperes in a loop using a 10- to 50-milliampere format).

The feedback voltage, e_{fb}, provides stability and balances the bridge. The proportion of feedback voltage supplied to the bridge determines the span of the instrument and is adjusted by the span potentiometer. The range that can be accommodated by the transmitter can be expanded to include values from a few millivolts to 400 volts through the use of voltage dividers that attach to the input terminals on the front case panel of the instrument.

For thermocouple input the transmitter is fitted with special modules that provide reference junction compensation. The modules are made and adjusted at the factory to match the type of thermocouple to be used and require no further adjustment. The compensating element is typically a nickel resistance temperature detector (RTD) whose resistance varies in accord with the difference between the reference junction and the standard reference temperature of 32°F.

Notice that if the current-to-pressure (I/P) transducer element in this transmitter is bypassed, the remainder constitutes an EMF/I (voltage-to-current) transducer or converter.

Resistance-to-Pneumatic Transmitter

It is a small step from an EMF-to-pneumatic (EMF/P) transmitter to a resistance-to-pneumatic instrument. The circuitry of the two transmitters is quite similar, as can be seen from the circuitry of a transmitter using a nickel RTD (fig. 11.10). A three-wire RTD is wired into one leg of the bridge, the third wire being a compensating line. Any change in the resistance of the RTD will cause a change in the voltage output of the bridge, and this output is applied to the chopping stage (FET) and differential amplifier. From there, action is identical to that of the EMF/P transmitter. Zero adjust, span, and stability are quite similar to the EMF/P. Components of the bridge differ somewhat due to the nonlinearity of the resistance-temperature characteristic of the RTD. The values of components in the bridge are so chosen that output from the bridge will be a linear representation of the temperature measurement.

In addition to the nickel RTD, a platinum RTD may be used but cannot be directly

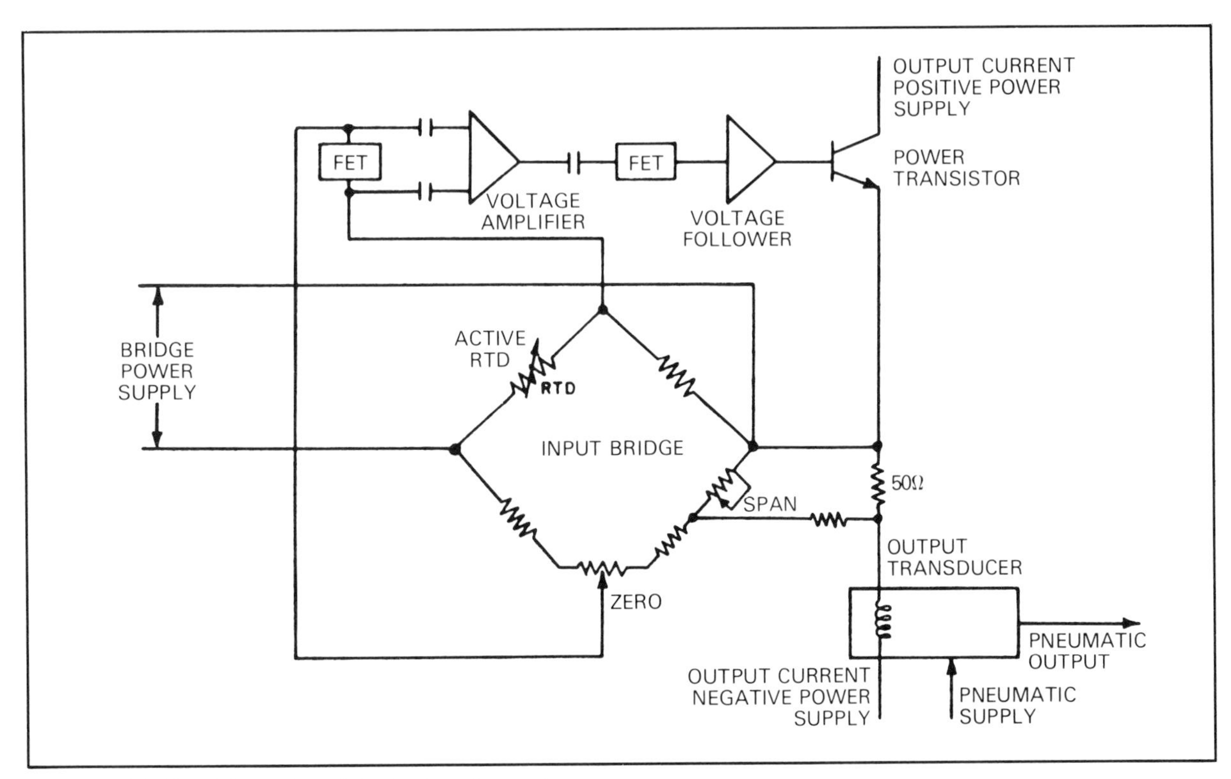

Figure 11.10. The simplified circuitry of a resistance-to-pressure transmitter (*Courtesy Foxboro*)

substituted for the nickel RTD. The circuitry is different and more complex. Platinum will be used in special cases where its high-temperature capability or its superior repeatability is required. It has the disadvantages of being significantly more expensive than nickel and is slightly less sensitive.

The nickel RTD can also be used to measure the temperature difference of two elements. Replacing the resistor in the appropriate leg (called the zero leg) of the bridge with a nickel RTD, as shown in figure 11.11, will enable the system to measure the temperature difference between the RTDs in the two legs. For this arrangement a difference temperature (ΔT) of zero represents a midscale bridge output. Temperature difference between the two elements will be indicated or transmitted as positive and negative values above and below $\Delta T = 0$. With nickel RTD elements, span can be as much as 600°F.

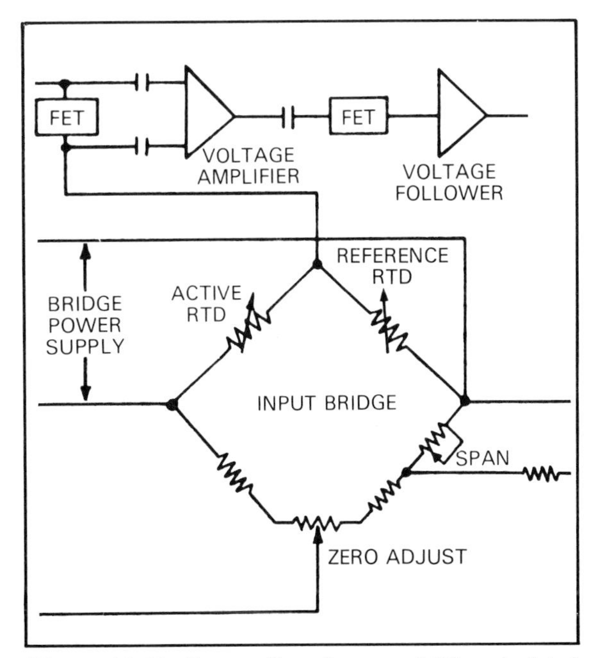

Figure 11.11. Connections for an additional nickel RTD element to provide temperature difference (ΔT) measurements (*Courtesy Foxboro*)

Electronic Differential Pressure Transmitter

The differential pressure cell or differential pressure transmitter discussed in chapter nine is a strictly *pneumatic* device capable of measuring the differential pressure (existing across an orifice plate, for example) and transmitting the analog of this pressure in terms of 3 to 15 psi pneumatic signals. A similar transmitter is an *electronic* device that measures differential pressure, but transmits signals in the 4- to 20-milliamp or 10- to 50-milliamp range. There are many similarities between this electronic transmitter (fig. 11.12) and the pneumatic one. High and

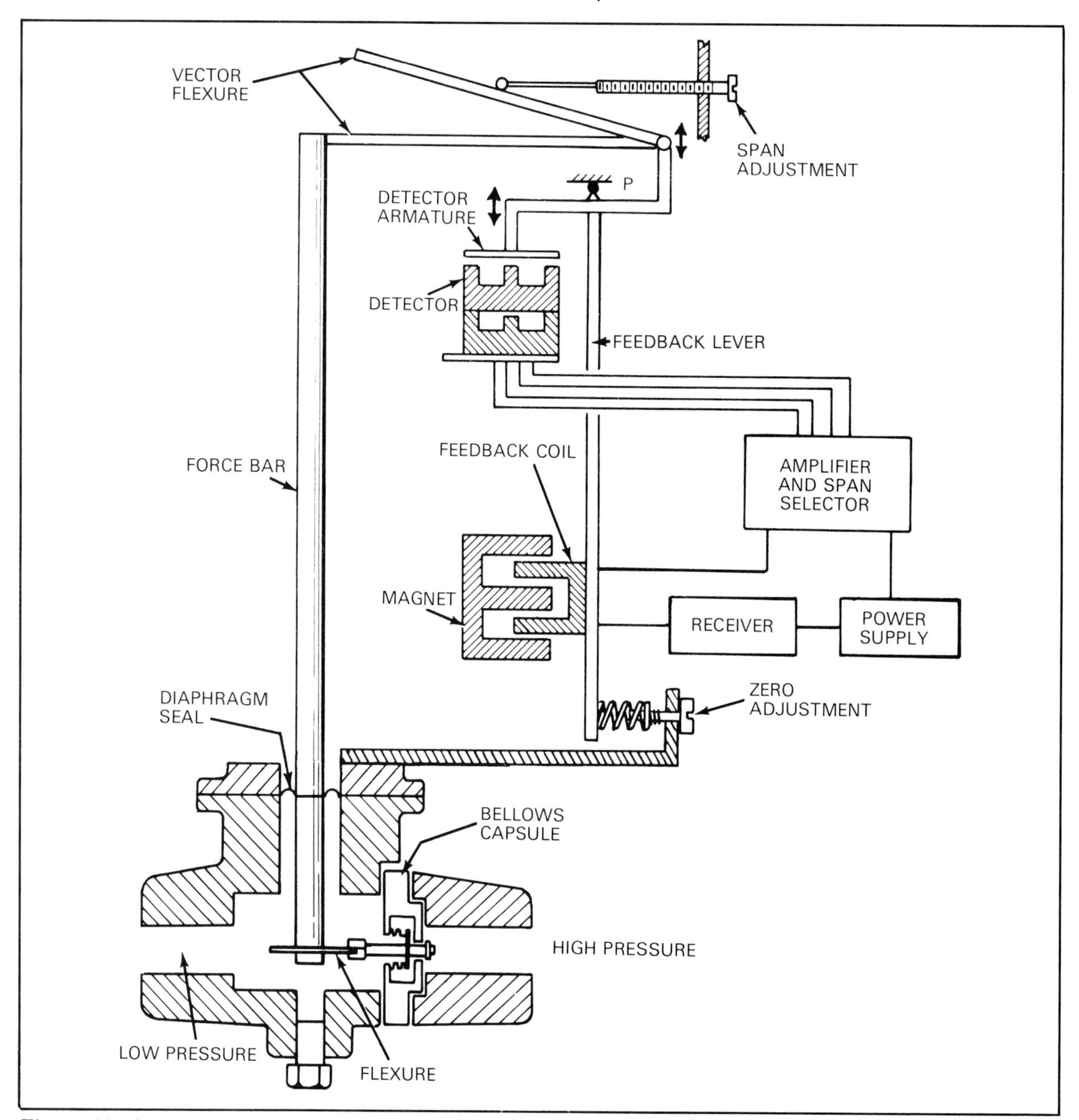

Figure 11.12. Schematic drawing of an electronic differential pressure transmitter (*Courtesy Foxboro*)

low pressures are applied to opposite sides of a bellows capsule, actuating a force bar that flexes at the diaphragm seal.

Notice that motion of the force bar is right and left in the drawing in figure 11.12. Any movement of the force bar will be transmitted through linkages to move the detector armature and the feedback lever. Should the differential pressure increase, the top of the force bar will move to the right, and vice versa.

It is important to realize that the detector armature and feedback lever are rigidly coupled and any motion of the force bar causes a pivoting of the feedback lever at point *P*. Should the differential pressure increase, the pivoting action will cause the detector armature to move away from the detector and the feedback coil to move toward a closer coupling with the permanent magnet.

Current in the feedback coil will cause a directed force that opposes motion of the detector armature, so equilibrium is quickly established. Current in the feedback coil is the 4- to 20-milliamp (10- to 50-milliamp) value that flows in the current loop composed of the power supply and receiver. The power supply will be the usual 24-volt unit for 4- to 20-milliamp loops, or 50 + volts for the 10- to 50-milliamp loops. The receiver block in the drawing represents the recorder, controller, or indicator at some remote point in the loop.

The electronic circuitry of this transmitter (fig. 11.13) is a reliable yet straightforward assembly using several transistors and inductive devices. Transistor *Q1* serves as an oscillator. The detector, which is essentially a transformer with variable reluctance coupling, receives its primary excitation (terminals *3* and *4*) from the oscillator and produces an output of 1 to 1.5 volts AC at terminals *1* and *2*. The output voltage depends on the position of the detector armature. The output from the detector is rectified by *CR3*

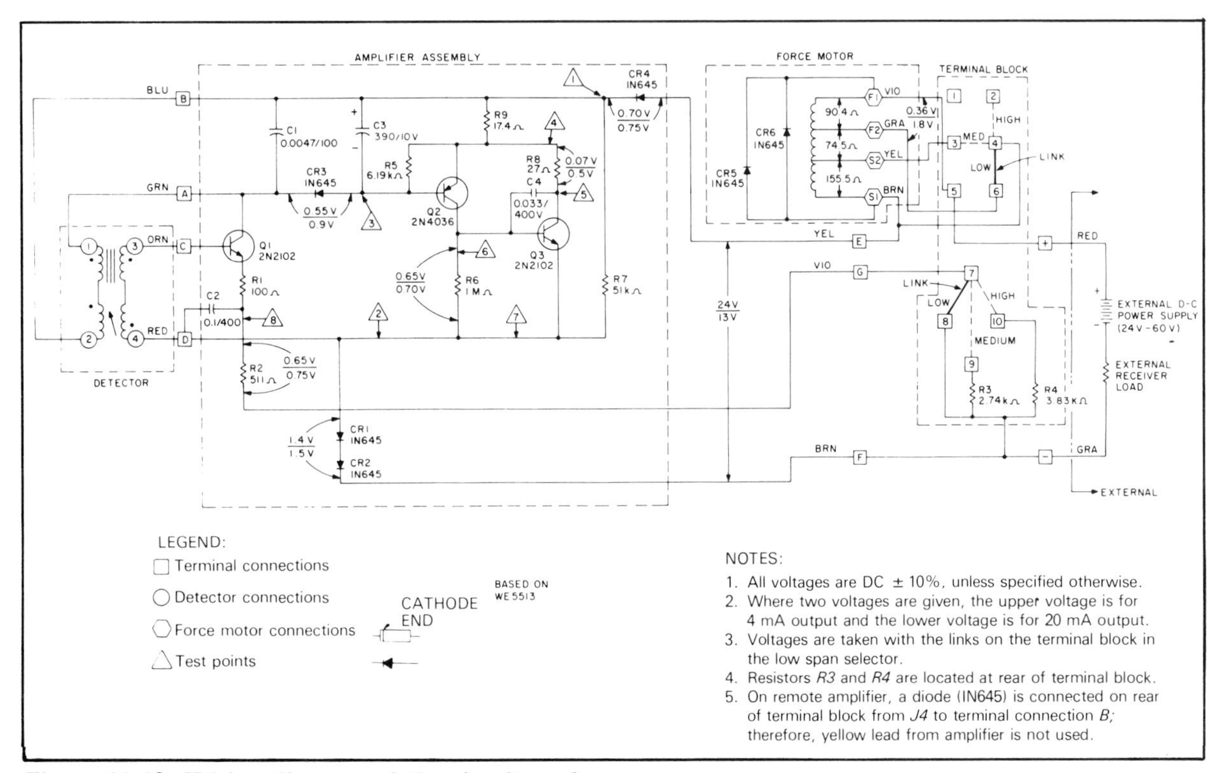

Figure 11.13. Wiring diagram of the circuitry of the electronic differential pressure transmitter of figure 11.12 (*Courtesy Foxboro*)

and biases transistor *Q2,* the amplifier stage. *Q2* in turn biases *Q3*, which is the current regulator that carries the loop current of 4 to 20 milliamps.

The feedback coil of figure 11.12 is identified as the force motor in the circuitry of figure 11.13. It consists of a coil divided into three sections, each section possessing a different magnetic capability for a given current flow. The selection of coil section, or combination of sections, provides a method of choosing a span range for the instrument. The selection is made by connecting a link between two of six terminals on the terminal block (fig. 11.14). In addition to this selection, another link corresponding to the low, medium, or high selection must also be made on the terminal block shown in the figure. This latter selection affects the voltage applied to the emitter of *Q1.*

The span adjustment screw shown in the drawing of figure 11.12 provides a vernier setting of the span and enables the user to read the setting with accuracy. The transmitter can be quickly recalibrated to previously established span ranges by merely changing positions of the links and repositioning the span screw to the appropriate setting. The appropriate settings will have been determined by previous calibrations and recorded on the instrument.

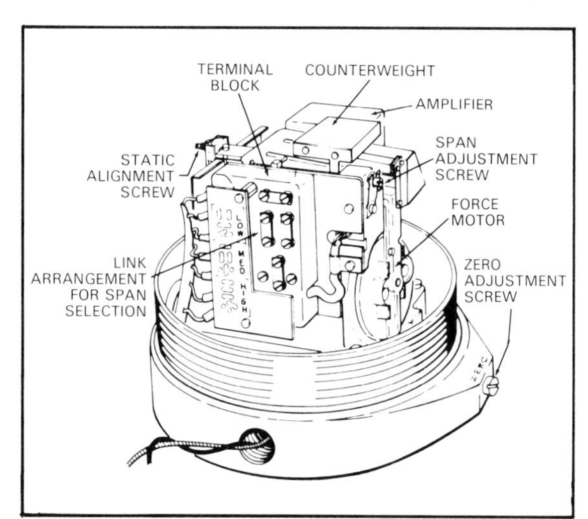

Figure 11.14. A view of the arrangement of components on the transmitter of figures 11.12 and 11.13 showing the various adjustment points. The counterweight provides a balance between the force motor and the detector armature. (*Courtesy Foxboro*)

Zero adjustment of the transmitter is made with the high- and low-pressure inputs to the bellows capsule open to the atmosphere, that is, with zero differential pressure on the instrument. With the control loop energized, the zero adjust screw should be set to produce 4 milliamps of current flow (or 10 milliamps for a 10- to 50-milliamp loop). The static adjustment should be made after the zero adjustment by allowing both pressure inputs to the bellows capsule to be brought to the highest static pressure expected in service. If the current output of 4 milliamps (or 10 mA) in the loop has not changed after 2 minutes, then the static adjustment is satisfactory. If the current changes more than desired, the static adjustment screw must be adjusted, following the manufacturer's instructions.

Converters

Pneumatic-to-Electronic Signal

One pneumatic-to-electronic, or air-to-current, converter is available as a rack-mounted instrument or packaged for field mounting (fig. 11.15). Basic characteristics of the instrument include inputs of 3 to 15 or 3 to 27 psi and outputs of 4 to 20 or 10 to 50 milliamps. A rack-mounted model is also available that produces an output of 0 to 10 volts. An elementary diagram of the converter circuitry is shown in figure 11.16.

The pneumatic input signal is applied to a bellows element. The active end of the bellows is linked to a closed loop, located between coils *L1* and *L2*. Movement of the loop varies the inductance of coils *L1* and *L2* that carry a 20-kilohertz signal from the oscillator. The detector is an arrangement of four diodes that rectify signals from *L1* and *L2*. Amplifier *1* is

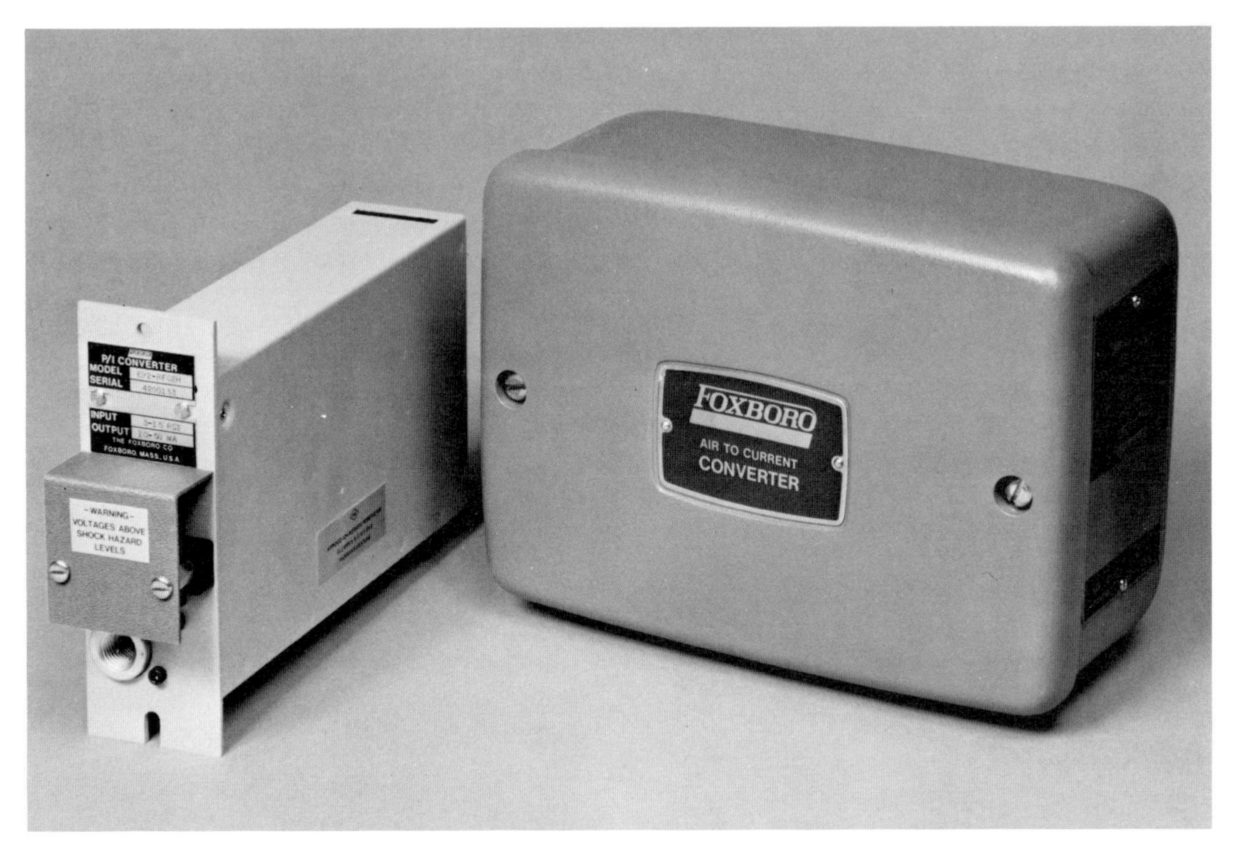

Figure 11.15. Two configurations of an air-to-current converter. *Left,* a version designed for mounting in a rack; *right,* a field package. (*Courtesy Foxboro*)

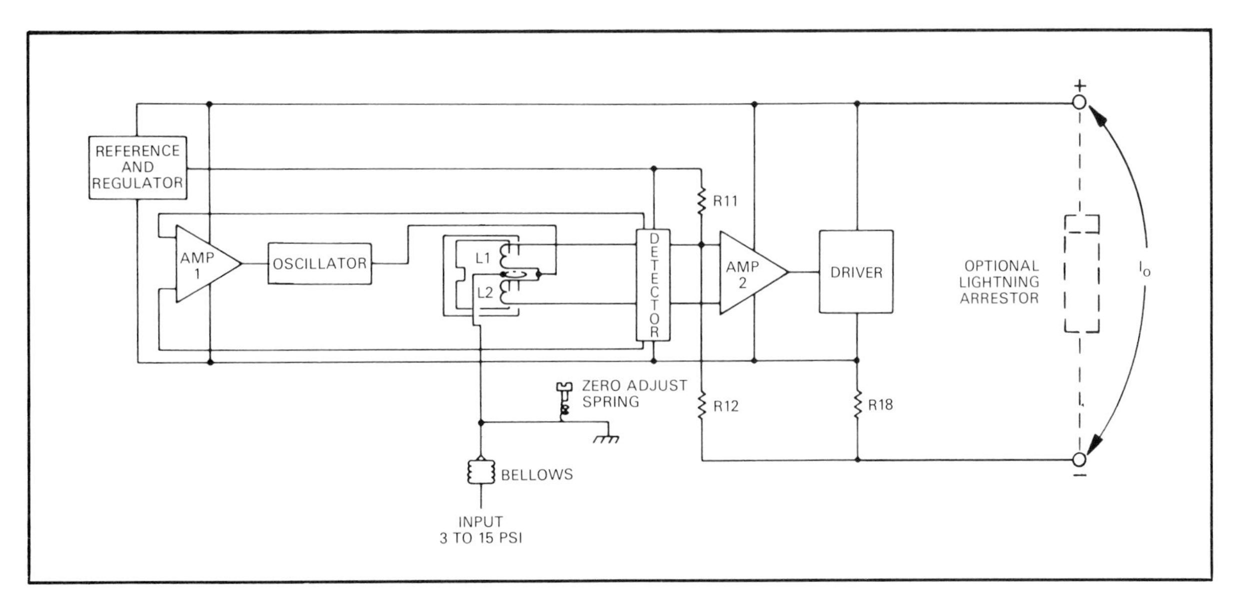

Figure 11.16. Simplified circuitry of the air-to-current converters of figure 11.15 (*Courtesy Foxboro*)

configured as a comparator and receives DC input signals from the detector.

With a pneumatic input of 9 psi (for a 3 to 15 psi format), the closed loop is located at the midpoint between coils *L1* and *L2*. For this condition the output current (I_o) of the converter should be 12 milliamps, the midpoint value in a 4- to 20-milliamp format. Also, for this condition the amplitude of the oscillator output is maximum.

Should the pneumatic input change from 9 psi, the closed loop will move away from the midpoint position between coils *L1* and *L2*, causing reduced amplitude in the output of the oscillator and unbalanced input to the detector. The design of the detector is such that should the pneumatic input go above 9 psi the DC signals to amplifiers *1* and *2* will be opposite in direction to those that will exist for an input below 9 psi.

Amplifier *1* causes the sum of the currents through *L1* and *L2* to be equal to a reference current by regulating the oscillator amplitude. The reference current is established by the regulator section whose principal components are three transistors, two diodes, and a Zener diode.

The noninverting input of amplifier *2* is initially biased by resistor *R11*. This input also contains a potentiometer, not shown in the drawing, that provides span adjustment. Resistor *R18* (100 ohms) is in the output current circuit. The voltage drop across it is effectively fed to the inverting input of amplifier *2*, thus aiding in stability.

The output of amplifier *2* provides base bias for the driver, a power transistor that regulates current flow in the 4- to 20-milliamp loop circuit in step with values of pneumatic input to the converter.

Calibration of the pneumatic-to-electronic converter is easily carried out, preferably in a shop equipped with capability for accurate measurement of pneumatic pressure and millivolts. The manufacturer's recommended procedures should be followed when making adjustments.

Millivolts-to-Volts Converter

Thermocouples and a few other devices commonly used in instrumentation produce millivolt signals in response to the variable they measure. It is often desirable to convert these millivolt signals to a proportional range of 1 to 5 volts. Special problems are encountered when this conversion is carried out.

Most millivolt sources will have a span of less than 100 millivolts. A platinum/platinum-15% rhodium thermocouple has a temperature range span of about 3,300°F, but its millivolt span for that range is less than 22 millivolts. The most sensitive thermocouples will have millivolt spans of less than 90 millivolts.

In view of the rather feeble input to the converter, great care is required to avoid several forms of disturbances that may influence the conversion process. For example, there is the risk of having the output signal feed back into some earlier stage of the conversion process, or spurious signals may find their way into the input section of the converter. Since the gain, or amplification, that must be provided to step the input signals from perhaps 5 or 6 millivolts to 3 or 4 volts is relatively large, special isolation and shielding precautions must be observed.

A simplified diagram (fig. 11.17) will be used to describe a millivolts-to-volts converter (shown out of its case in fig. 11.18). The unit has two channels, each capable of accepting independent inputs from thermocouples or other millivolt devices and producing independent outputs. A description of channel 1 will provide an understanding of the function of each stage.

Converter input. The input terminals for each channel provide for ranges that include elevated or suppressed zero. Elevated zero permits the conversion of negative millivolt signals. For example, a thermocouple produces a positive voltage at terminal *8* for temperature values above 32°F (or other reference temperature) and a negative voltage below that value. If there is a need to have the

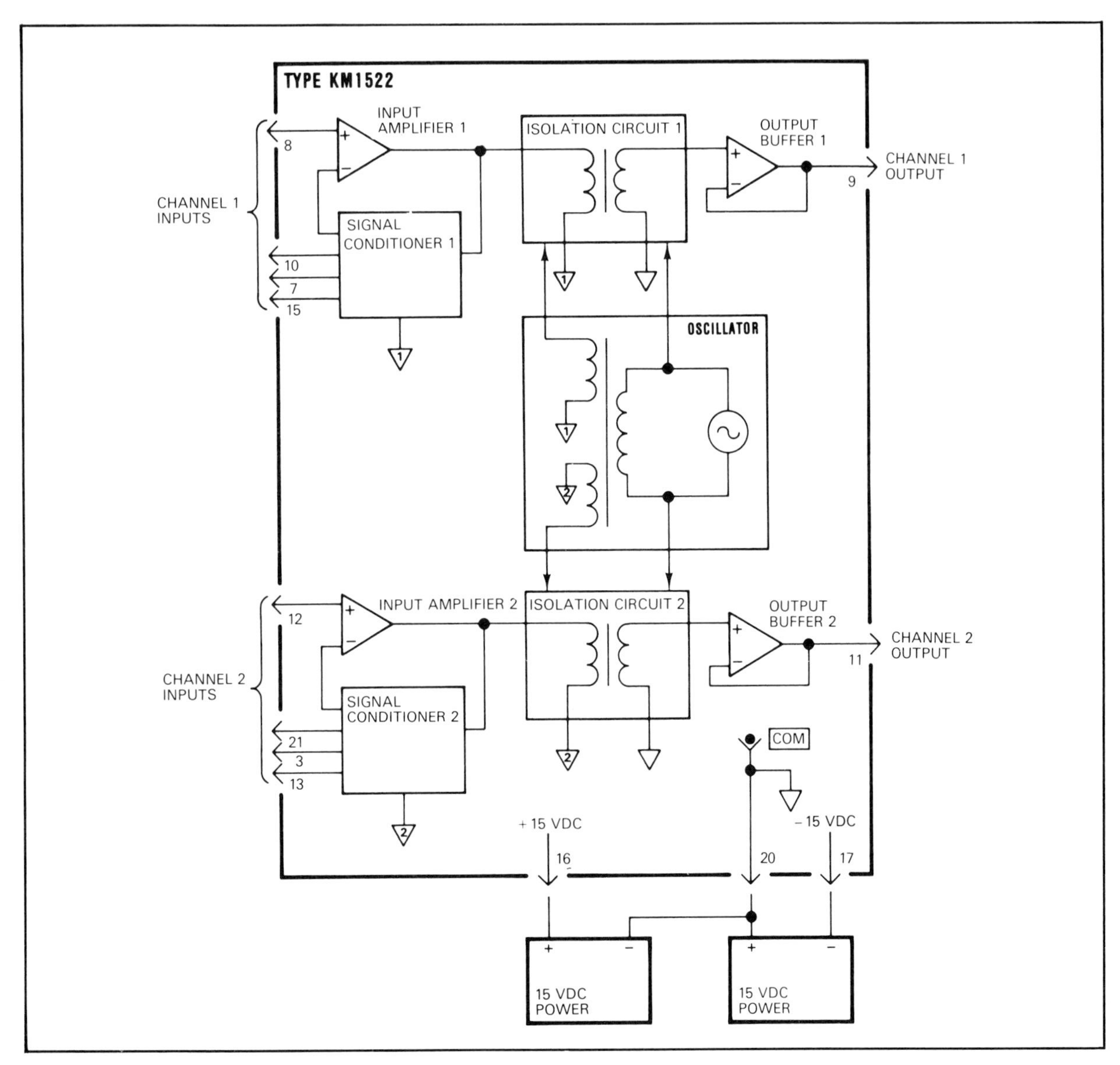

Figure 11.17. A block diagram of the circuitry of a millivolts-to-volts converter (*Courtesy Fisher Controls*)

converter respond to temperatures above and below the reference temperature, then an elevated zero must be provided. This elevated zero will enable the system to "go below zero" to negative values of millivolts. It is useful to point out, however, that the converter output will remain in the range of 1 to 5 volts DC. A suppressed zero means that the lowest millivolt signal the converter will respond to is some positive value. Assume the span is set for +10 to +80 millivolts. Should the millivolt input fall to, say, +5 millivolts, the converter output will be 1 volt, the same as for an input of +10 millivolts.

Terminal *8* is the nominal positive millivolt input. Terminal *7* is the negative input for suppressed zero range, and terminal *15* is for elevated zero range. Terminal *10* is for the negative lead from a thermocouple when no elevation or supression of zero is needed.

Figure 11.18. The millivolts-to-volts converter diagrammed in figure 11.17. Fine adjustments of zero and span are provided for each of the two channels on the front panel, as well as jacks for performing tests. (*Courtesy Fisher Controls*)

Signal conditioner and input amplifier. A signal conditioner is a device or circuit that acts on a signal, usually the signal from a primary element, to make it more suitable for further use or processing. The conditioning may be simple amplification, or it may be an effort to change a signal to a linear representation of the controlled variable. In the converter being described the signal conditioner contains switches and potentiometers for rough and fine zero and span adjustments and resistance networks to accommodate suppressed and elevated zero ranges.

Another function of the signal conditioner has to do with reference junction compensation when the millivolt input is from a thermocouple. A reference junction compensator is installed between the input terminals of the converter and the thermocouple. The compensator provides a millivolt signal that changes the same amount with ambient temperature as the millivolt signal at the reference junction, but polarities of the two are reversed, thus producing a constant junction voltage. The compensating signal contains two components: (1) a fixed voltage developed by the converter and (2) a variable voltage that follows resistance changes of the compensator due to varying ambient temperatures. The fixed voltage is selected by a selector switch in the signal conditioner and depends on the type of thermocouple in use. This fixed voltage can be varied over a narrow range by a compensation adjust potentiometer.

Input amplifier. The millivolt input from thermocouple or other source is applied directly to the noninverting input of the input amplifier where it is chopped (broken into a discontinuous series of pulses). The signal is then amplified and integrated to produce a 0- to 4-volt DC output, which is applied to the signal conditioner and isolation circuit. In the signal conditioner circuit the 0- to 4-volt output finds its way through an assemblage of resistors and switches, then back to the inverting input of the amplifier. The resistors, including the span adjustment potentiometer, combine to provide the negative feedback to the amplifier.

Isolation circuit. In the simple schematic the isolation circuit is shown as a transformer. It is more complex than that, but the transformer points up a part of the isolation that exists between the input and output ends of the converter. The transformer would imply that output from the input amplifier is other than direct current. A group of four field effect transistors (FETs) gated by square waves from the oscillator section enable the 0- to 4-volt DC output from the amplifier to be repeated on the secondary side of the isolation transformer. From there the DC signal is applied to the output buffer.

Output buffer. A buffer stage is usually one that provides isolation between two stages or sections of an electronic assembly. In the millivolt converter the output buffer furnishes a measure of isolation, and it does other service as well. In the diagram (fig. 11.17) the op amp is shown configured as a voltage follower providing unity gain and low impedance output. Not shown in the diagram is the method of adding 1 volt to the noninverting input to provide an output range of 1- to 5-volt DC. The 1 volt is obtained through a potentiometer operating across a 6.2-volt Zener diode.

Oscillator section. It was noted above that the oscillator section furnishes gating pulses to the isolation circuit. It is also a source of DC power to the input amplifier and isolation circuit. The oscillator consists of two transistors, plus a special transformer and various lesser components. Output of the oscillator (about 1,000 Hz) is transformer coupled to bridge rectifiers and filter sections to supply the input amplifier. This choice of power supply provides for further isolation by avoiding the risk of feedback that could occur in a common power supply arrangement.

Conclusion

Other converters, transmitters, and transducers are made and used, but those covered in this chapter should serve as a good introduction to the art of transforming various signals from one form to another, always retaining a meaningful relation between the beginning (input) and the end (output) of the transformation.

Manufacturers provide excellent technical information about their products. In many instances the manufacturer's technical manual describing a single model of converter or transducer may be more extensive than this entire chapter, for it will contain parts lists, installation instructions, maintenance and calibration procedures, wiring diagrams, and other drawings. These technical bulletins are highly recommended to the reader who wishes to pursue the subject matter.

CHAPTER 12

Recorders and Integrators

Recorders are devices of instrumentation that produce a permanent record of some variable in a system or process. In most instances the values of the variable are recorded on a time basis; that is, the value at any moment in a given period becomes a matter of record. Integrators are counters, actuated sometimes by ingenious mechanical contrivances. Integrators are literally devices that sum up "all the little bits."

Recorders are used extensively where temperature, pressure, voltage, current, humidity, and many other variables are important to a given system or process. On the other hand, integrators have no place in the measurement of temperature, pressure, or liquid level, but they are tremendously important in measuring the total quantity of material that is processed, delivered, consumed, or the like. Recorders are also used in flow measurement, where a permanent record forms an important element in the process of determining the total volumetric flow of material.

In many instances, the recorder is an integral part of a controller, and the same lever mechanism that actuates the nozzle-flapper assembly also drives a combination recording pen and indicator. The scale of the indicator is provided by a chart appropriately calibrated in values of the variable being measured. A clockwork mechanism drives the chart at some predetermined speed.

Types of Recorders

Recorders are divided broadly into two types—those having circular charts and those having strip charts. Circular charts are single sheets of paper in the form of disks, while strips are ribbonlike and of such length that they are usually wound on spools.

Another distinction that might be made between recorders is that some record a single variable and others are capable of handling two or more variables. For example, in flow measurement, both the differential head (in inches of water) and the pressure-tap pressure (in psi) can be conveniently recorded on the same chart.

Circular Chart Recorders

Circular chart recorders are the most popular recorders in use today. They have advantages that are reflected in the ease of changing charts and the fact that the charts are flat and easy to file. Circular chart recorders also adapt to multivariable recording more easily than other forms.

A principal disadvantage lies in the rather restricted width of the chart that is usable. For example, a circular chart of 12 inches in diameter offers a recording width of only 5 inches, while a strip chart only 6 inches wide offers this same recording width.

Single-variable recorders. Key components of a single-variable recorder are its measuring element, recording pen, pen arm, pen lifter, time indicator, chart plate, chart drive, and chart hub (fig. 12.1). The measuring element shown in figure 12.1 is a helical Bourdon tube, suitable for pressure or temperature measurement, but any other element capable of imparting motion to the lever system could be used just as well. The chart plate offers a flat, firm surface for the chart paper. The chart is held tightly to the chart hub by a hub

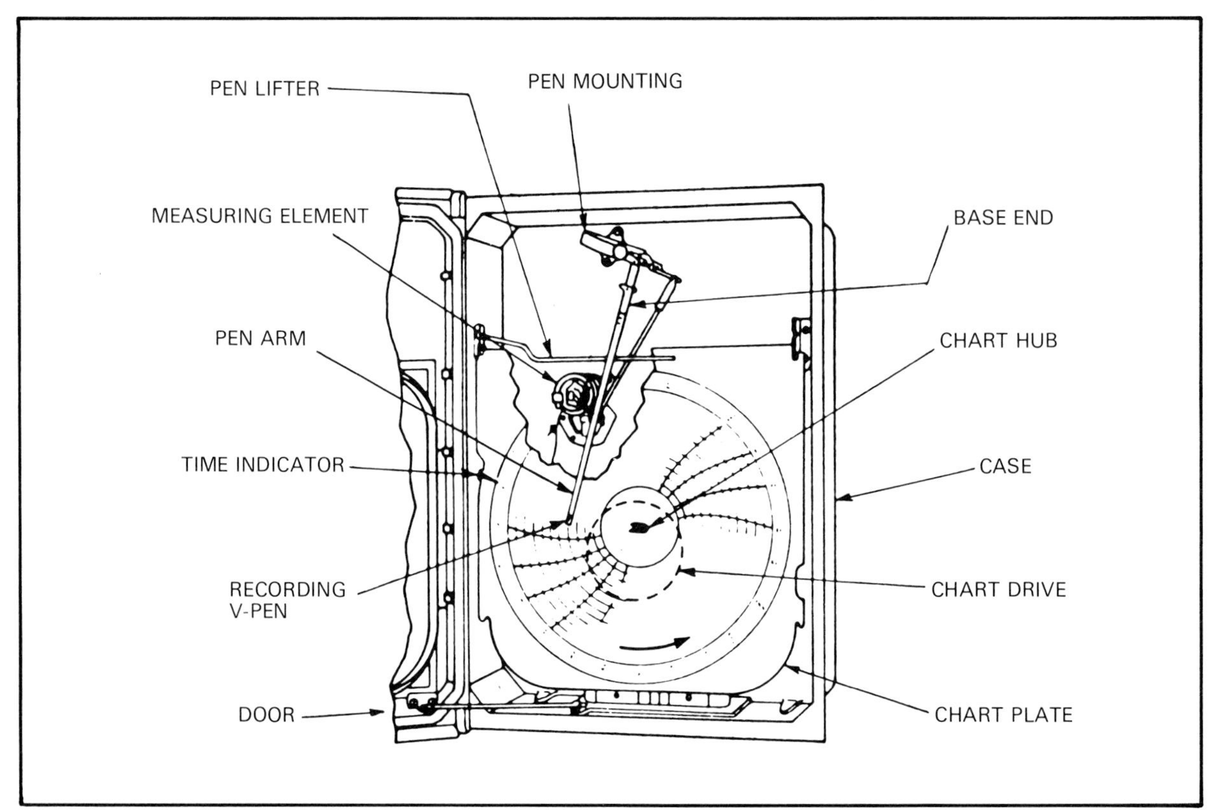

Figure 12.1. A single-pen, single-variable circular chart recorder (*Courtesy Bristol Babcock*)

nut. The pen lifter is used to hold the pen out of the way when installing or removing charts. The time indicator is a pointer that enables the operator to accurately position a new chart.

Multivariable recorders. Certain problems arise when recording more than one variable on a single chart. For example, if two pens are used and each is allowed free travel from one chart edge to the other, collision between the two pens is quite likely unless some means is used to allow the two pens to pass one another. For example, each pen on a recorder using four pens (fig. 12.2) has a radius of operation different from the other three. Each pen is attached to its own vertical shaft, and the shaft is driven by a measuring element and lever system.

The pens enjoy complete freedom from collision, but at the price of having to be offset from an exact time position of the chart. Sometimes this offset is not serious, and a typical example for a 24-hour, 12-inch chart shows a time difference of only a few minutes between the leading pen and the one with maximum time delay. The natural gas industry requires a 15-minute time lag between pens on 24-hour charts that are to be mechanically scanned.

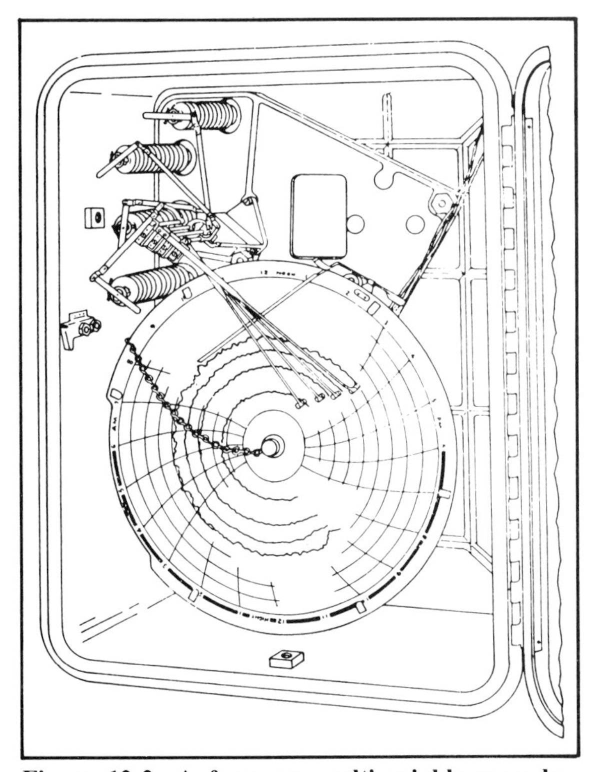

Figure 12.2. A four-pen multivariable recorder. Note the linkages that drive the pens from a common center (*Courtesy Bailey Meter*)

An arrangement with four pen arms rotating about a common center (fig. 12.3) provides a slightly smaller time difference between leading and lagging pens than the assembly shown in figure 12.2.

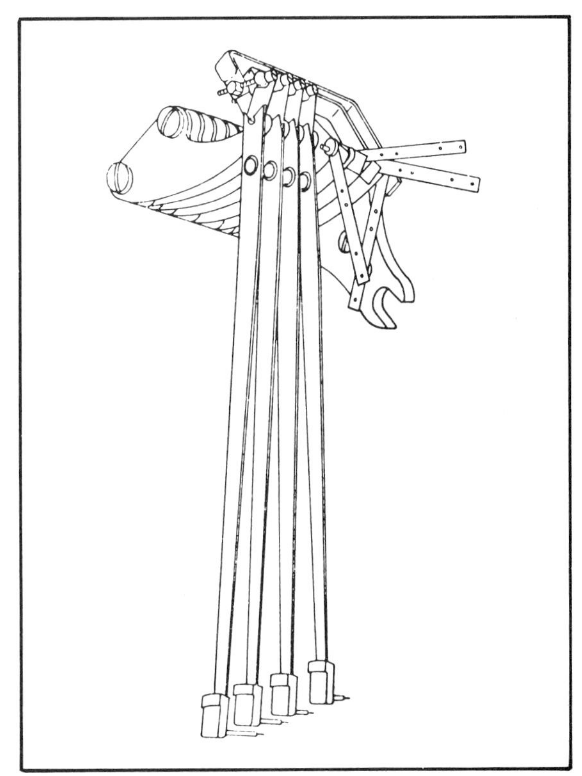

Figure 12.3. A detailed view of a four-pen assembly (*Courtesy Foxboro*)

Sometimes it is feasible to restrict the travel of each pen of a multivariable recorder to a certain zone of the chart. When a chart is divided into zones, the scale of values is obviously compressed in proportion to the number of zones required. This may not be serious if the system being monitored is stable enough that a legible and usable range of values can be incorporated in the zone. The

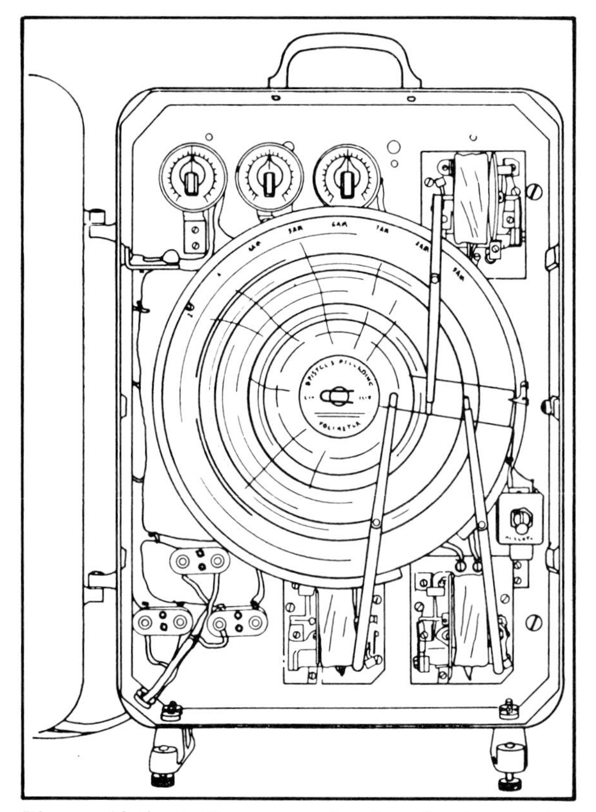

Figure 12.4. A three-pen circular chart recorder with each pen restricted to a limited portion of the chart (*Courtesy Bristol Babcock*)

zone system, of course, eliminates the problem of time lag between the pens (fig. 12.4).

A unique system for providing multiple recording on a single circular chart is the use of a single pen arm, recording up to six different tracks, each in a different color (fig. 12.5). Basis of the system is a switching and transfer mechanism that allows the single pen arm to respond to as many as six variables in sequential periods (fig. 12.5*A*). For example, variable *A* might be represented by red ink, and during its recording time a recording pen unit is tightly held to the pen arm by a strong permanent magnet. When the recording interval for variable *A* expires, the pen arm is lifted from the chart and swings off scale to the indexing and transfer assembly. The pen assembly with red ink lodges in a particular slot of the pen-wheel assembly. The pen wheel then rotates, pulling the red ink pen assembly away from the pen arm and replacing it with another pen, perhaps for variable *B*. The actuating mechanism for the pen arm is shifted to respond to variable *B*, and the arm moves out over the chart to a position determined by the value of variable *B*. After the pen for variable *B* is allowed to make its mark on the chart, the process is repeated, this time replacing the pen for variable *B* with one for variable *C*, and so forth.

Ink pads, six in all, rotate with the pen-wheel assembly, and pens are supplied with quantities of ink of the proper color as they rest on the pads (fig. 12.6*B*).

Strip Chart Recorders

Strip chart recorders are rarely found in the field where the mechanical clock driven, simple and rugged circular chart recorders seem to enjoy almost exclusive domain. Strip chart recorders are used extensively in laboratories and control centers since even the simplest strip chart recorder is a fairly complex instrument, and the more versatile—those specialized for particular applications—are apt to be quite intricate.

Advantages. Strip chart recorders possess several noteworthy advantages over the circular chart types. They are more economical of panel area space, making them desirable for central control rooms where panel area, rather than panel depth, becomes an important consideration.

Strip charts up to 120 feet long are practicable and can be confined in relatively small volume by use of spools. One form of strip chart begins as a roll, but as it passes beyond the recording pens, it gathers in folds in a bin. This format has advantages where there is frequent need to check on portions of the recording that have moved well past the visible area of the recorder. Also, when long strips are used, fairly high-speed recording (2 or 3 inches of chart paper per second) is possible.

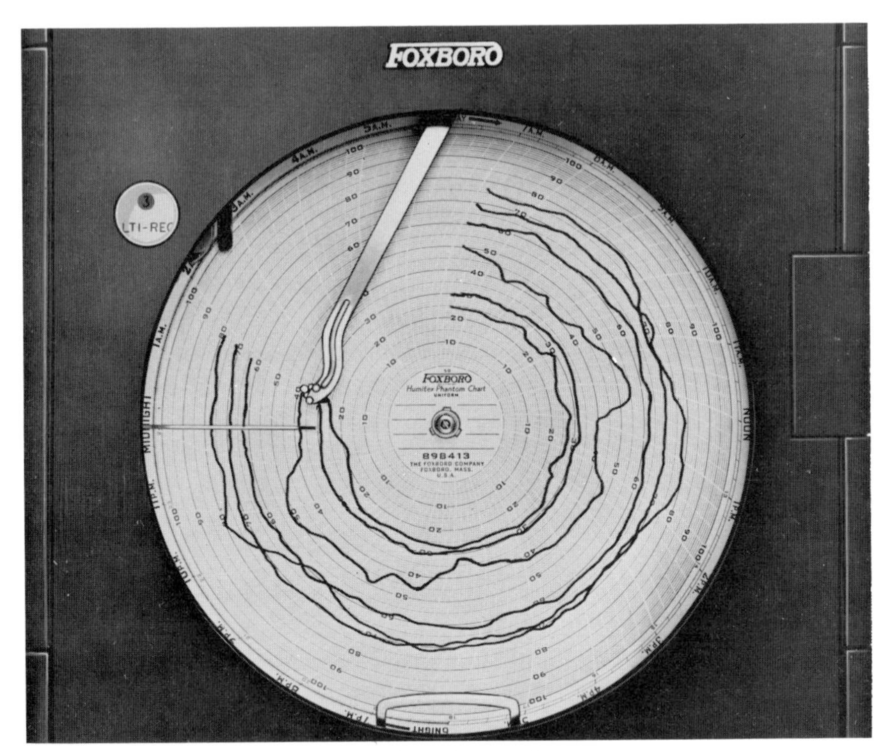

Figure 12.5. A pen-wheel recorder. On a slow-moving chart the traces will appear to be continuous. (*Courtesy Foxboro*)

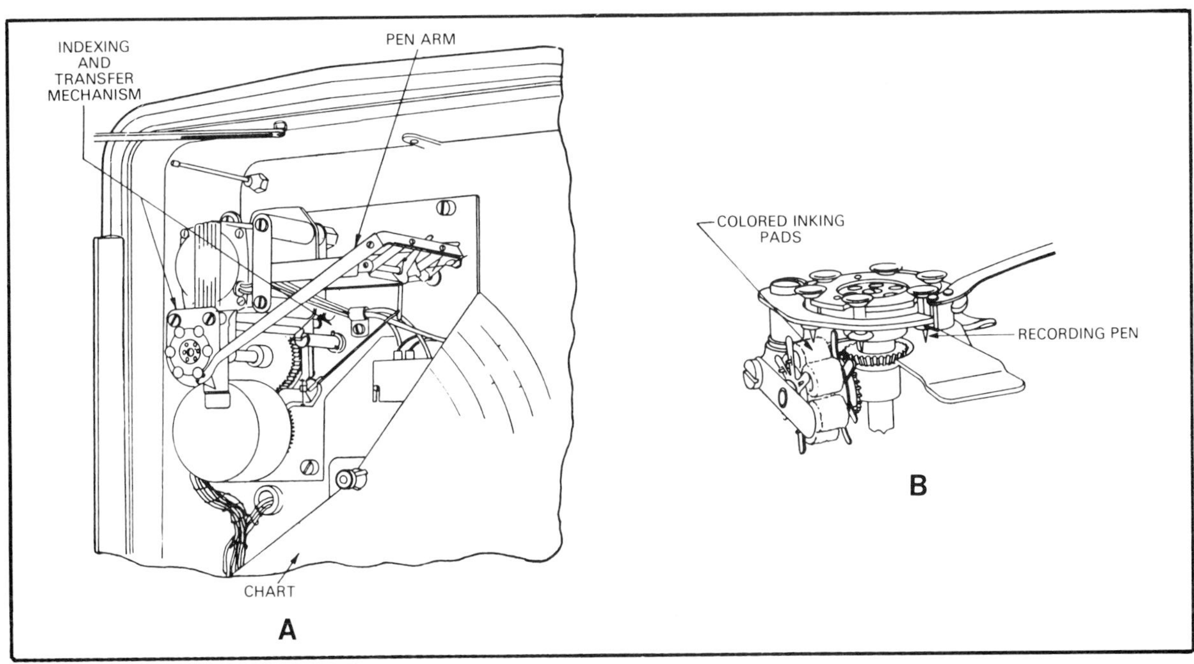

Figure 12.6. The pen-wheel mechanism of a multi-variable recorder. *A*, part of the mechanism that can record up to six variables, each using a different ink color; *B*, rotating ink pads that supply different colors of ink. (*Courtesy Foxboro*)

Data are recorded in rectangular coordinates or in curvilinear form on strip charts. A chart in rectangular coordinates is one for which the time basis is linear across the chart; that is, a given unit of time takes the same linear distance at the low end of measurement as it does at the high end. Also, the recording pen moves across the chart in a straight line, rather than in an arc. A curvilinear chart has uniform increments of time, but the recording pen swings in an arc, just as it did for the circular chart recorders.

A pneumatic strip chart recorder. A relatively recent development is the use of pneumatic mechanisms in strip chart recorders, although the use of electric systems is still prevalent. One type of pneumatic strip chart recorder (fig. 12.7) has a curvilinear chart: its recording pen follows an arc in its travel across the chart.

The measurement section of this recorder—the receiver—is designed to respond to a pneumatic signal of 3 to 15 psi. Such a signal might represent a range of values of temperature, pressure, or other variable whose normal units of measurement have been converted by a transducer to units of air pressure. The 50%, or mid-scale point, is the position taken by the recording pen or measurement indicator when a 9 psi signal is applied to the bellows element of the receiver section.

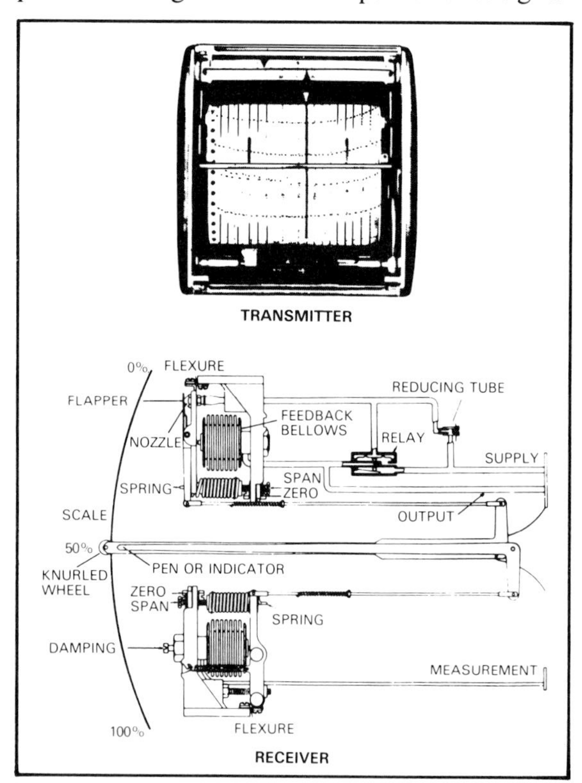

Figure 12.7. A strip chart recorder that produces a curvilinear trace (*Courtesy Foxboro*)

The transmitter section enables the recorder to perform as a remote control device for set-point adjustment. The set-point index contains a knurled wheel that serves to make fine adjustment and to prevent the index from drifting off the mark if subjected to vibration. Moving the set-point index varies the nozzle-flapper clearance, and the resulting change in nozzle back-pressure causes the air relay to change its output pressure. The air relay output is not only applied to the pneumatic-set device of the remote controller, but also to the local feedback bellows that acts to maintain the flapper-nozzle clearance in the throttling position.

Servomechanisms in Recorders

A servomechanism is a form of feedback control system in which the controlled variable is the *mechanical positioning* of a component such as a recording pen. The mechanical position of components has been a factor in previous studies—for example, the position of a valve plug or the position of a flapper with respect to a nozzle. But these instances of position were determined by the need to regulate the controlled variable and were simply coincidental.

The use of servomechanisms in strip chart recorders has a number of advantages, although their use will add to the cost and complexity of the units using them. As a rule, servomechanisms used in recorders provide for straight-line movement of the pen and indicator.

A Pneumatic Servomechanism

In the pneumatic curvilinear recorder (fig. 12.7), the pen swings about a rather long radius, thus reducing the arc it produces to a fairly flat form. Regardless of the length of radius of the arc, obtaining a flat recording line is not practical. However, there is a form of pneumatic servomechanism that causes the recording pen, or pointer, to follow a straight path and thus produce a trace in true rectangular coordinates (fig. 12.8*A*).

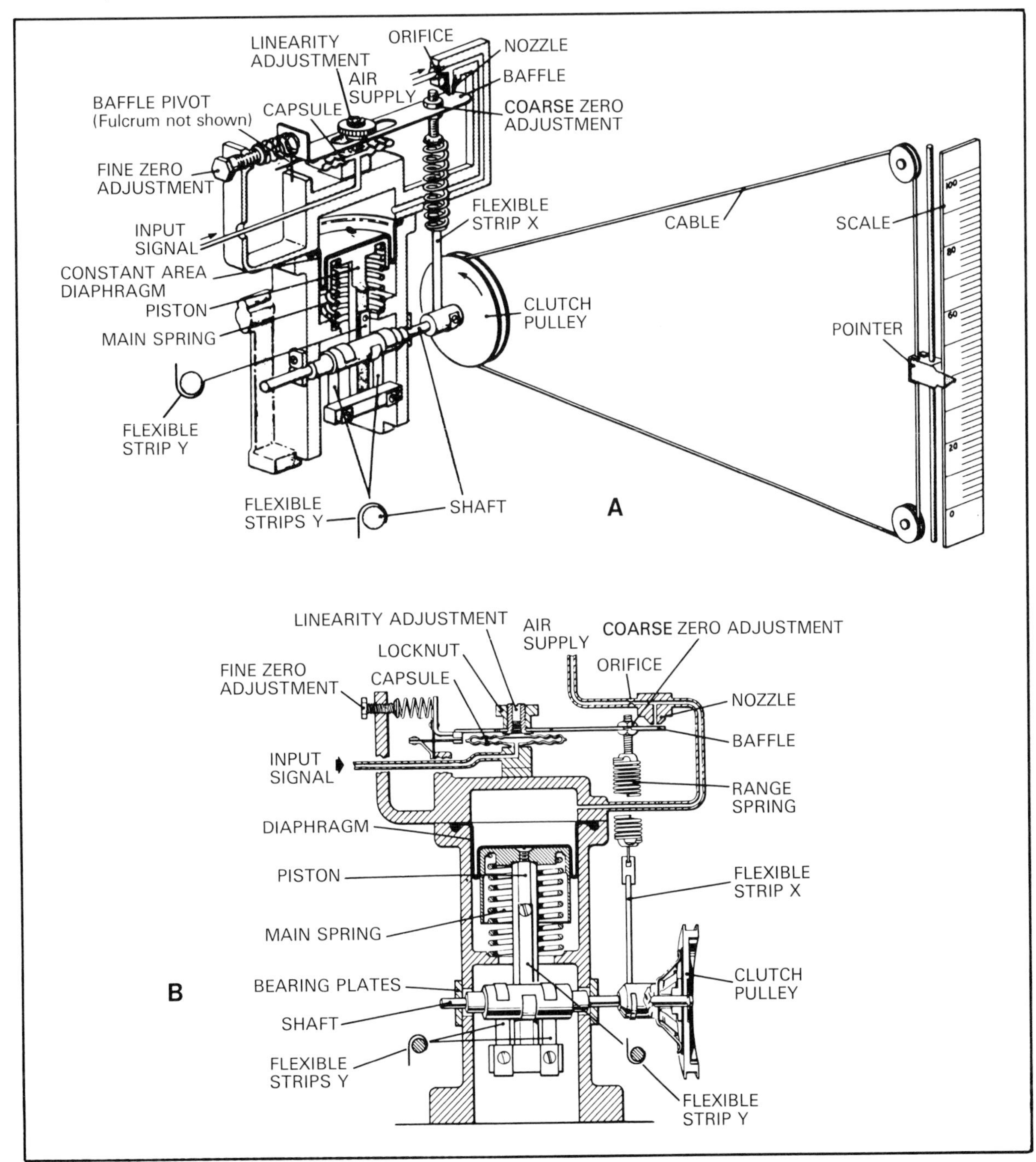

Figure 12.8. A straight-line strip recorder using a pneumatic servomechanism. *A*, diagram showing how pointer follows straight path; *B*, detail of clutch-pulley arrangement.

Control starts with a pneumatic signal applied to a diaphragm capsule. Expansion and contraction of the capsule positions a flapper-nozzle assembly, and the nozzle back-pressure is applied to a diaphragm cavity. Air pressure in the cavity tends to position the diaphragm and piston against the force of the main spring.

The clutch pulley is rotated by an arrangement of flexible strips, *Y,* two of them anchored to the main body of the device and the other attached to the piston-diaphragm-spring component (fig. 12.8*B*). Another flexible strip, *X,* is fastened between baffle and pulley hub. As the pulley rotates, it winds or unwinds the strip about its hub and varies the tension on the range spring. The mechanical feedback system performs the same function as that of a feedback bellows: it tends to counteract the displacement of the flapper caused by action of the capsule assembly. The linearity adjustment helps to attain a straight-line function between the pneumatic input signal and the recording pen and indicator.

Electronic Servo Systems

Control systems based on electronic components that use a voltage or current signal will almost invariably use a servomechanism to drive the recorder pens. A variety of such drive mechanisms is available, and the circuitry and principles of two such mechanisms, both intended to function with a 1- to 5-volt DC input signal, are described here.

Pen drive systems using bridge circuits. The use of bridge circuits is a common practice in electronic servo systems. One type of pen drive system uses a bridge composed of inductors, capacitors, and varactors (fig. 12.9). Varactors are diodes that possess significant capacitive reactance. Moreover, the diodes have the quality of experiencing appreciable change in capacitive reactance with change in current flow through them.

Circuits that incorporate inductors and capacitors, as in the bridge arrangement diagrammed in figure 12.9, require alternating current for operation. Coil L_o is energized with a 50-kilohertz signal that

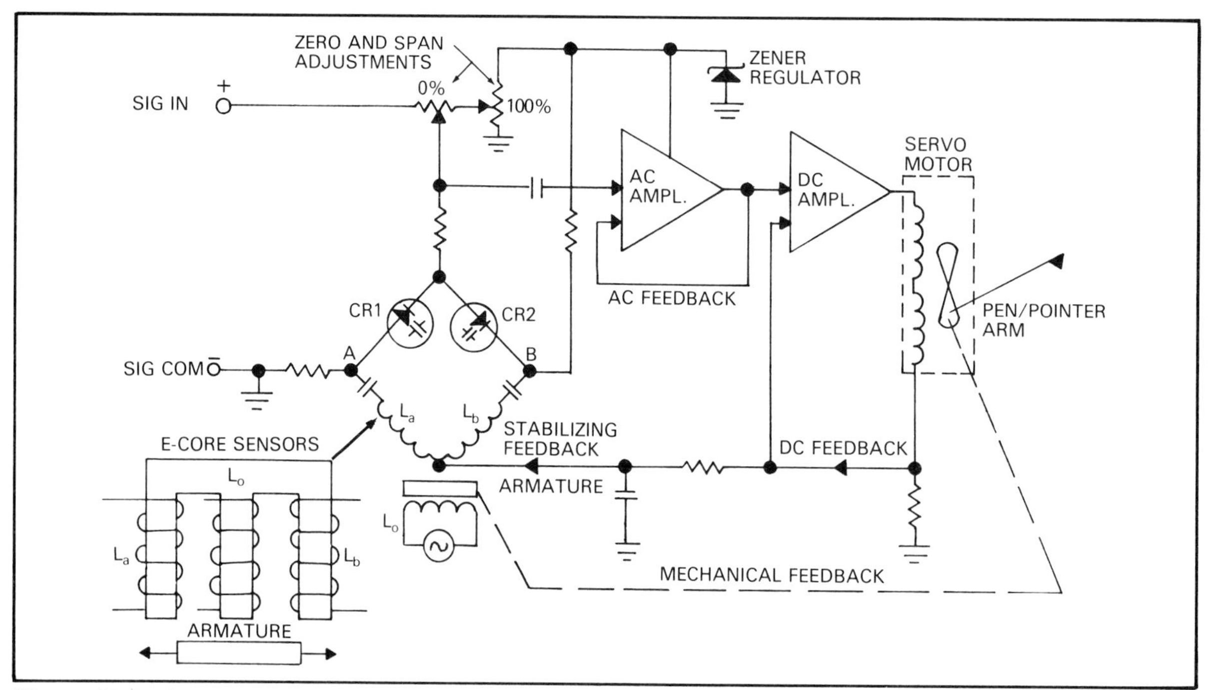

Figure 12.9. Servo system using a bridge circuit composed of inductors, capacitors, and varactors (*Adapted from Motorola Veritrak Circuitry*)

comes from an oscillator function in the AC amplifier section. Transformer action causes this AC voltage to be induced into coils L_a and L_b of the bridge. The position of the armature determines to a large extent the strength of the coupling between L_o and L_a and between L_o and L_b.

The 1- to 5-volt direct current representing the process variable is applied through the span-adjust potentiometer and other resistances to the bridge at the juncture of the varactors. Note that the opposite junction of the bridge receives a feedback signal from the servo motor circuit.

Balance of the bridge is achieved when the ratio of capacitive reactance of *CR*1 and *CR*2 equals the ratio of the inductive reactance of coils L_a and L_b:

$$CR1/CR2 = L_a/L_b.$$

For that condition a small AC voltage exists across the bridge. This small signal maintains the 50-kilohertz oscillation within the tuned amplifier loop.

Remembering that a changing voltage applied to a varactor diode changes its capacitive reactance, it is easy to understand that any change in the input signal, that is, any change in the process variable, will change the value of *CR*1/*CR*2 and upset the balance of the bridge. Note that for a change in input voltage, the reactance of one varactor will decline, while that of the other will increase. An unbalanced condition will cause a change in the AC voltage applied to the AC amplifier.

Output from the AC amplifier is rectified and fed to a two-stage DC amplifier that powers the servo motor. Motor torque resulting from increased current flow is opposed by the mechanical torque of a spring. In addition to driving the pen of the recorder, the servo motor also positions the armature associated with the *E*-core sensors. Positioning of the armature can cause the ratio of inductive reactances between L_a and L_b to equal the ratio of capacitive reactances between *CR*1 and *CR*2, thus balancing the bridge.

The servo motor (fig. 12.10) used in this pen drive system does not have an armature capable of complete rotation. The motor is a form of solenoid, with its armature fitted between two pole pieces. A torque spring maintains the armature at zero position until current flows in the field coils. Current flow will cause the armature to attempt to align itself with the magnetic field between the pole pieces. Linkages connect the armature to the recording pen.

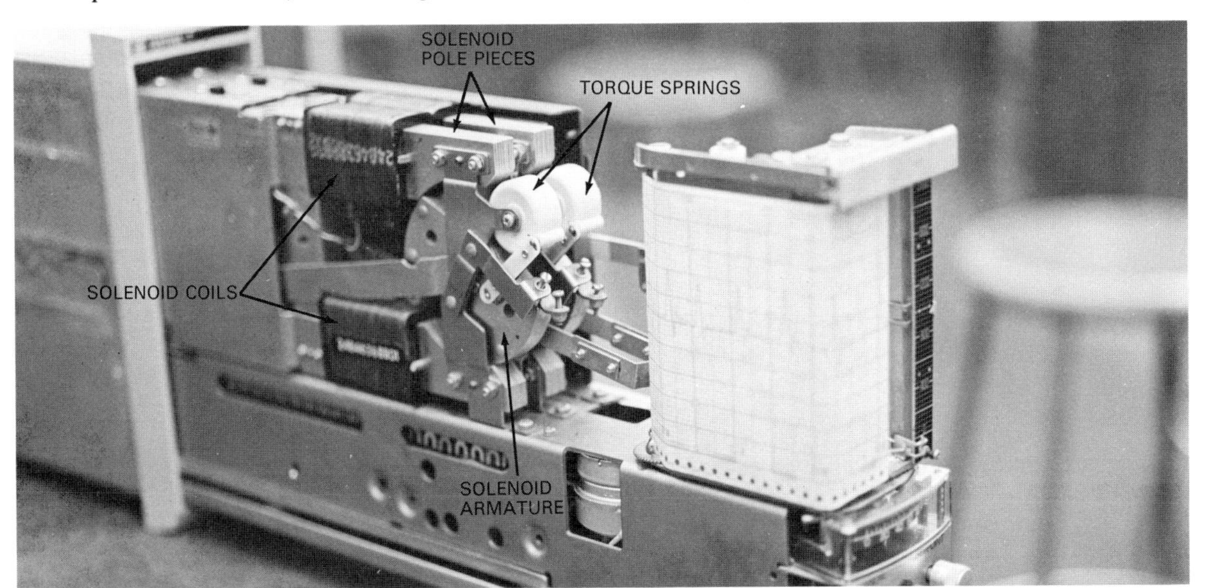

Figure 12.10. Servo motor in a two-pen recorder (*Courtesy Motorola Instrumentation and Control*)

The Zener diode establishes accurate reference voltages at points *A* and *B* of the bridge (fig. 12.9), and the input signal from the controlled variable is measured in relation to these constant reference voltages. Several feedback paths assure stable operation. Each amplifier section has a feedback loop, and the bridge circuit receives a DC stabilizing feedback signal from the servo motor drive circuit.

Pen drive systems using slidewire elements. One type of effective pen drive servo uses an op amp, power transistors, a small reversible DC motor, and a slidewire element (fig. 12.11). The op amp *IC*1 is configured as a differential amplifier, and its output provides drive signals to the two power transistors that control the servo motor. This compact system is commonly used in recorders capable of housing two or more pen drives.

The differential amplifier (discussed in chapter 4, *Automatic Control*) has the characteristic that if equal voltages are applied to its two inputs, its output will be zero. The noninverting input receives the 1- to 5-volt DC signal that is the analog of the process variable. This input is also connected to the regulated +18-volt DC supply through a high-resistance circuit so that a minimum of 3 to 4 volts is present at this input. The inverting input receives a voltage signal from the slidewire resistance element. The sliding contact of this element can have a voltage of 0 to something over 6 volts, depending to some extent on the span control setting. Another source of voltage for the inverting input comes from the zero-adjust potentiometer which is in a path between the +18-volt DC supply and signal ground.

The servo motor for this system has two separate field windings that are wound in opposition to one another. Then when both windings are energized equally, they neutralize one another and the motor's armature is at a standstill. When energizing currents of the two windings differ, the motor will develop torque in proportion to this difference and the direction of rotation will be determined according to which winding has the heaviest current flow.

The power transistors control current flow through the field windings of the motor. The transistors are arranged in push-pull format

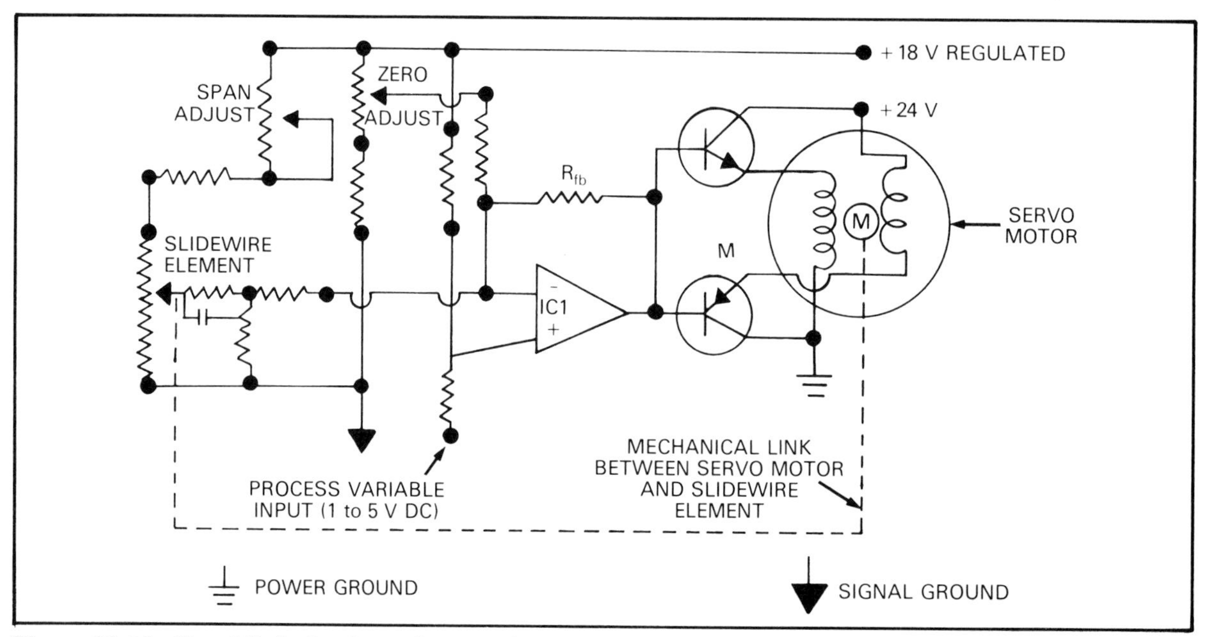

Figure 12.11. Simplified circuitry of a pen drive servo system that uses a slidewire element, an OP AMP, two transistors, and a special servo motor (*Adapted from Bristol Babcock*)

so that as current flow through one increases, current flow through the other declines. Current flow through the transistors is controlled by the voltage applied to their base elements. The base elements are connected together and connected to the output of the op amp, so the op amp controls current flow through the transistors. Remember, a differential amplifier can have a positive or negative output voltage, depending on whether voltage at the inverting input is less than or greater than that at the noninverting input.

The output from the op amp differential amplifier will be positive if the analog voltage of the process variable exceeds the reference voltage from the slidewire element. If the analog voltage falls below the reference voltage, the reverse is true. Since one transistor is NPN and the other PNP, any output from the op amp will cause one transistor to pass more current and the other correspondingly less current. The motor armature will move in one direction or the other. Rotation of the armature will simultaneously drive the pen and slidewire moving contact. The pen will indicate the new value of the process variable, and the slidewire contact will adjust the reference voltage to match the analog value of the process variable.

The zero-adjust and span-adjust settings are made using a precision source of 1- to 5-volt DC. With 1 volt applied to the process variable input, the recorder pen should read 0%, or other minimum value on the strip chart. If the reading is not within 0.5% of the correct value, the zero-adjust control should be varied to obtain such reading. Next, 3 volts is applied to the input and the pen checked to see whether it has moved to the midway, or 50%, mark on the chart. Then, 5 volts is applied to the input and the pen checked against the 100% mark on the chart. If the readings are not within 0.5% of the correct value, the span-adjust control must be reset. The pen is correctly set for 0%, using the zero adjust to achieve this setting. Then with a 5-volt input the pen position is checked. If it is off more than 0.5%, the span adjust is used to bring it to the correct position. Finally, the performance is checked again at 0% and at 50% marks. Minor touch-up of the controls may be necessary in order to make the pen track the input signals correctly.

Servo system with an alarm circuit. Recorders generally have circuits that provide for visual and audible alarms when process variables get out of bounds. A highly simplified alarm circuit (fig. 12.12) that can be

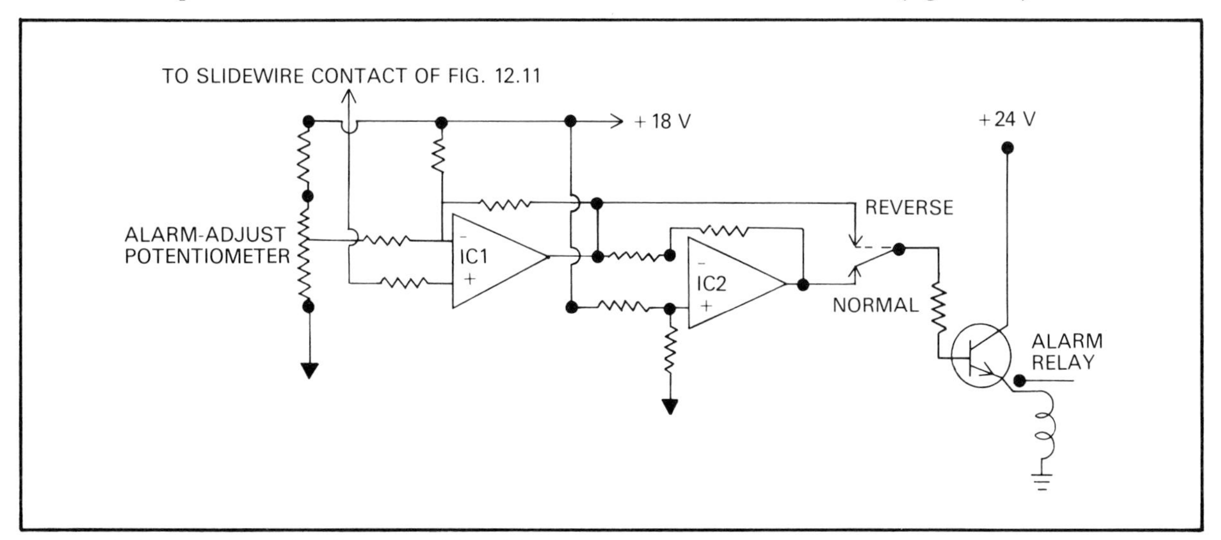

Figure 12.12. An alarm circuit compatible with the slidewire servo mechanism shown in figure 12.11 (*Adapted from Bristol Babcock*)

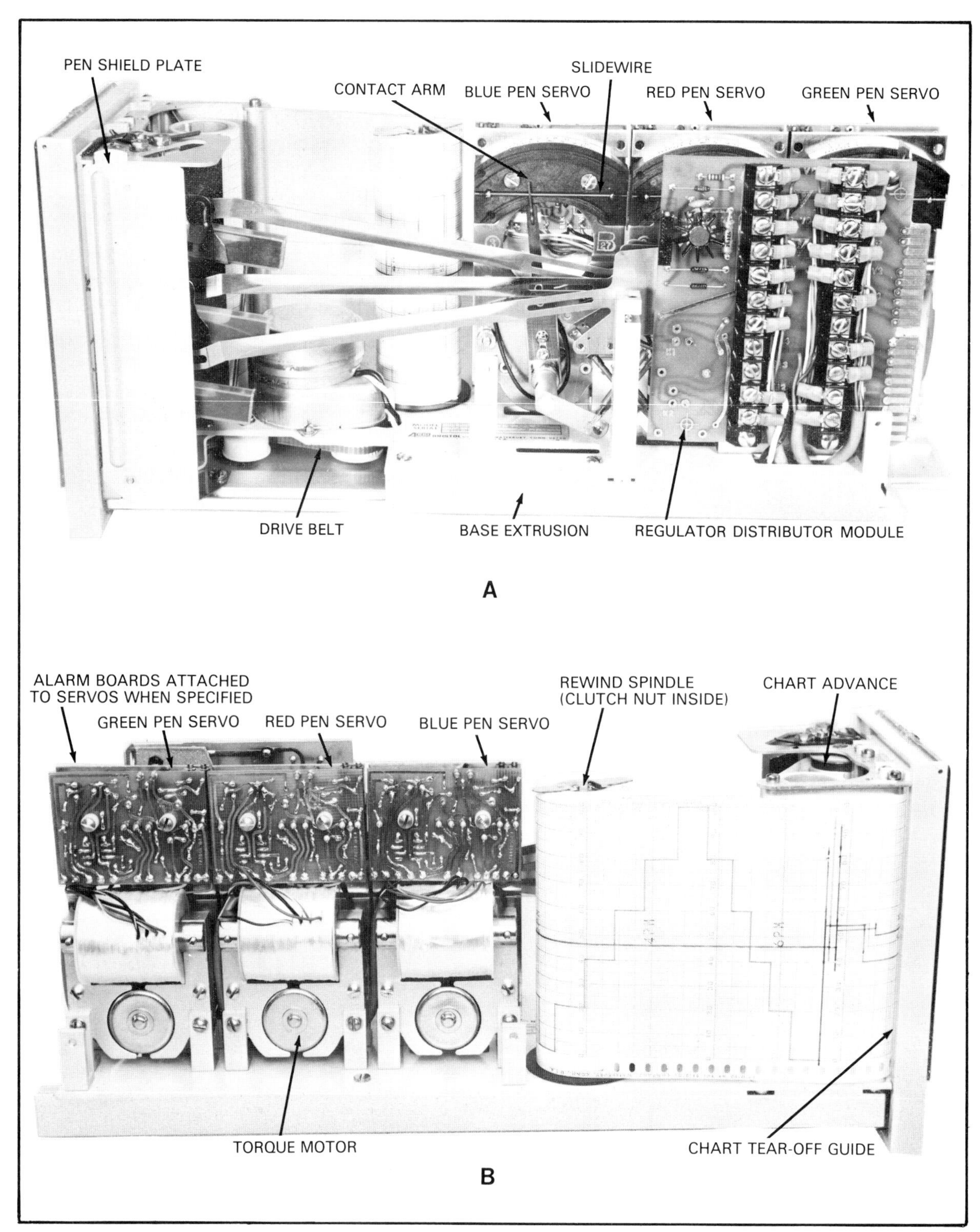

Figure 12.13. A three-pen recorder with shroud removed, using refined versions of the circuitry in figures 12.11 and 12.12. *A,* right side of chassis; *B,* left side of chassis. (*Courtesy Bristol Babcock*)

associated with the slidewire servo system described above consists of two op amps, a driver transistor, and a relay.

The inverting input of *IC*1 can be adjusted in the range of 0 to 6.5 volts by the alarm-adjust potentiometer. The noninverting input of *IC*1 receives its input from the moving contact of the slidewire element of figure 12.11. With the slidewire reference voltage less than the alarm-adjust voltage, the ouput from *IC*1 will be negative and the ouput from *IC*2 will be positive. For this condition and with the normal/reverse selector in the *normal* position, a positive bias is applied to the base of the relay drive transistor, Q_1, causing it to conduct and energize the alarm coil. The contacts of the relay in this instance are arranged so that an energized relay coil is a *no alarm* condition. Note that changing the selector to *reverse* changes the setup so that an energized relay coil is an *alarm* condition. With the normal/reverse selector in the *reverse* position *IC*2, the inverter op amp, is bypassed, and bias for the driver transistor comes from the output of *IC*1.

The three-pen recorder in figure 12.13 uses servo systems and alarm circuits. Actual circuitry of this recorder contains many refinements in the form of additional electronic components, but principles of operation are the same as those described.

Modern recorders make extensive use of solid state electronics and benefit from permanent lubrication. The three-pen recorder of figure 12.14 uses disposable markers, contains a DC power supply to drive 4- to 20-milliamp control loops, and is available with a fourteen-speed stepping motor chart drive system.

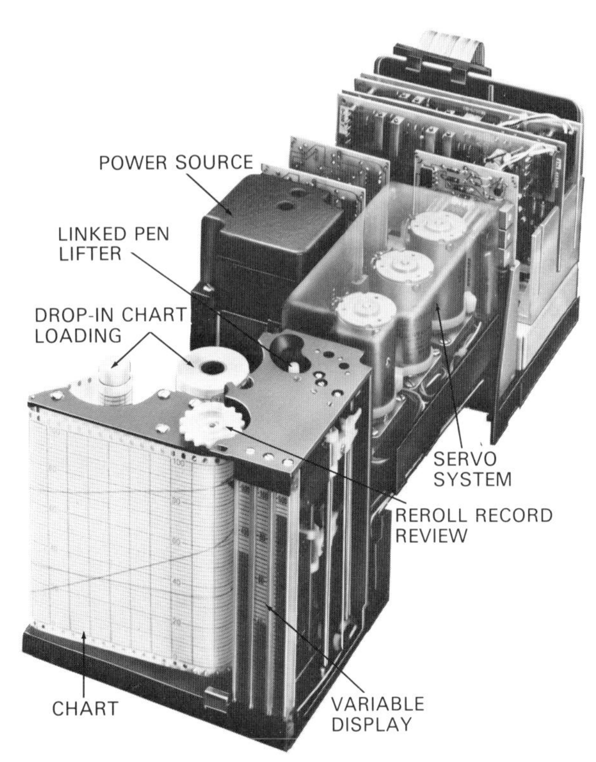

Figure 12.14. A modern strip chart recorder with three pens and novel display scales. Three servo motors and conductive plastic slidewire elements are contained in a single enclosed housing. (*Courtesy Leeds & Northrup Instruments*)

Chart Drive Mechanisms

Drive mechanisms include not only the clockwork device that causes the chart to move with respect to the recording pen, but also special gearing for changing speed of the drive, rewind features on strip chart recorders, and other contrivances that are related in some way with movement of the chart.

Driving Motors

The device that furnishes the driving power for the chart may be any of several forms of clockwork mechanisms: (1) a hand-wound clockwork mechanism driven by a spiral spring; (2) an electric synchronous motor, similar to that found in electric clocks; (3) electric-wind spring drive; or (4) other less popular forms such as pneumatic-powered drives.

Most circular chart recorders use the hand-wound mechanical clock motor, while synchronous motors are commonly found in strip chart recorders. The driving mechanism for circular charts must possess enough energy to provide at least one revolution without rewinding. The advent of chart-changing features has made it necessary to have clock motors that enable the recorder to produce a score or more circular charts without attention. One such product (fig. 12.15) has a stack of charts placed over a special spindle. After one revolution, the device automatically lifts the pens from the chart, releases the completed chart, and lets the pens back down on a new chart. This action takes place in a fraction of a second. The completed charts drop into a special container below the main body of the recorder.

Strip chart recorders require drives that produce a comparatively large amount of energy because of the great length of the strips. Although spring-powered drives are available for strip charts, most are driven by synchronous electric motors. Such drives have superiority where reliable electrical service is available.

Figure 12.15. A chart-changing mechanism used in a typical field installation for measuring natural gas (*Courtesy Mullins Manufacturing*)

Pneumatic Chart Drive Systems

Pneumatic chart drive systems have much merit in some installations, for example, when twenty to thirty recorders are concentrated in a control room. Another installation that might benefit from a pneumatic chart drive system is one in a hazardous location where the intrinsic safety requirements would impose much added expense if electric drive motors were used.

The control room installation, with its twenty to thirty recorders, points up the desirability of using pneumatic chart drives. Needed for such an installation are one chart drive motor for each recorder (fig. 12.16) and

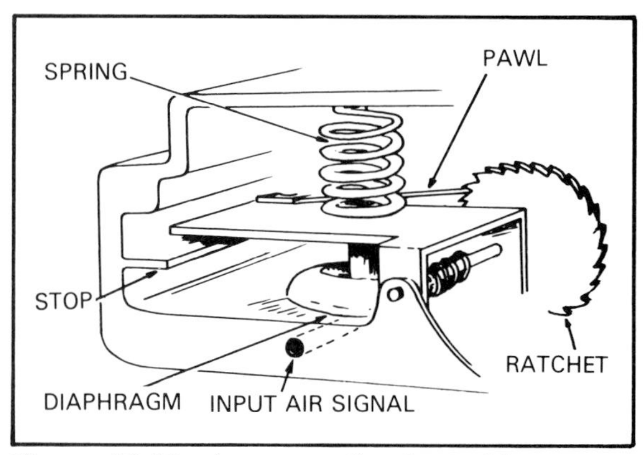

Figure 12.16. A pneumatic chart drive motor (*Courtesy Foxboro*)

just one master impulse unit (fig. 12.17). The simplicity of the drive motor is quite clear, and the impulse unit is not a very complex nor expensive component.

The master impulse unit is essentially a combination of a one rpm, 60-hertz synchronous motor, a four-lobe cam, cam follower, flapper-nozzle assembly, and a standard air relay. The motor drives the cam.

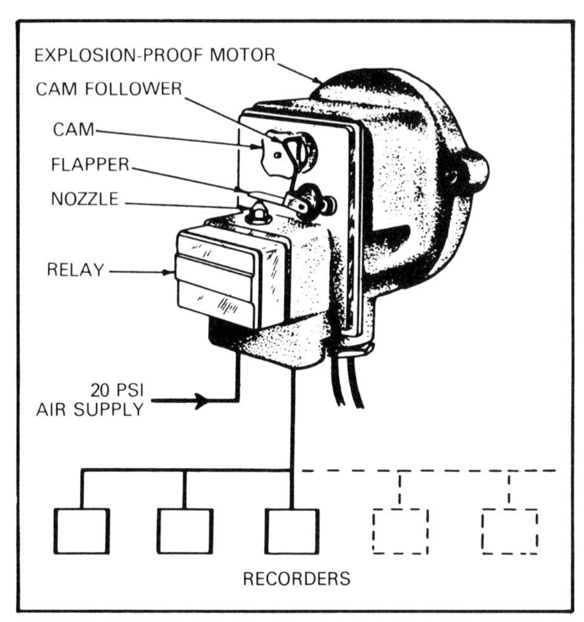

Figure 12.17. A master impulse unit, or transmitter (*Courtesy Foxboro*)

The cam follower is attached to the flapper. As the cam rotates, the cam follower causes the flapper to alternately cover and uncover the nozzle four times each minute. During the time the nozzle is covered, the air relay sends a signal, a pulse of 20 psi, to each of the recorders associated with the system.

The receivers, or chart drive motors, consist mainly of a diaphragm, ratchet, pawl, and mounting hardware. A small fraction of a revolution of the ratchet is made each time a pulse of air reaches the diaphragm. A shaft from the ratchet drives the sprocket shaft of the chart drive mechanism.

Such a system provides an economical approach to chart drive mechanisms, especially where more than three or four recorders are in close proximity.

Multispeed Drives

Having available two or more chart speeds for any recorder is desirable unless the recorder is made for a specified purpose and is always used for that purpose. Most chart drives are capable of only one speed, although many do

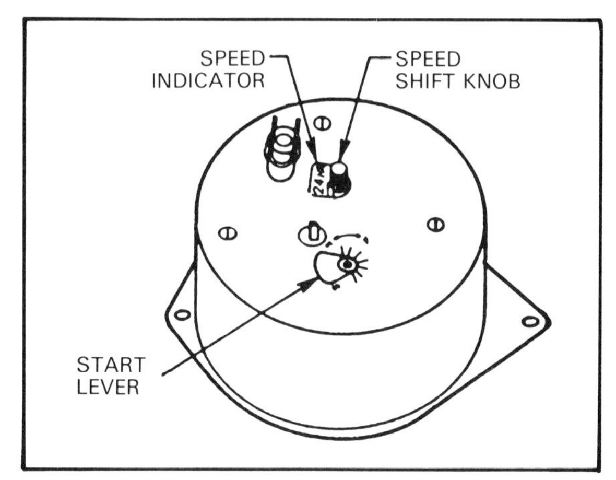

Figure 12.18. Two-speed clock drive with speed shift knob (*Courtesy General Time Corp.*)

possess the option of two-speed drives. In most cases a two-speed drive is shifted from one speed to the other by means of a shift lever or knob (fig. 12.18). Some recorder manufacturers provide gear sets for changing speed, each gear set being easily placed into service. Figure 12.19 shows a clockwork mechanism with a gear set installed and additional gear sets that can give the recorder a range of several speeds from two hours to seven days.

Some variable-speed chart drive systems employ a separate drive motor for each speed. This approach is not uncommon where large recorders are used, those having chart widths of 12 inches or more, for example. Each chart speed is selected by an interlocking device or switch. As many as five drive motors may be

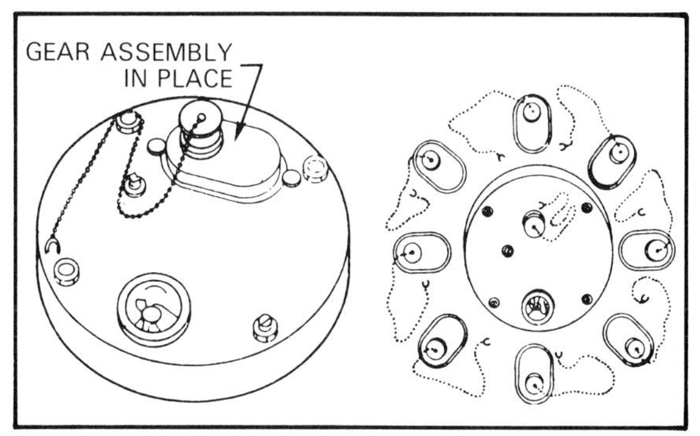

Figure 12.19. Multiple-speed gear sets for circular chart drive (*Courtesy Rockwell Manufacturing*)

Figure 12.20. Large five-speed strip chart recorder (*Courtesy Leeds & Northrup Instruments*)

used in a recorder, one for each chart speed. In the five-speed recorder shown in figure 12.20, the gearing ratio between motors and chart drive shaft can be changed quickly to adapt to either 50-hertz or 60-hertz power source.

When a large variety of chart speeds is desirable, recorders that use *stepping motors* in the drive mechanism can be used. As its name would imply, a stepping motor is one that rotates not steadily, but in distinct start-and-stop fashion. A step may amount to 15 degrees of rotation, with the stepping action so rapid that it gives the illusion of continuous motion.

Several types of stepping motors are available, ranging from the relatively simple and inexpensive solenoid and ratchet type, through the pulse or square wave operated permanent magnet types, to the variable reluctance type. Quality and reliability of stepping motors seem to increase with their complexity and cost. Solid state electronics has helped make stepping motors an effective choice for multispeed service.

Recorder Marking Systems

Chart recorders typically use pens and ink to produce the record of the variable being monitored. Perhaps the greatest single source of difficulty in a recorder is a failure of the pen to produce a legible trace on the chart. Certainly this problem plagued recorders in the past and is still a serious problem where old equipment is in use. Great progress has been made in recent years to provide components of marking systems that perform reliably for considerable periods of time. Much credit for improved performance is due to improved inks and disposable pen-ink combinations.

Much of the difficulty with any marking system can be avoided by closely following maintenance instructions issued by the manufacturer. This applies especially to older equipment.

V-Pens and Box Pens

Older models of circular chart recorders usually employed V-pens (fig. 12.21). The pen

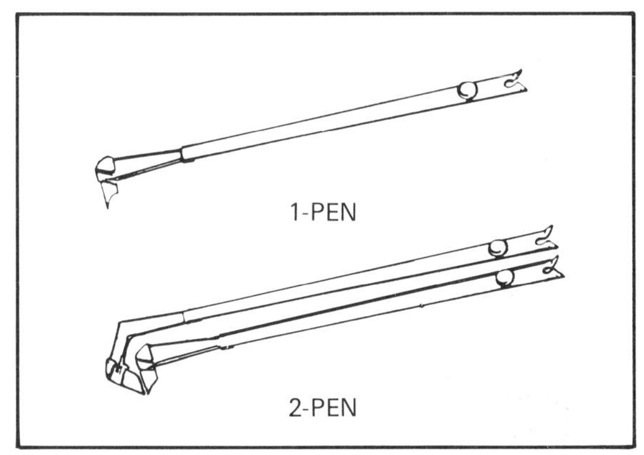

Figure 12.21. Several forms of V-pens for use on circular charts

gets its name from its shape, although the shape of most of them resembles a thin pyramid. V-pens are simple devices made of pressed thin metal that either slip on the end of the pen arm or fit into a socket formed at the end of the arm. Such pens hold about one drop of ink, so they should be checked closely each time a chart is changed. Care should be exercised to avoid overfilling.

A V-pen that has become fouled by the buildup of a residue when the ink has evaporated can usually be cleaned by flushing with water or fresh ink. In all instances of

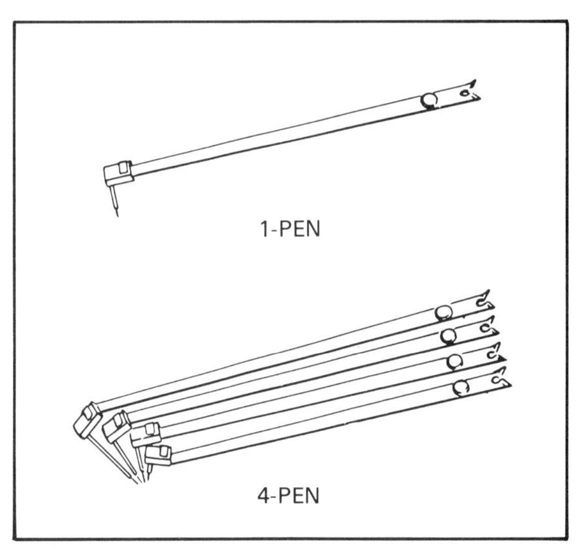

Figure 12.22. Box pens used on circular chart recorders

fouled pens, the manufacturer's instructions should be followed if available.

Box pens (fig. 12.22) have been popular for use in circular chart recorders, particularly because the fast movement of the pen arm might displace the ink from a V-pen. The writing end of a box pen is a capillary-like tube supplied with ink from the boxlike reservoir. Box pens may be filled from the point, using special ink and equipment supplied by the manufacturer.

Older models of strip chart recorders were usually fitted with ink reservoirs, although some laboratory-type instruments had fountain pens or ball-point pens. These inking methods met the need for the greater amount of ink needed on strip charts.

Disposable Pen-Ink Combinations

The use of V-pens and box pens causes a troublesome maintenance problem. Such pens require frequent refilling, a tedious task and one that takes time and care. The ink in these pens is prone to rapid evaporation, especially in hot and dry climates. Disposable pens are well on the way to totally replacing the older model marking devices.

Figure 12.23. A modern strip chart recorder that uses long-lasting disposable markers (*Courtesy Leeds & Northrup Instruments*)

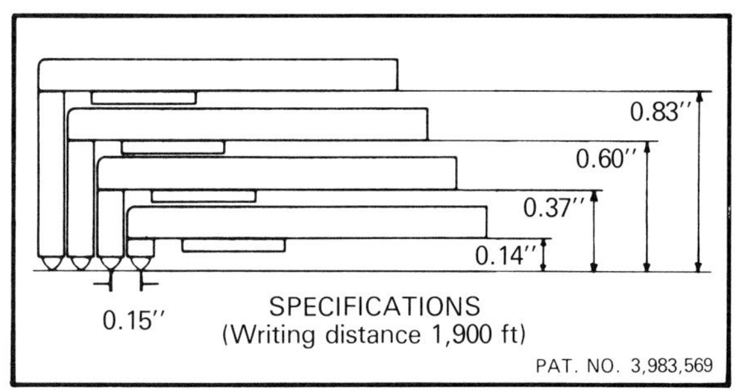

Figure 12.25. Snap-on disposable pens for multiple circular chart recorder (*Courtesy Graphic Controls*)

Modern strip chart recorders are fitted with the disposable pen-ink cartridges (fig. 12.23), and in many cases the older models of strip chart recorders can be retrofitted with this modern concept. The disposable cartridges are quite compact, but contain enough ink to draw a trace that is 2,500 feet long.

A sectional view of a disposable pen (fig. 12.24) shows the construction of a common type used widely today. It consists of a molded plastic body, ink reservoir, and fiber writing nib. An array of four disposable pens (fig. 12.25) shows vertical and radial separation normal for a multiple-variable circular chart recorder. Using the proper cartridge in each of the four positions is clearly necessary. The proper choice of ink color is also important in multiple-variable recorders.

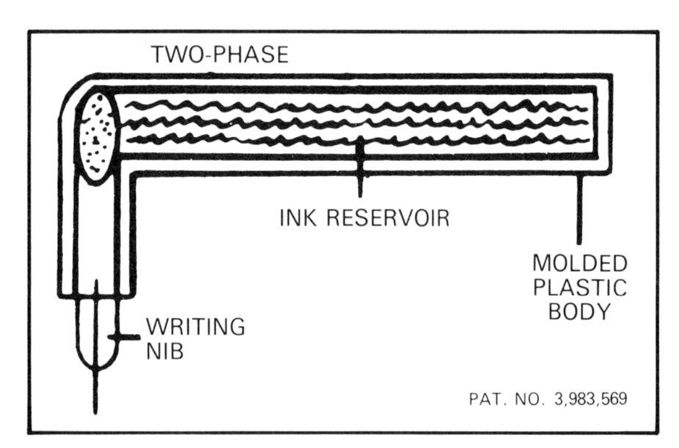

Figure 12.24. Lightweight disposable recorder pen (*Courtesy Graphic Controls*)

Disposable pen-ink cartridges are readily available in five colors, and each contains enough ink to run off nearly one thousand 12-inch circular charts. The shelf life of such cartridges is rated at one year. Considering their cost and ease of replacement, disposable pens represent a sensible approach to economical operation.

Styli and Other Trace Forms

In some situations recording with ink is not satisfactory for one reason or another. Several other methods are available that are suitable for environments or demands that rule out the use of ink tracing. One form uses a photosensitive paper and a fine traveling light beam that responds to the measured variable and traces an exposure path on the sensitive paper. This trace form is used in oscillographic recordings.

A purely mechanical device for tracing on chart paper involves the use of a sharp stylus (needle) that literally scratches a trace line in a soft coating material that is applied to the paper. The coating material will be black in most instances and applied to a white paper backing. Thus, the trace will appear as a white line on a black or gray background.

Some recorders use a heated stylus that causes a change in color of a thermal-sensitive coating applied to a paper backing. The stylus, frequently of sapphire or other jewel-like substance, is maintained at a fairly high temperature (perhaps 300°F) by a special electric heating element that fits around the upper body of the stylus.

Solid State Devices

During the past few years recorder manufacturers have been quick to take advantage of the opportunity to adapt their instruments to the technology surrounding solid state devices. It is now common to find recording instruments that contain microprocessors that enable the operator to easily program a recorder to accomplish a vast assortment of functions. As an example, one such recorder (fig. 12.26) accepts inputs in the form of

Figure 12.26. A multivariable, microprocessor-based recorder (*Courtesy Honeywell*)

volts, millivolts, milliamperes, and the output of resistance temperature detectors. It contains a dot-matrix printer that produces letters as well as numerals and graphic characters. A battery-powered clock enables it to produce time lines across the uniformly spaced 100-division chart. It records or prints in six colors, although a one-color version is available that uses pressure-sensitive chart paper.

The versatility of such modern recorders enables them to serve just about any need, but they are usually employed in laboratories or special situations that require this great versatility. Most recording needs in the petroleum industry are served quite well by recorders that have a limited capability. These traditional recorders are effective instruments known for reliability and are easy to maintain by following the manufacturers' instructions.

Integrators

Integrators find wide application in connection with instruments that measure the flow of something—liquids, gases, even electricity. Of course, they are also used to measure miles or other units of distance and units of time.

Integrators are devices that count, and the results of the count are indicated by a direct-reading counter or a dial counter. Automobile mileage indicators are direct-reading devices, while most gas meters and electric kilowatt-hour meters are dial types. In addition to the counters, this section will explore the methods of driving them.

Some of the driving devices for counters are quite simple. The nutating-piston meter is an example as it requires only a gear train attached between the nutating pin and a counter. A similar situation exists for the system associated with a mass flow meter, where the basic requirement is a gear train between the counter and an axle driven by motion about the major axis of a gyroscope assembly. Other devices take more complex forms of coupling between the measured variable and the counter.

Almost all counters are actuated by having rotary motion applied to an input shaft. The rotary motion may be intermittent, as from a ratchet-and-pawl device, or it may be continuous.

Intermittent Integrators

One form of an intermittent integrator associated with a flow recorder is shown in figure 12.27. Its counter is actuated by a periodic (once-a-minute) motion of a lever called a striker arm. Striker arm *B* is attached to drive shaft *C*, which in turn is geared to the counter unit. Once each minute the chart drive motor moves the striker arm from its normal *up* position to a downward position that is determined by the position of cam *A*. Movement of the striker arm is accomplished through gears *K*, *L*, and *M* and lever *J*. The position of cam *A* is controlled by movement of the pen arm through lever *R*.

The striker arm *B* does not turn shaft *F* in its downward motion due to a clutch arrangement between shafts *C* and *F*. However, in its upward motion toward the normal horizontal position, striker arm *B* rotates shaft *F* and consequently operates the counter.

The position of cam *A* is determined by the pen arm, as there is a link between the two mechanisms. However, for 55 seconds of each minute integrator cam *A* is locked into a set position by brake *H*, during which time a change in pen-arm position will be reflected by cam *A*. In order to move during this period, the pen arm must overcome the force of the pickup springs between the pen bracket and link *R*. The fact that measuring elements are active for only about 5 seconds during each minute, as far as the integrator is concerned, is not considered significantly adverse to acceptable accuracy.

The range of the instrument shown in figure 12.27 is changed by changing the ratio

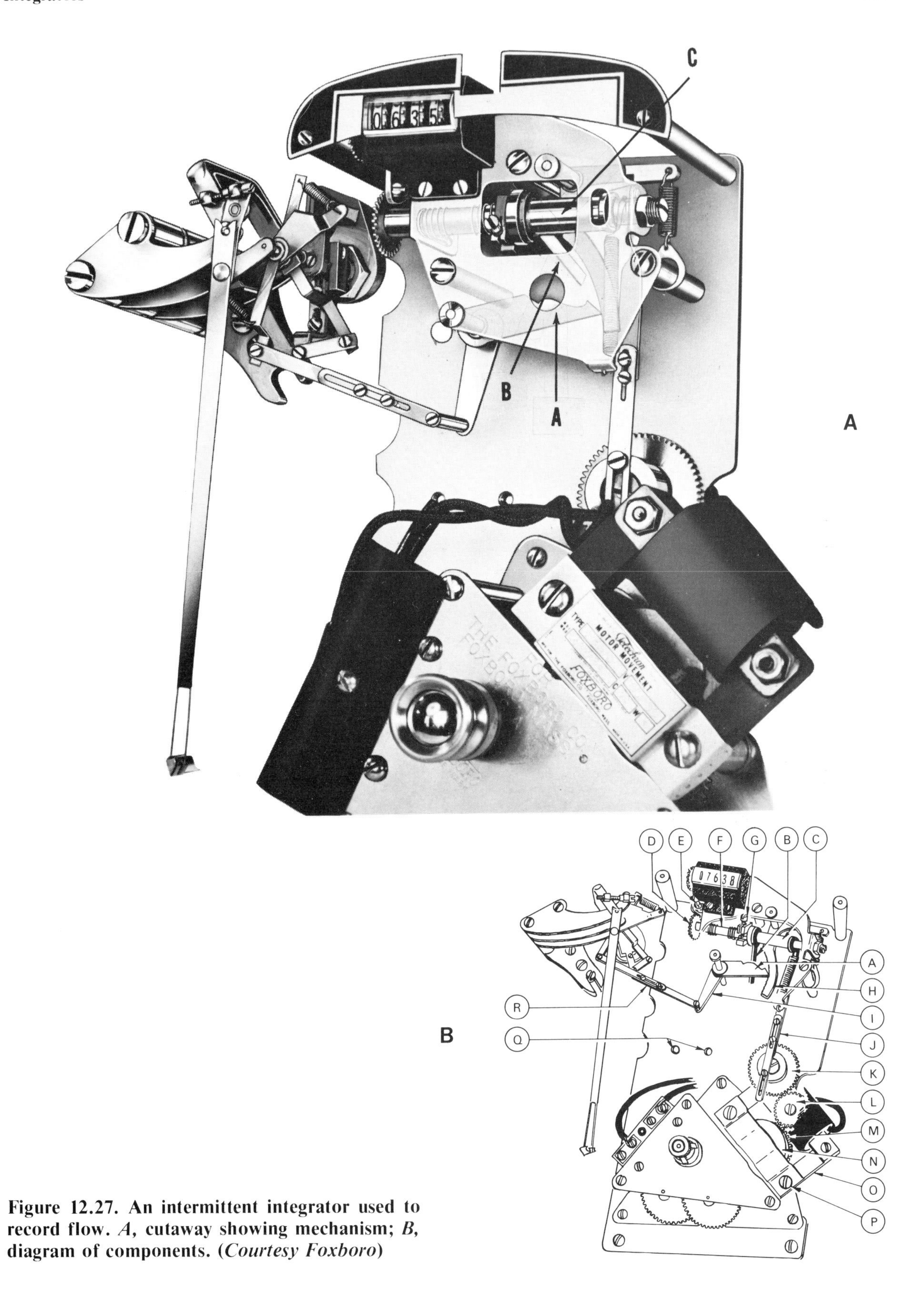

Figure 12.27. An intermittent integrator used to record flow. *A*, cutaway showing mechanism; *B*, diagram of components. (*Courtesy Foxboro*)

of the gearing between the counter and shaft *F*. This is done by substituting for gear *D*. These gears are available with as few as six and as many as fifty teeth.

The pen arm and cam *A* of the recorder are positioned by the differential pressure that exists across an orifice plate or other differential pressure device. Recalling the flow equation, quantity of flow as calculated from such devices is a square root function; that is, volume of flow is a function of the radical quantity $\sqrt{hp}$, where h is the differential pressure in inches of water, and p is the pressure-tap pressure in psi. As a result of the square root function, cam *A* has a special shape so that for each position of the pen arm, the amount of rotation permitted the striker arm represents a true value of the actual volume of flow. A quite similar integrator mechanism has cam *A* shaped for a linear relation between actual flow volume and pen-arm position.

Another form of intermittent integrator mechanism has a synchronous motor that drives a heart-shaped cam at a constant speed of two rpm (fig. 12.28). A friction clutch between the cam and the escape wheel causes the latter to be revolved in unison with the cam unless the pawl engages the teeth of the wheel.

The counter is driven by a gear train, which is connected to the escape wheel. Therefore, the counter input shaft is turned at a constant speed except when the pawl engages teeth on the escape wheel. Whether or not the pawl is engaged with the escape wheel is a function of two things: (1) the position of the roller arm and (2) the position of the flow arm. Position of the flow arm is determined by the value of the variable input to the integrator mechanism—differential pressure across the orifice plate, for example. Position of the roller arm varies (1) as the cam rotates, because a roller attached to the arm rides on the cam's working surface; and (2) as a function of the flow-arm position, because the left end of the roller arm is pivoted on the flow arm.

For zero input at the flow arm, the pawl is pushed so far toward the escape wheel that at no point during a revolution of the heart-shaped cam will the pawl be disengaged from the escape wheel teeth. As input to the flow arm increases to values above zero, the pawl will be retracted from the escape wheel. The

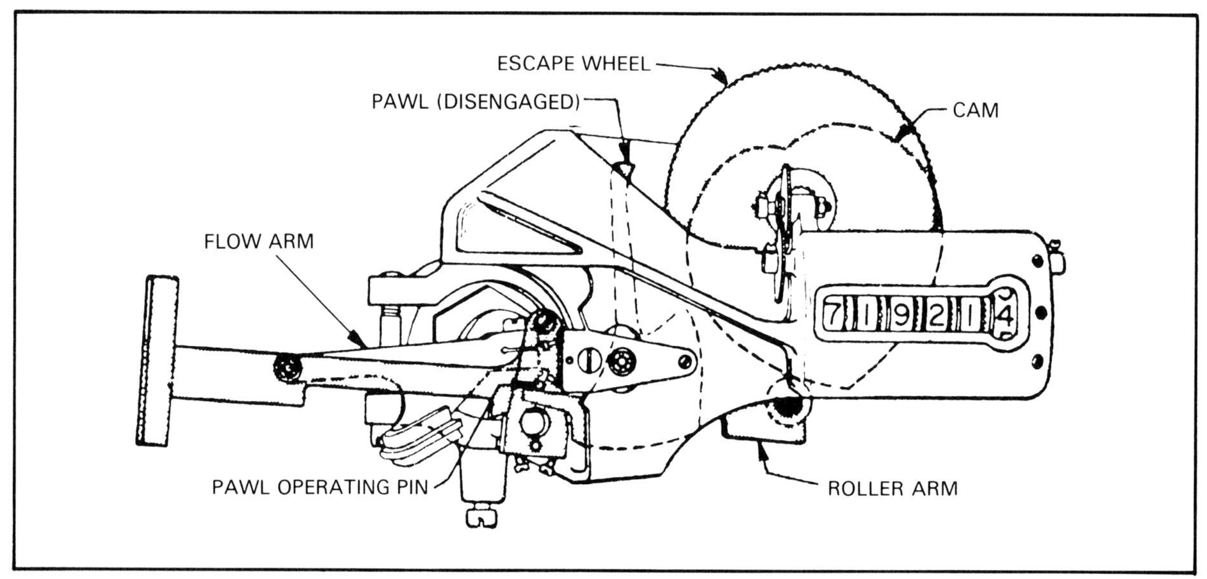

Figure 12.28. Integrator with a heart-shaped cam rotated by a synchronous motor (*Courtesy Bailey Meter*)

amount of the retraction will be in proportion to the magnitude of input: as it increases, the escape wheel will be able to turn increasing portions of each revolution made by the cam.

Continuous Integrators

One continuous-type integrator has a pneumatic-powered mechanism (fig. 12.29*A*)

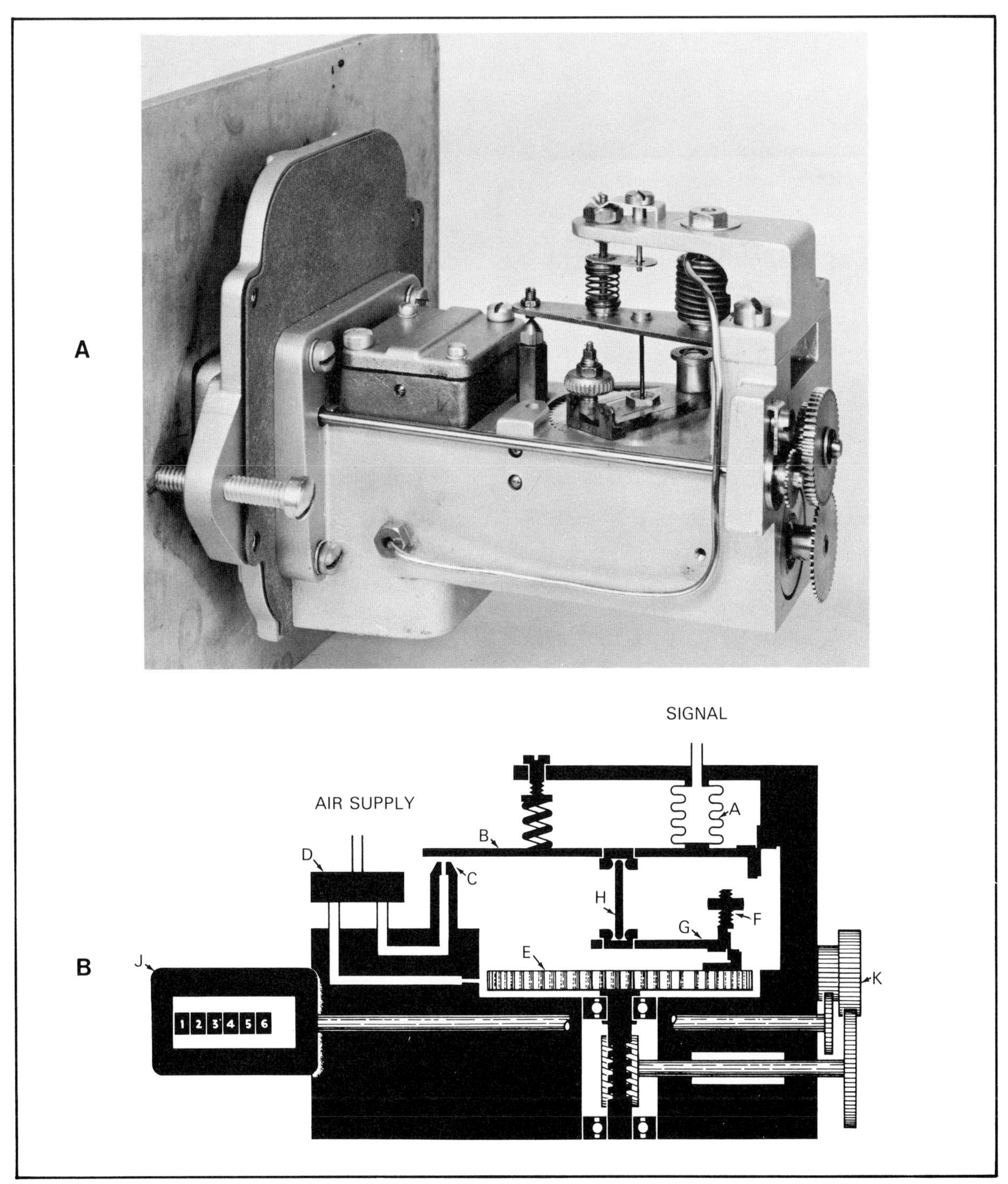

Figure 12.29. A continuous-type integrator designed for commercial use. *A*, pneumatic-powered mechanism; *B*, diagram of components. (*Courtesy Foxboro*)

and contains many fundamental features such as the nozzle-flapper combination and the bellows input element. This integrator is very satisfactory for operation from the 3 to 15 psi air signals that could be received from a differential pressure cell transmitter (discussed in chapter 9).

The input signal to the integrator (fig. 12.29*B*) is applied to bellows *A*, and increasing the pressure to this bellows tends to narrow the separation of the nozzle *C* and flapper *B*. The air relay *D* has its air output controlled by nozzle back-pressure, and the relay output is fed to a jet that drives turbine motor *E* at a speed that is proportional to air relay output pressure.

Rotation of the turbine wheel *E* drives the counter *J* through a system of shafts and a gear train *K*. It also causes another effect through centrifugal force acting on weight *F*, which is attached to bell crank *G*. The bell crank is flexure-pivoted near the outer edge of turbine wheel *E*.

Centrifugal force on *F* is transferred through the bell crank to thrust pin *H*, and the force tends to push flapper *B* away from nozzle *C*. The centrifugal force acting on weight *F*, and therefore on thrust pin *H*, is proportional to the square of the turbine speed:

$$F_c = M_F \times L_F \times \omega^2$$

where

F_c = centrifugal force due to uniform angular velocity;

M_F = mass of weight *F*;

L_F = distance of weight *F* from axis of turbine wheel;

ω = angular velocity, in radians per second.

Since the 3 to 15 psi signal pressure is proportional to the square of the rate of flow, the turbine speed is directly proportional to the rate of flow. Disregarding possible error due to friction or other losses, a proper gear ratio between turbine wheel and counter will produce an exact accounting of total volumetric flow.

Summary

Integrators and recorders are devices that totalize and produce a permanent record of variables in a system. A permanent record is desirable for a number of reasons. Such a record might be used to determine the amount of steam, gas, or water used or delivered; to account for variations in product quality; and to serve as operating guides for proper performance of equipment and personnel.

Numerous types and models of recorders and integrators are in use today. A few types are discussed here to give a general understanding of the overall characteristics. Making use of these basics and the appropriate manufacturers' literature should assure the successful operation and maintenance of these straightforward, but delicate, instruments.

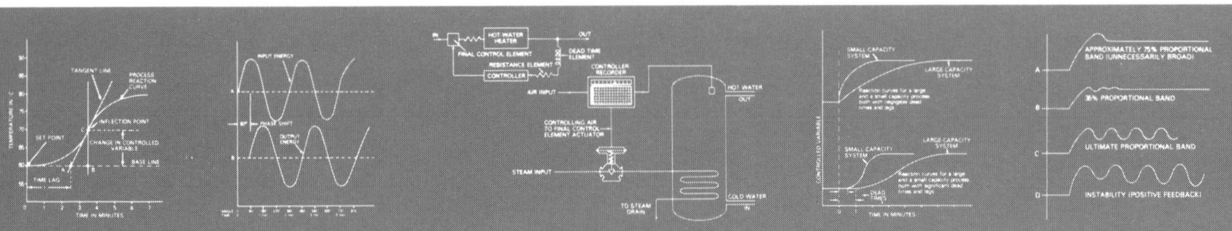

CHAPTER

13

Adjusting Automatic Controllers to a Process

In previous chapters the variables that need to be measured and controlled, their units of measurement, and the individual components needed to accomplish the measuring and controlling were discussed. There was also discussion of how any given system that is to be controlled possesses important characteristics that might set it apart from another system. Common characteristics include capacity, resistance, and response lag among the measuring and controlling elements. These characteristics influence the reaction rate of a control system.

This chapter will deal extensively with the techniques and the information needed to adjust a control system to the particular process to which it is associated. The ultimate goal is to fine-tune a controller so that it maintains the controlled variable as near the set point as practicable, yet not so sensitive that the system becomes unstable in the event of a minor upset. The topics of stability, phase shift, and process reaction rate will be covered in good detail, as will acceptable methods for achieving good control characteristics through proper adjustments of the three modes of control.

Process Reaction Rate

The process reaction rate is significant to the proper design of automatic control systems. As described earlier, the performance curve for a process reaction is obtained by disabling the automatic features of a system and introducing a step-change in energy input by either opening or closing the final control element a very small amount. This step-change must take place while the system is stable at the set point, and the change must be virtually instantaneous. A recorder or other means is used to obtain the reaction of the controlled variable to the step-change.

The curves of figure 13.1 are similar to those of figures 1.13 and 1.14, except that pairs of similar curves have been combined to form single figures. Briefly, the following remarks can be made concerning the curves and the processes they represent.

1. Small-capacity processes have steeper response curves for a given energy input or output change and, therefore, reach new levels of stability more rapidly than large-capacity processes.
2. Dead time is the lapse in time between the moment the step-change is made and the moment the measuring means (or recorder) first begins to reflect the change. The location of the primary element relative to the point at which the energy change is taking place has a pronounced effect on the dead time of a system.
3. Lag in the process is shown as a concave shape occurring immediately after dead time. Lag is the slowing down of the effect of the energy change as reflected by the measuring means or recorder. Lag is caused by resistance elements in the process or the control system.
4. The *reaction rate* of a process is indicated by the *slope of its reaction curve* at the inflection point (fig. 13.9). The slope can be determined with good accuracy by doing the following:
   a. Draw a line tangent to the curve at the inflection point and extend this line to the base line. The base line is the set-point line.
   b. Draw a line from the point of inflection to the base line, making it perpendicular.
   c. The slope of a curve at any point is the slope of its tangent at that point. To obtain the slope of the tangent, divide the length of line *BC*, as shown on figure 13.9, by the length of line *AB*. This amounts to expressing the reaction rate as the change in value of the controlled variable per unit time.
5. The effective time lag is determined by the point at which the tangent line crosses the base line, as shown in figure 13.9.

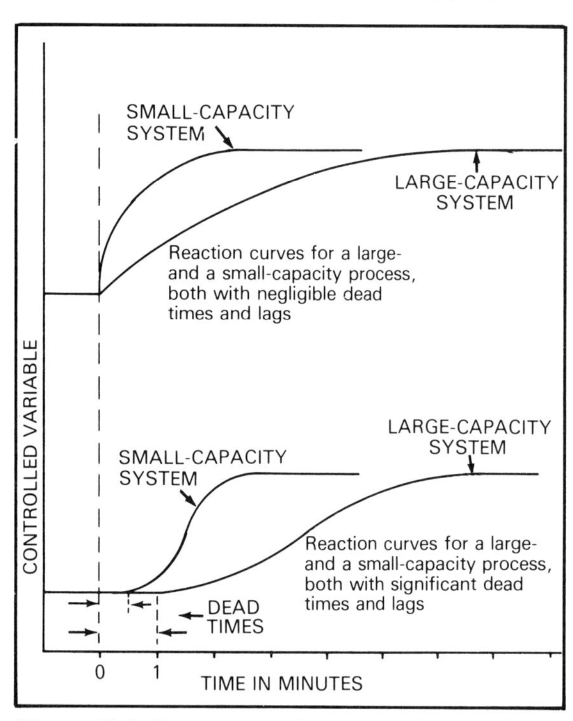

Figure 13.1. Process reaction curves for small- and large-capacity systems

An ideal process or system has no lags or dead times. Such a system is never achieved, although some systems approach such perfection. A system or process having virtually no time lag is usually easy to control, because any change, however small, is instantly detected, and immediate countermeasures are applied to exactly offset the change. Control problems arise when time lags exist; unless these factors are taken into consideration, the corrective countermeasures will be applied to the process at the wrong times and probably in the wrong amounts. In such instances, the corrective measures are not properly "phased" for best results. It is conceivable that if dead time and lags are sufficiently severe, the countermeasures might be so poorly phased and of such quantity that instead of correcting the deviation they will cause it to become greater.

Stability of Control

Stability of control refers to the effect the controller has on the controlled variable. Controllers can be either stable or unstable, depending on the characteristics of the system. An unstable system, of course, is apt to cause the controlled variable to fluctuate more widely than no controller at all. Understanding stability and instability will make it easier to adjust a given controller to a process or system.

A closed-loop control system, which is the sort most commonly used, depends on the *feedback principle* to achieve the balance between the input and output of a process. The feedback principle calls for taking a little of the energy from the output and feeding it back to the input of the system. A simple hot-water system with a feedback-type control loop (fig. 13.2) can be used to clarify several things. When the temperature of the outgoing water exceeds the set-point value, the energy output is greater than desired. Normal corrective action in this case requires a *reduced* energy *input*. If the situation is reversed, that is, if the output temperature (and energy) is below that desired, the input energy needs to be *increased*. The main point here is that the feedback energy must be applied to the input in such a way that it is *out of phase* with the

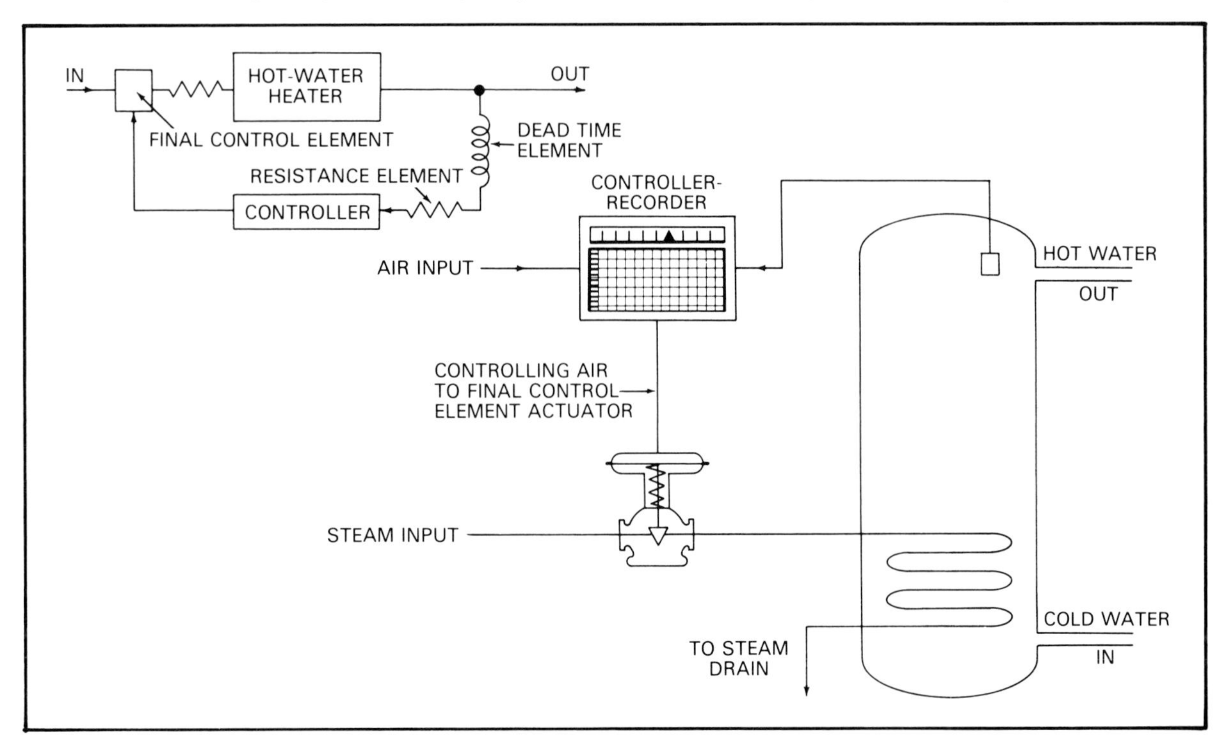

Figure 13.2. A hot-water system with closed-loop control

output: as the output energy rises above normal, the input energy must be *reduced;* as the output energy falls *below* normal, the input energy must be *increased.*

Ideally, the corrections applied at the input of a process should be 180° out of phase with the deviations in the output, but due to dead time and lags, such exactness is not practicable. Many practical systems have time lags small enough to permit an almost ideal situation. Of course, achieving a complete and exact phase reversal at the input is easy to accomplish by merely reversing the action of the final control element. In the hot-water system (fig. 13.2), for example, the steam control valve moves toward the *closed* position as the water temperature increases, although earlier studies pointed out that final control elements are available in such forms as air-to-open and air-to-close. Substituting one of these valves for the other will cause a complete reversal of the action, or a 180° phase shift. It is important to realize that any problems caused by dead time and lags have not been solved. It has merely been shown that a 180° phase shift can be obtained quite easily. The real problems will arise from trying to offset phase shifts of somewhat more or less than 180°. Then, the needs for a feedback control system can be expressed in a few words: the corrections applied at the input to the process must be *negative feedback,* or of *opposite sign* (positive or negative), to the output deviation.

Phase Shift and Natural Frequency

The terms *phase* and *phase shift* always imply the existence of two or more variable quantities whose simultaneous values are being compared. In this case the quantities are the input and the output energies of a controlled process. Time is an all-important factor in measuring phase (although phase difference is usually expressed in degrees or radians), and that is why time lags in a controlled system become vital considerations.

All feedback systems have *natural frequencies.* This is true whether the system is an electronic oscillator or an energy loop. The natural frequency of a closed loop is the frequency at which the system would like to oscillate, that is, rove between maximum and minimum values at a rate determined by the phase shifts between input and output.

The sine waves depicted in figure 13.3 show phase shift and gain between input and output energy. Sine wave *A* represents energy that is being fed into the hot-water system of figure 13.2. This sine wave form of energy input could be achieved possibly by a special device that controlled air to the pneumatic actuator in such rhythm as to cause a sine wave input. The automatic feature of the controller would be disconnected for this study, of course. Sine wave *B* represents the output energy from the process, but it does not reflect the same amplitude (height above and below the reference line) as the input wave. This is

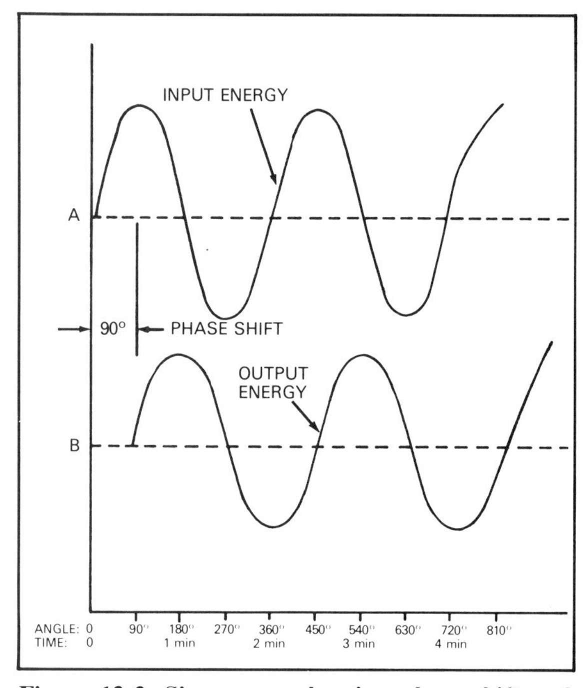

Figure 13.3. Sine waves showing phase shift and gain between input and output energy

accounted for by the fact that time lags in the system prevent the output variations from ever catching up with those occurring at the input and by the additional fact that energy is simply lost in the process due to heat leakage through the walls and other areas.

An important point here is that the system puts out less energy than it takes in, and it has a *gain* (amplification) of less than 1.0. Another equally important point is the fact that deviations in the output energy do not occur in step with the deviations of the input energy. In fact, they occur somewhat later—about 90° later—resulting in a *phase shift* between input and output.

A cycle of change is 360°. Note that the sine wave representing input energy begins at the reference line, rises to a maximum peak, declines to a minimum below the line, then rises back to the reference line. This completes one cycle, or 360°, and a new cycle then begins. The frequency of the wave is the number of cycles that occur in a unit of time. The time base for figure 13.3 is in minutes and degrees, and it can be seen that one cycle takes 2 minutes. The frequency, then, is one-half cycle per minute, or thirty cycles per hour. Note the relation between time and degrees.

The 90° phase shift indicated in figure 13.3 between the values of the input energy and output energy is caused by dead time and lags. When the controller is reconnected, additional lags that amount to another 90° might be picked up (in the final control element actuator, in the response of the primary element, etc.). If this 180° phase shift is combined with the built-in 180° phase shift of the controller, it is easy to suspect that the system will be unstable at the frequency of one-half cycle per minute, because of the 360° phase shift. A phase shift of 360° between input and output of any system produces *positive feedback*. Positive feedback means that the signal sent from the output will be in the same direction as the action of the controller. If the controller is feeding in more energy than needed to sustain the output demands, positive feedback will tend to support and aggravate this undesirable condition. Negative feedback is needed in control systems, but positive feedback is generally a no-no.

For any feedback control system in which the phase shift between the output energy and the input energy is 360°, there is danger of severe instability. If the overall gain of the system is greater than 1.0 at the frequency at which the phase shift is 360°, the controlled variable will oscillate in an increasing amplitude above and below the set point. If the overall gain is less than 1.0, the system will eventually settle out at the set point. For an overall gain of 1, the controlled variable will oscillate above and below the set point to a constant amplitude.

What about this *gain* in the overall loop? Normally, less energy will be delivered at the output than is injected at the input (fig. 13.3). However, if some of the output energy is used as an in-phase signal to boost the input energy, the gain of the overall system can, in effect, be made greater than 1. Of course, energy will still be lost in the process, but the amplification factor, or gain, can be greater than unity.

The gain of a control system is determined by the amount of energy lost in the system and the sensitivity of the controller. Controller sensitivity is adjustable and thus plays a key role in adjusting the controller settings. The sensitivity of a proportional controller can be varied by adjusting the proportional band—a narrow band being more sensitive than a broad one, and so on.

In the ideal system—one having no time lags at all—the feedback signal would always be 180° out of phase with the input energy. In any practical system, however, time lags always exist, and for any time lag there is always a natural frequency for which the feedback signal reaches the input side of the process in phase with the input energy. Assuring that the feedback signal is small enough to produce a gain of less than unity at the natural frequency of the process is the primary concern here.

Before studying how to adjust automatic controllers, understand that the matter of stability has only been touched upon. For example, the natural frequency of oscillation for a process will in most cases change for different load situations in a given process. Ordinarily the change in either frequency or load for a given process will not be so severe as to cause trouble, but a little margin must always be allowed for possible deviations beyond the normal operating conditions. Most critical processes are capable of being tuned to a very fine degree for optimum control, but trying to achieve even finer control might result in a control system that can "go over the brink" and cause a serious upset of the process, with consequent loss of time, material, product, and so forth.

Making Controller Adjustments

Three basic types of controller adjustments are (1) cut-and-try, or trial-and-error, adjustments; (2) calculated settings based on a concept called ultimate sensitivity; and (3) calculated settings based on the process reaction curve. Each method has its merits to recommend it for certain situations.

The cut-and-try method implies a haphazard approach to adjustment, although it need not be haphazard. Very frequently one of the other two methods is used to achieve a close approach to the best possible sort of control, then the cut-and-try method is used to add the fine touch to make the controller work at its best. Such cut-and-try techniques produce near-perfect results when carried out by a technician who is familiar with the process and who has had considerable experience in making adjustments.

The use of settings based on the concept of ultimate sensitivity, which was developed by two gentlemen named Ziegler and Nichols, has gained wide popularity. It provides a means for arriving at good controller performance by making calculations from information easily gained after introducing a small "upset" in the process.

Using calculations made with information provided by the process reaction curve produces excellent results. With this information and very slight additional adjustments based on the cut-and-try technique, a fine degree of control can be achieved.

Adjusting the Proportional Mode

Many controllers have proportional band adjustment dials that are graduated directly in percentages or proportional band factors. Other controllers have dials graduated in arbitrary numbers, with a chart or other information source showing the relation between numbers on the dial and the percentages of proportional band. The latter system is sometimes used in similar controllers that have two or more ranges of sensitivity. For example, one company's controllers use dials graduated from 1 to 150. In the broad-band or standard controller this range of numbers corresponds to proportional bands of 1% to 207%, while in a narrow-band version of the controller the proportional bands range from 1% to only 14%. Another controller uses a proportional band adjustment called a *sensitivity* adjustment, and its scale is graduated accordingly (fig. 13.4). A conversion scale to the right of the dial shows the relation between the sensitivity dial setting and the percentage of proportional band. Regardless of the sort of graduations the manufacturer has chosen to use on the dials of its controllers, it is always useful to know the relation between the dial setting and the actual proportional band percentage when making adjustments.

The automatic reset rate dial shown in figure 13.4 is graduated directly in *repeats per minute,* and this is quite common practice,

Figure 13.4. A three-mode recorder/controller (*Courtesy Taylor Instrument*)

although some controllers have arbitrary scales on their reset dials, as noted above. Dials for setting the rate of response, or the pre-act time, as noted in figure 13.4, are usually graduated in *minutes,* although some controllers have arbitrary scales.

In making the adjustment to the proportional band, or sensitivity of the controller, other needs will also be accomplished. For example, the frequency of oscillation of the system will be found, and this frequency value will be used to determine the *period* of the system, which is needed for later adjustments.

In many cases, connecting a sensitive high-speed recorder in parallel with the controller while making adjustments is beneficial. The speedy and sensitive recorder will show much greater detail than most circular chart recorders that are normally associated with controllers, and this will be a help in making accurate calculations and noting the frequency, or period, of the system. However, this detail is not absolutely essential.

Adjustments are begun by having the process operating at its normal set point and in a stable condition. At this time reset and rate functions will be locked out. It is important that this lock-out occur when no reset action is in effect in the controller; otherwise, efforts will go awry. Lock-out of reset and rate functions is accomplished by turning the reset dial to its lowest reset value and the rate dial to the position for slowest rate response. In effect, a proportional controller now exists, and the proportional band is arbitrarily set at about 75%.

To adjust the system, a small upset is introduced in the process by changing the set point on the controller to a slightly different value—one that will not cause significant economic loss due to ruining the product. Then the following steps are performed, or observations are made:

1. Observe the reaction of the process to the set-point change. Allow the system to settle out fully. For many processes that

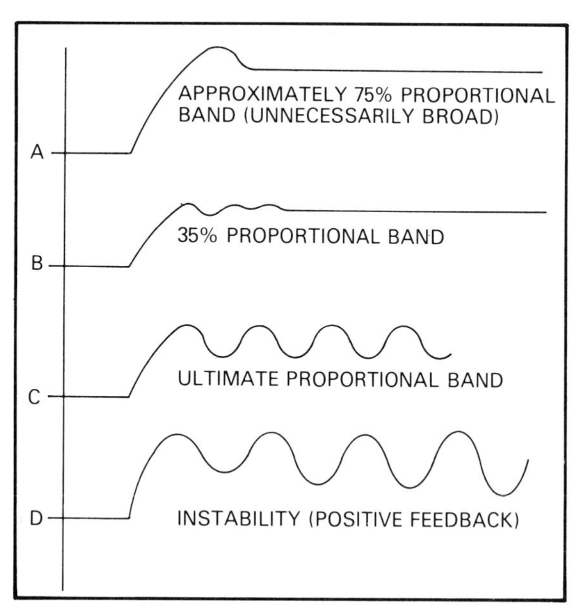

Figure 13.5. Performance curves for the proportional mode of a controller

require the transfer of heat between large capacities, this may take a long time—many minutes or even hours. Where rate of flow or pressure is the variable, the settling out will be comparatively rapid (curve *A* in fig. 13.5).

2. Return the process to stability at the set point, then narrow the proportional band to a value of about 35% and again change the set point as in step 1 above, being sure to use a small deviation. If the process again settles out, restore the variable to stability at the set-point value (curve *B* in fig. 13.5).
3. By systematically reducing the proportional band to half its previous percentage value and introducing a minor upset, it will be observed that the *offset* in the controlled variable becomes less and less as the proportional band is narrowed, but the number of oscillations that occur before the system settles out becomes greater. Eventually a point of adjustment will be reached at which the controlled variable does one of two things:
   a. Oscillates with small but constant amplitude about the new offset value (curve *C* in fig. 13.5).
   b. Oscillates about the new offset value with oscillations that are increasingly larger in amplitude until a limit is reached (curve *D* in fig. 13.5).

   If the first (curve *C*) occurs, the *ultimate* proportional band has been reached. It is the point of adjustment of feedback (and sensitivity) at which the gain around the control loop is unity. If the latter (curve *D*) occurs, the feedback is so great and so phased as to create a gain around the control loop that is greater than unity, so the proportional band is broadened somewhat and rechecked. The ultimate proportional band (PB_u), expressed in percentage, is the goal of the adjustment.
4. When the ultimate proportional band is determined, two things are noted:
   a. The percentage of proportional band.
   b. The *period* of the oscillation about the offset value of the controlled variable. The period of an oscillation is the reciprocal of its frequency; that is, if the controlled variable is oscillating at the rate of one-half cycle per minute, then its period is 2 minutes. The period, then, is merely the time required for one complete cycle. In this case, it is called the *ultimate period* (P_u) and is usually expressed in minutes (fig. 13.6).

Once the ultimate proportional band, PB_u, is known, the proportional band setting for best performance of the control system can be selected. For most processes it is $2PB_u$, that is,

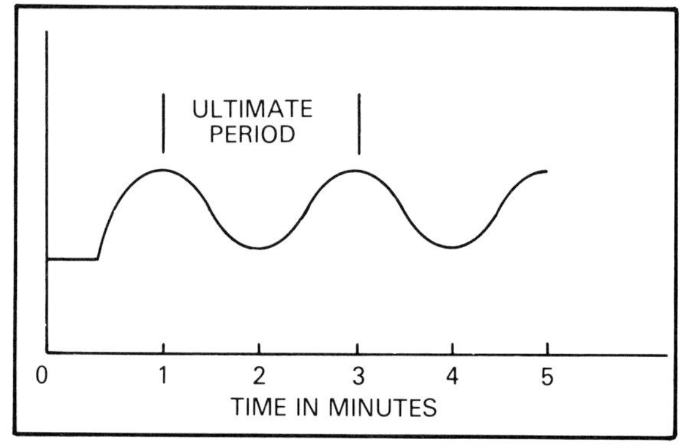

Figure 13.6. The period of the oscillation, or time required for one complete cycle

twice the percentage value as was noted for constant amplitude oscillation. This choice will give a sort of control that will cause the amplitude of each cycle about the offset point to be one-fourth the value of the preceding cycle. Such action will be satisfactory for the great majority of processes that are suited to proportional control. A smaller offset and smaller amplitude of oscillation can be obtained by narrowing the proportional band. However, for the narrowed band, oscillation will continue for a longer period of time, so it is a matter of choice—fewer oscillations of a given amplitude or a greater number of oscillations of somewhat smaller amplitude. Adjustments for narrowing the proportional band below the $2PB_u$ value should be done by cut-and-try method, slowly and carefully arriving at a band width that produces the performance desired, always keeping in mind the danger of instability that arises from too great a sensitivity.

Tuning the Reset Mode

From the measurements made in connection with choosing the proper proportional band the ultimate period, P_u, of the process is known. A quick and fairly appropriate setting for the reset dial can be made by using the *reciprocal* of the ultimate period. For example, if the period is 2 minutes, the reset dial should be set to 0.5 repeat per minute, or its equivalent.

According to the ultimate sensitivity method of carrying out adjustments, a proportional plus reset controller can achieve its best performance with the following settings:

Percent proportional band = $2.2PB_u$.
Reset rate, repeats per minute = $1.2/P_u$.

Such settings will provide the sort of damping that will produce the 0.25 amplitude ratio that exists between a given cycle and the preceding cycle.

After the adjustments are made, a study of the control characteristics as shown by the recorder will indicate whether optimum results are being obtained (fig. 13.7). Should the standard adjustments produce results that are other than optimum, slight readjustment by the cut-and-try method will bring about the desired characteristic form.

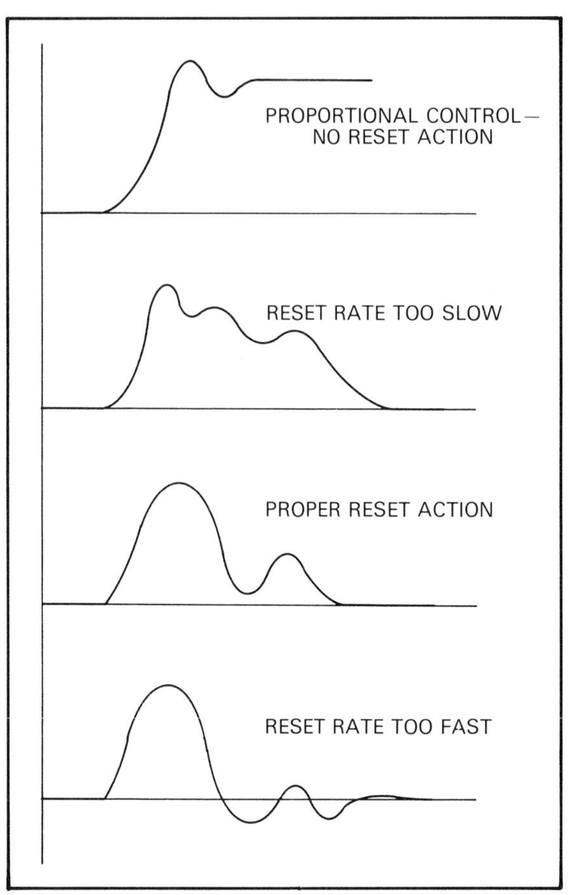

Figure 13.7. Effects of reset rate settings on the control characteristics

Tuning the Rate Mode

The rate mode of a controller is a function of how fast the controlled variable is changing with respect to the set-point value. Setting of the rate dial is determined by information gathered in preparing the initial adjustments—ultimate proportional band and ultimate period. A controller that has a rate mode adjustment, as well as adjustments for proportional band and reset rate, will work best if the proportional band is narrowed slightly

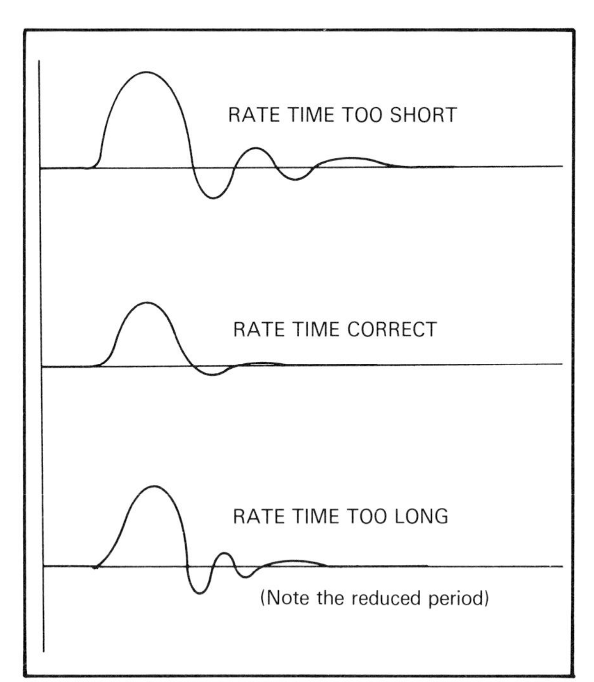

Figure 13.8. Effects of rate settings for a three-mode controller

and the reset rate increased slightly from what these settings would be for a two-mode controller. Based on the ultimate sensitivity method, the following settings are recommended for a three-mode controller:

Proportional band (%) = $1.6PB_u$.
Reset (repeats/minute) = $2/P_u$.
Rate time (minutes) = $P_u/8$.

These settings are made, and then the control characteristic studied after a minor upset in the system. Referring to the curves of figure 13.8 and comparing them with the actual curve shown by the recorder will indicate whether optimum settings have been achieved. Remember, the goal here is minimum deviation from the set-point value and a rapid but practical return to stability at the set point.

Adjustments Using the Process Reaction Curve

The process reaction curve is obtained by using the proportional mode only and introducing a *step-change* into the input or output energy of a process and carefully plotting the response of the controlled variable against a time base. The step-change for purposes here (1) is quite small; (2) is accurately known—that is, 5% of the total possible change in the final control element; and (3) is a very quick, almost instantaneous, change.

A high-speed, sensitive recorder capable of producing a graph in *rectangular coordinates* is of enormous value in obtaining the reaction curve. In one way or another a reaction curve is obtained or prepared in rectangular coordinates—either by using the proper kind of recorder or by transposing tabulated or recorded values to the right kind of coordinate paper. Having prepared the process reaction curve in this way, the following is done in much the same way as stated in the introductory section of this chapter. Refer to figure 13.9.

1. Draw a tangent line to the inflection point.
2. Draw a line perpendicular to the time base and have it intersect the inflection point on the curve.
3. The reaction *rate* of the process is equal to the slope of the reaction curve at the inflection point. The reaction rate, R, is found by dividing the length of line BC by the length of line AB.
4. The total time lag, t, is determined by measuring from time *0* to the point at which the tangent line crosses the base line.

As noted earlier in the chapter, reaction rate R is the rate of change in the controlled variable per unit time; that is

$$R = \frac{\text{change in controlled variable}}{\text{time (in minutes)}} .$$

If the controlled variable is temperature, then R might be expressed as "degrees Celsius per minute," or simply as °C/min. The time lag factor t is usually expressed in minutes, although it might sometimes be seconds or hours.

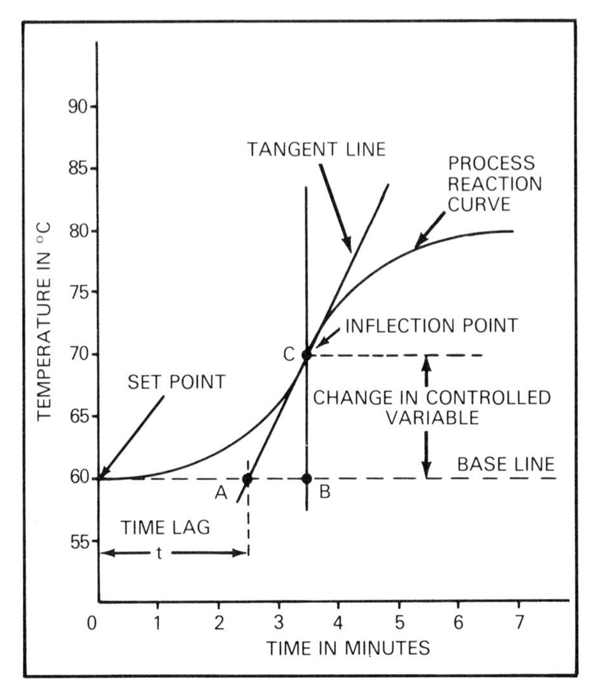

Figure 13.9. Method of deriving reaction rate and time lag

With information about the reaction rate R and time lag t, the following empirical equations can be applied to obtain proper settings for a process.

For controllers with proportional mode only:

Proportional band (%) $= 100Rt/\Delta V$

where ΔV is the change in the final control element to cause the step-change. ΔV is expressed as a percentage of the *total* change possible. For example, with a pneumatic actuator that operates in the range of 3–15 psi, a 1 psi change in pressure is $\frac{1}{12}$, or 8.3%, of the total change possible.

For controllers with two modes, proportional plus reset:

Proportional band (%) $= 110Rt/\Delta V$.

Reset rate (repeats/minute) $= 0.3/t$(min).

For three-mode controllers, proportional plus reset plus rate:

Proportional band (%) $= 83Rt/\Delta V$.

Reset rate (repeats/minute) $= 0.5/t$.

Rate time (minutes) $= 0.5t$.

Solving a three-mode problem using data from figure 13.9 demonstrates the use of the above empirical equations. Assume that the step-change was caused by increasing the air pressure on the final control element actuator from 7 to 10 psi—a 3 psi change, equivalent to 25% of the overall range of the unit.

From the graph the time lag t is established to be approximately 2.4 minutes (the time represented by the distance from time *0* to point A on the base line); the change in controlled variable (its value at the inflection point of the curve, minus its value at time *0*) is 10°C; and the length of the line AB is 1.1 minutes. From this information the reaction rate R is determined as follows:

$$R = \frac{\text{change in variable}}{\text{change in time}}$$
$$= 10°\text{C}/1.1 \text{ min}$$
$$= 9.09°\text{C/min}.$$

This shows that for purposes here the temperature is changing at a rate of slightly more than 9°C per minute.

For a three-mode controller the proportional band is set according to the equation,

$$\text{proportional band} = 83Rt/\Delta V$$
$$= \frac{83 \times 9.09 \times 2.4}{0.25}$$
$$= 0.724, \text{ or } 72.4\%,$$

and the reset rate is set according to the equation,

$$\text{reset rate} = 0.5/t$$
$$= \frac{0.5}{2.4}$$
$$= 0.2 \text{ repeats/min (approx.)}.$$

The rate mode adjustment is made according to the equation,

$$\text{rate time} = 0.5t$$
$$= 0.5 \times 2.4$$
$$= 1.2 \text{ min}.$$

Once these adjustments are made to the controller, it will be wise to study the process reaction to slight upsets and note whether it will be worthwhile to make additional fine adjustments by the cut-and-try method. A study of the curves shown in figure 13.8, as was done for the ultimate sensitivity method, will aid in determining whether or not the various modes are producing the very best performance.

Trouble-shooting Excessive Settings of the Modes

It will be recalled that the period of a process is determined in preparation for making controller adjustments. If, after all the adjustments are completed on a three-mode controller, the controlled variable is found to be oscillating, or cycling, excessively, it can probably be determined which of the modes—proportional, reset, or rate—is causing the trouble (figs. 13.7 and 13.8). The period of cycling is well defined in each curve. Notice particularly that the curve drawn to show too long a time for the rate mode has a significantly shorter period than the curve showing too short a rate time.

Knowing the period of the cycling as determined with just the proportional mode in action, the source of difficulty can quickly be determined by observing the period of the system with all modes in action. The following observations will be useful:

1. If the period of cycling is the *same* as that determined for the ultimate period, then the proportional band is too narrow and is causing the system to cycle excessively.
2. If the period of cycling is *less* than the ultimate period, that is, the cycling is more rapid, then the rate mode is set for too long a time.
3. If the period of cycling is *greater* than the ultimate period, that is, the cycling is slower, then the reset rate is too fast.

Once the period of the excessive cycling is compared with the ultimate period and the mode causing the cycling is determined, it is a simple matter to make one or more very slight corrections to the setting in order to bring about the optimum form of control.

Conclusion

Familiarity with equipment and process, which is a matter of experience coupled with a serious interest in what is going on, is just as important to keeping a control system in tip-top form as any other consideration.

Almost all systems of control that possess more than two or three related control adjustments are accompanied by literature furnished by the design engineers or the engineering firm that did the installation work. Routine sort of installations—those that possess no unique control problems or features—usually have standard items of control equipment for which the manufacturer has provided adequate instructions for maintenance and adjustment. In all cases, the instructions that accompany control equipment and deal with its adjustment to the system or process, should be studied, consulted, and followed when adjustments seem in order.

As a final note, adjustments should never be attempted without prior authority of persons responsible for seeing that specifications of the process product are adhered to.

APPENDIX

A

The International System of Units (SI)

The metric system of measurement is strange to most English-speaking peoples, although nearly all such peoples live in nations committed to adopting SI. SI is a universally accepted version of the metric system.

Nearly all who have studied science in Canada and the United States have developed an appreciation for the metric system. Its use of a single multiplier, the number 10^n, to express larger and smaller quantities simplifies calculations. The multiplier is used with positive or negative exponents (n) to achieve larger or smaller values, respectively.

SI is attractive for another reason. Only a few quantities are needed to express all the dimensions of measurement. The unit of length is the *metre.* Larger and smaller values of length can be expressed by multiples and submultiples of the metre – kilometre and millimetre are examples. Other unrelated measures for length do not exist in SI. Consider the conventional system: inch, foot, yard, rod, fathom, chain, mile, and league are all used to measure length. None of these units bears a sensible relation to any of the others. The same clumsy situation exists for the measurement of mass and force.

The metric system has enjoyed wide acceptance over the nearly two hundred years since its development, but it was tampered with, and several versions were introduced to meet alleged needs. Two subsystems were in common use: the MKS and CGS systems, short terms for metre-kilogram-second and centimetre-gram-second. These systems had their attendant derived units – dyne and erg as units of force and energy in the CGS system, and newton and joule for corresponding units in the MKS system. Various forms of the calorie were introduced as units of heat energy.

SI, while retaining the fine points of the metric system, establishes a few fundamental and supplementary units that will meet all the needs of science, industry, and commerce. It rules out the use of terms that tend to deviate from the basic units. The *litre,* for example, should preferably be replaced by the *cubic decimetre,* a volume unit of equal size. The use of the word *litre* as the name for a common volume unit is so entrenched in the metric system that it will continue to be used despite efforts to the contrary.

TABLE A.1
SI UNITS

Base Units

Quantity	*Unit Name*	*Unit Symbol*	*Dimensional Symbol*
Length	metre	m	L
Mass	kilogram	kg	M
Time	second	s	T
Electric current	ampere	A	I
Temperature	kelvin	K	θ
Amount of substance	mole	mol	Mol
Luminous intensity	candela	cd	–

Supplementary Units

Quantity	*Unit Name*	*Unit Symbol*
Plane angle	radian	rad
Solid angle	steradian	sr

Some Derived Units with Special Names

Quantity	*Unit Name*	*Unit Symbol*	
Frequency	hertz	Hz	1/s
Force	newton	N	kg•m/s^2
Pressure, stress	pascal	Pa	N/m^2
Energy, work, quantity of heat	joule	J	N•m
Power	watt	W	J/s
Electric charge	coulomb	C	A•s
Electric potential	volt	V	W/A
Electric resistance	ohm	Ω	V/A
Electric capacitance	farad	F	C/V

Other Derived Units

Quantity	*Unit Name*	
Acceleration, angular	radian per second squared	rad/s^2
Acceleration, linear	metre per second squared	m/s^2
Area	square metre	m^2
Density	kilogram per cubic metre	kg/m^3
Moment of force	newton metre	N•m
Permeability (electricity)	henry per metre	H/m
Thermal capacity (entropy)	joule per kelvin	J/K
Thermal conductivity	watt per metre kelvin	W/(m•K)
Viscosity, dynamic	pascal second	Pa•s
Velocity, angular	radian per second	rad/s
Velocity, linear	metre per second	m/s
Volume	cubic metre	m^3

The Dimensions, Units, and Symbols of SI

Table A.1 contains the seven basic units, the two supplementary units, and a number of derived units in SI. The basic unit for thermodynamic temperature is the *kelvin* (K). It is commonly used for scientific work. A more common temperature scale for everyday use (weather reporting, cooking, room temperature, etc.) is called *Celsius* (°C). It is equivalent to the metric scale, centigrade, having 0°C and 100°C occurring at the freezing and boiling points of water, respectively. The degree symbol (°) is always used with Celsius (C) but never with kelvin (K). Celsius is not a part of SI, but its use is accepted.

The supplementary units relate to angles, and they represent a great departure from the average person's knowledge and use of angles. The measurement of angles in radians has a good foundation in mathematics and comes about quite naturally. However, the use of degrees to express angle size will continue for many years, just as the use of *litre* will.

Table A.1 contains the quantity name, the unit name, the unit symbol, and the dimensional symbol of several quantities. Notice that the unit names contain all lowercase letters. Also notice that the symbols for the units may be lowercase letters, uppercase letters, a combination of the two (Pa, Hz), or a letter from the Greek alphabet (ohm, Ω). *The forms of the symbols are of vital importance, and it is essential to learn them well.*

The column headed *Dimensional Symbol* contains symbols for the basic quantities. These symbols are useful in expressing the dimensions of a quantity without specifying the measurement units involved. For example, force has dimensions of mass (M), length (L), and time (T). Using dimensional symbols, force equals $M \times L/T^2$, regardless of the base units involved. In SI the unit of force is the newton (N), and it is equal to kg•m/s².

The unit of mass is the kilogram. It is the only base unit containing a prefix. When expressing larger or smaller quantities of mass, the *gram* is used with a prefix.

Prefixes and Multipliers

In the conventional system of measurements, inches are used to measure small distances and miles to measure large distances. The inch is divided into quarters, sixteenths, sixty-fourths, and mils (for 0.001 inch). Intermediate measurements are in feet, yards, rods, and other units. A poor mathematical relationship exists among these length units, and some of them are plagued with ambiguity. For example, there are three kinds of mile.

SI insists on the metre (m) as a length unit. Through the use of prefixes applied to the metre, values of length from extremely small to extremely large can be easily expressed. Table A.2 lists the prefixes and multipliers to be used with SI, and they represent values ranging from 10^{-18} to 10^{18}. The small values

TABLE A.2
LIST OF PREFIXES

Value	*Prefix*	*Symbol*
10^{18}	exa	E
10^{15}	peta	P
10^{12}	tera	T
10^{9}	giga	G
10^{6}	mega	M
10^{3}	kilo	k
10^{2}	hecto	h
10^{1}	deca	da
10^{-1}	deci	d
10^{-2}	centi	c
10^{-3}	milli	m
10^{-6}	micro	μ
10^{-9}	nano	n
10^{-12}	pico	p
10^{-15}	femto	f
10^{-18}	atto	a

are capable of expressing atomic sizes, and the upper values meet most needs for great sizes. A light-year, for example, is about 5.68 trillion miles. In SI a light-year can be stated as about 9.5 petametres, or 9.5 Pm. Stated in powers of ten, it is 9.5×10^{15} m.

The kilogram is never given a prefix. A quantity of 1,000 kilograms is not properly termed a kilokilogram. It is a megagram (Mg), and a particularly small quantity of mass is not a microkilogram, but a milligram (mg).

Note the Greek letter *mu* (μ) used for the 10^{-6} prefix. It has enjoyed this use for many years—for example, μfd (microfarad) to express electrical capacity.

The most common prefixes, and thus those likely to be memorized first, are M, k, m, c, and μ. Once again, the form of the prefixes is important. Some are capitalized letters, some are lowercase, and one is a Greek letter.

Rules and Comments

The style for writing SI expressions is strict. It is well worth the trouble to learn it correctly, because proper use will avoid misinterpretation. The symbols for quantity units and prefixes have been thoughtfully chosen, but failure to use them correctly can lead to serious error. Assume that an FM broadcast station operates on a frequency of 101 megahertz. In symbolic form that is 101 MHz. The error caused by substituting a lowercase *m* for the uppercase *M* changes the statement to 101 millihertz! Such an error is so gross, of course, that it would be detected immediately.

Writing SI Units

All SI units written out in full should be in lowercase letters. The non-SI unit of temperature, Celsius, is always spelled with a capital *C*.

When writing out derived units, density and velocity, for examples, use the word *per* instead of a solidus or slash (/): write *metre per second* and *kilogram per cubic metre,* not *metre/second* and *kilogram/cubic metre.*

Symbols for units are not followed by a period unless they are at the end of a sentence. Remember they are symbols, not abbreviations. Symbols do not change form in the plural—for example, 1 km, 10 km, or 100 km.

The symbols of a derived unit formed by division may be written with the use of a solidus between the symbols in the numerator and those in the denominator (50 m/s; 15 kg/m³). They may also be written using negative exponents for the appropriate units ($50\ \text{m}\bullet\text{s}^{-1}$; $15\ \text{kg}\bullet\text{m}^{-3}$).

The symbols of a derived unit formed by multiplication of two or more units have the multiplication indicated by a dot (•) placed between the symbols, preferably above the line, thus N•m; Pa•s. The dot has the same meaning if it is placed on the line, however. To avoid confusion, the multiplication symbol to be used in conjunction with numerals is an $\times$, not a dot. Examples of use are—

$4\ \text{kg} \times 9.803\ \text{m/s}^2 = 39.21\ \text{N}$ and

$1.4 \times 10^4\ \text{Pa}\bullet\text{s} = 14\ \text{kPa}\bullet\text{s}$.

When more than one factor is contained in the denominator of a derived unit, the factors are enclosed in parentheses. For example, W•s/(kg•m). Factors in the numerator do not need to be parenthesized.

Prefixes are not used in the denominator. The single exception is the use of *kilo* with kilogram. Otherwise, the term is arranged through the use of a single appropriate prefix in the numerator. Mg/m^3, *not* mg/(mm)^3; m/s, *not* mm/ms; and mol/kg are correct examples.

A space is left between numerals and the first letter of a prefix or symbol. Examples are 25 km and 15 N, *not* 25km and 15N.

Writing Numerals

In text and tables, absolute numerical values of less than one (1) should have a zero (0) preceding the decimal point, which is a dot in

English-speaking countries. In many countries the comma is used as a decimal point. To avoid confusion, the use of spaces instead of commas to break up long numbers is recommended. For example, write *35 241.404 2* instead of *35,241.4042.* Arrange the digits in groups of three beginning at the decimal point and proceeding both ways.

Using Prefixes

Prefix symbols are printed in Roman type with no spacing between prefix and the unit symbol, thus, Mg, km, kPa. Never affix more than one prefix to a unit; for example, use gigahertz (GHz), *not* kilomegahertz (kMHz). Do not apply a prefix to *kilogram.* Apply the appropriate prefix to *gram* instead.

A prefix symbol is combined with the unit symbol it precedes to form a new symbol. This new symbol can be raised to a positive or negative power and can be combined with other symbols to form compound symbols. Examples:

$$1\ \mathrm{km}^3 = 1\ (\mathrm{km})^3 = (10^3\mathrm{m})^3 = 10^9\ \mathrm{m}^3$$

$$1\ \mathrm{ms}^{-1} = 1\ (10^{-3}\ \mathrm{s})^{-1} = 10^3\ \mathrm{s}^{-1}$$

Only one prefix should be used in forming decimal multiples or submultiples of derived units. This prefix should be attached to a unit in the numerator. As noted earlier, an exception is when the base unit *kilogram* appears in the denominator. Derived units with degrees of angle or degrees Celsius may also be treated exceptionally, as these degrees cannot be prefixed.

The prefix for a given unit should be one that provides a convenient numerical range, although in computations some thought should be given to expressing units for a given quantity with the same prefix. Normally, however, a range of 0.1 to 1 000 is desirable. Here is an example. Suppose a mass of 30 kg is accelerated at the rate of 15 m/s². What is the force acting on the mass? Force is the product of mass and acceleration and in dimensional symbols is

$$F = M \times L/T^2,$$

where

F = force;
M = mass;
L = length;
T = time.

Mass (M) in this case is 30 kg, length (L) is 15 m, and time (T) is 1 s², or unity. This can be written as:

$$F = 30\ \mathrm{kg} \times 15\ \mathrm{m} = 450\ \mathrm{N},$$
$$F = 30\ \mathrm{kg} \times 0.015\ \mathrm{km} = 0.45\ \mathrm{kN}.$$

Comparing SI with Conventional Units

For some years to come a great deal of time will be absorbed in converting conventional units to SI equivalents. Many of these conversions will be straightforward and trouble-free, requiring only the application of a multiplier constant. Statute miles are converted to kilometres by multiplying the number of miles by 1.609. The reciprocal of this number (0.621) used as a multiplier converts kilometres to miles. Common volume measurements in the conventional system—gallons and quarts—can be expressed in litres with close approximation. A U.S. quart is about 0.946 litre; a Canadian quart is 1.14 litres, and the respective gallons are just four times these values in litres—3.78 litres per U.S. gallon, 4.55 litres per Canadian gallon.

Many commodities are sold or exchanged on the basis of weight, or more correctly, mass. In SI the kilogram is the common unit for expressing the quantity of butter, meat products, or other solid foods or household goods. Roughly, the kilogram is equal to 2.2 pounds of mass.

A source of much confusion is the fact that conventional systems have used the pound as a measure of mass and of force. As units of mass, the kilogram and the pound have a simple relation as stated above: 1 kilogram equals about 2.2 pounds of mass. The conversion between units of force in the conventional

TABLE A.3
CONVERSIONS

To Convert from Conventional Units	*To SI Units*	*Multiply by Conversion Factor*
barrel (42 U.S. gallons)	metre^3 (m^3)	1.589×10^{-1}
British thermal unit	joule (J)	1.055×10^{3}
calorie	joule (J)	4.186
centimetre of mercury at 0°C	pascal (Pa)	1.333×10^{3}
centimetre of water at 4°C	pascal (Pa)	98.064
centipoise	pascal second (Pa•s)	10^{-3}
centistoke	metre^2 per second (m^2/s)	10^{-6}
circular mil	metre^2 (m^2)	5.067×10^{-10}
degree (angle)	radian (rad)	1.745×10^{-2}
degree Celsius (°C)	kelvin (K)	$T_K = T_C + 273.15$
degree Fahrenheit (°F)	degree Celsius (°C)	$T_C = (T_F - 32)/1.8$
degree Fahrenheit (°F)	kelvin (K)	$T_K - (T_F + 459.67)/1.8$
electronvolt	joule (J)	1.602×10^{-19}
fluid ounce (U.S.)	metre^3 (m^3)	2.957×10^{-5}
foot	metre (m)	0.308
foot cubed (ft^3)	metre^3 (m^3)	2.831×10^{-2}
foot squared (ft^2)	metre^2 (m^2)	9.29×10^{-2}
foot-pound-force	joule (J)	1.355
gallon (Canadian)	metre^3 (m^3)	4.546×10^{-3}
gallon (U.S. liquid)	metre^3 (m^3)	3.785×10^{-3}
horsepower (550 ft-lb/s)	watt (W)	746
inch	metre (m)	2.54×10^{-2}
inch of mercury column (32°F)	pascal (Pa)	3.386×10^{3}
inch of water column (39.2°F)	pascal (Pa)	2.491×10^{2}
kilowatt-hour	joule (J)	3.60×10^{6}
mile (U.S. statute)	kilometre (km)	1.609
ounce-force (avoirdupois)	newton (N)	0.278
ounce-mass (avoirdupois)	kilogram (kg)	2.834×10^{-2}
ounce-fluid (U.S.)	metre^3 (m^3)	2.957×10^{-5}
pound-force (avoirdupois)	newton (N)	4.448
pound-mass (avoirdupois)	kilogram (kg)	0.453 5
pound per square inch (psi)	pascal (Pa)	6.894×10^{3}
quart (U.S. liquid)	metre^3 (m^3)	9.463×10^{-4}
quart (Canadian)	metre^3 (m^3)	1.136×10^{-3}
square inch	metre^2 (m^2)	6.451×10^{-4}
square foot	metre^2	9.290×10^{-2}
ton (U.S. short, 2 000 lb)	tonne (t)	0.907 18
ton (U.K. long, 2 240 lb)	tonne (t)	1.016
yard	metre (m)	0.9144

system and SI can be a troublesome task. The difficulty extends into and complicates the conversions of energy and pressure between the systems. What is the equivalent of a foot-pound of work in SI? Based on the definition of force, a pound-force equals the product of pound-mass and acceleration. Gravitational acceleration is about 32.16 feet per second squared, or 9.8 m/s^2. A pound of mass is approximately 0.453 kg; a foot is 0.305 m.

$$\begin{aligned} \text{ft-lb} &= \frac{1 \text{ lb (mass)} \times 32.16}{1} \times 1 \text{ ft} \\ &= \frac{0.453 \text{ kg} \times 9.8}{1} \times 0.305 \text{ m} \\ &= 1.35 \text{ N}\bullet\text{m} \\ &= 1.35 \text{ joule} \end{aligned}$$

Conversions between the two systems are best carried out through the use of tables, because the number of factors that enter into the calculations for some of the more involved conversions, as in the example above, increases the chance for error. Table A.3 is a table of common conversions. Many of the pamphlets and texts listed in the bibliography contain extensive conversion tables.

APPENDIX

B

Symbols for Instrumentation and Electronics

Symbols serve a useful function. They are a form of shorthand, because they are simple and can be sketched quickly. Symbols, done properly, are like the proverbial picture—worth a thousand words.

The instrumentation symbols shown here are those recognized by the process control industry. Component identification (letters and numbers) are usually included in the symbols to designate complete instrumentation application. Not many of these symbols are used in this text. Generally, components are represented by drawings that closely resemble the actual equipment. However, electrical and electronic components are represented extensively by conventional symbols throughout the text.

In addition to these symbols, many Greek letters are used throughout the text to represent units of measure, mathematical numbers, or other factors.

Instrumentation Symbols

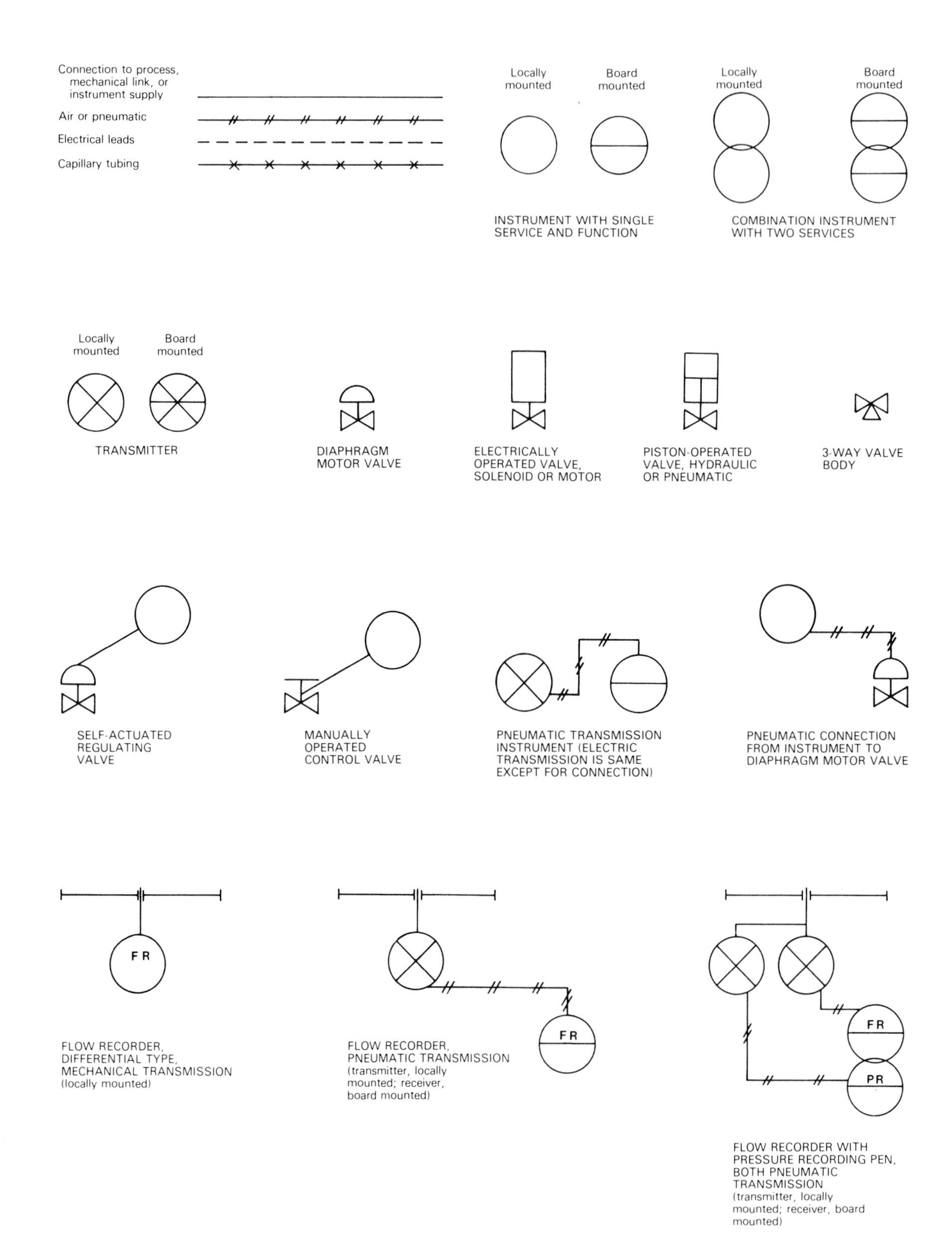

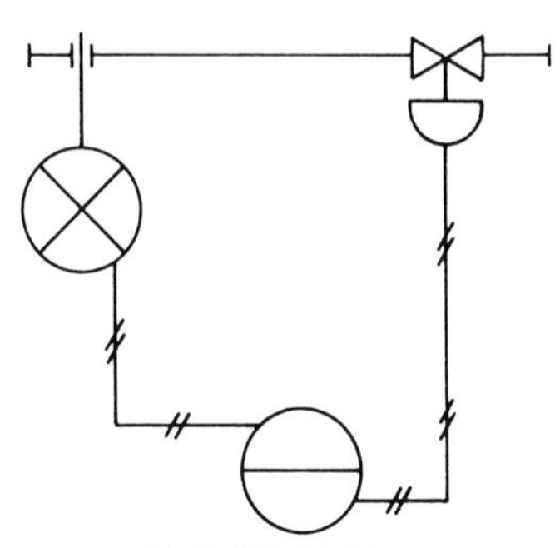

FLOW RECORDING CONTROLLER, PNEUMATIC TRANSMISSION (transmitter, locally mounted; receiver, board mounted)

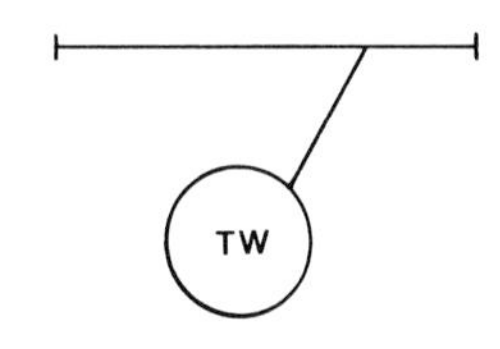

TEMPERATURE WELL

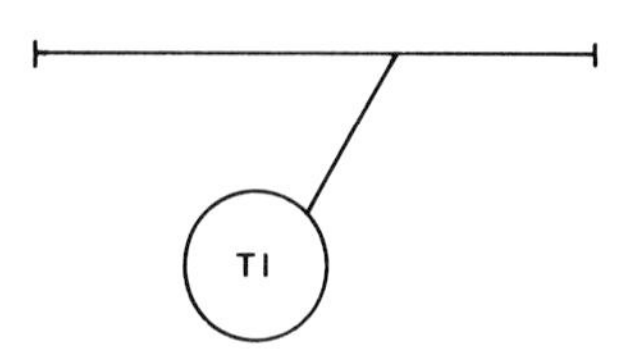

TEMPERATURE INDICATOR OR THERMOMETER (locally mounted)

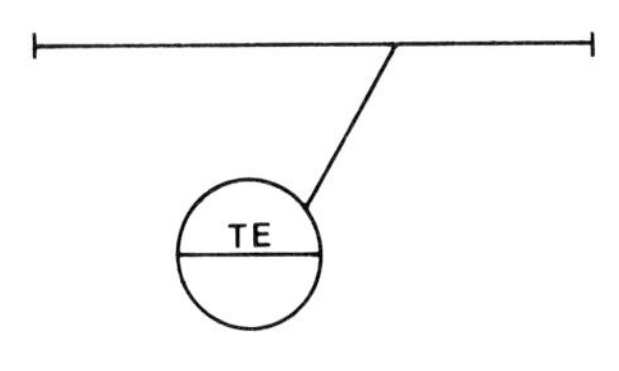

TEMPERATURE ELEMENT WITHOUT INSTRUMENT CONNECTION

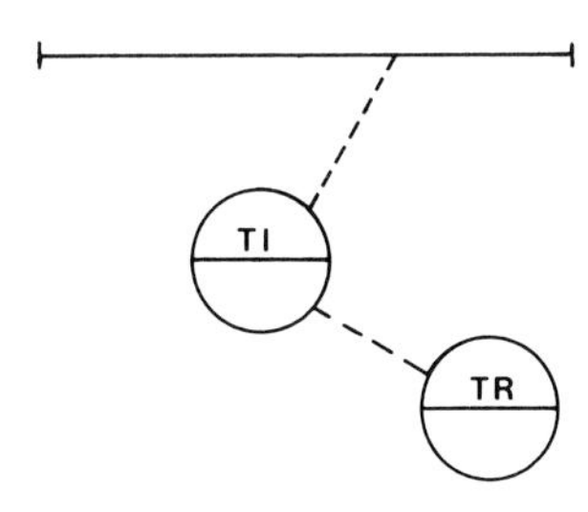

TEMPERATURE INDICATING AND RECORDING POINT CONNECTED TO MULTIPOINT INSTRUMENTS (board mounted)

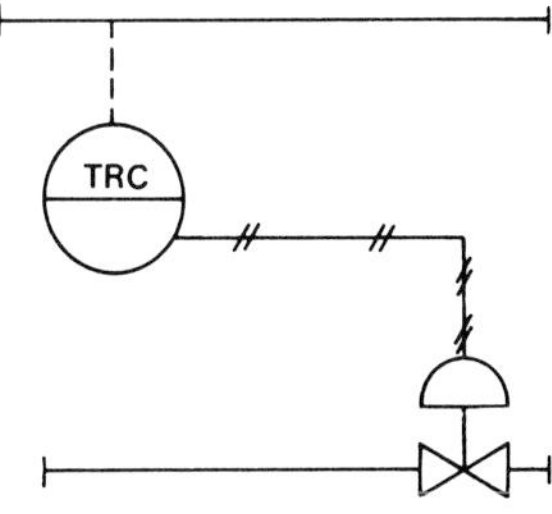

TEMPERATURE RECORDING CONTROLLER FOR ELECTRIC MEASUREMENT (board mounted)

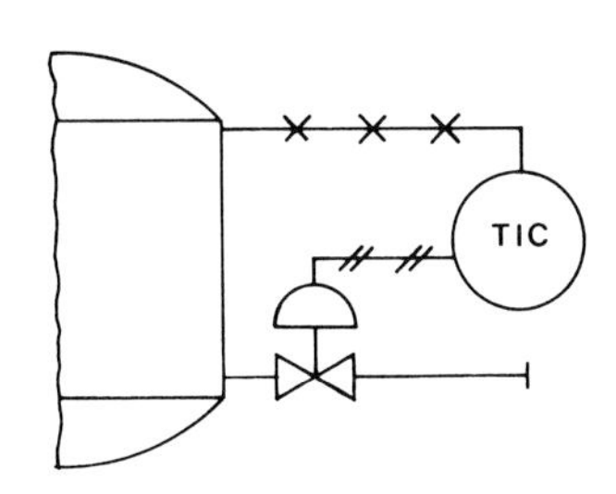

TEMPERATURE INDICATING CONTROLLER, FILLED SYSTEM (locally mounted)

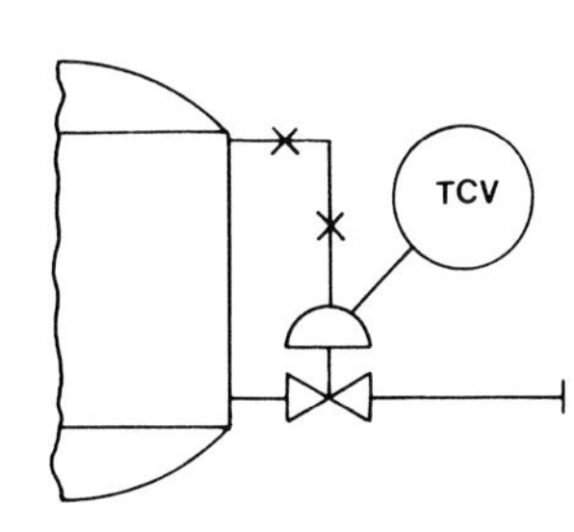

TEMPERATURE CONTROLLER, SELF-ACTUATED

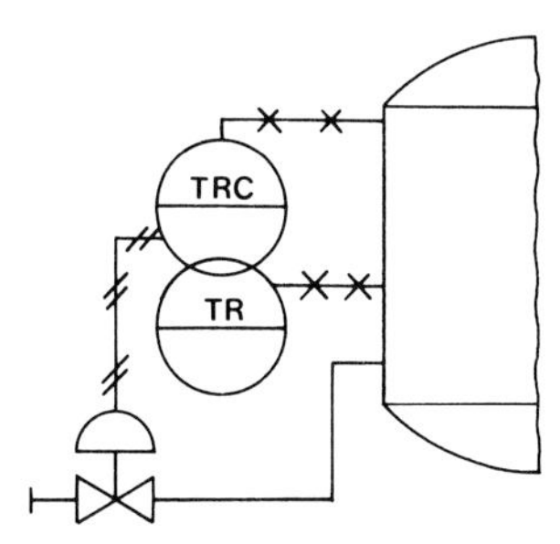

TEMPERATURE RECORDING CONTROLLER AND RECORDER, COMBINED INSTRUMENT (board mounted)

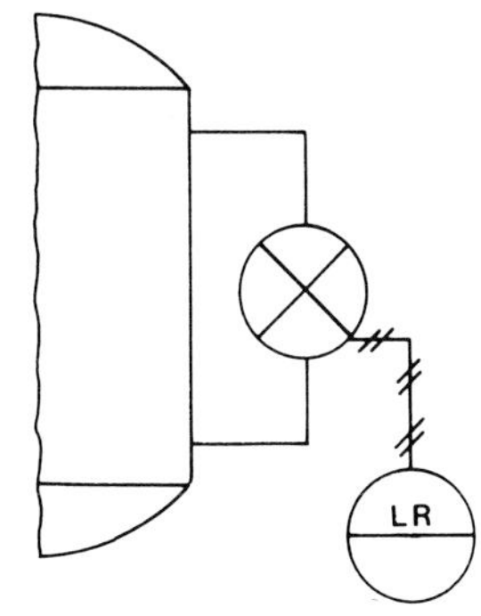

LEVEL RECORDER, PNEUMATIC TRANSMISSION WITH RECEIVER (board mounted); EXTERNAL TYPE TRANSMITTER

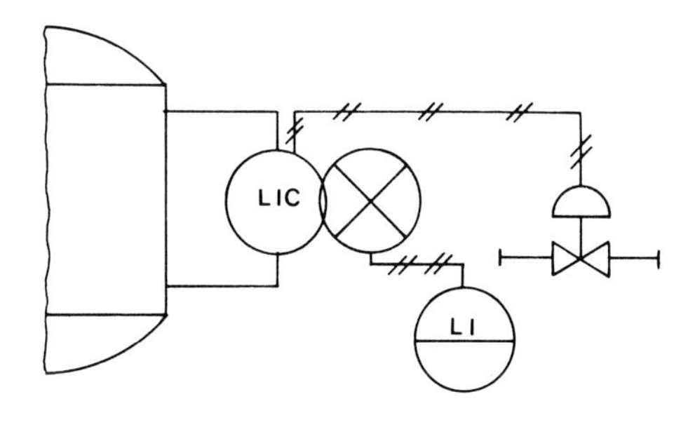

LEVEL INDICATING CONTROLLER AND TRANSMITTER COMBINED WITH LEVEL INDICATING RECEIVER (board mounted)

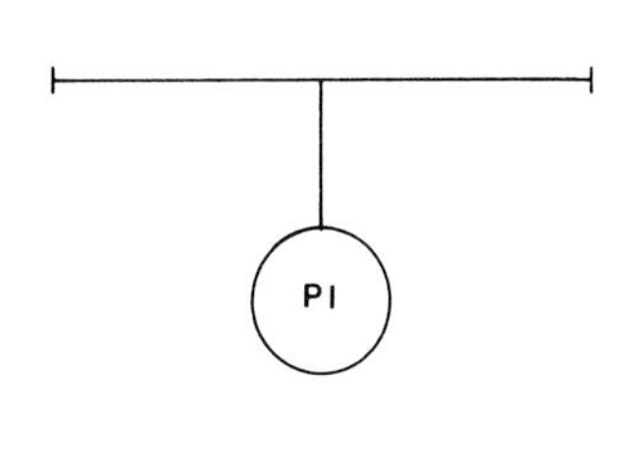

PRESSURE INDICATOR (locally mounted)

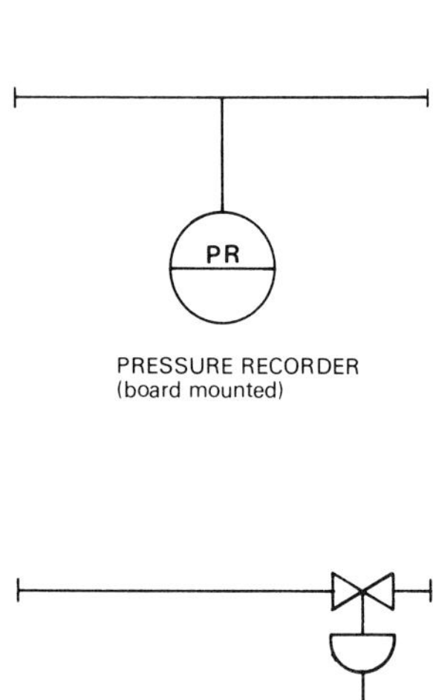

PRESSURE RECORDER
(board mounted)

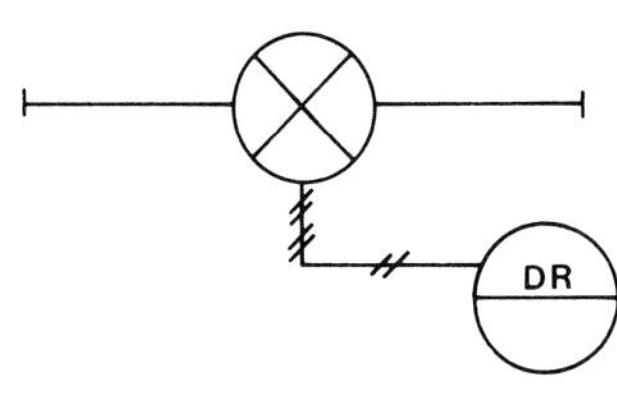

DENSITY RECORDER,
PNEUMATIC TRANSMISSION
(board mounted)

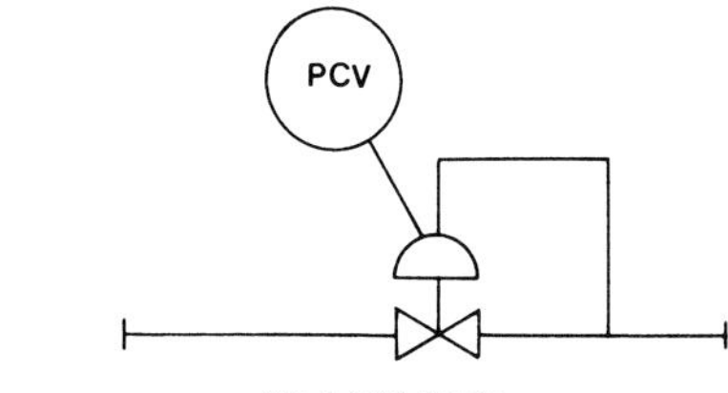

SELF-ACTUATED
(INTEGRAL) PRESSURE
REGULATING VALVE

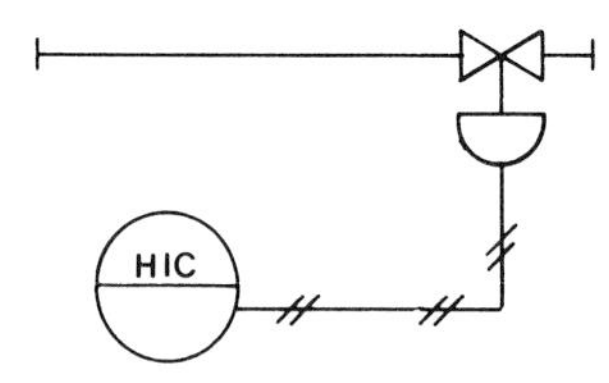

HAND-ACTUATED PNEUMATIC
CONTROLLER WITH INDICATION
(board mounted)

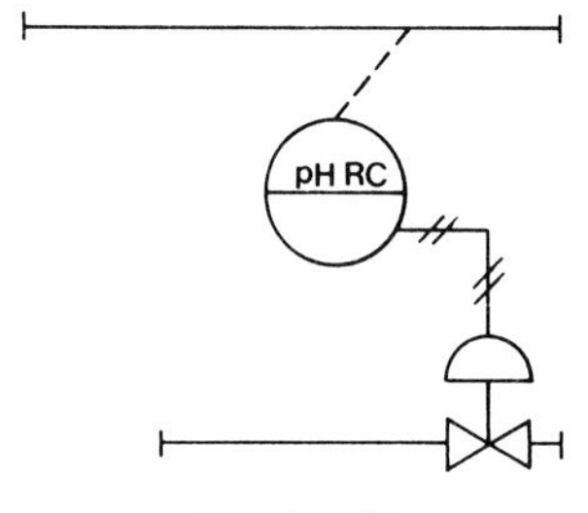

pH RECORDING
CONTROLLER
(board mounted)

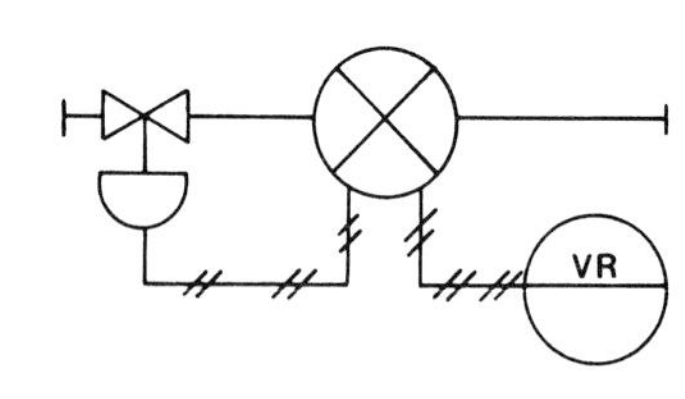

VISCOSITY RECORDER,
PNEUMATIC TRANSMISSION
(board mounted)

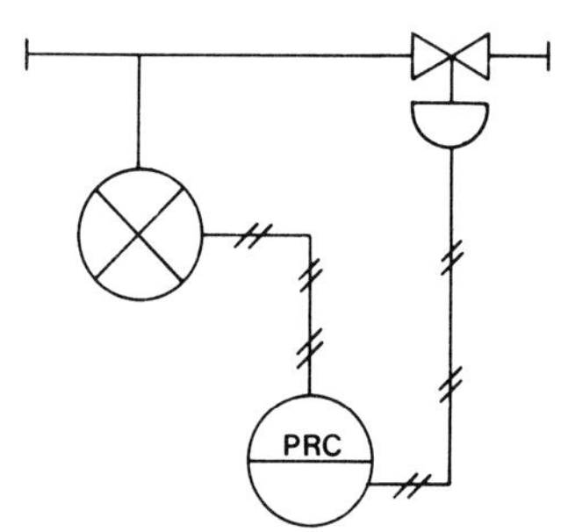

PRESSURE RECORDING
CONTROLLER, PNEUMATIC
TRANSMISSION, WITH
RECEIVER
(board mounted)

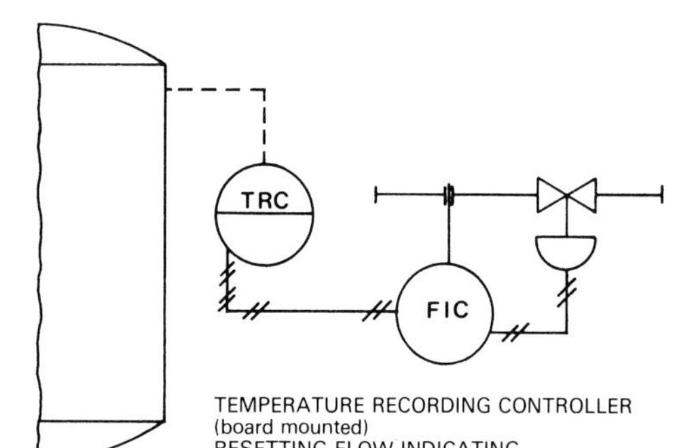

TEMPERATURE RECORDING CONTROLLER
(board mounted)
RESETTING FLOW INDICATING
CONTROLLER (locally
mounted)

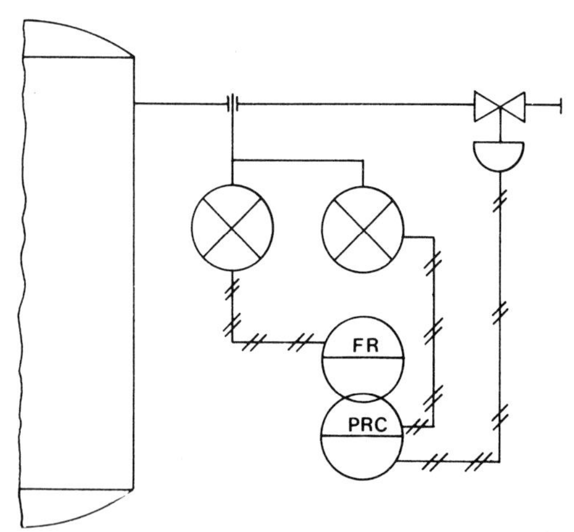

PRESSURE RECORDING CONTROLLER
WITH FLOW RECORD, PNEUMATIC
TRANSMISSION, COMBINED
RECEIVER (board mounted)

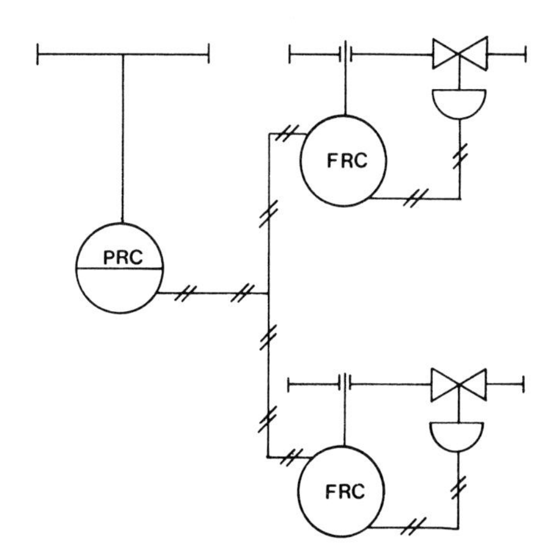

PRESSURE RECORDING CONTROLLER
(board mounted)
RESETTING FLOW RECORDING
CONTROLLERS (locally
mounted)

Electrical Symbols

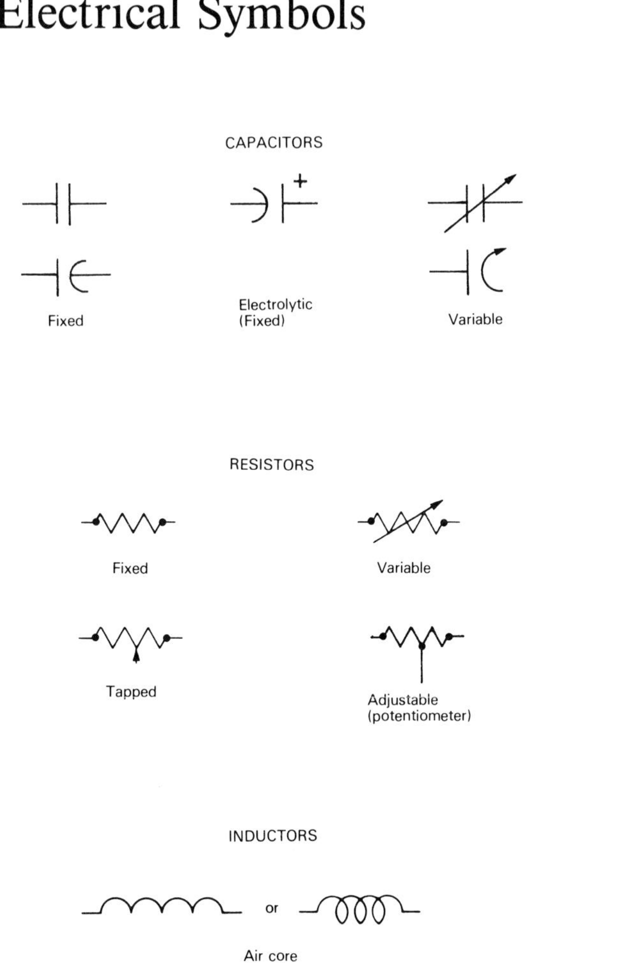

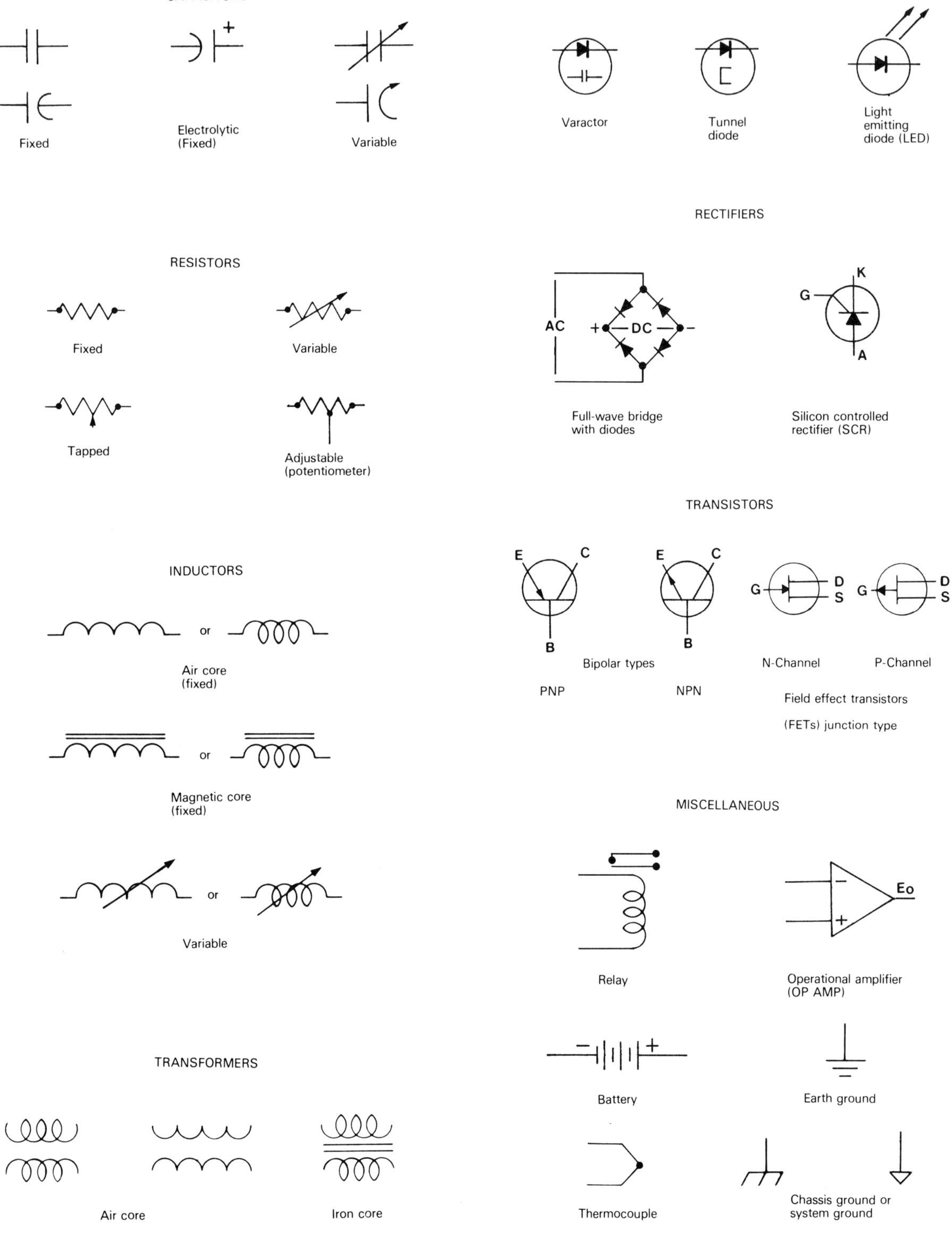

Greek Alphabet

Α	α	Alpha	Angles, coefficients, attenuation constant, absorption factor, area
Β	β ϐ	Beta	Angles, coefficients, phase constant
Γ	γ	Gamma	Complex propagation constant (cap), specific gravity, angles, electrical conductivity, propagation constant
Δ	δ	Delta	Increment or decrement (cap or small), determinant (cap), permittivity (cap), density, angles
Ε	ϵ	Epsilon	Dielectric constant, permittivity, base of natural logarithms, electric intensity
Ζ	ζ	Zeta	Coordinates, coefficients
Η	η	Eta	Intrinsic impedance, efficiency, surface charge density, hysteresis, coordinates
Θ	ϑ θ	Theta	Angular phase displacement, time constant, reluctance, angles
Ι	ι	Iota	Unit vector
Κ	κ	Kappa	Susceptibility, coupling coefficient
Λ	λ	Lambda	Permeance (cap), wavelength, attenuation constant
Μ	μ	Mu	Permeability, amplification factor, prefix micro
Ν	ν	Nu	Reluctivity, frequency
Ξ	ξ	Xi	Coordinates
Ο	ο	Omicron	
Π	π	Pi	3.1416
Ρ	ρ	Rho	Resistivity, volume charge density, coordinates
Σ	σ	Sigma	Summation (cap), surface charge density, complex propagation constant, electrical conductivity, leakage coefficient
Τ	τ	Tau	Time constant, volume resistivity, time-phase displacement, transmission factor, density
Υ	υ	Upsilon	
Φ	ϕ φ	Phi	Scalar potential (cap), magnetic flux, angles
Χ	χ	Chi	Electric susceptibility, angles
Ψ	ψ	Psi	Dielectric flux, phase difference, coordinates, angles
Ω	ω	Omega	Resistance in ohms (cap), solid angle (cap), angular velocity

Bibliography

Books

Addison-Wesley. *Consumer's Power; Fundamentals of Electricity.* 2 vols. Reading, Massachusetts: Addison-Wesley Publishing, 1966.

Berger, Bill D., and Anderson, Ken E. *Plant Operations Training Series.* 3 vols. Tulsa: PennWell Books, 1979.

Canadian Standards Association. *Canadian Metric Practices Guide.* Rexdale, Ontario: Canadian Standards Association, 1979.

Chemical Rubber Company. *Handbook of Chemistry and Physics.* 58th ed. Cleveland: CRC Press, 1977–78.

Considine, Doublas M., ed. *Process Instruments and Controls Handbook.* 2nd ed. New York: McGraw-Hill, 1974.

Faulkenberry, Luces M. *An Introduction to Operational Amplifiers with Linear IC Applications.* 2nd ed. New York: John Wiley & Sons, 1982.

Johnson, Curtis D. *Process Control Instrumentation Technology.* 2nd ed. New York: John Wiley & Sons, 1982. An advanced-level text.

Jung, Walter G. *IC OP-AMP Cookbook.* 2nd ed. Indianapolis: Howard W. Sams, 1980.

Kirk, Franklin W., and Rimboi, Nicholas R. *Instrumentation.* 3rd ed. Chicago: American Technical Society, 1974.

Palmer, J. F. *The International System of Units Handbook.* Huntsville, Alabama: Brown Engineering, 19--.

Tobey, Gene E; Graeme, Jerald G.; and Huelsman, Lawrence P., eds. *Operational Amplifiers: Design and Application.* New York: McGraw-Hill, 1971.

Williams, H. B.; Andrew, William G.; and Zoss, Leslie M. *Applied Instrumentation in the Process Industries.* 4 vols. Houston: Gulf Publishing, 1979–1982. Advanced-level texts.

Self-Study Programs

Vocational Training Series.
E. I. DuPont de Nemours & Company, Inc.
F & F Department, Applied Technology Division
Clayton Building, Concord Plaza
Wilmington, DE 19810

About fifteen titles appropriate to instrumentation.

API PROFIT Series of Programed Learning Courses.
Howell Training Company
5201 Langfield Road
Houston, TX 77040

Numerous titles appropriate to measurement and control in the petroleum industry.

Petroleum Learning Programs, Ltd.
400 FM 1960 West, Suite 260
Houston, TX 77090

Numerous self-study titles appropriate to petroleum industry instrumentation.

TPC Training Systems
Technical Publishing Company
1301 South Grove Avenue
Barrington, IL 60010

Self-study programs in *Instrumentation & Process Control, Maintenance Fundamentals, Electrical Maintenance,* and *Industrial Electronics.*

Manufacturers' Training Materials

The Foxboro Company
Educational Services
Foxboro, MA 02035

Numerous audiovisual programs relating to instrumentation; also short courses.

Honeywell
Education Center
1100 Virginia Drive
Fort Washington, PA 19034

A wide selection of courses in industrial instrumentation and control technology.

National Photographic Laboratories, Inc.
1926 West Gray
Houston, TX 77019

The *API Target Series* provides for complete classroom instruction using films, student workbooks, standard tests, instructor's manual, and answer keys; multimedia courses in instrumentation.